Bibliographia zoologiae et geologiae

A General Catalogue of All Books, Tracts, and Memoirs on Zoology and Geology

VOLUME 1

LOUIS AGASSIZ
EDITED BY H.E. STRICKLAND

CAMBRIDGE
UNIVERSITY PRESS

CAMBRIDGE UNIVERSITY PRESS

Cambridge, New York, Melbourne, Madrid, Cape Town,
Singapore, São Paolo, Delhi, Mexico City

Published in the United States of America by Cambridge University Press, New York

www.cambridge.org
Information on this title: www.cambridge.org/9781108062503

© in this compilation Cambridge University Press 2013

This edition first published 1848
This digitally printed version 2013

ISBN 978-1-108-06250-3 Paperback

CAMBRIDGE LIBRARY COLLECTION

Books of enduring scholarly value

Physical Sciences

From ancient times, humans have tried to understand the workings of the world around them. The roots of modern physical science go back to the very earliest mechanical devices such as levers and rollers, the mixing of paints and dyes, and the importance of the heavenly bodies in early religious observance and navigation. The physical sciences as we know them today began to emerge as independent academic subjects during the early modern period, in the work of Newton and other 'natural philosophers', and numerous sub-disciplines developed during the centuries that followed. This part of the Cambridge Library Collection is devoted to landmark publications in this area which will be of interest to historians of science concerned with individual scientists, particular discoveries, and advances in scientific method, or with the establishment and development of scientific institutions around the world.

Bibliographia zoologiae et geologiae

Born in Switzerland, Louis Agassiz (1807–73) distinguished himself as one of the most capable and industrious naturalists of the nineteenth century, working in fields as diverse as ichthyology and glaciology. In the late 1840s, he moved to North America, where he became a professor of zoology at Harvard and established the Museum of Comparative Zoology. His extensive bibliography of all known works relating to zoology and geology, which he had compiled for private use, was revised and substantially expanded by the English naturalist Hugh Edwin Strickland (1811–53) and published by the Ray Society in four volumes between 1848 and 1854. As such, it stands as the fullest record of the existing scientific literature just prior to the publication of Darwin's *On the Origin of Species*. Volume 1 (1848) provides a global list of all relevant periodicals before beginning the principal list of works, arranged alphabetically by author, ranging here from Aalborg to Bywater.

Cambridge University Press has long been a pioneer in the reissuing of out-of-print titles from its own backlist, producing digital reprints of books that are still sought after by scholars and students but could not be reprinted economically using traditional technology. The Cambridge Library Collection extends this activity to a wider range of books which are still of importance to researchers and professionals, either for the source material they contain, or as landmarks in the history of their academic discipline.

Drawing from the world-renowned collections in the Cambridge University Library and other partner libraries, and guided by the advice of experts in each subject area, Cambridge University Press is using state-of-the-art scanning machines in its own Printing House to capture the content of each book selected for inclusion. The files are processed to give a consistently clear, crisp image, and the books finished to the high quality standard for which the Press is recognised around the world. The latest print-on-demand technology ensures that the books will remain available indefinitely, and that orders for single or multiple copies can quickly be supplied.

The Cambridge Library Collection brings back to life books of enduring scholarly value (including out-of-copyright works originally issued by other publishers) across a wide range of disciplines in the humanities and social sciences and in science and technology.

THE

R A Y   S O C I E T Y.

INSTITUTED MDCCCXLIV.

W.Bagg

LONDON.

MDCCCXLVIII.

BIBLIOGRAPHIA
ZOOLOGIÆ ET GEOLOGIÆ.

A

GENERAL CATALOGUE

OF ALL

BOOKS, TRACTS, AND MEMOIRS

ON

ZOOLOGY AND GEOLOGY.

By Prof. LOUIS AGASSIZ,

CORR. MEMB. BRIT. ASSOC. ADV. SC. &c.

CORRECTED, ENLARGED, AND EDITED

By H. E. STRICKLAND, M.A., F.G.S. &c.

VOL. I.

CONTAINING PERIODICALS,
AND THE ALPHABETICAL LIST FROM A TO BYW.

LONDON:
PRINTED FOR THE RAY SOCIETY.
1848.

PRINTED BY
RICHARD AND JOHN EDWARD TAYLOR,
RED LION COURT, FLEET STREET.

ALERE FLAMMAM.

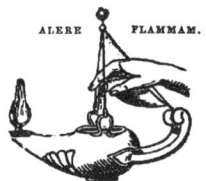

CONTENTS.

a 2

PREFACE BY THE EDITOR.

THE voluminous work which the RAY SOCIETY has here undertaken to publish, has been mainly compiled by Professor Agassiz, during the leisure moments of a life of almost incessant scientific research. Originally prepared for his own private use as an index to the numerous works on Zoology and Geology which he had occasion to consult, the collection gradually increased upon his hands, and the importance of its publication as an aid to other scientific investigators became daily more evident. Not trusting however to the means within his own reach, he established an extensive correspondence with naturalists in all parts of Europe, from whom he received an immense addition of materials on the scientific bibliography of their respective countries, and he was thus enabled to prepare a catalogue,—not indeed a complete one, for that can never be attained in practice, but far more perfect than any existing,—of all known works and detached memoirs on Zoology and Geology.

The utility of such a work to the scientific student is sufficiently evident, yet its technical nature rendered its publication a matter of very doubtful policy to one who could ill afford to sustain a probable loss. The work might therefore have been indefinitely delayed, or never have seen the light at all, were it not for the intervention of the RAY SOCIETY, whose Council were so convinced of its value, that they resolved on proposing such terms to M. Agassiz as induced him to make over to the Society the whole of his materials. Having been requested to act as Editor of the work, and having been in the habit of compiling for my own use a similar, though less extensive, Catalogue of Zoo-bibliography, I am enabled to make numerous additions and corrections to the labours of M. Agassiz and his coadjutors. I have also consulted various published lists, and have derived much assistance from the very useful " Verzeichniss der Bücher über Naturgeschichte," lately published by Engelmann. To Mr. J. E. Gray also I am indebted for the use of a MS. catalogue of entomological works compiled some years ago by Mr. Bennett and himself. I am well aware that a Catalogue which is intended to include all that has ever been published on so extensive a subject, must necessarily be very defective both in completeness and accuracy ; still, so far as it goes, there can be no doubt of its utility ; and if the work be found, as I trust it will, to be by far more extensive and correct than any of its predecessors, the RAY SOCIETY will have no cause to regret the undertaking.

The greatest difficulty which has been felt in making this compilation, has been to define the limits by which to circumscribe it. So intimately connected are the several sciences, that bibliographers well know the difficulty of making a satisfactory classification of books according to their subject-matter, and the same impediment is equally felt in making lists of works which shall exhibit the whole, and nothing but the whole, of any given department of knowledge. In the present case, the primary object of M. Agassiz was to insert in his Catalogue the titles of every work and essay on ZOOLOGY, as a science of observation and of classification. The Zoology of past epochs, as well as of the present, necessarily formed an element in this design, and hence it was necessary to include such Geological treatises as contained any information on the structure or distribution of organic remains. But this definition embraced, in fact, nearly every existing geological work, and the labour of examining all such writings to ascertain whether they did or did not contain palæontological matter would have been so great, that it was thought better to extend the limits a little wider so as to include all works on descriptive Geology, whether they referred to organic or to inorganic phænomena.

The first portion of the work contains a list, arranged geographically, of all the known periodical or miscellaneous publications which contain treatises on Zoology or Geology. As however it was in many cases impossible to make a personal examination of all such works, on account of their extreme rarity and limited circulation, it is probable that some of them may not in fact contain matter relevant to our purpose. This seems to be especially the case with the periodicals of France and Italy, enumerated hereafter, many of which appear from their titles to be either of a popular nature, or to be remotely connected with our subject-matter. We have however preferred to err on the side of excess rather than of defect, especially as it may interest the student in other branches of science, to find here recorded the titles of many valuable periodicals whose existence was previously unknown to him. Nor must it be supposed that some Journals, whose titles have no apparent reference to Zoology or Geology, are here inserted without good reason. Many a valuable fact relating to those subjects lies buried in periodicals specially devoted to Chemistry, Meteorology and other physical sciences, and in the majority of instances the compilers have themselves ascertained such to be the case before they ventured to insert a title into their list.

Under *Periodical Works* are included all serial publications which contain scientific treatises by a variety of hands, such as Transactions of Societies, Journals, Magazines and Encyclopædias. The several treatises of which these periodicals consist, as far as they relate to our subject-matter, are registered alphabetically under the author's name in the second part of the work. This vast undertaking can at present be only partially executed, from the difficulty before alluded to, of obtaining access to all the periodicals enumerated in the first part. So far however as the compilers have been able to perform this part of their project they have done it conscientiously, taking such periodicals as were within their reach, and faithfully extracting the titles of the memoirs which they contain. The titles of all detached and independent works are also inserted in the alphabetical series, so that under the name of each author will be found a complete list of all his writings in these departments of natural history.

As already stated, this compilation is professedly limited to *scientific Zoology and Geology*, yet it must be admitted that M. Agassiz has inserted, rather too indiscriminately in my opinion, many titles of works which are only remotely connected with those subjects. Had I been the sole compiler instead of the editor of the work, I should, for the sake of rendering it less bulky, have excluded all merely popular and elementary essays, all treatises on Human and Morbid Anatomy, on Ethnography, on Domestication of Animals and the Veterinary Art, on pure Mineralogy and Crystallography, on Mining and Metallurgy, and all popular Voyages and Travels. Too many of such works have however crept into the Catalogue, and I, as editor, did not, except occasionally, feel justified in excluding them. To have erased a work from the list merely because its *title appeared* irrelevant to our especial subjects would have been highly improper, and to have consulted the original work in all doubtful cases would not only have been most difficult and laborious, but with regard to many rare and inaccessible publications it would have been impossible. The list furnished by M. Agassiz is therefore substantially produced here, but it has undergone a great amount of careful revision and correction. Several thousand additional titles, selected with great care, have also been added, either by myself or under my immediate inspection. These are principally extracted from British scientific periodicals, with whose contents continental naturalists are too imperfectly acquainted. Many of these titles had been already extracted at second-hand by M. Agassiz from French and German journals where they are translated, and the works of many British writers consequently appeared in a foreign dress. In such cases the original English title has been substituted, and a reference made to the foreign work where the translation appears. In some instances however I have not had an opportunity of verifying such titles with the originals, and they consequently still remain in the list in their exotic form.

In the case of very short or unimportant communications, a certain amount of discretionary power was indispensable. Our popular " Magazines" of Natural History teem with trifling notices, often anonymous, sometimes brief and indefinite, sometimes wordy and inflated, but which do not contain a single fact of scientific importance. To have recorded all such in our list would have added bulk but not value, and they have therefore been in general omitted. But I have always endeavoured to render due justice to every properly authenticated statement, however brief, of a new or important observation in Zoology or Geology.

Wherever it has been practicable, the lists of each author's writings have been forwarded to him for correction before sending them to the press, and the works of living British

naturalists have thus been accurately registered up to the
time of publication.

I am fully prepared to admit that the Catalogue would have
been rendered more useful if the works of each author had
been *chronologically* arranged. To a certain extent this has
been done; but the mere *alphabetical* arrangement of such a
mass of authors' names was no slight undertaking, and to
have cut up and classified each book-title in the order of time
would have involved an amount of labour, both mechanical
and critical, which it was impossible to spare for such an ob
ject. Those therefore who consult this work must be content
to find here a full list of the zoo-geo-logical writings of each
author, but must not expect any systematic arrangement of
those writings, either as to date or subject.

It will be seen that in many instances I have been unable
to ascertain the Christian names of the authors recorded in
the alphabetical list.   This is owing to the singular practice,
followed by many persons on the continent of Europe, of
signing their surnames only, omitting both the Christian
name and its initials.   I have never been able to discover the
motive for a custom so productive of inconvenience and con-
fusion.

The Council of the RAY SOCIETY propose, when this publication is completed, to print in a Supplement all the additional titles of books or memoirs on Zoology or Geology
which have been inadvertently omitted during its progress.
There still remain many periodical works, both British and
foreign, whose contents have never been looked at with a
view to this publication; and of those which have been so
consulted, the later volumes have in some cases escaped notice.   To render the remaining volumes of the Bibliographia,
and its proposed Supplement, as complete as possible, the
Council earnestly request the cooperation of the Members of

the Ray Society and of the public. Any person who has access to periodical works, whether British or foreign, which appear to have been unnoticed by the compilers of the present Catalogue, will confer a boon by extracting the titles of the zoological and geological papers which they contain, referring to the volume and page of the original work, in the form adopted in this volume. The titles of such Books also as have been omitted in our list, or have been published subsequently, especially those of rare provincial and otherwise inaccessible tracts and pamphlets on Zoology or Geology, will be exceedingly acceptable. All such titles should be written *on one side of the paper only*, and may be addressed to the Secretary, Dr. E. Lankester, 22 Old Burlington Street, London.

I wish it had been in my power to have given an exact list of those periodicals, the titles of whose contents have been wholly or partly extracted into this work, so that those who are disposed to aid in the above object might at once see whether any given periodical had already been made use of or not. Unfortunately M. Agassiz omitted to keep such a list; and to enumerate those which have been consulted by myself would have only partially remedied this defect. By referring however to a few of the individual titles, as entered in this volume, it will easily be seen whether the work in which they are contained has been systematically consulted or not; and the List of the Abbreviations employed will also assist this object.

A person who was willing to devote his life to compilation might no doubt have made this work more convenient for some purposes by arranging it as a *classed* instead of an *alphabetical* catalogue. Such a task was not, however, practicable in the present case, but every one may to a certain extent make it into a classed catalogue for himself, by attaching distinctive marks in the margin to those titles which relate to his own especial departments of scientific research.

It only remains for me to deprecate criticism as to the many errors unavoidable in a work of this kind;—a work of considerable difficulty in itself, and undertaken on my part for the RAY SOCIETY as a labour of love, in the midst of other avocations. And when it is stated that the matter supplied to the printer was written in eight or nine different languages, and in a variety of handwritings, some of them very illegible, and replete with erasures and corrections, due credit will be given to Messrs. R. and J. E. Taylor for the manner in which the printing has been executed.

H. E. STRICKLAND.

Oxford, April 1848.

LIST OF ABBREVIATIONS USED IN PART II.

N.B. In the following list of the contractions employed in the alphabetical portion of this work, I have endeavoured, as far as I am able, to identify the periodical and other works to which these contractions refer. It must not however be supposed that all the periodicals, here referred to, have been systematically consulted, and the title of every appropriate memoir in them extracted. Many of these titles have been obtained at second hand from works where they are reprinted, and the original sources have not in every case been accessible. This circumstance will also explain the fact (which I cannot but regret) that so many different forms of abbreviation have been occasionally used to indicate one and the same publication. This was perhaps unavoidable in a work which was compiled by a variety of hands ; and to have rectified these diversities, and produced the desired uniformity, would have been a task of immense labour. I must also add, that some of the works to which references are made in Part II. are omitted in this list, but they will in general be found under the author's name in the alphabetical series.

The first column in the list contains the abbreviated forms, arranged alphabetically ; the second gives the place of publication, and the No. attached to the work in Part I. ; and the last column refers the reader to the page in Part I. where the title of the periodical will be found at full length.— H. E. S.

Abbreviated form.	Place of publication and No.	Page of Part I.
Abh. Baiersch. Akad........................	München, 1	29
Abh. Berl. Ak.	Berlin, 3	11
Abh. einer Privatges. in Böhm.	} Prag, 3	32
Abh. Privatges. Böhm.		
Ac. Pét.	St. Pétersbourg, 1–6	6
Acad. Sc. Par.	Paris, 2	49
Act. Acad. Nat. Cur.	} Nürnberg, 5	31
Act. Nat. Cur.		
Act. Bonon.	Bologna, 1, 2	64
Act. Erudit.	Leipzig, 2	26
Act. Hafn.	Kiöbenhavn, 1	2

Abbreviated form.	Place of publication and No.	Page of Part I.
Act. Harl.	Haarlem, 1	9
Act. Helv.	} Basel, 1	58
Act. Helvet.		
Act. Holm.	Stockholm, 1	4
Act. Œcon. Soc. Reg. Dan.		
Act. Nidr.	Kiöbenhavn, 23	3
Act. Upsal............................	Upsala, 1, 2	4
Act. Vliss.	Middelburg, 1..............	10
Actes de la Soc. d'Hist. nat. Par..........	Paris, 17	50
Allg. Anz. d. Deutsch.		
Am. Ac.		
Amœn. Ac..............................	} *Vide* LINNÆUS, in Part II.	
Amœnit. Acad.		
Amœnit. Academic.		
Am. J.		
Amer. J.	} New York, 9	84
Amer. Journ............................		
Analyst	London, 52	79
Anat. Transalp.		
Ann. Agric. Regno d' Ital.		
Ann. d' Agric. R. d' Ital.	} Milano, 9	67
Ann. d' Agr. R. d' Ital.		
Ann. and Mag. N. H.		
Ann. a. Mag. N. Hist.	} London, 48................	78
Ann. of Nat. Hist........................		
Ann. Chim. et Phys........................	Leipzig, 23	27
Ann. d'Anat.	} Paris, 63	53
Ann. d'Anat. et de Phys.		
Ann. d'Auv.	Clermont-Ferrand, 1	43
Ann. d. Erdk.		
Ann. de Chim.		
Ann. der Wetterauischen Gesellschaft...	Frankfurt am Main, 3 ...	19
Ann. des Mines	} Paris, 48	52
Ann. Min.		
Ann. di St. nat............................	} Bologna, 11	64
Ann. di St. Nat. Bol.		
Ann. Gén. Sc. Phys........................	Bruxelles, 12	37
Ann. Lyc. N. York	} New York, 5	84
Ann. Lyc. N. Y.		
Ann. med.-chir. di Roma	Roma, 4	71
Ann. Med. Stran.		
Ann. Phys. et Chim........................	Leipzig, 23	27
Ann. Provenç. d'Agr.		
Ann. R. Lomb.-Ven........................		
Ann. Sc. Regn. Lomb.-Ven.	} Milano, 20	67
Ann. Sc. R. Lomb.-Ven.		
Ann. Sc. Nat.		
Ann. Sc. n.	} Paris, 52	52
Ann. des Sciences nat.		

Abbreviated form.	Place of publication and No.	Page of Part I.
Ann. Sc. nat. Bologna	} Bologna, 13..................	64
Ann. Sc. nat. Bol......................		
Ann. Soc. Ent. Fr.	} Paris, 28	50
Ann. Soc. Entom. Fr.		
Ann. Soc. Linn. Par.	Paris, 25	50
Ann. Soc. Minér. Jena		
Ann. un. Medic..............................	Milano, 12	67
Antol...	} Firenze, 15	66
Antologia		
Arch. f. Physiol.............................	Berlin, 31	13
Arch. Gén. Méd.	Paris, 50	52
Arch. hist. et stat. Rhône	}	
Arch. hist. et stat. du Rhône		
Asiat. Research..............................	} Calcutta, 1	81
Asiat. Res..................................		
Athen...	London, 60..................	79
Att. Accad. Fisiocrit. di Siena		
Atti Accad. Sc. Sien.......................		
Atti di Siena	} Siena, 1	71
Atti Accad. Sien.		
Att. Istit. d' Ital.		
Att. Soc. Gioen. Catan.	} Catania, 1	64
Att. Soc. Gioen. di Catan.		
Atti Instit. Sc. nat. di Napoli	Napoli, 4.....................	68
Atti R. Accad. di Torino	Torino, 3.....................	71
Ballenst. Arch.	} Quedlinburg, 1	33
Ball. Arch.................................		
Beitr. zu Gesch. der Erfindungen		
Berl. Beschäft.		
Besch. Berl. Naturf.......................		
Besch. d. Berl. Naturf......	} Berlin, 8	12
Beschäft. der Berlin Gesellsch. Natur-forsch. Freunde		
Berl. Mag.	Berlin, 19	12
Berl. Samml.	Berlin, 6	11
Bibl. Ent.	*Vide* GORY, in Part II.	
Bibl. Ital.	Milano, 11	67
Bibl. Med.-chir...............................		
Bibl. Univ. Gen.	Genf, 5	60
Biogr. Un.		
Bost. Soc.	} Boston, 5.....................	83
Bost. Soc. N. Hist.		
Brehm, Ornis.	*Vide* BREHM, in Part II.	
Brewst. Edinb. Journ. of Science		
Brewst. Journ. Sc.........................	} Edinburgh, 11	75
Brewst. Journ. Science....................		
Brit. Assoc.	London, 34..................	78
Bull. Ac. Brux.	Bruxelles, 4	37

Abbreviated form.	Place of publication and No.	Page of Part I.
Bull. Ac. Pet.		
Bull. Ac. Petersb.	St. Pétersbourg, 11	6
Bull. Acad. Petersb.		
Bull. Petersb.		
Bull. d. Sc.	Paris, 14, 15	50
Bull. Phys. Ac. Pétersb.	St. Pétersbourg, 12	6
Bull. Sc. med. di Bol.	Bologna, 14	64
Bull. Soc. med. Bol.		
Bull. Soc. Géol.	Paris, 26	50
Bull. Soc. Linn. Bord.	Bordeaux, 5	41
Bull. Soc. Med. Bol.	Bologna, 14	64
Bull. Soc. Natur. Mosc.	Moscow, 5	5
Bull. Natur. Mosc.		
Bull. Soc. Phil.	Paris, 14	50
Bull. Soc. Philom.		
Bull. Un. Sc.	Paris, 51	52
Büsch. Mag.		
Comm. Ac. Petr.	St. Petersbourg, 1	6
Comm. Acad. Petrop.		
Comm. Bonon.	Bologna, 2	64
Comment. Bonon.		
Comm. Gœtt.		
Comment. Gœtt.	Göttingen, 1	21
Comm. recent. Gœtt.		
Comm. Litter. Norimb.	Nürnberg, 12	31
Commerc. Norimberg.		
C. r.		
C. R.	Paris, 8	49
Comptes Rendus		
Compt. Rend.		
Congr. Sc. de Pise	Pisa, 1	70
Contr. Macl. Lyc.	Philadelphia, 7	85
Corresp. Bl. Würt. Landw. Ver.	Stuttgart, 8	34
Danske Vidensk.	Kiöbenhavn, 2, 3, 4	2
Denkschr. Allg. Schweiz. Ges.	Neuchatel, 4	62
Denkschrift. der Academ. der Wissensch. zu München	München, 3	30
Dict. class. d'H. n.	Paris, 74	53
Dict. Sc. Nat.	Paris, 73	53
Dublin Phil. Journ.	Dublin, 8	74
Ed. J. of Sc.	Edinburgh, 11	75
Edinb. Encyclopædia	Edinburgh, 16	75
Edinb. Journ. Nat. and Geogr. Sc.	Edinburgh, 12	75
Edinb. Journ.		
Edinb. New Phil. Journ.	Edinburgh, 10	75
Ed. N. Phil. J.		
Edinb. Phil. Journ.	Edinburgh, 9	75
Edinb. Versuch. u. Bemerk.	Altenburg, 1	10

Abbreviated form.	Place of publication and No.	Page of Part I.
Ent. Mag.	} London, 51..................	79
Entom. Mag.		
Entom. Soc.	London, 31..................	77
Ephem. Ac. Nat. Cur.	} Nürnberg, 4	31
Ephem. Nat. Cur.		
Eph. scient. e litt. per la Sicil.	Palermo, 4	69
Erdm. Journ. f. prakt. Chemie		
Féruss. Bull.	Paris, 51	52
Field Naturalist's Mag.	London, 49..................	79
Fror. Notiz.	Weimar, 5	35
Fuessl. N. Mag.		
Fuessl. N. Mag. d. Ent.	} Zürich, 4.....................	63
Fuess. N. Ent. Mag.		
Genootsch. te Vliss.	Middelburg, 1..............	10
Germ. Mag. Ent.	Halle, 13.....................	23
Giorn. Arcad...................................	Roma, 3	71
Giorn. d' Ital..................................	Venezia, 9	72
Giorn. di Grisell.		
Giorn. enc......................................	{ Firenze, 12	65
	{ Napoli, 14	69
Giorn. Ligust.	Genova, 4	66
Giorn. Sc. e Lett. Torin.	Torino, 10	71
Giorn. Sc. Med. Tor.	Torino, 21	72
Giorn. Soc. d' Incorrag. Milano	Milano, 2	67
Gotheb. Wetensk..............................	Götheborg, 1	1
Gött. gel. Anz.	Göttingen, 6	21
Gött. Mag......................................	} Göttingen, 9	21
Gœtt. Mag.....................................		
Grundig, Nat. u. Kunstgesch...........	*Vide* GRUNDIG, in Part II.	
Guérin, Mag. Entom.	} Paris, 55	52
Guér. Mag. de Zool.......................		
Hall. Naturf.	Halle, 6	22
Hamb. Mag.	Hamburg, 1	23
Hann. Mag.	Hannover, 2	24
Hertha		
Heus. Zeitschr.	} Eisenach, 2.................	18
Heusing. Zeitschr..........................		
Holl. Mag......................................		
Höpfn. Mag.	Zürich, 6.....................	63
Ill. Mag.	Braunschweig, 2............	15
Ind. Rev.	Calcutta, 9	82
Instit..	Paris, 59	52
Isis	Jena, 5	25
James. N. Philos. Journ.	Edinburgh, 10...............	75
J. A. S. B......................................	} Calcutta, 4	81
Journ. Asiat. Soc. Bengal		
J. Acad. Philad...............................		
Journ. Ac. Philad............................	} Philadelphia, 3	84
Journ. Acad. Philad.........................		

Abbreviated form.	Place of publication and No.	Page of Part I.
J. d'Hist. nat.	Paris, 44	51
J. d. Mines	Paris, 47	52
J. Sc. Agr. et Arts B.-Rhin	Strasbourg, 5	56
Journ. Agr. Soc. Engl.	London, 35	78
Journ. d. Mines	Paris, 47	52
Journ. d. Sav.	Paris, 32	50
Journ. de Méd.		
Journ. de Phys.	} Paris, 41	51
J. de Phys.		
Journ. f. prakt. Chem.		
Journ. Geol. Soc.	London, 20	77
Journ. Hebd.		
Journ. Highland Soc.	Edinburgh, 6	75
Journ. Roy. As. Soc.	London, 23	77
Journ. Roy. Inst.	London, 22	77
Journal of the Geolog. Soc. of Dublin	Dublin, 5	74
Karst. Arch.	Berlin, 30	13
Kröy. Tidskr.	Kiöbenhavn, 18	3
L'Institut	Paris, 59	52
Leipz. Mag.	Leipzig, 14	27
L. u. Br.		
L. u. Br. N. Jahrb.	} Heidelberg, 6	25
Leonh. u. Br. N. Jahrb.		
Leonh. Tasch.		
Leonh. Zeitschr.	Heidelberg, 4	24
Leop. Ac. d. Naturf.	Nürnberg, 6	31
Licht. u. Voigt's Mag.	} Gotha, 4	20
Lichtenberg's und Voigt's Mag.		
Lond. Lit. Gaz.	} London, 59	79
Lond. Liter. Gaz.		
L. and Ed. Phil. Mag.	} London, 43	78
Lond. and Edinb. Phil. Mag.		
Lond. Med. and Phys. Journ.		
Lyc. of New York	New York, 5	84
Madras Journ. Lit.	Madras, 1	82
Mag. Berl. Naturf. Fr.	Berlin, 12	12
Mag. d' Ital. Literat.	Weimar, 2	35
Mag. Encycl.	Paris, 46	51
Mag. for Naturv.	Christiania, 2	1
Mag. Hannov.	Hannover, 2	24
Mag. Nat. Hist.	} London, 46	78
Mag. of Nat. Hist.		
Mag. Zool.	Paris, 55	52
Mag. Zool. and Bot.	London, 47	78
Meck. Arch.	} Halle, 14, 15	23
Meck. Arch. f. Anat. u. Physiol.		
Med. Jahrb. d. Oesterr. St.	Wien, 7	35
Med. Journ.	Marburg, 3	29
Med. Sil. Satyr.		

Abbreviated form.	Place of publication and No.	Page of Part I.
Mél. Soc. Roy. Turin	Torino, 2	71
Mém. Ac. Brux.	} Bruxelles, 1	36
Mém. Acad. Brux.		
Mém. Ac. d'Aix	Aix, 2	39
Mém. Acad. Dijon	Dijon, 3	43
Mém. Acad. Pét.		
Mém. Acad. St. Pét.	} St. Pétersbourg, 5, 6	6
Mém. Ac. Pétersb.		
Mém. Acad. Sc. Par.	} Paris, 1	49
Mém. Acad. des Sc.		
Mém. Accad. Tor.	} Torino, 3	71
Mem. Accad. Torin.		
Mém. Acad. Tur.		
Mém. Ac. Tur.	} Torino, 4	71
Mém. Tur.		
Mém. de la Soc. d'Hist. Nat. de Paris		
Mém. Soc. d'Hist. Nat.		
Mém. Soc. d'H. N.	} Paris, 18	50
Mém. Soc. d'Hist. Nat. Par.		
Mém. de la Soc. Méd. d'Émulation		
Mem. di Lucca	} Lucca, 2	66
Mem. di Fis. e d' Ist. nat. Lucc.		
Mém. Flessing.	Middelburg, 1	10
Mem. I. R. Istit. Lombardo-Ven.	} Milano, 3	67
Mem. Istit. Lomb.-Ven.		
Mém. Laus.	Lausanne, 1	61
Mem. Lit. and Phil. Soc. Manchester	} Manchester, 1	80
Mem. Lit. and Scient. Soc. Manchester		
Mém. Mus.	Paris, 21	50
Mém. Natur. Mosc.	Moscow, 2	5
Mém. prés. à l'Acad.	Paris, 7	49
Mém. prés. Ac. Pétersb.	St. Pétersbourg, 8	6
Mém. Sav. étr.	Paris, 5	49
Mem. Sien.	Siena, 1	71
Mém. Soc. Acad. Sav.	Chambery, 1	65
Mém. Soc. Géol. Fr.	Paris, 27	50
Mém. Soc. H. Nat. Strasb.	Strasbourg, 7	56
Mem. Soc. Ital.	Verona, 1	73
Mém. Soc. Linn. Calv.	} Caen, 4	42
Mém. Soc. Linn. Norm.		
Mém. Soc. Med. d'Obs.		
Mém. Soc. Neuch.	} Neuchatel, 1	61
Mém. Soc. sc. nat. Neuch.		
Mém. Soc. Phys. et Hist. Nat.		
Mém. Soc. Roy. Méd.		
Mém. Stockh.	} Stockholm, 1	4
Mém. Ac. Stockh.		
Mem. Wern. Soc.	Edinburgh, 4	74
Mineral. Belustig.		

Abbreviated form.	Place of publication and No.	Page of Part I.
Misc. Nat. Cur.	} Nürnberg, 1	30
Miscel. Acad. Nat. Cur.		
Moll. Mag.		
Müll. Arch.	Berlin, 31	13
Münch. Acad.	München, 3	30
Mus. Helvet.	Bern, 8	59
N. Abh. Schw. Acad.	Stockholm, 2	4
N. Act. Ac. Petr.	St. Pétersbourg, 4	6
N. Act. Holm.	Stockholm, 2	4
N. Act. Leop. Ac.		
N. Act. Nat. Cur.	} Nürnberg, 6	31
N. Acta Ac. Nat. Cur.		
N. Act. Soc. R. Upsal.		
N. Act. Upsal.	} Upsala, 2	4
N. Act. Ups.		
N. Ann. Mus.	Paris, 22	50
N. Ann. Sc. Nat.		
N. Ann. Sc. Nat. Bol.	} Bologna, 13	64
N. Ann. Sc. n. di Bol.		
N. Bull. Soc. Phil.	} Paris, 15	50
N. Bull. Soc. Philom.		
N. Comm. Bonon.	Bologna, 3	64
N. Comment. Gœtt.	Göttingen, 1	21
N. Ed. J. of Sc.	Edinburgh, 11	75
N. Edinb. Philos. Journ.	Edinburgh, 10	75
N. Giorn. Lett.	Pisa, 3	70
N. Hamb. Mag.	Hamburg, 1	23
N. Mém. Ac. Berl.	Berlin, 2	11
N. Mém. Brux.	Bruxelles, 3	37
N. Mém. Natur. Mosc.	Moscow, 3	5
N. Mém. Soc. Helv. Sc. n.	Neuchatel, 4	62
N. Miscell. Lucch.	Lucca, 4	66
N. Oekon. Nachr. Schles. Ges.		
N. Nachr. d. Schles. Ges.	}	
Neue Oekon. Nachr. Ges. in Schles.		
N. Schr. d. Jenaer Miner. Soc.	}	
N. Schr. d. Miner. Ges. in Jena		
N. Schr. Naturf. Ges.		
N. Schr. Naturf. Ges. Halle.	} Halle, 12	23
N. Schr. Nat. Ges. Halle.		
N. Schr. Hall. Naturf. Ges.		
N. Schw. Akad.	} Stockholm, 2	4
N. Schwed. Akad.		
Nachr. d. Schles. Ges.		
Nat. Verh. Maatsch. d. Vet. Harl.		
Nat. Verh. d. Vet.	} Haarlem, 2	9
Natuurk. Verh. Holl. M. d. Vet.		
Natuurk. Verh. Harlem		
Natur. de Mosc.	Moscow, 2	5

Abbreviated form.	Place of publication and No.	Page of Part I.
Naturf.	} Leipzig, 4	26
Naturforscher		
Neue Nordische Beytr.	St. Petersbourg, 17	7
Nov. Act. Ae. Leop.		
Nov. Act. Ac. Leop. Car.		
Nov. Act. Nat. Cur.	>Nürnberg, 6	31
Nov. Act. Phys. Med. Acad. Cæs. Leop. Carol. Nat. Curios.		
Nouv. Mém. de Dijon	Dijon, 4	43
Oekon. Nachr. Ges. Schles.		
Opusc. Scelt.	} Milano, 6	67
Opusc. Sc.		
Ornis	Vide BREHM, Part II.	
Par. Verh.	Paris, 3	49
Phil. Mag. ser. 1.	London, 41	78
Phil. Mag. ser. 2.	} London, 42	78
Phil. Mag. a. Ann. of Philos.		
Phil. Mag. ser. 3.	} London, 43	78
Phil. Mag. and Journ.		
Phil. Tr.	} London, 1	76
Phil. Trans.		
Physiogr. Selskap. Aarsber. Lund	Lund, 3	4
Pogg. Ann.	Berlin, 36	13
Prag. Abh.	Prag, 1	32
Prize Essays, Highland Soc.	Edinburgh, 5	75
Proc. Bost. Soc. N. H.	Boston, 5	83
Proc. Com. Zool. Soc.	} London, 27	77
Pr. Com. Zool. Soc.		
Proc. Geol. Soc.	London, 19	77
Proc. R. Irish Acad.	} Dublin, 2	74
Proc. Roy. Irish Acad.		
Proc. Roy. Soc.	London, 14	77
Pr. Zool. Soc.	} London, 28	77
Proc. Zool. Soc. Lond.		
Quart. Journ. Geol. Soc.	London, 20	77
Quart. Journ. Sc.	} London, 21	77
Quart. Journ. of Sc.		
Rec. Gen. Sc.	London, 54	79
Rep. Brit. Assoc.	London, 34	78
Rev. Zool.	} Paris, 56	52
Rev. Zool. Soc. Cuv.		
Revue Franç.		
Roux, Journ. Med.		
Rozier, Journ. de Phys.	} Paris, 39	51
Roz. Journ. de Phys.		
Russ. Samml. f. Naturw. u. Heilk.	Riga, 2	6
Scelta de Opusc. interess.	Milano, 5	67
Schles. Ges.		
Schles. Gesellsch.	}Breslau, 1, 2	15
Schles. Patr. Ges.		

Abbreviated form.	Place of publication and No.	Page of Part I.
Schr. Berl. Naturf.		
Schr. Berl. Naturf. Fr.		
Schrift. der Gesellsch. Naturforsch. Freunde	Berlin, 9	12
Schr. d. Berlin. Ges. Naturf. Freunde ...		
Schröt. Journ.	Weimar, 1, 3	35
Schw. Akad.		
Schwed. Ak. Abh.	Stockholm, 1	4
Schwed. Abh.		
Schweig. Journ. f. Chem. u. Phys.	Nürnberg, 14, 15	31
Schw. Seid. N. Jahrb. Phys.		
Schweigg. N. Jahrb.		
Schweitz. Arch. f. Statist.		
Silb. Rev. Ent.	Strasbourg, 8	56
Silberm. Rev. Ent.		
Sill. Am. J.	New York, 9	84
Sill. Am. Journ.		
Skr. Kiöb. Selsk.	Kiöbenhavn, 1, 2, 3	2
Skr. Kiöbenh. Selsk.		
Skrivt. af Naturhist. Selsk.	Kiöbenhavn, 9	2
Soc. d'Agric. Seine et Oise	Versailles, 2	58
Soc. d'Hist. nat. Bord.	Bordeaux, 1	41
Soc. d'Hist. nat. Par	Parıs, 16, 17, 18	50
Soc. Ital. d. Sc.	Modena, 1, 2	68
Soc. Ital. in Modena		
Tidskr. f. Naturw.	Kiöbenhavn, 17	3
Tiedem. Zeitschr.	Darmstadt, 1	16
Todd's Cyclop. of Anat. and Physiol. ...	London, 74	80
Tr. Am. Phil. Soc.		
Trans. Am. Phil. Soc.	Philadelphia, 1	84
Transact. Americ.		
Tr. Geol. S. Lond.	London, 18	77
Trans. Geol. Soc.		
Tr. Lin. Soc. Lond.		
Trans. Linn. Soc. Lond.	London, 16	77
Trans. Linn. Soc.		
Tr. Liter. a. Hist. Soc. Queb.	Quebec, 1	85
Tr. Phil. Soc. of Manch.	Manchester, 1	80
Tr. R. Soc. Edinb.	Edinburgh, 2	74
Trans. Roy. Soc. Edinb.		
Trans. Cambr. Phil. Soc.	Cambridge, 1	74
Trans. Entom. Soc.	London, 31	77
Trans. Geol. Soc. Cornw	London, 77	80
Trans. Med. and Phys. Soc. Calcutta ...	Calcutta, 7	81
Trans. Med. et Phys. Soc. Calcutta ...		
Trans. Microsc. Soc.	London, 33	78
Trans. Nat. Hist. Soc. Newcastle	Newcastle, 2	80
Trans. R. I. Acad.	Dublin, 1	74
Trans. R. Irish Acad.		
Trans. Zool. Soc.	London, 26	77

Abbreviated form.	Place of publication and No.	Page of Part I.
Trondh. Selsk.	Kiöbenhavn, 23	3
Ungrisches Magazin	Pressburg, 1	33
Uytg. Verh.	Amsterdam, 13	8
Van der Hoev. Tijdsch.	Leyden, 4	9
Van der M. Rec.		
Verh. Kurland. Ges. f. Litt. u. K.		
Verh. Maatsch. te Haarl.		
Verhand. Maatsch. te Haarl.	Haarlem, 1	9
Verhand. Maatsch. Haarl.		
Verh. Teyler's Genoots...................		
Verhandl. der Ges. Naturforsch. Fr. in Berlin	Berlin, 13	12
Vet. Ac.................................		
Vet. Acad................................		
Vet. Acad. Handl........................	Stockholm, 1	4
Vet. Acad. Handling.		
Vetensk. Ac. Handl......................		
Vet. Acad. N. Handl.	Stockholm, 2	4
Vid. Selsk. Naturvidensk. Afhand.	Kiöbenhavn, 4	2
Voigt's Mag.............................	Gotha, 4	20
Walk. Ent. Mag.	London, 51	79
Wiedem. Arch.	Berlin, 25	13
Wiedemann, Zool. Mag....................	Kiel, 2.....................	2
Wiegm. Arch.............................	Berlin, 33	13
Wittenb. Wochenbl.	Wittenberg, 1..............	36
Zeitschr. f. Anthrop.		
Zeitschr. f. Mineral.	Frankfurt am Main, 15 ... Heidelberg, 4	19 24
Zeitschr. f. organ. Phys.	Eisenach, 2	18
Zeitschr. f. die organ. Phys..............		
Zool. J.	London, 45	78
Zool. Journ.		

BIBLIOGRAPHIA

ZOOLOGIÆ ET GEOLOGIÆ.

PARS PRIMA,

ACTA SOCIETATUM, DIARIA, ET TRACTATUUM SYLLOGAS CONTINENS.

SCANDINAVIA ET DANIA.

(Vide etiam GREIFSWALD, 2; NÜRNBERG, 13.)

CHRISTIANIA.

1. Bidrag till Kundskab over Naturvidenskaberne. Christiania, 1813.
2. Magazin for Naturvidenskaberne, udg. af det *Physiographiske Selskab i Christiania.* 10 vols. 8°. Christiania, 1823–1831. 2den Række, 2 vols. 1832–1836.
3. Nyt Magazin for Naturvidenskaberne, udg. af det *Physiographiske Forening i Christiania.* 8°. Christiania, 1836–1846.
4. Norsk Magazin for Laege-videnskaben, herausgegeben von dem *ärztlichen Vereine in Christiania.* 1841–1842.

GÖTHEBORG (Gottenburg).

1. *Götheborgska Vetenskaps och Vitterhets samhällets* Skrifter. 8°. Götheb. 1778–1808, 4 fasc.
2. *Götheborgska Vetenskaps och Vitterhets* Nya Handlingar 1819–22, 5 Hefte.

3. Gothenburgisches Magazin ; Suecico idiomate edidit *D. J. Roseen* 1759, et continuavit titulo: Gothenburgisches Wochenblatt, *J. Gothenius* ; 1765.

4. Samling af Rön och Uptakter i senare tider uti Physik. Gothemb. 1781, 8°.

5. Förhandlingar vid *Skandinaviska Naturforskarnes och Läkarnes* hälna möte i Götheborg, År 1839. Götheb. 1840. 8°.

KIEL (Kilia).

1. Beyträge zur Naturkunde, v. *F. Weber.* 2 vols. 8°. Kiel, 1805–1810.

2. Zoologisches Magazin, v. *C. R. W. Wiedemann.* 2 vols. 8°. Kiel u. Altona, 1817–1823.

KIÖBENHAVN (Hafnia, Copenhagen, Kopenhagen).

1. Acta *Societatis Scientiarum Regiæ Hafniensis.* (Dan.) Skrifter som udi det *Kiöbenhavnske Selkskab af Lærdoms og Videnskabs Elskere* ere fremlagte og oplæste. Hafniæ, 1745–79, 12 vols. 4°.

2. Nye Samling af det *Kong. Danske Videnskabernes Selskabs* Skrifter. Kiöbenhavn, 1781–99, 5 vols. 4°.

3. Det *Kongel. Danske Videnskabers Selskabs* Skrifter for Aarene 1800–12. Kiöbenhavn, 1801–18, 6 vols. 4°.

4. Det *Kongelige Danske Videnskabernes Selskabs* naturvidenskabelige og mathematiske Afhandlinger. Kiöbenhavn, 1824–43, 10 vols. 4°.

5. Physikalische, chemische, naturhistorische u. mathematische Abhandlungen aus der neuen Sammlung der Schriften der *Königl. dänischen Gesellschaft der Wissenschaften.* Aus dem Dänischen von *P. Scheel* u. *K. F. Degen.* 2 vols. 8°. Copenhagen, 1798–1803.

6. Schriften der physikalischen Classe der *Königlich Dänischen Gesellschaft der Wissenschaften* in Kopenhagen. Von *C. G. Rafn.* Aus dem Dänischen. 3 vols. 8°. Copenhagen, 1801–1805.

7. Oversigt over det *Kongelige Danske Videnskabernes Selskabs* Forhandlinger og dets Medlemmers Arbeider ; af *H. C. Örsted.* Kiöb. 1815–1843.

8. Acta medica et philosophica Hafniensia ann. 1671–1679 ; a *Th. Bartholini.* 5 vols. 4°. Hafn. 1673–1680.

9. Skrivter af *Naturhistorie-Selskabet.* Kiöbenhavn, 1790–1810. 6 vols. 8°.

10. Schriften der *naturforschenden Gesellschaft in Kopenhagen.* Aus dem Dänischen. 8°. Kopenhagen, 1793.

11. Kleine Abhandlungen einiger Gelehrten in Schweden über verschiedene in die Physik, Chemie und Mineralogie laufende Materien. Aus dem Schwed. übers. Kopenhagen u. Leipzig. 1766–68; 2 Th. 8°.

12. Neue Sammlung kleiner Abhandlungen einiger Gelehrten in Schweden über Naturgeschichte. 8°. Kopenhagen, 1774.

13. Schwedisches Magazin, oder gesammelte Schriften der grössten Gelehrten in Schweden für die Liebhaber der Arzneywissenschaft, der Naturgeschichte, Chemie u. Oekonomie; aus dem Schwed. übers. v. *J. K. Weber*, Copenh., 1768, 1770, 2 vols. 8°.

14. Neue Sammlung verschiedener Schriften der grössten Gelehrten in Schweden, für die Liebhaber der Naturgeschichte, &c. Kopenh., 1775, 8°.

15. Nordisches Archiv für Natur u. Arzneiwissenschaft u. Chirurgie, v. *C. H. Pfaff, P. Scheel* u. *K. A. Rudolphi.* 4 vols. 8°. Kopenhagen, 1799–1805

16. *Veterinær Selskabets* Skrifter. Kiöbenhavn, 1808–1818, 3 vols. 8°.

17. Tidskrift for Naturvidenskaberne; af *H. C. Örsted, J. W. Hornemann, J. Reinhardt.* Kiöbenhavn, 1822–28. 5 vols. 8°.

18. Naturhistorisk Tidskrift; udgivet af *H. Kröyer.* Kiöbenhavn. 4 vols. 8°. 1837–1843.—Ny Række, 1844–1847.

19. Bibliothek for Laeger. Kiöbenhavn, 1825–1843.

20. Ugeskrift for Laeger. Kiöbenhavn, 1839–1843.

21. Forhandlinger ved de *Skandinaviske Naturforskeres Forsamling* i Kjöbenhavn, 1840; 8°. Kjöbenhavn, 1842.

22. Rit pèsz Islenska Laerdöms (Scripta *Societatis Scientiarum Islandicæ*). Copenh., 1780–82. 2 vols. 8°.

23. Det *Trondhiemske Selskabs* Skrifter. 3 vols. 12mo. Kiöbenhavn, 1761–1765.—(Contin.) Det *Kongelige Norske Videnskabers Selskabs* Skrifter. 2 vols. 12mo. Kiöbenhavn, 1768–1774.

24. Schriften der *Drontheimischen Gesellschaft.* Kopenh. u. Leipz. 1765–70; 4 vols. 8°. fig.—(Contin.) Der *Königl. Norwegischen Gesellschaft der Wissenschaften* Schriften, aus dem Dänischen übers. 4 vols. Kopenh. 1770, 8°.

25. Nye Samling af det *Kongl. Norske Videnskabers Selskabs* Skrifter. Kiöbenhavn, 1784–88, 2 vols. 4°. Nyeste Samling, Kiöbenhavn, 1798, 8°

26. Skrifter i det 19de Aarhundrede af det *Norske Videnskabers-Selskab i Trondhjem.* 3 vols. Kiöbenhavn et Trondhjem, 1817–1832.

LUND.

1. Physiographiska Sällskapets Handlingar. Lund. 1776, 8º.
2. Physiographiska Sällskapets Magazin. Lund. 1781, 8º.
3. Physiographiska Sällskapets Årsberättelse. Lund. 1823–24, 8º.
4. Physiographiska Sällskapets Tidskrift. 1 vol. 8º. Lund. 1837–1838.

STOCKHOLM (Holmia).

(Vide etiam UPSALA, 1 ; BONN, 2 ; LEIPZIG, 30, 31 ; STRAL-SUND, 3 ; PARIS, 79.)

1. *Kongliga Svenska Vetenskaps Akademiens* Handlingar. 40 vols. 8º. Stockholm 1739–1779.
2. *Kongliga Svenska Vetenskaps Akademiens* Nya Handlingar. 33 vols. 8º. Stockholm, 1780–1812.—Handlingar för 1813–1844, 34 vols. 8º. Stockholm, 1813–1846.
3. Register öfver *Kongl. Vetenskaps Acad.* Handlingar. Stockh. 1831–37, af *A. T. Ståhl,* 8º.
4. Öfversigt af *Kongl. Vetenskaps Academiens* Förhandlingar, 8º. Stockholm, 1844.
5. Stockholmisches Magazin, darinnen kleine Schwedische Schriften, welche die Geschichte, Staatsklugheit u. Naturforschung betreffen, etc., mitgetheilt werden. Stockh., 1754–56, 8º. 3 vols.
6. Tidskrift för Jägar och Naturforskar. Stockholm, 1832–34, 3 vols. 8º.
7. Förhandlingar vid de *Skandinaviske Naturforskarnes* tredje Möte i Stockholm, d. 13–19 Juli, 1842, 8º. Stockh., 1843.

TRONDHIEM (Nidrosia, Drontheim).

(Vide KIÖBENHAVN, 23, 24, 25, 26.)

UPSALA (Upsalia, Upsal.)

(Vide etiam PARIS, 79.)

1. Acta Litteraria Sueciæ. Rostochii, Lipsiæ, Upsaliæ et Stock-holmiæ, 1720–1739, 4 vols. 4º.
2. Acta *Regiæ Societatis Scientiarum Upsaliensis,* ann. 1740–1750. Stockholmiæ, 1744–1751, 5 vols. 4º. Nova Acta, Upsalæ, 1733–1842, 12 vols. 4º.
3. Acta Medicorum Suecicorum, s. Sylloge Observationum et Ca-suum rariorum, præsertim in Historià naturali, Praxi medicâ et chirurgicâ. Ups. 1783, 8º.

ROSSIA ET POLONIA.

DORPAT.

(Vide etiam BERLIN, 47.)

1. Beiträge zur Naturkunde aus den Ostsee-Provinzen Russlands herausgegeben von *C. H. Pander*. Dorpat. 1820.

HELSINGFORS.

1. Acta *Societatis Scientiarum Fennicæ*, Helsingfors, 1838.

KASAN (Casanum).

1. Kasanskii Vestnick, 1819.
2. Zavolgeskii Mouravei, 8°. Kasan, 1831.
3. Annales *Universitatis Casanensis*, 8°. Casan.

MITTAU (Mitau).

1. Jahresverhandlungen der *Kurländischen Gesellschaft für Litteratur und Kunst*. Mittau, 1819, 4°.
2. Sendungen der *Kurländischen Gesellschaft für Litteratur und Kunst*. Mittau, 4°. 1840–1845.
3. Die Quatember, Zeitschrift für naturwissenschaftliche, geschichtliche, philologische, literarische u. gemischte Gegenstände; von *E. v. Trautvetter*. Mittau, 8°. 1829.

Moscow (Mosquia, Moscou, Moscau).

1. Journal de la *Société Impériale des Naturalistes de Moscou*. 4°. 1805.
2. Mémoires de la *Société Impériale des Naturalistes de Moscou*. 4 vols. 4°. Moscou, 1806–1823.
3. Nouveaux Mémoires de la *Société Impériale des Naturalistes de Moscou*. Moscou, 1829–41, 1 vol. 4°.
4. Rapport périodique de la *Société Impériale des Naturalistes de Moscou*, 1809.
5. Bulletin de la *Société Impériale des Naturalistes de Moscou*. 18 vols. 8°. Mosc. 1829–1846.
6. Acta *Societatis Physico-medicæ Mosquensis*, 1805.
7. Commentationes *Societatis Physico-Medicæ* apud Universitatem Mosquensem institutæ. Vol. i. 4°. Mosc. 1808.
8. Nouveau Magasin d'Histoire naturelle, de *Dirgereski*. Mosc. 1824–29.
9. Outcheniia Zapiski *Imp. Moskorskago Ouniversiteta*, 8°. 1833.

RIGA.

1. Physikalische u. medicinische Abhandlungen der *Kaiserlichen Akademie der Wissenschaften in Petersburg*. Aus der Lateinischen von *J. Lor*, u. *C. Mümler*, 3 vols. 8°. Riga, 1782–85.

2. Russische Sammlungen für Naturwissenschaften u. Heilkunst, v. *A. Crichton, J. Rehmann u. K. F. Burdach*. 2 vols. 8°. Riga u. Leipzig, 1815–1818.

ST. PETERSBOURG (Petropolis, Petersburg).

1. Commentarii *Academiæ Scientiarum Imperialis Petropolitanæ*. 14 vols. Petropoli, 4°. 1728–1751.

2. Novi Commentarii *Academiæ Scientiarum Imperialis Petropolitanæ*. 20 vols. Petrop. 4°. 1750–76.

3. Acta *Academiæ Scientiarum Imperialis Petropolitanæ*. 6 tom. Petrop. 1778–86, 4°.

4. Nova Acta *Academiæ Scientiarum Imperialis Petropolitanæ*. 15 vols. 4°. Petrop. 1787–1806.

5. Mémoires de l'*Académie Impériale des Sciences à Pétersbourg*. (Sciences naturelles.) 11 vols. 4°. Pétersb. 1809.

6. Mémoires de l'*Académie Impériale des Sciences de St. Pétersbourg*. 6me série. Sciences mathématiques physiques et naturelles. 4°. Petersburg, 1830–1846.

7. Sermones in primo solemni *Academiæ Scientiarum Petropolitanæ* conventu recitati. Petropoli, 1725, 4°.

8. Mémoires présentés à l'*Académie Impériale des Sciences à St. Pétersbourg*, pour répondre à la question minéralogique proposé pour le prix de 1785. St. Pétersbourg, 1786, 4°.

9. Recueils des actes des Séances publiques de l'*Académie Impériale des Sciences de St. Pétersbourg*, 4°. St. Pétersbourg, 1827–1846.

10. Mémoires présentés à l'*Académie Impériale des Sciences de St. Pétersbourg*, par divers savans. 4°. Pétersbourg. 6 vols. 1830–1846.

11. Bulletin scientifique, publié par l'*Académie Impériale des Sciences de St. Pétersbourg*, 10 vols. 4°. 1836–1841.

12. Bulletin Physico-Mathématique de l'*Académie Impériale des Sciences de St. Pétersbourg*, 4°. 1842–1846.

13. Annuaire du Journal des Mines de Russie, pour les années 1835, 36, 37, et 38, avec une introduction. 5 vols. 8°. Pétersbourg, 1840.

14. Annuaire Magnétique et Météorologique du corps des Ingénieurs des Mines de Russie, ou Recueil d'Observations Magnétiques et Météorologiques faites dans l'étendue de l'empire de Russie. Par *A. T. Kupffner*, année 1841.

15. Mémoires de la *Société Minéralogique Impériale de St. Péters-bourg.*

16. Actes (en Russe) de la *Société Impér. Minéralogique* de St Pé-tersbourg, 1801–30, 8°.

17. Neue nordische Beyträge für physicalisch. und geograph. Erd- und Volkerbeschreibung Naturgeschichte und Oeconomie, v. *P. S. Pallas.* Petersb. und Leipz., 1781–1796, 7 vols. 8°.

18. Journal des Mines, par *Gornoi.* Pétersb. 1825.

19. Beiträge zur Kenntniss des Russischen Reiches u. der angrenzen-den Länder Asiens. Von *K. E. v. Baer* u. *G. v. Helmersen,* 8°. Petersburg, 1839–1845.

WARSCHAU (Varsovia, Warsaw).

1. Vermischte Abhandlungen der *Warschauer Gesellschaft zur Be-förderung der Naturkunde, Oekonomie, Manufacturen,* etc. Warsch. u. Dresd., 1768. 8°.

2. Slavoniana. Varsov. 1829.

3. Magazin für Heilkunde u. Naturwissenschaft in Pohlen, v. *L. Leo.* 4 vols. 8°. Warschau, 1828.

WILNA (Vilna).

1. Universitas et *Academia Wilnensis* instaurata ac nomine *Scholæ Principis* insignata. Vilnæ, 1781, fol.

HOLLANDIA.

(Vide etiam HADAMAR, 1.)

AMSTERDAM (Amstelodamum).

(Vide etiam HAARLEM, 1, 4; Paris, 1, 10, 68.)

1. Verhandelingen der erste Klasse van het *Hollandsch Instituut van Wetenschappen, Letterkund een schoone Kunsten te Amsterdam.* Amst. 1812–1825, 7 vols. 4°.

2. Nieuwe Verhandlingen der erste Klasse van het *Koningl. Neder-landsche Instituut van Wetenschappen, Letterkunde en schoone Kunsten te Amsterdam.* 10 vols. 4°. Amsterd. 1827–1844.

3. Verhandelingen der tweede Klasse van het *Koningl. Neder-landsche Instituut.* Amsterd. 1812–1826, 5 vols. 4°. Nieuwe Verhandel. Amst. 1827–1840, 9 vols.

4. Commentationes Latinæ tertiæ classis *Instituti Regii Belgici.* Amstelod. 1818–1836, 6 vols. 4°.

5. Prijs Verhandelingen der derde Klasse van het *Koningl. Neder-landsche Instituut.* Amsterd. 1822, 1 vol. 4°.

6. Proces-verbaal van de algemeene vergaderingen van het *Koningl. Nederlandsche Instituut,* gehouden te Amsterdam. 4°. Amst. 1810–1827.

7. Byvoegsel tot het proces-verbaal van de twintigste algemeene vergadering van het *Koningl. Nederl. Instituut,* gehouden den 27 Augustus, 1827. 4°.

8. Gedenkschriften in de hedendaagsche talen van de derde Klasse van het *Koningl. Nederlandsche Instituut.* 1ste deel, 1817 ; 3de deel, 1826 ; 5de deel, 1836. 3 vols. 4°.

9. Werken der *Hollandsche Maatschappy van fraaije Kunsten en Wetenschappen.* 8 vols. 8°. Amsterd. en Leyde, 1810–1830.

10. Natuur en Konstkabinet, of Kabinet der natuurlyke Historien, Wetenschappen, Konsten en Handwerken, auct. *W. van Ranouw.* Amst. 1719–27. 8°. 8 vols.—T. ix continet Indicem a *R. Vander Meetsch,* Amst. 1732.

11. Vytgeleezene Natuurkundige Verhandelingen. Amst. 1735 et seq. 8°.

12. Natuurkundigeaanmerkingen,waarneemingenen ondervindingen van de *Koningklyke Societeit van London.* Door *P. Le Clercq.* 2 vols. 8°. Amsterdam, 1735.

13. Vytgezockte Verhandelingen nyt de nieuweste Werken van de Societeten der Wetenskapen in Europa. Amsterd. 1764. 8°.

14. Geneez-heel en natuurkundige Verlustigungen, getrokken uit verscheide Werken, over deeze Wetenschappen in Europa uitgekommen of nog uitkomende. Door een Genootschap. Amst. 1766. 8°.

15. Natuurkundige Verhandelingen, of Verzameling van Stükken der Natuurkunde, Geneeskunde, en natuurlyke Historie betreffende, getrokken uit de geachte Werken van Engelsche, Fransche en Hoogduitsche Schryvers, etc. vertaald. Amst. 1767. 8°.

16. Tijdschrift voor natuurkundige Wetenschappen en Kunsten, door *C. G. G. Reinwardt.* 2 vols. 8°. Amsterdam, 1810, 1811.

17. Bydragen tot de natuurkundige Wetenschappen, verzameld door *H. C. Van Hall, W. Vrolik* en *G. J. Mulder.* 7 vols. 8°. Amsterd. 1826–1832.

18. Magasin des Sciences Philosophiques et Naturelles, par cahiers. 8°. Amsterd.

GRONINGEN (Groninga).

1. Annales *Academiæ Groninganæ* (1815–1828), 13 vols. 4°. Groningæ, 1817–1835.

HAARLEM.

(Vide etiam ALTENBURG, 2 ; LEIPZIG, 35, 37.)

1. Verhandelingen van de *Hollandsche Maatschappy der Weten-schappen te Haarlem*. Haarl. en Amsterd. 1654–1798, 43 vols. 8°.

2. Natuurkundige Verhandelingen van de *Hollandsche Maatschappy de Wetenschappen te Haarlem*. Haarl. 1799–1839, 24 vols. 8°.

3. Registre ofte hoofdraaklyke inhoud der Verhandelingen die in de twaalf eerste deelen van de *Hollandsche Maatschappy der We-tenschappen te Haarlem* voorkomen; door *J. F. Martinet*. 8°. Haarl. 1772.

4. Beredeneerd Register ofte hoofdzaaklyke inhoud der Verhande-lingen die in de 28 deelen van de *Hollandsche Maatschappy der Wetenschappen* voorkomen; door *J. F. Martinet*. 8°. Haarl. en Amsterd. 1793.

5 Wijs gerige Verhandlingen van de *Maatschappij der Weten-schappen te Haarlem*. 1ste deel, Amsterd. 1811. 8°. 2de deel, Haarl. 1822. 8°.

6. Natuurkundige Verhandelingen van de *Hollandsche Maatschappije der Wetenschappen te Haarlem*. Tweede verzameling. Haarl. 1841. 1 vol. 4°.

HAGEN (Haga Comitum, s'Gravenhage, the Hague, la Haye).

1. Verhandelingen van de *Natuur en Geneeskundige Corresponden-ten Societeit* in de Vereenigde Nederlanden te Hagen 1784–1798. 8 vols. 8°.

2. L'Europe savante, depuis 1718 à 1720. La Haye, 1718–1720. 12 vols. 12°.

3. Mémoires pour servir à l'Histoire Naturelle des Animaux et des Plantes. Par *C. Perrault, Dodart*, &c. 4°. La Haye, 1731.

4. Bibliothèque des Sciences et des Arts. La Haye, 1754–1778. 50 vols. 8°.

5. Ephemeriden der *Natuurkundige Wetenschappen*. 8°. Hage, 1834.

LEYDEN (Lugdunum Batavorum, Leide).

(Vide etiam AMSTERDAM, 9.)

1. Annales *Academiæ Lugduno-Batavæ*. 4°. Leyden, 1816–1838.

2. Museum Academicum *Academiæ Lugduno-Batavæ*; ed. *G. Sandifor*. Lugd. Bat. 1827, fol.

3. Genees-Natuur en Huis-houdkundig Kabinet, uitgeg. door *T. V. van Engelen*. Leyd. 1779.

4. Tijdschrift voor natuurlike Geschiedenis en Physiologie. Van *J. van der Hœven* en *W. H. Vriese*. 8°. Leyden en Utrecht, 1834–1846.

5. Natuur en Scheikundig Archief. Van *G. J. Mulder* en *W. Wenckebach.* 4 vols. 8°. Leyden en Rotterdam, 1836.

6. Bulletin des Sciences Physiques et Naturelles en Neerlande, par *F. A. W. Miquel, G. J. Mulder*, et *W. Wenckebach.* 4°. Leyden, 1838. Rotterdam, 1839. Utrecht, 1840.

MIDDELBURG.

1. Verhandelingen uitgegeven door het *Zeeuwsche Genootschap der Wetenschappen* te Vlissingen. Middelburg, 1769–1792. 15 vols. 8°.

2. Nieuwe Verhandelingen van het *Zeeuwsche Genootschap der Wetenschappen.* Middelburg, 1807–1835. 5 vols. 8°.

3. Nieuwe Werken van het *Zeeuwsche Genootschap der Wetenschappen.* 8°. Middelburg, 1836.

ROTTERDAM.

(Vide etiam LEYDEN, 5, 6.)

1. Verhandelingen van het *Bataafsch Genootschap der præfondervindelyke Wysbegeerte te Rotterdam.* 12 vols. 4°. Rotterd. 1774–1798. Nieuwe Verhandelingen, 10 vols. 4°. Rotterdam, 1800–1846.

UTRECHT (Trajectus ad Rhenum, Utregt).

(Vide etiam LEYDEN, 4, 6.)

1. Annales *Academiæ Rheno-Trajectinæ.* 4 vols. 8°. Traj. ad Rhen. 1817–1820. Supplementum Annalium, Traj. ad Rhen. 1818, 8°. Annales, 17 vols. 8°. Traj. ad Rhen. 1822–1837.

2. Verhandelingen van het provinciaal *Utregtsch Genootschap van Kunsten en Wetenschappen.* Utregt, 1781–1801. 10 vols. 8°.

3. Nieuwe Verhandelingen, &c. 8°. 1802–1846.

VLISSINGEN (Flushing, Flessingue).

(Vide MIDDELBURG, 1, 2, 3 ; Giessen, 1.)

GERMANIA.

ALTENBURG.

1. Neue Versuche u. Bemerkungen der *Gesellschaft zu Edinburg.* Altenb. 1756–1762, 7 vols. 8°.

2. Abhandlungen der *Holländischen Gesellschaft der Wissenschaften zu Harlem.* Aus dem Holländischen von *A. G. Kästner.* 2 vols. 8°. Altenburg, 1758.

3. Mittheilungen aus dem Osterlande. Von dem Kunst- u. Handwerks-Vereine, der *naturforschenden u. der pomologischen Gesellschaft zu Altenburg.* 8°. Altenburg, 1837–1846

ALTONA.

(Vide etiam KIEL, 2.)

1. Nordische Beyträge zum Wachsthume der Naturgeschichte, Haushaltungskunst, Handlung, etc. Altona, 1756–58. 8°. 2 vols.

2. Astronomische Nachrichten, von *H. C. Schumacher.* 17 vols. 4°. Altona, 1836–1841.

ANNABERG.

1. Obererzgebürgisches Journal, oder Sammlung von allerhand in die hiesige Naturwissenschaft, etc. merkwürdigen Abhandlungen; von *J. Chr. Themel.* Annab. u. Freiberg, 1747–53. 12 vols. 8°.

AUGSBURG (Augusta Vindelicorum).

1. Der aus dem Reiche der Wissenschaften wohl versuchte Referendarius, oder auserlesene Sammlungen von allerhand wichtiger Abhandlungen, Schriften u. Versuchen aus der Naturlehre, Arzneywissenschaft, etc. Von *J. A. E. Maschenbauer.* Augsb. 1767 et seq. 12 Th. 4°.

BERLIN (Berolinum).

(Vide etiam BONN, 1; GOTHA, 5; HALLE, 14; PARIS, 80.)

1. Miscellanea Berolinensia ad incrementum scientiarum ex scriptis Societati exhibitis edita; 8 vols. 4°. Berol. 1710–1744.

2. Histoire et Mémoires de l'*Académie Royale des Sciences de Berlin,* de 1745 à 1769; 25 vols. 4°. Berl. 1746–1771. Nouveaux Mémoires, de 1770 à 1786; 17 vols. 4°. Berl. 1772–1788. Mémoires depuis 1786 jusqu'à 1804; 14 vols. 4°. Berl. 1762–1807.

3. Abhandlungen der physikalischen Klasse der *Königlich-Preussischen Akademie der Wissenschaften.* 13 vols. 4to. Berlin, 1815–1832. (Fortsetz.) Physikalische Abhandlungen der *Königlichen Akademie der Wissenschaften* zu Berlin. 4°. Berlin, 1832–1846.

4. Dissertations, qui ont remporté les prix adjugés par l'*Académie Royale des Sciences et Belles-Lettres de Prusse en* 1766. Berlin, 1767. 4°.

5. Histoire de l'*Académie Royale des Sciences et Belles-Lettres,* depuis son origine jusqu'à présent. Berlin, 1788. 4°.

6. Sammlung der deutschen Abhandlungen, welche in der *Königl. Akademie der Wissenschaften zu Berlin* vorgelesen worden. (1788–1834.) Berl. 1793–1836, 14 vols. 4°.

7. Bericht über die zur Bekanntmachung geeigneten Verhandlungen der *Königl. Preussischen Akademie der Wissenschaften zu Berlin.* Berl. 1836–1844, 8°.

8. Beschäftigungen der *Berlinischen Gesellschaft Naturforschender Freunde.* Berl. 1775–1779, 4 vols. 8°.

9. Schriften der *Berlinischen Gesellschaft Naturforschender Freunde.* 6 vols. 8°. Berl. 1780–1785.

10. Beobachtungen und Entdeckungen aus der Naturkunde, von der *Gesellschaft Naturforschender Freunde zu Berlin.* 5 vols. 8°. 1787–1794.

11. Neue Schriften der *Gesellschaft Naturforschender Freunde zu Berlin.* 4°. 1795–1805, 4 vols.

12. Magazin der *Gesellschaft Naturforschender Freunde zu Berlin,* für die neuesten Entdeckungen in der gesammten Naturkunde. Berl. 1807–1818, 8 vols. 4°.

13. Verhandlungen der *Gesellschaft Naturforschender Freunde in Berlin.* 1819–1829, Berlin.

14. Zweifaches Universalregister über die bisherigen Schriften der *Gesellschaft Naturforschender Freunde.* 8°. Berlin, 1794.

15. Mittheilungen aus den Verhandlungen der *Gesellschaft Naturforschender Freunde zu Berlin.* 8°. 1836–1839, 3 vols.

16. Berlinische wöchentliche Berichte der merkwürdigsten Begebenheiten des Reichs der Wissenschaften und Künste. 1749. 8°.

17. Monathliche Beyträge zur Naturkunde, von *J. D. Denso.* Berl. 1752, 8°. 6 vols. Fortgesetzte Beyträge, 1765, 6 vols.

18. Gemeinnütziges Natur- u. Kunst-Magazin, oder Abhandlungen zur Beförderung der Naturkunde, der Künste, Manufacturen und Fabriken. Berl. 1763–67, 8°. 3 vols.

19. Berlinisches Magazin, oder gesammelte Schriften u. Nachrichten für die Liebhaber der Arzneywissenschaft, Naturgeschichte und der angenehmen Wissenschaften überhaupt. Berl. 1765–67, 4 vols. 8°. Neues Berlinisches Magazin, 1782 et seq.

20. Berlinische Sammlungen zur Beförderung der Arzneywissenschaft, Naturgeschichte, etc. Berl. 1768–77, 10 vols. 8°.

21. Auszüge aus den neuesten u. besten Dissertat. der mehrsten Akademien die zur Naturgeschichte, Arzneywissenschaft, Chemie u. Physik gehören. 6 vols. 8°. Berlin u. Stralsund, 1769–1772.

22. Mannichfaltigkeiten, eine gemeinnützige Wochenschrift. Berl. 1770 et seq. 8°. 4 vols. (Fortsetz.) Neue Mannichfaltigkeiten etc. Berl. 1774, et seq. 4 vols. Neueste Mannichfaltigkeiten etc. Berl. 1778 et seq. 4 vols. Allerneueste Mannichfaltigkeiten, etc. Berl. 1782 et seq.

23. **Abbildung** u. Beschreibung naturhistorischer Gegenstände. 17 vols. 8°. Berlin, 1795–1802.

24. Journal der praktischen Heilkunde; von *Hufeland*. 8° Berl. 1795. Fortgesetzt v. *Osann* u. *Busse*.

25. Archiv für Zoologie und Zootomie; von *C. R. W. Wiedemann*. 4 vols. 8°. Berl. 1800–1804.

26. Repertorium des neuesten und wissenswürdigsten aus der gesammten Naturkunde. Herausg. von *Heinr. Gust. Flörke*. 8°. Berlin, 1811, 1813.

27. Bulletin des neuesten und wissenswürdigsten aus der Naturwissenschaft, von *Sigism. Fried. Hermstädt*. 15 vols. 8°. Berlin, 1809–1813.

28. Museum des neuesten u. wissenswürdigsten aus dem Gebiete der Naturwissenschaft, der Künste, der Fabriken u. s. w., v. *S. F. Hermstädt*. 15 vols. 8°. Berlin, 1814–1818.

29. Magazin für die gesammte Heilkunde; von *Rust*. 8°. Berl. 1817. Fortgesetzt von *Eck*.

30. Archiv für Mineralogie, Geognosie, Bergbau u. Hüttenkunde; v. *Karsten* u. *von Dechen*. Berl. 15 vols. 8°. 1829–1841.

31. Archiv für Anatomie, Physiologie und wissenschaftliche Medizin; von *J. Müller*. 8°. Berl. 1834–1844.

32. Jahrbücher der Insectenkunde, mit besonderer Rucksicht auf die Sammlung im Konigl. Museum zu Berlin, v. *F. Klug*. Berlin, 1834, 8°.

33. Archiv für Naturgeschichte; von *Wiegmann*. 8°. Berl. 1835 –1841, 14 vols. Fortgesetzt von *W. F. Erichson*. 1841–1847.

34. Repertorium für Anatomie und Physiologie; von *G. Valentin*. Berl. 1837. Bern u. St. Gallen, 183 .

35. Jahrbuch für die Leistungen der gesammten Heilkunde; v. *J. J. Sachs*. Berl. 8°. 1838–1839. Leipz. 1840, 2 vols. 8°.

36. Annalen der Physik u. Chemie; von *Poggendorf*. 8°. Berl.

37. Repertorium der Physik; von *Dove* u. *Moser*. 8°. Berl.

38. Medizinische Zeitung des *Vereins für Heilkunde in Preussen*. 4°. Berl.

39. Wochenschrift für die gesammte Heilkunde; von *Casper*. 8°. Berl.

40. Magazin für die gesammte Thierheilkunde; von *Gurlt* u. *Hertwig*. 8°. Berl.

41. Berliner medizinische Centralzeitung; von *J. J. Sachs*. 4°. Berl.

42. Encyklopädisches Wörterbuch der medicinischen Wissenschaften; v. *Busch, Gräfe, Hufeland*, etc. 36 vols. 8°. Berl.

43. Amtlicher Bericht über die Versammlung *deutscher Natur-forscher und Aerzte* zu Berlin, erstattet von *Alex. v. Humboldt* und *H. Lichtenstein.* Berlin, 1829, 4°.

44. Stralsundisches Magazin, oder Sammlung auserlesener Neuig-keiten zur Aufnahme der Naturlehre, Arzneywissenschaft, u. Haushaltungskunst. Berl. u. Strals. 2 vols. 8°. 1767–1776.

45. Wittenbergisches Magazin für die Liebhaber der philosophi-schen u. schönen Wissenschaften ; v.*J. J. Ebert.* Berl. 1781, 8°.

46. Schwedische Annalen der Medecin u. Naturgeschichte, v. *K. A. Rudolphi.* 8°. Berlin, 1799–1800.

47. Naturwissenschaftliche Abhandlungen aus Dorpat. Berlin, 8°. 1823.

48. Mittheiluugen über Wien in naturwissenschaftlicher u. ärtz-licher Beziehung. Berlin, 1832.

49. Archiv für die wissenschaftliche Kunde von Russland; v. *A. Ermann.* Berl. 8°. 1841–1846.

50. Linnæa Entomologica, Zeitschrift herausgegeben von dem *Entomologischen Vereine in Stettin.* 8°. Berlin, 1846.

BONN (Bonna).

(Vide etiam NÜRNBERG, 6 ; BRESLAU, 9.)

1. Horæ Physicæ Berolinenses, collectæ ex symbolis virorum doc-torum. Edi curavit *C. G. Nees ab Esenbeck.* fol. Bonn, 1820.

2. Jahresberichte der *schwedischen Akademie der Wissenschaften* übers. von *J. Müller.* 8°. Bonn, 1826–1828.

3. Dreizehnte Versammlung *deutscher Naturforscher und Aerzte* in Bonn, 1835.

4. Organ für die gesammte Heilkunde. Herausgeg. von der *nie-derrheinischen Gesellschaft für Natur- u. Heilkunde in Bonn.* 1841. Bonn, 8°.

5. Verhandlungen des *Naturhistorischen Vereins der preussischen Rheinlande.* Von *L. C. Marcquart.* 8°. Bonn. 1844–1846.

6. Indische Bibliothek ; von *A. W. Schlegel.* 8°. Bonn.

BRANDENBURG.

1. Abhandlungen zur Naturgeschichte, Chemie, Anatomie, Medicin u. Physik, aus den Schriften des Instituts der Künste u. Wissen-schaften zu Bologna. Herausg. v. *N. G. Leske.* Brand. 1781–1782, 2 vols. 8°.

BRAUNSCHWEIG (Brunswick).

1. Vermischte Bibliothek, oder Auszüge aus verschiedenen zur

Arzneygelehrtheit, Chymie, Naturkunde, Œconomie gehörigen academischen Schriften, mit Anmerkungen begleitet; von *C. L. Neuenhahn.* Braunschw. 1758, 8°. 2 vols.

2. Magazin für Insektenkunde, von *J. K. W. Illiger.* Braunschw. 1802–1807, 6 vols.

3. Amtlicher Bericht über die neunzehnte Versammlung *deutscher Naturforscher und Aerzte* zu Braunschweig, 1841, erstattet von *F. K. v. Strombeck* und *Dr. Mansfeld.* Braunschweig, 1842, 4°.

BREMEN (Brema, Breme).

(Vide etiam GÖTTINGEN, 4.)

1. Bremisches Magazin, zur Ausbreitung der Wissenschaften, Künste und Jugend, mehrentheils aus den Englischen Monathsschriften gesammelt. Hann. u. Brem. 1756-65, 8°. 7 vols. Neues Bremisches Magazin, etc. 2 vols. Brem. 1766–1769.

2. Amerikanische Annalen der Arzneikunde, Naturgeschichte, Chemie u. Physik, v. *J. A. Albers.* 8°. Bremen, 1802–1803.

3. Amtlicher Bericht über die 22ste Versammlung *deutscher Naturforscher u. Aerzte,* zu Bremen, 1844. 4°. Bremen.

BRESLAU.

1. Uebersicht der Arbeiten und Veränderungen der *schlesischen Gesellschaft für vaterländische Cultur.* Bresl. 1838–1843, 8°.

2. Verhandlungen der *schlesischen Gesellschaft für vaterl. Cultur.* 4°.

3. Beiträge zur Entomologie, besonders in Bezug auf die schlesische Fauna, verf. u. herausg. von den Mitgliedern der Entomol. Section der *schlesischen Gesellschaft für vaterländische Kultur.* Bresl. u. Leipz. 8°.

4. Jahresberichte der Entomologischen Section der *schlesischen Gesellschaft für vaterländische Kultur* zu Breslau. 8°. Breslau, 1839–1843.

5. Sammlung von Natur- und Medicin-Geschichten, etc. 21 vols. 4. Breslau, 1718–1726.

6. Der *Königl. Academie der Wissenschaften in Paris* physische Abhandlungen, übers. v. *B. Ad. von Steinwehr.* Bresl. 1748–1759, 8°. Derselben anatomische, etc. Abhandlungen, übers. v. demselben; Bresl. 1749–1760, 8°. 9 vols.

7. Mineralogische Jahres-Hefte; von *Glocker.* Bresl. 8°. 1833.

8. Amtlicher Bericht über die 11te Versammlung *deutscher Naturforscher und Aerzte* zu Breslau, 1833; von *J. Wendt* und *A. W. Otto.* Breslau, 1834, 4°.

9. Verhandlungen der *Kaiserlich Leopoldisch-Karolinischen Akademie der Naturforscher.* 19 vols. Breslau und Bonn, 8°.

BROCKHAUSEN.

(Vide DÜSSELDORF, 1.)

BRÜNN.

1. Mittheilungen der *mährisch-schlesischen Gesellschaft zur Beförderung des Ackerbaues, der Natur- und Landeskunde* in Brünn. 4°. Brünn, 1840–1843.

CARLSRUHE.

1. Carlsruher nützliche Sammlungen, oder Abhandlungen aus allen Theilen der Wissenschaften. 1 Bd. Carlsr. 1759, 8°.

CASSEL.

(Vide etiam MARBURG, 1.)

1. Meteorologische und Naturhistorische Annalen ; herausg. von *Al. Th. Nahl.* Cassel, 1842, 8°.

CÖLN (Colonia Agrippina, Köln, Cologne).

1. Kölnisches Encyclopädisches Journal, herausg. v. *J. P. Eichhoff.* Jahrg. Köln, 1779, 8°.

DANZIG (Dantzig).

1. Versuche und Abhandlungen der *Naturforschenden Gesellschaft in Danzig.* 3 Theile. Danzig, 1747–1756, 4°.

2. Neue Sammlung von Versuchen und Abhandlungen der *Naturforschenden Gesellschaft in Danzig.* 1 vol. 4°. 1778.

3. Neueste Schriften der *Naturforschenden Gesellschaft in Danzig.* 4 vols. 4°. Danzig, 1820–1843. Halle, 1826. Königsberg, 1840.

DARMSTADT.

1. Zeitschrift für Physiologie, v. *F. Tiedemann, G. R. Treviranus* u. *L. C. Treviranus.* 5 vols. 8°. Darmstadt u. Heidelberg, 1824–1835.

DESSAU.

(Vide etiam LEIPZIG, 14.)

1. Abhandlungen der *Hallischen Naturforschenden Gesellschaft.* Dessau, 1783, 1 vol.

2. Neue monatliche Beiträge zur Naturkunde. 5 vols. 8°. Dessau, 1782.

3. Verhandlungen des *Naturhistorischen Vereins für Anhalt* in Dessau. 8°. Dessau, 1844.

DRESDEN (Dresda).

(Vide etiam LEIPZIG, 41 ; PRAG, 1.)

1. Sammlung physikalischer Aufsätze, besonders die Böhmischen Naturgeschichte betreffend, von einer *Gesellschaft Böhmischer Gelehrten.* Von *J. Mayer* u. *F. A. Reuss.* 5 vols. 8°. Dresden, 1791–1799.

2. Schriften der *Naturforschenden Gesellschaft zu Jena.* 8°. Dresden, 1802.

3. Zeitschrift für Natur- und Heilkunde ; herausgegeben von den Professoren der *chirurgisch-medizinischen Akademie* zu Dresden. 5 vols. 8°. Dresd. 1819–1828.

4. Neue Zeitschrift für Natur- u. Heilkunde. 1 vol. 8°. Dresden, 1829–1830.

5. Auszüge aus den Protokollen der *Gesellschaft für Natur- u. Heilkunde in Dresden.* 8°. 1832–34.

6. Verhandlungen der *Wandergesellschaft Sächsischen Landwirthe u. Naturforscher*, 8°. Dresden u. Leipzig, 1837–1840.

7. Dresdenisches Magazin, oder Ausarbeitungen u. Nachrichten zum Behuf der Naturlehre, Arzney- Sitten- u. schönen Wissenschaften. Dresd. 1760, 1765. 2 vols. 8°.

8. Medicinische und Physische Nachrichten. 1 vol. Dresd. u. Warsch. 1762, 8°.

9. Sammlung aus der Naturgeschichte, Oekonomie, etc. Dresd. 1774, 8°.

10. Abbildungen aus dem Thier- u. Pflanzenreiche. 4°. Dresden, 1812.

11. Allgemeine Deutsche Naturhistorische Zeitung, von *C. T. Sachse.* 8°. Dresden, 1846.

DÜSSELDORF.

1. Nachricht von dem Fortgange der *Westphälischen Naturforschenden Gesellschaft zu Brockhausen.* 8°. Düsseldorf, 1798–1799.

2. Neue Schriften der *Gesellschaft Naturforschender Freunde Westphalens.* 2 vols. 4°. Düsseldorf, 1798. Berlin, 1805.

EISENACH.

1. Der Zoologe, oder compendiose Bibliothek des Wissenswürdigsten aus der Thiergeschichte. 8°. Eisenach u. Halle, 1795–1797.

2. Zeitschrift für die organische Physik, von *C. F. Heusinger.*
3 vols. 8°. Eisenach, 1827–1829.

EMDEN.

1. Jahresberichte der *Naturforschenden Gesellschaft in Emden.* 8°.
Emden, 1841–1843. 1853.

ERFURT (Erfurtum).

1. Acta *Academiæ Electoralis Moguntinæ Scientiarum, quæ Erfurti
est.* 12 vols. 4°. Erfurti, 1757–1796.

2. Nova Acta *Academiæ Electoralis Moguntinæ Scientiarum utilium,
quæ Erfurti est.* 4 vols. 8°. Erfurti, 1799–1809.

3. Neue Physikalische Abhandlungen der *Akademie nützlicher
Wissenschaften zu Erfurt.* 8°. Erfurt, 1806.

4. Abhandlungen der Akademie gemeinnütziger Wissenschaften zu
Erfurt. 4°. 1828.

5. Bericht über die 1–3 Versammlung des *naturwissenschaftlichen
Vereins für Thüringen.* Erfurt, 1842–44.

6. Miscellanea Phys. Med. Mathematica ; oder Nachricht von physikal. und medicinischen, auch dahin gehörigen Kunst- u. Litteraturgeschichten, welche 1728 u. 1729 in Deutschland u. andern
Reichen sich zugetragen. Erf. 1732, 1733, 4°.

7. Notizen aus dem Gebiete der Natur- und Heilkunde. 4°. Erfurt,
1824.

ERLANGEN.

(Vide etiam NÜRNBERG, 6 ; FRANKFURT A. M. 2.)

1. Erlangische gelehrte Anzeigen, darinnen kurze und zur Verbesserung der Wissenschaften ausgearbeitete Materien befindlich, auf
das Jahr 1744. Erlang. 4°.

2. Zeitschrift der Geschichte der gesämmten Naturlehre, von *K. W.
G. Kastner.* 8°. Erlang. 1828.

3. Amtlicher Bericht über die 18te Versammlung der *Gesellschaft
deutscher Naturforscher u. Aerzte* zu Erlangen, 1840. Erstattet
von *J. M. Leupoldt* u. *L. Stromeyer.* Erlang. 1841, 4°.

FRANKFURT AM MAIN (Francofurtum, Francfort).

1. Miscellanea Curiosa *Ac. Cæs. Leop. Car.* (vide *Nürnberg,* 1, 4.)

2. Abhandlungen der *Phusicalischen Medicinischen Societät* zu Erlangen. 2 vols. 4°. Frankf. 1810.

3. Annalen der *Wetterauischen Gesellschaft für die gesammte Natur-kunde.* 4°. Frankf. 1809–1814. 3 vols.

4. Neue Annalen der *Wetterauischen Gesellschaft für die gesammte Naturkunde.* Frankf. 1818–20. 1 vol. 4°.

5. Jahrbuch zur Verbreitung naturwissenschaftlicher Kenntnisse ver-anstaltet vom *physikalischen Vereine zu Frankfurt.* 8°. Frankfort, 1831.

6. Museum Senkenbergianum. Abhandlungen aus dem Gebiete der beschreibenden Naturgeschichte, von Mitgliedern der *Senchen-bergischen Naturforschenden Gesellschaft* in Frankfurt. Frankf. 1833–1846. 3 vols. 4°.

7. Vermischte Schriften aus der Naturwissenschaft, Chymie und Arzneygelahrtheit, v. *Fr. Cartheuser.* Francf. 1759, 8°.

8. Neue Auszüge aus den besten ausländischen Wochen- und Mo-nathsschriften, Frankf. 1765–67 ; Hanau 1771. 4 vols. 8°.

9. Beyträge zur Rechtsgelahrtheit, Œconomie, Polizey, Naturge-schichte, etc., besonders von Hessen ; v. *G. Conr. Stockhausen.* Frankf. 1769, 8°.

10. Fragmente zur Arzney- und Naturkunde und Geschichte von *D. Berchelmann.* Frankf. 1780–81. 2 vols. 8°.

11. Hessische Beiträge zur Gelehrsamkeit u. Kunst. 2 vols. 8°. Frankf. 1784–1787.

12. Journal für Liebhaber der Entomologie, v. *G. Scriba.* Frankf. 1790–93, 3 vols. 8°.

13. Entomologische Hefte, enthaltend Beyträge zur weitern Kennt-niss u. Aufklärung der Insectengeschichte, von *Hoffmann, Koch, Linz, u. J. P. Müller.* 8°. Frankfurt, 1803.

14. Journal für Naturwissenschaft u. Medicin, v. *F. J. Schelver.* 8°. Frankfurt, 1810.

15. Zeitschrift für Mineralogie, v. *K. C. v. Leonhard.* Frankf. 1825, 8°.

FRANKFURT AN DER ODER.

1. Neues Nordisches Archiv. Verfasst von einer *Gesellschaft Nord-ischer Gelehrten.* 1 vol. 8°. Frankfurt a. d. O. 1807.

FREIBERG.

1. Altes und Neues aus dem Erzgebirge, nebst ausführlichen Nachrichten von denen im Erzgebirge herauskommenden neuen Schriften, 1747–49. Freiberg, 8°.

2. Obererzgebirgisches Journal, oder Sammlung von allerhand in die hiesige Naturwissenschaft überhaupt, v. *J. C. Themel.* 8°. Freiberg, 1748–1751.

3. Magazin für die Oryktographie von Sachsen; v. *Freiesleben.* Freib. 1831.

4. Jahrbuch für den Berg- und Hütten-Mann; herausg. von der *K. Berg.-Akademie zu Freiberg.* 8°. Freib. 1833.

5. Berg- und Hüttenmännische Zeitung mit besonderer Berücksichtigung der Mineralogie und Geologie, von *Carl Hartmann.* Freiberg. 1842–1844. 3 vols. 4°.

FREIBURG IM BREISGAU.

1. Bericht über die sechzehnte Versammlung *deutscher Naturforscher u. Aerzte* zu Freiburg, 1838. Verfasst von *Dr. F. S. Leuckart.* Freib. 1839, 8°.

GIESSEN.

1. Abhandlungen der *Seeländischen Gesellschaft der Wissenschaften zu Vlissingen;* übers. mit Anmerkungen v. *A. Böhm.* 1 vol. Giessen, 1775, 8°.

2. Rheinisches Magazin zur Erweiterung der Naturkunde; von *Borkhausen* u. *Brahm.* 8°. Giessen, 1793.

3. Abhandlungen u. Untersuchungen aus dem Gebiete der Naturwissenschaft, v. *J. B. Wetter.* 8°. Giessen, 1839.

GÖRLITZ.

1. Abhandlungen der *Naturforschenden Gesellschaft zu Görlitz.* Görlitz, vols. 8°. 1827–1842.— 59.

2. Lausitzisches Magazin, oder Sammlung verschiedener Abhandlungen u. Nachrichten zum Behuf der Naturkunst- Welt- u. Vaterlandsgeschichte. Görlitz, 1769, 4°.

3. Der Naturforscher, oder Abhandlung über ausgewählte Gegenstände aus dem Reiche der Natur. 8°. Görlitz, 1795.

GOTHA.

1. Allgemeines Historisches Magazin zu Beförderung der Erdbeschreibung u. der Natur- Staats- u. Kirchengeschichte. Gotha, 8°. 1762.

2. Gothaisches Magazin der Künste und Wissenschaften. Gotha, 1776, 8°.

3. Gothaisches gemeinnütziges Wochenblatt. I quart. 1779, 4°.

4. Magazin für das Neueste aus der Physik und Naturgeschichte. Von *L. Christ. Lichtenberg.* Fortg. von *J. H. Voigt.* 8°. Gotha, 1781–1799, 12 vols.

5. Physikalische und Medicinische Abhandlungen der *Königl. Akad.*

der Wissenschaften zu Berlin, aus dem Latein. u. Franz. übersetzt von *J. L. C. Mümler.* Gotha, 1781–1786, 4 vols. 8°.

6. Journal der Erfindungen, Zweifel u. Widersprüche in der gesammten Natur- u. Arzneywissenschaft. Von *A. F. Hecker.* 11
vols. 8°. Gotha, 1792–1809.

7. Neuestes Journal der Erfindungen, Zweifel u. Widersprüche in
der gesammten Natur- u. Arzneywissenschaft. Von *J. C. G.
Jörg, J. C. A. Heinroth* u. a. 2 vols. 8°. Gotha, 1810–1813.

GÖTTINGEN (Gottinga).

1. Commentarii *Societatis Regiæ Scientiarum Göttingensis,* ab anno
1751 ad 1754. Götting. 1752–1755, 4 vols. 4°. Novi Commentarii (1769–1777). Gött. 1771–1778, 8 vols. 4°. Commentationes (1778–1807). Gött. 1779–1808, 16 vols. 4°. Commentationes recentiores (1808–1837). Gött. 1811–1841, 8 vols. 4°.

2. Abhandlungen der *Königlichen Gesellschaft der Wissenschaften
zu Göttingen,* 1 Band. 4°. Göttingen, 1838–1841. Abhandlungen der Physikalischen Klasse. Göttingen, 1843.

3. *S. Chr. Holmanni* Commentationum in *Reg. Scientiarum Societate
Göttingensi* recensitarum, Sylloge. Gött. 1762, 4°. Ejusdem
Sylloge altera. Lips. 1775. Gött. 1785, 4°.

4. *J. D. Michaelis* Syntagma Commentationum *Societatis Göttingensis.* Gött. 1759–1767, 2 vols. 4°. Ejusdem Commentationes
Societati per ann. 1758–1762 oblatæ. Bremæ, 1763; ed. 2ª. 1774,
4°. Eædem per ann. 1763–1768. Bremæ, 1769, 4°.

5. Deutsche Schriften von der *Königl. Soc. der Wissenschaften zu
Göttingen,* 1 Bd. Gött. 1771, 8°.

6. Göttingische gelehrte Anzeigen, unter der Aufsicht der *Königl.
Gesellschaft der Wissenschaften.* Göttingen, 1805–1841, 3 vols.
8°.

7. Studien des Göttingischen *Vereins Bergmannischer Freunde.* 8°.
Göttingen, 1824–1841.

8. Physikalisch-ökonomische Bibliothek, worinnen von den neuesten
Büchern, welche die Naturgeschichte. Naturlehre, u. die Land- u.
Stadtwirthschaft betreffen, zuverlässige u. vollständige Nachrichten
ertheilet werden; von *J. Beckmann.* Gött. 1770–1785, 8°.
13 vols.

9. Göttingisches Magazin der Wissenschaften und Litteratur, herausg.
v. *G. Chr. Lichtenberg* u. *G. Forster.* 3 vols. 8°. Gött. 1780–
1785.

10. Magazin für allgemeine Natur- u. Thiergeschichte, v. *C. F. A.
Müller.* 8°. Göttingen, 1788.

11. Annalen der Naturgeschichte, v. *H. F. Link.* 8°. Göttingen,
1791.

12. Göttingisches Journal der Naturwissenschaften, v. *J. F. Gmelin.*
1 vol. 8°. Göttingen, 1797–1798.

13. Dissertationes Academicæ Upsaliæ habitæ sub præsidio *Carol.*
Petr. Thunberg. 2 volumina, 8°. Göttingæ, 1799 et 1800.

14. Freihefte für wissenschaftliche Natur- und Heilkunde; von
Krauss. 8°. Gött. 1836.

GRÄTZ.

1. Recueil du *Johanneum de Grätz*, en Styrie. 4°.

2. Steyermarkische Zeitschrift : N. S. Grätz. 1834–18 .

3. Amtlicher Bericht der 21sten Versammlung *deutscher Naturfor-*
scher u. Aerzte. 1843, 4°. Grätz, 1844.

GREIFSWALD.

1. Greifswaldische Academische Zeitschrift. Greifsw. 1822–23, 8°.
2 vols.

2. Archiv skandinavischer Beiträge zur Naturgeschichte. Von
C. F. Hornschuch. 8°. Greifswald, 1845.

HADAMAR.

1. Journal für die neueste holländische medicinische u. naturhistorische Literatur, v. *S. J. L. Döring* u. *G. Salomon.* 8°. Hadamar, 1802.

HALLE (Hala Saxonum).

(Vide etiam DANZIG, 3 ; DESSAU, 1 ; EISENACH, 1.)

1. Gelehrte Anzeigen, welche vormals denen wöchentlichen Hallischen Anzeigen einverleibt worden, von *Joh. Peter Ludewig.*
3 vols. Halle, 1743–1745, 4°.

2. Amusemens Philologiques, ou Mélange agréable de diverses
Pièces concernant l'histoire des personnes célèbres, la morale, la
mythologie, l'histoire naturelle ; par *Choffin.* Halle, 1749, 8°.

3. Das Reich der Natur und Sitten. Halle, 1751–62, 12 vols. 8°.

4. *Academiæ Sacri Romani Imperii Leopoldino-Carolinæ Naturæ*
Curiosorum Historia (par *A. El. Büchner*). Halæ Magdeburgi,
1754, 4°.

5. Beyträge zur Beförderung der Naturkunde. Halle, 1774, 8°.

6. Der Naturforscher, eine physikalische Wochenschrift. 30 vols.
8°. Halle, 1774–1804.

7. Archiv für die Physiologie, v. *J. C. Reil.* Fortges. v. *J. H. F.*
Autenrieth. 12 vols. 8°. Halle, 1796–1815.

8. Annalen der Physik, von *L. W. Gilbert.* 76 vols. 8°. Halle, 1799–1824.

9. Zeitschrift für organische Physik, v. *F. J. Schelver.* 8°. Halle, 1802.

10. Verhandlungen der *Gesellschaft zur Beförderung der Natur-kunde u. Industrie Schlesiens.* 8°. Halle, 1806.

11. Jahrbuch der *Naturforschenden Gesellschaft zu Halle.*

12. Neue Schriften der *Naturforschenden Gesellschaft zu Halle.* Halle, 1809–17, 8°.

13. Magazin der Entomologie, von *Germar* u. *Zinken.* Halle, 1813–21, 4 vols. 8°.

14. Deutsches Archiv für die Physiologie, v. *J. F. Meckel.* Halle u. Berl. 8 vols. 8°. 1815–1823.

15. Archiv für Anatomie und Physiologie, v. *J. F. Meckel.* 6 vols. 8°. Halle, 1826–1832.

HAMBURG.

1. Hamburgisches Magazin, oder gesammelte Schriften zum Unterricht und Vergnügen aus der Naturforschung und den angenehmen Wissenschaften. Hamb. und Leipzig, 1747–1767, 8°. 27 vols. Neues Hamburg. Magaz. oder Fortsetzung gesammelter Schriften aus der Naturforschung, etc. Hamb. und Leipzig, 1767–84, 20 vols. 8°.

2. Gesellschaftliche Erzählungen für die Liebhaber der Naturlehre, der Haushaltungswissenschaft, etc. Hamb. 1753–57, 4 vols. 8°. Neue gesellschaftliche Erzählungen, etc. Leipz. 1758 et seq. 4 vols. 8°.

3. Der physicalische und œconomische Patriot, oder Bemerkungen aus der Naturhistorie, etc. Hamb. u. Leipz. 1756, 4°. 4 vols. (2te Aufl.) ibid. 1765, 8°.

4. Gemeinnützige Nachrichten aus dem Reiche der Wissenschaften und Künste. Hamb. 1768 und 1769, 4°.

5. Hamburgisches Wochenblatt, eine moralisch-physikalisch-œconomische Wochenschrift. Hamb. 1768, 8°. 3 Th.

6. Verhandlungen und Schriften der *Hamburgischen Gesellschaft, zur Beförderung der Künste und nützlichen Gewerben.* 7 vols. Hamb. 1792–1807, 8°.

7. Mittheilungen aus den Verhandlungen der *Naturwissenschaftlichen Gesellschaft in Hamburg.* 8°. Hamburg, 1846.

8. Zeitschrift der in- und ausländischen Medizin; von *Fricke, Dieffenbach* u. *Oppenheim.* 8°. Hamb.

9. Abhandlungen der *Königl. Schwedischen Akademie der Wissenschaften* (1739–1755), aus dem Schwed. übersetzt durch *A. G.*

Kästner. Hamb. u. Leipz. 1749–1783, 41 vols. 8º. Neue Abhandlungen (1780), Leipz. 1784–1794, 12 vols.

10. Französische Annalen für die allgemeine Naturgeschichte Physik, Chemie, Physiologie u. ihre gemeinnützigen Anwendungen, v. *C. H. Pfaff* u. *J. Friedländer.* 9 vols. 8º. Hamburg u. Leipzig, 1802–1803.

11. Annalen des *Nationalmuseums der Naturgeschichte zu Paris.* Uebers. mit Anm. v. *J. J. Bernhardi.* Hamb. 1803–1804, 2 vols. 4º.

12. Amtlicher Bericht über die 9te Versammlung *deutscher Naturforscher und Aerzte* in Hamburg, 1830, von *J. H. Bartels* und *J. C. G. Fricke.* Hamburg, 1831, 4º.

HANAU.

(Vide FRANKFURT A. M. 9.) 3.

HANNOVER (Hanover).

(Vide etiam BREMEN, 1.)

1. Hannöverische Annalen für die gesammte Heilkunde. Herausg. von *Holscher.* Hanov. 8º.

2. Hannöverische Anzeigen (unter folgenden Titeln):—

Sammlung kleiner Ausführungen aus verschiedenen Wissenschaften. Hannov. 1750–54, 4 vols. 4º.

—— nützliche Sammlungen, 1755–58, 4 vols.

—— Hannöverische Beyträge zum Nutzen u. Vergnügen. 1759–62.

—— Hannöverisches Magazin, worin kleine Abhandlungen, etc., so die Verbesserung des Nahrungstandes, Physik, etc., betreffen, aufbewahret werden. Hanov. 1763. 4º.

HEIDELBERG.

(Vide etiam DARMSTADT, 1.)

1. Medizinische Annalen; von *Puchelt, Chelius* u. *Nägele.* 8º. Heidelb.

2. Amtlicher Bericht über die achte Versammlung *deutscher Naturforscher und Aerzte* in Heidelberg 1829, von *F. Tiedemann und L. Gmelin.* Heidelberg, 1829, 4º.

3. Annalen der Pharmacie, von *J. Liebig* und *F. Wöhler.* Heidelberg. 8º.

4. Zeitschrift für Mineralogie, Neue Folge, Jahrg. 1829. Von *K. C. v. Leonhard.*

5. Jahrbuch für Mineralogie, Geognosie, Geologie und Petrefacten-

kunde, 1, 2 u. 3 Jahrg. Mit Steindrucktafeln. Von *Leonhard* u. *Bronn*. Heidelberg, 8°. 1830–32.

6. Neues Jahrbuch für Mineralogie, Geognosie, Geologie und Petrefactenkunde; von *Leonhard* u. *Bronn*. 8°. Heidelberg u. Stuttgart, 1833–1844.

HERBORN.

(Vide MARBURG, 4.)

JENA.

(Vide etiam DRESDEN, 2; NÜRNBERG, 1.)

1. Natur- und Kunstkabinet, oder Sammlung nützlicher Nachrichten zu Beförderung der Naturkunde und Manufacturen. Jena, 1755, 8°.

2. Nachricht von der Gründung einer *Naturforschenden Gesellschaft zu Jena*. Von *A. J. G. K. Batsch*. 8°. Jena, 1793.

3. Magazin für den neuesten Zustand der Naturkunde, von *J. H. Voigt*. 12 vols. 8°. Jena u. Weimar, 1797–1806.

4. Nachricht von dem Fortgange der *Naturforschenden Gesellschaft zu Jena*. 8°. Jena, 1802.

5. Isis, oder Encyclopädische Zeitung; von *L. Oken*. 4°. Jena, u. Leipzig, 1817–1847.

6. Entomologisches Archiv v. *T. Thon*. Jena, 1827–29, 4°.

7. Amtlicher Bericht über die 17te Versammlung *deutscher Naturforscher und Aerzte* in Jena 1836, von *D. G. Kieser* und *J. C. Zenker*. Weimar, 1837, 4°.

8. Archiv für die gesammte Medicin; v. *Häser*. Jena, 8°.

INSBRUCK.

1. Beiträge zur Geschichte, Statistik, Naturkunde, etc., von *Pfaundler*, *Roggel* und *du Mersi*. Inspruck, 1825–1834, 8 vols. 8°.

2. Neue Zeitschrift des *Ferdinandeums für Tyrol und Voralberg*. 8°. Inspr. 1834–1845.

KÖNIGSBERG.

(Vide etiam DANZIG, 3; LEIPZIG, 42.)

1. Der Preussische Sammler zur Kenntniss der Naturgeschichte, etc. Königsb. 1774, 8°. 2 vols.

2. Königsberger Archiv für Naturwissenschaft und Mathematik. Königsberg, 1811–12, 8°. 2 vols.

3. Königsberger naturwissenschaftliche Unterhaltungen. 8°. Königsberg, 1844.

LANDAU.

1. Erster Jahresbericht der *Pollichia*, eines naturwissenschaftlichen Vereins der bayerschen Pfalz. 8°. Landau, 1843.

LAUTERN.

(Vide MANNHEIM, 3.)

LEIPZIG (Lipsia).

(Vide etiam KIÖBENHAVN, 11; UPSALA, 1; RIGA, 2; ST. PETERSBOURG, 17; BERLIN, 35; DRESDEN, 6; GÖTTINGEN, 3; HAMBURG, 1, 2, 3, 9, 10; NÜRNBERG, 1, 4.)

1. Miscellanea curiosa *Ac. Cæs. Leop. Car.* (vide NÜRNBERG).

2. Acta Eruditorum Lipsiensia, ab anno 1682 ad ann. 1731. 50 vols. 4°.—Nova Acta, 1732–1776, 43 vols.—Supplementa, 1692–1734, 10 vols. 4°.—Nova Supplementa, 1735–1757, 8 vols. 4°.—Indices generales Actorum, etc., 1693–1745, 6 vols. 4°.

3. Commentarii de rebus in Scientiâ naturali et Medicinâ gestis. Lipsiæ, 1752–1806, 37 vols. 8°.—Supplementa, 1763–1796, 3 vols. 8°. et Suppl. I. Decadis IV. 8°.—Indices Decad. I.–III. 1770–1793, 3 vols. 8°.

4. Der Naturforscher. Leipz. 1747–48, 8°.

5. Œconomisch-physicalische Abhandlungen von *P. v. Hohenthal.* Leipz. 1751–63, 20 vols.

6. Allgemeines Magazin der Natur, Kunst und Wissenschaften. Leipz. 1753–67, 12 vols. 8°.

7. Neue Gesellschaftliche Erzählungen für die Liebhaber der Naturlehre, der Haushaltungswissenschaft, etc. Leipz. 1758 et seq. 4 vols. 8°.

8. Leipziger Intelligenzblatt, 4°. 1763.

9. Gemeinnütziger Vorrath auserlesener Aufsätze zur Beförderung der Haushaltungswissenschaft, Künste, Manufacturen, Fabriken, Arzneygelahrtheit u. Naturkunde; von *D. Krünitz.* Leipz. 1767–68, 3 vols. 8°.

10. Allgemeines Repertorium der gesammten deutschen medicinisch-chirurgischen Journalistik, v. *Kleinert*, und fortgesetzt v. *Neumeister.* Leipz. 8°.

11. Sammlungen nützlicher u. angenehmer Gegenstände aus allen Theilen der Naturgeschichte, Arzneywissenschaft u. Haushaltungskunst, v. *Fr. Xav. v. Wasserberg.* Leipz. 1773, 8°.

12. Sammlungen zur Physik und Naturgeschichte, von einigen Liebhabern dieser Wissenschaften. Leipz. 1779–1792, 4 vols. 8°.

13. Beiträge zur Naturgeschichte u. Bergpolizeywissenschaft; v. *Fr. Gottl. Glaser.* Leipz. 1780, 8°.

14. Leipziger Magazin zur Naturkunde, Mathematik u. Œkonomie, herausg. v. *C. B. Funke, N. G. Leske* u. *C. F. Hindenburg.* Leipz. u. Dessau, 1781–87, 8°.

15. Neue Litteratur und Beyträge zur Kenntniss der Naturgeschichte; von *J. S. Schröter.* Leipz. 1784, 8°.

16. Archiv zur neuern Geschichte, Geographie, Natur- u. Menschenkenntniss; v. *J. Bernoulli.* Leipz. 1785, 8°.

17. Neues Journal der Physik, von *F. A. C. Gren.* 4 vols. 8°. Leipz. 1795–97.

18. Zoologisches Archiv. Von *F. A. A. Meyer.* 2 vols. 8°. Leipzig. 1796.

19. Archiv für die systematische Naturgeschichte v *F. Weber* u. *D. M. H. Mohr.* 8°. Leipzig, 1804.

20. Jahrbücher der gesammten in- und ausländischen Medicin; von *C. C. Schmidt.* 4°. Leipz.

21. Acta *Societatis Naturæ Scrutatorum Lipsiensis.* 1 vol. 4°. Leipzig, 1822.

22. Allgemeine Encyclopädie der Wissenschaften, von *J. S. Ersch,* und *J. G. Gruber.* 4°. Leipzig, 1818–1846.

23. Annalen der Physik und Chemie, von *J. C. Poggendorf.* 53 vols. 8°. Leipzig, 1834–41.

24. Pfennig- Encyclopädie der Anatomie, oder bildliche Darstellung der gesammten Anatomie, nach *Rosenmüller, Loder, Bell, Gordon, Bock,* etc. Leipz. 4°.

25. Zeitschrift für die Entomologie; v. *E. F. Germar.* Leipz. 8°. 1838–1846.

26. Encyclopädie der gesammten Medicin. Im Vereine mit mehreren Aerzten herausg. v. *C. G. Schmidt.* Leipz. 4°. 1841.

27. Acta philosophica *Societatis Regiæ in Anglia* annorum 1656–1659. Lipsiæ, 1675, 4°.

28. Abhandlungen zur Naturgeschichte, Physik und Œkonomie, aus den philosophischen Transaktionen und Sammlungen. Gesammelt und mit einigen Anmerkungen übersetzt von *Nath. Gottfr. Leske.* 1 vol. 4°. Leipzig, 1779–1780.

29. Sammlung von Natur und Medicin, wie auch hierzu gehörigen Kunst- und Litteraturgeschichten, so sich von 1717–1726, in Schlesien und ándern Ländern begeben, in 38 Versuchen, nebst 4 Supplementen u. Universalregistern. Leipz. u. Budiss. 1718–1730, 4°.

30. Neue Abhandlungen der *K. Schwed. Akademie der Wissenschaften* aus der Naturlehre, Haushaltungskunst und Mechanik. Aus dem

Schwedischen von *Abr. Gotth. Kästner* und *Joachim Dietrich Brandis.* 13 Theile, nebst einem zwiefachen Universal-Register. Leipzig, 1784–1794.

31. Die vorzüglichsten Vorlesungen welche in der *Königlichen Schwedischen Akademie der Wissenschaften* zu Stockholm gehalten worden sind. Aus dem Schwedischen von *D. O. G. Gröning.* 2 vols. 8°. Leipzig, 1794–1795.

32. Auserlesene Abhandlungen welche an die *Königliche Akademie der Wissenschaften zu Paris* von einigen Gelehrten eingesendet worden. Aus dem Französischen übers. von *F. W. Beer.* 2 vols. 8°. Leipzig, 1752–54.

33. Abhandlungen zur Naturgeschichte der Thiere u. Pflanzen welche ehemals der *Königl. Französischen Akademie der Wissenschaften* vorgetragen wurden, von *Perrault, Charres,* u. *Dodert.* 3 vols. 4°. Leipzig, 1757–58.

34. *Sammlung brauchbarer Abhandlungen aus Roziers Beobachtungen über die Natur und Kunst,* mit Anmerk. v. *Wünsch.* Leipz. 1775–76, 2 vols. 8°.

35. Abhandlungen aus der Naturgeschichte praktischer Arzneykunde u. Chirurgie, aus den Schriften der *Harlemer* u. andern holländischen *Gesellschaften.* 2 vols. 8°. Leipzig, 1775, 1776.

36. Abhandlungen der *Gesellschaft der Künste u. Wissenschaften in Batavia,* aus dem Holländischen übers. u. mit Anmerkungen u. Zusätzen versehen. 8°. Leipzig, 1782.

37. Naturhistorische Abhandlungen der *Batavischen Gesellschaft der Wissenschaften zu Harlem* ; a. d. Holländ. von *Halem.* Leipz. 1802, 1 vol.

38. Italiänische Bibliothek, oder Sammlung der merkwürdigsten kleinen Abhandlungen zur Naturgeschichte, Œkonomie u. Fabrikwesen aus den neuesten Italiänischen Monathschriften, v. *J. J. Volkmann.* Leipz. 1778–79, 2 vols. 8°.

39. Physikalische u. philosophische Abhandlungen der *Gesellschaft der Wissenschaften zu Manchester.* Aus dem Englischen von *E. B. G. Hebenstreit.* 2 vols. 8°. Leipzig, 1788.

40. Auserlesene Abhandlungen für Aerzte, Naturforscher u. Psychologen ; aus den Schriften der *literar-philosophischen Gesellschaft zu Manchester.* Aus dem Englischen von *A. W. Schwenger.* 8°. Leipzig, 1795.

41. Auswahl aus den Schriften der unter *Werner's* Mitwirkung gestifteten *Gesellschaft für Mineralogie zu Dresden.* 2 vols. 8°. Leipz. 1818.

42. Berichte der *König. Anatom. Anstalt zu Königsberg.* 2 vols. Leipz. 1819–40.

43. Beiträge zur Entomologie, besonders in Bezug auf die Schlesische

Fauna; v. den Mitgliedern der Entomol. Sect. der *Schlesischen Gesellschaft für vaterländische Cultur.* 8°. Leipz. 1829.

44. Dresdner naturwissenschaftliches Jahrbuch, von *A. Petzholdt.* 8°. Leipzig, 1846.

45. Abhandlungen bei Begründung der *König. Sachsischen Gesellschaft der Wissenschaften* am Tage der 200-jährigen Geburtsfeier Leibnitzens. 4°. Leipzig, 1846.

MAINZ (Moguntia, Mayence, Mentz).

1. Recueil des Mémoires et Actes de la Société des Sciences et Arts du Mont Tonnère. Mayence, 1804, 8°.

2. Amtlicher Bericht über die zwanzigste Versammlung der Gesellschaft *deutscher Naturforscher und Aerzte* zu Mainz im 1842, von *Gräser* und *Bruch.* Mainz, 1843, 4°.

MANNHEIM.

1. Historia et Commentationes *Academiæ Electoral. Scientiarum et elegantior. Litterarum Theodoro-Palatinæ.* 11 vols. 4°. Mannh. 1766–1794.

2. Beiträge zur Sittenlehre, Œconomie, Arzneywissenschaft, Naturlehre u. Geschichte, aus den westlichen Gegenden Deutschlands. Manh. 1770 and 1772, 2 St. 8°.

3. Bemerkungen der *Kurpfälzischen physikalisch-ökonomischen Gesellschaft*; vom Jahr 1770–1783. 15 Theile. Mannheim und Lautern. 1771–1785, 4°.

4. Vorlesungen der *Kurpfälzigen physikalisch-ökonomischen Gesellschaft.* 5 vols. 8°. Mannheim, 1785–1790. Jahresbericht des *Mannheimer Vereins für Naturkunde,* 8°. Mannheim, 1834–1843.

MARBURG.

1. Schriften der *Gesellschaft für Beförderung der gesammten Naturwissenschaften zu Marburg.* 4 vols. 8°. 1823–1839. Cassel, 1832.

2. Neues Journal für Œkonomie, Naturgeschichte und Chemie, von *E. Mönch.* 1 vol. 8°. Marburg, 1794.

3. Neues medicinisches u. physikalisches Journal, v. *E. G. Baldinger.* 3 vols. 8°. Marburg, 1797–1802.

4. Kritisches Repertorium der auf in- u. ausländischen Lehranstalten herausgekommenen Probe- u. Einladungsschriften aus dem Gebiete der Arzneygelahrtheit u. Naturkunde, von *S. J. L. Döring.* 4°. Herborn u. Marburg, 1803.

MÜNCHEN (Monacum, Munich).

1. Abhandlungen der *Chur-Baierischen Akademie der Wissenschaften*

Münch. 1763–1776, 10 vols. 4°.—Neue philosophische Abhandlungen. München, 1778–1797, 4°. 7 vols.

2. Physikalische Abhandlungen der *Königlich. Baierischen Akademie der Wissenschaften*. 2 vols. 8°. München, 1803 und 1806.

3. Denkschriften der *Königl.Baierischen Akademie der Wissenschaften zu München*, für die Jahre 1808–1824. Münch. u. Sulzb. 1809–1825, 9 vols. 4°.

4. Abhandlungen der Mathematisch-physikalischen Klasse der *Königl. Bayerischen Akademie der Wissenschaften*. 3 vols. 4°. München, 1832.

5. Gelehrte Anzeigen, herausgegeben von Mitgliedern der *Königl. Baierischen Akademie der Wissenschaften*. 11 vols. 4°. Münch. 1835–1840.

6. Almanach der *Königlichen bayerischen Akademie der Wissenschaften* zu München. München, 1843, 8°.

7. Baierische Beyträge zur schönen u. nützlichen Litteratur, 1779, et seq.

8. Abhandlungen einer Privatgesellschaft von Naturforschern und Oekonomen in Oberdeutschland. Herausg. von *Franz von Paula Schrank*. 1ster Theil. München, 1792, 8°.

9. Faunus, Zeitschrift für Zoologie u. Vergleichende Anatomie; v. *J. Gistl.* Münch. 1832–35, 2 vols. 8°. Neue Folge. 8°. Münch. 1837.

10. Jahrbücher des *Aerztlichen Vereins in München*. 8°. 1835.

NAUMBURG.

1. Archiv der Naturgeschichte oder Sammlung belehrender Abbildungen aus dem Thierreiche. Von *W. Thienemann* u. *Thou.* 4°. Naumburg. 1824–1829.

NEUSTADT.

1. Zweiter Jahresbericht der *Pollichia*. 8°. Neust. a. d. Hardt. 1844.

NORDHAUSEN.

1. Bericht des *Naturwissenschaftlichen Vereins des Harzes*. 4°. Nordhausen, 1840–1842.

NÜRNBERG (Berga, Norimberga, Nuremberg).

(Vide etiam WIEN, 3.)

1. Miscellanea curiosa medico-physica *Academiæ Cæsareo-Leopoldinæ Naturæ Curiosorum*. Decuria prima. Ann. 1670–1679. 6 vols. Lipsiæ, Jenæ, Francofurti, Vratislaviæ et Bergæ. 1670–1680. 4°.—Decuria II. Ann. 1682–1691. 10 vols. Norimbergæ.

1683–1692. 4º.—Decuria III. Ann. 1694–1705. 7 vols. Lipsiæ et Francofurti, 1694–1706. 4º.

2. Index rerum notabilium Dec. I. et II. Ephemeridum *Academiæ Naturæ Curiosorum*, ab anno 1670 usque ad annum 1692 (à *J. P. Wurfbain*). 4º. Norimb. 1695.

3. Index generalis et absolutus Miscellaneorum curiosorum, sive Ephemeridum, etc. *Academiæ Naturæ Curiosorum*, Dec. I. et II. Norimb. 1695. 4º. Index generalis et absolutissimus. Dec. III. Francof. 1713, 4º.

4. *Academiæ Cæsareo-Leopoldinæ Naturæ Curiosorum* Ephemerides, sive Observationum medico-physicarum, etc. Cent. I. VIII. Francof. et Lips. 1712–1722, 4º.

5. Acta physico-medica *Academiæ Cæs. Leop. Carol. Naturæ Curiosorum*, exhibentia Ephemerides, sive observationes, historias et experimenta à celeberrimis viris habita, singulari studio collecta. Norimb. 1727–1754. 10 vols. 4º.

6. Nova Acta physico-medica *Academiæ Cæsareæ Leopoldino-Carolinæ Naturæ Curiosorum*, exhibentia Ephemerides. Norimb. 1757–1791, 4º. 8 vols. cum Appendice. Bonnæ, Erlang. et Vratisl. 1821–1846, 13 vols.

7. Synopsis Observationum medicarum et physicarum quas Decuriæ III. ac Centuriæ X. Ephemeridum *Naturæ Curiosorum* ab ann. 1670–1722 publicatarum, continent; à *W. Andr. Kellner*. Norimb. 1739, 2 vols. 4º.

8. Der *Röm. Kaiserl. Akademie der Naturforscher* auserlesene medicinisch- chirurgisch- anatomisch- chymisch und botanische Abhandlungen, aus dem Latein. übersetzt. 20 vols. 4º. Nürnb. 1755–1771.

9. Fränkische *Acta Eruditorum* in diesem Kreise vorgefallne Curiosa und Merkwürdigkeiten in sich haltend. Nürnb. 1726–1732.

10. Fränkische Sammlungen von Anmerkungen aus der Naturlehre Arzneygelahrtheit, Œconomie u. denen damit verwandten Wissenschaften. Nürnb. 1756–65, 8 vols.

11. Gemeinnütziges Fränkisches Magazin, oder Sammlung merkwürdiger nützlicher Grundsätze u. Erfahrungen aus der Naturlehre, Naturgeschichte, etc. von *Heppe*. Nürnb. 1779, 8º.

12. Commercium litterarium ad Rei medicæ et Scientiæ naturalis incrementum institutum. Norimb. 1731–45, 15 vol. 4º.

13. Neues Schwedisches Magazin kleiner Abhandlungen, welche in die Natur- und Haushaltungskunde einschlagen; besorgt v. *T. C. D. Schreber*. 1 vol. Nürnb. 1783, 8º.

14. Journal für Chemie und Physik. In Verbindung mit *Bernhardi, Buchholz, Crell, Gehlen* u. m. a. herausg. von *J. S. C. Schweigger*. 30 vols. Nürnberg, 1811–1820, 8º.

15. Neues Journal für Chemie und Physik. Herausgegeben von Dr. *Schweigger* und Dr. *Meinecke.* Neue Reihe. 24 vols. Nürnberg, 1821–1829, 8°.

16. Archiv für die gesammte Naturlehre von Dr. *K. W. G. Kastner.* 18 vols. 8°. Nürnberg, 1824–1829.

17. Archiv für Chemie und Meteorologie. von Dr. *K. W. G. Kastner.* VIII. Bde 8°. Nürnberg, 1830–1834.

18. Jahrbücher der Mineralogie, Geologie, Berg- u. Hüttenkunde; von *C. Hartmann.* Nürnb. 8°. 1833.

19. Repertorium der Pharmacie von *Buchner.* Nürnberg, 8°.

PRAG (Prague).

1. Abhandlungen der *Königl. Böhmischen Gesellschaft der Wissenschaften* in Prag, auf die Jahre 1785–1788. 5 Bände. Prag u. Dresden, 1785–1789. 4°.

2. Neuere Abhandlungen der *Königl. Böhmischen Gesellschaft der Wissenschaften.* II. Theile. Wien und Prag, 1791 u. 1795. 4°. 7 vols. 8°. Prag, 1804–1822. Neue Folge, 4°. Prag, 1827–1845. Neue physikalische Belustigungen, mit Kupf. 1sten Bandes 1ste und 2te Abth. Prag, 1770, 8°.

3. Abhandlungen einer Privatgesellschaft in Böhmen zur Aufnahme der Mathematik, der vaterländischen Geschichte und der Naturgeschichte, zum Druck befördert von *Ignaz Edlen von Born.* 6 Theile. Prag, 1775–1784, 8°.

4. Monatschrift der *Gesellschaft des vaterländischen Museums in Böhmen.* Von *F. Palacky.* 8°. Prag, 1827–1829.

5. Verhandlungen der *Gesellschaft des vaterländischen Museums in Böhmen.* 8°. Prag, 1833–1844.

6. Jahrbücher des *Böhmischen Museums für Natur- u. Länderkunde, Geschichte, Kunst u. Literatur.* Von *F. Palacky.* 2 vols. 8°. Prag, 1830–31.

7. Bericht über die Versammlung *deutscher Naturforscher u. Aerzte* in Prag (1837). Von *C. v. Sternberg* u. *J. V. v. Krombholz.* Prag, 4°. 1838.

8. Berichte über die Verhandlungen der königlichen *Böhmischen Gesellschaft der Wissenschaften* in den Sektionsversammlungen, von 1840–1841. 4°. Prag, 1842.

9. Beiträge zur gesammten Natur- und Heilwissenschaft; von *Weitenweber.* 8°. Prag.

10. Hesperus, ein Nationalblatt für gebildete Leser; von *C. C. André.* 4°. Prag.

11. Magnetische und Meteorologische Beobachtungen zu Prag. Von *Karl Kreil.* 3 Jahrgang.

PRESSBURG.

1. Ungrisches Magazin, oder Beiträge zur Ungrischen Geschichte, Geographie, Naturwissenschaft, etc. Von *K. G. v. Windisch.* 4 vols. 8°. Pressburg. 1781–1788.

PYRMONT.

1. Amtlicher Bericht der siebenzehnten Versammlung *deutscher Naturforscher und Aerzte* in Pyrmont 1839.

2. Zeitschrift für Malakozoologie. Herausg. von Dr. *Karl Theodor Menke.* 8°. Pyrmont, 1844–1846.

QUEDLINBURG.

1. Archiv für die neuesten Entdeckungen aus der Urwelt; v. *J. G. J. Ballenstedt.* Quedl. 1819 et seq.

REGENSBURG (Ratisbona, Ratisbon).

1. Denkschriften der *Königl. Bayrischen botanischen Gesellschaft* zu Regensburg. Regensburg, 1822–1841, 3 vols. 4°.

ROSTOCH (Rostochium, Rostock).

(Vide UPSALA, 1.)

STETTIN.

(Vide etiam BERLIN, 50.)

1. Entomologische Zeitung, herausgegeben von dem *Entomologischen Vereine* zu Stettin. Redig. v. *W. Schmidt.* 1840–46, 7 vols. 8°.

STRALSUND.

(Vide etiam BERLIN, 21, 44.)

1. Stralsundisches Magazin oder Sammlung auserlesener Neuigkeiten, zur Aufnahme der Naturlehre, Arzneiwissenschaft und Haushaltungskunst: mit Kupf. II. Theile. Berlin und Stralsund, 1767–1776, 8°.

2. Neues Magazin für Liebhaber der Entomologie, v. *D. H. Schneider.* Strals. 1791–94, 8°. 5 Hefte.

3. Die schwedischen Naturforscher *K. P. Thunberg* u. *J. W. Dalman.* Aus den Abhandl. der *Akademie der Wissenschaften zu Stockholm.* Uebers. von *G. Mohnike.* 8°. Stralsund, 1831.

STUTTGART.

(Vide etiam HEIDELBERG, 6.)

1. *Selecta physico-œconomica,* oder Sammlungen von allerhand zur

Naturforschung u. Haushaltungskunst gehörigen Begebenheiten. Stuttg. 1750, etc. 3 vols. 8°.

2. Stuttgardter allgemeines Magazin. 1752–68, 8°.

3. Physikalisch-œconomische Realzeitung. Stuttg. 1755–56, 4°.— Physikalisch-œconom. Wochenschrift, 1757, 2 vols.—Physikal.-œconom. Auszüge aus den neuesten u. besten Schriften, die zur Naturlehre, etc. gehören. Stuttg. 1758–70, 10 vols. 8°.

4. Etwas für Alle, oder Neue Stuttgardter Realzeitung, 1765–66, 8°.

5. Schwäbisches Magazin von gelehrten Sachen, 1779 u. 1780. Stuttg.

6. Medicinisches Correspondenzblatt des *Würtembergischen ärztlichen Vereins* : v. *Blumhardt, Duvernoy* u. *Seeger.* 4°. Stuttg.

7. Denkschriften der *vaterländischen Gesellschaft der Aerzte u. Naturforscher Schwabens.* 8°. Stuttgart, 1805.

8. Correspondenzblatt des *Würtembergischen landwirthschaftlichen Vereins.* 8°. Stuttgart, 1822.

9. Naturwissenschaftliche Abhandlungen. Stuttgart und Tübingen, 2 vols. 8°. 1826–1828.

10. Schriften des *Würtembergischen naturhistorischen Reisevereins.* 8°. Stuttgart, 1834.

11. Amtlicher Bericht über die 12te. Versammlung *deutscher Naturforscher und Aerzte* zu Stuttgart, 1834, von C. v. KIELMEYER und G. JÄGER. Stuttgart, 1835, 4°.

12. Paläontologische Collectaneen, hauptsächlich als beliebiges Ergänzungsheft zum N. Jahrbuch für Mineralogie, etc. von. *H. G. Bronn.* Stuttg. 1843, 8°.

13. Jahreshefte des *Vereins für Vaterländische Naturkunde in Würtemberg.* Von *H. von Mohl, T. Plieninger Fehling, W. Menzel, F. Krauss.* 8°. Stuttgart. 1845.

14. Medicinische Vierteljahrsschrift, von *Roser* und *Wunderlich.* Stuttgart. 8°.

TÜBINGEN.

(Vide etiam STUTTGART, 9.)

1. Denkschriften der *Vaterländischen Gesellschaft der Aerzte und Naturforscher Schwabens.* Tübing. 1 vol. 8°. 1805.

2. Tübinger Blätter für Naturwissenschaften u. Arzneikunde. v. *J. H. F. Autenrieth* u. *J. G. F. v. Bohnenberger.* 3 vols. 8°. Tübingen, 1815–1817.

3. Naturwissenschaftliche Abhandlungen, herausgegeben von einer *Gesellschaft in Würtemberg.* 8°. Tübingen, 1826.

WEIMAR.

(Vide etiam JENA, 3, 7.)

1. Journal für die Liebhaber des Steinreiches u. der Konchyliologie v. *J. S. Schröter.* Weimar, 1774–84, 6 vols. 8°.

2. Magazin der Italiänischen Litteratur u. Künste; von *C. J. Jagemann.* Weimar, 1780, 8°.

3. Journal für die Litteratur u. Kenntniss der Naturgeschichte, sonderlich der Conchylien u. Steine; v. *J. S. Schröter.* Weimar, 1782, 8°.

4. Zoologische Annalen; v. *F. A. A. Meyer.* Weimar, 1794, 1 vol. 8°.

5. Notizen aus dem Gebiete der Natur u. Heilkunde: v. *L. F. Froriep.* 50 vols. 4°. Weimar, 1821–1836.
   Neue Notizen aus dem Gebiete der Natur und Heilkunde. Von *L. von Froriep* und *R. Froriep.* 4°. Weimar, 1837–1846.

6. Wörterbuch der Naturgeschichte dem gegenwärtigen Stande der Botanik, Mineralogie, u. Zoologie angemessen, mit Atlas. 11 vols. 8°. Weimar, 1824–1837.

7. Naturhistorischer synoptischer Atlas. fol. Weimar, 1833–43.

8. Fortschritte der Geographie u. Naturgeschichte. Von *L. F. v. Froriep.* 4°. Weimar, 1846.

WIEN (Vindobona, Vienne, Vienna).

(Vide etiam BERLIN, 48 ; PRAG, 2.)

1. Annalen des *Wiener Museums der Naturgeschichte,* herausg. von der Direction desselben. 2 vols. 4°. Wien, 1835–40.

2. Zoologische Abhandlungen aus den Annalen des *Wiener Museums der Naturgeschichte.* 2 vols. 4°. Wien, 1841.

3. Kosmographische Nachrichten und Sammlungen auf das Jahr 1748. Wien und Nürnberg, 1750, 4°.

4. Physikalische Arbeiten der einträchtigen Freunde in Wien. Herausg. von *Ign. Edeln von Born.* 2 vols. 4to. Wien, 1783–1788.

5. Zeitschrift für Physik u. verwandte Wissenschaften; von *Baumgärtner* u. *Ettinghausen.* 8°. Wien, 1832.

6. Amtlicher Bericht über die 10te Versammlung *deutscher Naturforscher und Aerzte* zu Wien, 1832, von *Frhr. v. Jacquin* und *J. J. Littrow.* Wien, 1833, 4°.

7. Medizinische Jahrbücher der K. K. Oesterreichischen Staaten. Herausg. von *Raimann* und *Rosas.* 8°. Wien.

8. Oesterreichische medicinische Wochenschrift; als Ergänzungs-

blatt der Medicinischen Jahrbücher der K. K. Oesterr. Staaten. Wien, 8º.

WIESBADEN.

1. Jahrbücher des *Vereins für Naturkunde im Herzogthum Nassau.* Von *C. Thomä.* 1 vol. 8º. Wiesbaden, 1844.

WITTENBERG.

(Vide etiam BERLIN, 45.)

1. Wittenbergisches Wochenblatt, zum Aufnehmen der Naturkunde u. des ökonomischen Gewerbes, von *J. D. Titius.* Wittenb 1769 et seq. 14 vols. 4º. (Vol. XV.) Aufsätze und Wahrneh- mungen über die Witterung, Haushaltungskunde, das Gewerbe, Naturkenntniss, etc.

2. Philosophical Transactions, reprinted according to the London edition, for 1751–1765 ; 9 vols. 4º. Wittenberg, 1775.

WÜRZBURG.

1. Neue Jahrbücher der *königlichen philosophischen medicinischen Gesellschaft zu Würtzburg.* Von *J. B. Friedreich.* 8º. Würz- burg, 1830.

WRATISLAW (Vratislavia).

(Vide Nürnberg, 1, 6.)

BELGIUM.

ANVERS (Antwerpen, Antwerp).

1. Mémoires et Observations de la *Société de Médecine d'Anvers.* Anv. 1836–38, 2 vols. 8º.

2. Annales de la *Société de Médecine d'Anvers,* 1838–42, 3 vols. 4º.

BRUGES.

1. Annales de la *Société des Sciences naturelles de Bruges,* 1839–42, 3 vols. 8º.

2. Bulletin de la *Société des Sciences naturelles de Bruges.* 8º. Bruges, 1841–42.

3. Annales de la *Société Médico-Chirurgicale de Bruges.* 1840–43, 3 vols. 8º.

BRUXELLES (Bruxellæ, Brussels).

1. Mémoires de l'*Académie impériale et royale des Sciences et Belles- Lettres de Bruxelles.* Brux. 1777–1788, 5 vols. 4º.

2. Mémoires sur les questions proposées par l'*Académie de Bruxelles*, qui ont remporté les prix en 1769 et ann. suiv. jusqu'en 1788, et dès lors jusqu'en 1817. 4°. Brux. 1769–1818.

3. Nouveaux Mémoires de l'*Académie royale des Sciences et Belles-Lettres de Bruxelles*, 1820–1846, 4°.

4. Bulletin des séances de l'*Académie royale des Sciences et Belles-Lettres de Bruxelles*. Brux. 1833–1846, 10 vols. 8°.

5. Mémoires couronnés par l'*Académie royale des Sciences et Belles-Lettres de Bruxelles*, de 1817 à 1836. Brux. 1818–1846, 15 vols. 4°.

6. Annuaire de l'*Académie royale de Bruxelles*, 1835–1843, 9 vols. 18°.

7. Annuaire de l'*Observatoire royal de Bruxelles*. Bruxelles, 1833–1843, 10 vols. 18°.

8. Annales de l'*Observatoire royal de Bruxelles*, publiées par *M. A. Quetelet*. Bruxelles, 1834–1837, 2 vols. 4°.

9. Bulletin de l'*Académie royale de Médecine de Belgique* à Bruxelles, 1842–43, 2 vols. 8°.

10. Annales de la *Société des Sciences naturelles et médicales de Bruxelles*, 1834–41, 6 vols. 8°.

11. Bulletin de la *Société des Sciences naturelles et médicales de Bruxelles*, 1842, 8°.

12. Annales générales des Sciences physiques et naturelles, par *MM. Bory de St. Vincent, Drapiez*, et *Van Mons*. Brux. 1819–21, 8 vols. 8°.

13. Dictionnaire classique des Sciences naturelles. 10 vols. 8°. Bruxelles, 1828–1845.

14. Encyclographie des Sciences médicales, publiée par une Société de Médecins, sous la direction de *Fl. Cunier*. Brux. 4°. 1832, et suiv.

15. Mémoires de la *Société de Médecine de Gand*. Brux. 1835–36, 8°.

16. Annales *Academiæ Lovaniensis*. 4 vols. 4°. Bruxelles, 1821–1823.

GAND (Gandavum, Ghent).

(Vide etiam BRUXELLES, 15.)

1. Annales *Academiæ Gandavensis*, ann. 1817–1826, 6 vols. 4°.

2. Bulletins de la *Société de Médecine de Gand*. 1835–40, 6 cah. 8°.

3. Annales de la *Société de Médecine de Gand*. 1837–41, 9 vols. 8°.

4. Annales Belgiques des Sciences, Arts et Littérature. Gand, 1818–24, 13 vols. 8°.

5. Messager des Sciences et des Arts. Gand, 1823–30, 6 vols. 8°. —(N. Sér.), 1832–43, 8°. 10 vols.

6. Belgish Museum, uitgegeven door M. WILLEMS, 1ste—4ste deel, 1837–1841.

LIÈGE (Leodium, Luttich).

1. Annales *Academiæ Leodiensis*, anni 1819–1823. Leodii, 1822–1824, 4 vols.

2. Procès-verbal de la Séance publique du 25 Avril 1821 de la *Société libre d'émulation et d'encouragement pour les sciences et les arts*, établie a Liège. Liège, 1821, 1 vol. 8°.

3. Procès-verbal de la Séance publique tenue le 12 Juin 1828. Liège, 1828, 1 vol. 8°.

4. Mémoires de la *Société royale des Sciences de Liège.* 8°. Liège, 1843.

5. Journal encyclopédique, par une Société de gens de lettres ; dirigé par *P. Rousseau.* Liège, 1756–1759, et *Bouillon*, 1760–1791, 288 vols. 12°.

6. Bibliographie académique Belge, ou Répertoire systématique et analytique des Mémoires, etc. publiés jusqu'à ce jour par l'ancienne et la nouvelle *Académie de Bruxelles ;* par *P. Namur.* Liège, 1838, 8°.

LOUVAIN (Lovanium).

(Vide etiam BRUXELLES, 16.)

1. Annuaires de l'*Université catholique de Louvain.* Années, 1837–1842. Louvain, 5 vols. 18°.

MONS.

1. Mémoires et publications de la *Société des Sciences, des Arts et des Lettres du Hainaut.* 8°. Mons.

2. Compte rendu des travaux de la *Société des Sciences, des Arts et les Lettres du Hainaut,* 1840–41, par *J. B. Bivort.* 8°.

GALLIA.

ABBEVILLE.

1. Mémoires de la *Société d'émulation d'Abbeville.* 1834, 8°.

2. Bulletin de la *Société Linnéenne du nord de la France.* 8°. Abbeville, 1840.

AIX (Aquæ Sextiæ).

1. Séances publiques de la *Société académique d'Aix.* 1825.

3. Neue Alpina v. *J. R. Steinmüller.* Winterthur, 1821–1827, 2 vols. 8°.

ZUG.

(Vide AARAU, 1.)

ZÜRICH.

(Vide etiam AARAU, 1 ; CHUR, 1.)

1. Der *Schweizerischen Gesellschaft in Bern,* Sammlung von land-wirthschaftlichen Dingen; oder Abhandlungen u. Beobachtungen durch die *Œkonomische Gesellschaft zu Bern* gesammlet. (Franç.) Mémoires et Observations concernant l'Economie Rurale, par une Société établie à Berne en Suisse. Zürich, 1760–73.

2. Abhandlungen der *Naturforschenden Gesellschaft in Zürich.* Zürich, 1761–66, 3 vols. 8°.

3. Verhandlungen der *Helvetischen Gesellschaft in Schinznach.* Zürich, 1764, 8°.

4. Magazin für Liebhaber der Entomologie, von *J. C. Fuessly.* Zür. u. Winterth. 1778 u. 1779, 2 vols. 8°.—Neues Magazin, etc. Zürich, 1782–1786, 3 vols.

5. Archiv der Insectengeschichte, von *J. Casp. Fuessly.* Zür. u. Winterth. 1781–1786, 8 vols. 4°. (Franç.) Zür. et Wint. 1794.

6. Magazin für die Naturkunde Helvetiens. Herausgeg. von. *Albr. Höpfner.* 8°. Zürich, 1787–1789, 4 vols.

7. Eröffnungsrede der Jahresversammlung der *Allg. Schw. Gesellsch. für die gesammten Naturwissenschaften,* v. *Usteri.* Zürich, 1817, 8°.

8. Bericht über die Verhandlungen der *Naturforschenden Gesellschaft in Zürich.* 8°. Zürich, 1826–1838.

9. Verhandlungen der *Schweizerischen Naturforschenden Gesellschaft* bei ihrer Versammlung zu Zürich, 1841. Zürich, 8°. 1827 u. 1841.

10. Denkschriften der *Allgemeinen Schweizerischen Gesellschaft für die gesammten Naturwissenschaften.* Zürich, 1829–1833, 4°.

11. Schweizerische Zeitschrift für Natur- und Heilkunde; herausg. von *F. v. Pommer.* 8°. Zürich, 1835–1841, 6 vols.

12. Mittheilungen aus der theoretischen Erdkunde; von *Frobel* u. *Heer.* 8°. Zürich, 1836.

2. Recueil de Mémoires des *amis des Sciences, Lettres, Agriculture et Arts*. 1823–1827.

AJACCIO.

1. Journal scientifique et littéraire de la Corse. 1823.

AGEN, Dep. du Lot et Garonne.

1. Bulletin des travaux de la *Société d'Agriculture, Sciences et Belles-Lettres d'Agen*. 1804, 8°.
2. Séances publiques de la *Société centrale d'Agriculture, Sciences et Belles-Lettres d'Agen*. 1821, 8°.

ALBY.

1. Journal de la *Société d'Agriculture du Dép. du Tarn*. Alby, 1820, 8°.

ANGOULÊME, Dép. de la Charente.

1. Annales de la *Société d'Agriculture, Sciences et Belles-Lettres de la Charente*. Angoulème, 1819, 8°.

AMIENS.

1. Mémoires de l'*Académie des Sciences, Agriculture, Commerce, Belles-Lettres et Arts du département de la Somme*. Amiens, 1835–1841, 8°.
2. Collection de rapports analytiques des travaux de l'*Académie d'Amiens*. 1805, 4°.

ANGERS.

1. Bulletin de la *Société Industrielle d'Angers et du département de Maine et Loire*. Angers, 1830-1842, 8°.
2. Mémoires de la *Société d'Agriculture, Sciences et Arts d'Angers*. 8°. 1831–1840, 4 vols.
3. Seánce générale et annuelle des *Sociétés d'Agriculture industrielle et de Médecine d'Angers*. 8°. Angers, 1835.
4. Travaux du *Comité horticole-agricole de Maine et Loire*. Angers, 1839–1841, 8°. 2 vols.

ARRAS, Dép. du Pas de Calais.

1. Mémoires de la *Société royale d'Arras* pour l'encouragement des Sciences, Lettres et Arts. Arras, 1821, 8°.
2. Bulletin de la *Société royale d'Arras* pour l'encouragement des Sciences, Lettres et Arts. Arras, 1825, 8°.

AUCH.

1. Travaux de l'*Athénéum du Gers*. An 12 (1805).

AURILLAC.

1. Bulletin de la *Société d'Agriculture du Cantal*. Aurillac, 1821.

AUXERRE.

(Vide DIJON, 2.)

AVIGNON.

1. Mélanges curieux et interessans de divers objets relatifs à la Physique, à la Médecine et à l'Histoire naturelle, par *Haguenot*. Avign. 1771, 12º.

BAYEUX.

(Vide etiam CAEN.)

1. Mémoires de la *Société d'Agriculture, des Sciences, Arts et Belles-Lettres de Bayeux*. 1842, 8º.

BEAUVAIS.

1. Bulletin de la *Société agricole et industrielle du Département de l'Oise*. Nos. 1–11, 8º. 1834–1835.

BESANÇON.

1. Rapport des Travaux de la *Société libre d'Agriculture, des Arts et du Commerce du Doubs*. Besançon, An VII.

2. Mémoires et Rapports de la *Société d'Agriculture, des Arts et du Commerce du Doubs*. Besançon, 1820, 8º.

3. Séances publiques de l'*Académie royale des Sciences, Inscriptions et Belles-Lettres de Besançon*. Besançon, 1807, 8º.

4. Comptes-rendus des travaux de l'*Académie royale des Sciences, Inscriptions et Belles-Lettres de Besançon*. Besançon, 1829.

5. Analyse des travaux de l'*Académie royale des Sciences, Inscriptions et Belles-Lettres de Besançon*. Besançon, 1834.

6. Mémoires et Comptes-rendus de la *Société libre d'émulation du Doubs*. Besançon, 1841–42, 8º.

BÉZIERS.

1. Recueil de Lettres, Mémoires et autres pièces pour servir à l'Histoire de l'*Académie des Sciences et Belles-Lettres de Béziers*. Béz. 1736, 4º.

BLOIS, Dép. du Loir et Cher.

1. Procès-verbal des séances générales de la *Société royale d'Agriculture de Loir et Cher*. Blois, 1824, 8°.

2. Mémoires de la *Société des Sciences et Lettres de Blois*. 1834, 8°.

3. Congrès scientifique de France ; 4me session tenue à Blois en 1836.

BORDEAUX.

1. Journal de la *Société de Santé et d'Histoire naturelle de Bordeaux*. Par *J. F. Capelle* et *D. Villars*. 3 vols. 8°. Bordeaux, 1797–1798.

2. Rapports des travaux de l'*Académie royale des Sciences, Belles-Lettres et Arts de Bordeaux*. Bordeaux, 1825, 8°.

3. Actes de l'*Académie royale des Sciences, Belles-Lettres et Arts de Bordeaux*. Bordeaux, 1839–1843, 8°.

4. Séances publiques de l'*Académie royale des Sciences, Belles-Lettres et Arts de Bordeaux*. Bordeaux, 1839, 8°.

5. Bulletin d'Histoire naturelle de la *Société Linnéenne de Bordeaux*. 10 vols. 8°. Bordeaux, 1826–1838.

6. Actes de la *Société Linnéenne de Bordeaux*. Bordeaux, 1839–1844, 12 vols. 8°.

7. L'Ami des Champs, journal d'Agriculture, de Botanique, et Bulletin littéraire du département de la Gironde. 8°. Bordeaux, 1833.

BOULOGNE SUR MER (Bononia).

1. Procès-verbal de la séance publique de la Société d'Agriculture, Commerce et Arts de Boulogne sur Mer. 1818, 8°.

BOURG.

1. Comptes-rendus des travaux de la *Société départementale d'émulation et d'Agriculture de l'Ain*. Bourg, 1808.

2. Journal d'Agriculture, de Littérature et des Arts de la *Société départementale d'émulation de l'Ain*. Bourg, 1810.

3. Exposé historique et statistique des travaux de la *Société départementale d'émulation et d'Agriculture de l'Ain*. Bourg, 1821, 8°.

BOURGES.

1. Bulletin de la *Société d'Agriculture du Cher*. Bourges, 1830.

BRIVES, Dép. de Corrèze.

1. Annales de la *Société d'Agriculture, Sciences, Arts et Belles-Lettres de la Corrèze*. Brives, 1825, 8°.

CAEN.

(Vide etiam PARIS, 82, 83.)

1. Mémoires analytiques des travaux de l'*Académie royale des Sciences, Belles-Lettres et Arts de Caen.* 1754–1760.

2. Rapports annuels de l'*Académie de Caen.* 1805–1809.

3. Séances publiques de l'*Académie royale des Sciences, Belles-Lettres et Arts de Caen.* 1811.

4. Mémoires de la *Société Linnéenne du Calvados.* Caen, 1824–1838, 6 vols. 4°.

5. Mémoires de l'*Académie des Sciences, Belles-Lettres et Arts de Caen.* 1825.

6. Précis des travaux de la *Société royale d'Agriculture et du Commerce de Caen.* 1827, 8°.

7. Rapport sur les mémoires de la *Société royale d'Agriculture et du Commerce de Caen.* 1827, 8°.

8. Séances publiques de la *Société Linnéenne de Normandie.* 8°. Caen, 1835–1837.

9. Séance du 19 Janvier 1838, de la *Société royale d'Agriculture et de Commerce de Caen.* 8°.

10. Revue Normande, rédigée par une Société de Savans et de Littérateurs de Rouen, de Caen et des principales villes de la Normandie.

CAHORS.

1. Bulletin de la *Société agricole et industrielle du Lot.* Cahors, 1841, 8°.

CAMBRAI.

1. Mémoires de la *Société d'émulation de Cambrai.* Cambrai, 1817, 8°.

CARCASSONNE.

1. Journal de la *Société d'Agriculture de Carcassonne.* 1833, 8°.

CHALONS sur Marne.

1. Procès-verbal des séances de la *Société d'Agriculture, des Sciences et des Arts du département de la Marne.* Châlons sur Marne.

2. Compte-rendu et Sommaire des travaux de la *Société d'Agriculture, des Sciences et des Arts du département de la Marne.* 1807.

3. Séance publique de la *Société d'Agriculture, des Sciences et des Arts du département de la Marne.* 1811.

4. Journal du *comice agricole de l'arrondissement de Châlons sur Marne.* 1826.

CHARTRES.

1. Mémoires de la *Société d'Agriculture de l'Eure et Loir.* Chartres, 1826, 8°.

CHATEAUROUX.

1. Comptes-rendus des travaux de la *Société libre d'émulation du département de l'Indre.* Châteauroux, An X. 8°.

2. Séances publiques de la *Société libre d'émulation du département de l'Indre.* Châteauroux, An XI.

3. Ephémérides de la *Société d'Agriculture du département de l'Indre.* Châteauroux, 1805–1829.

CLERMONT FERRAND.

1. Annales scientifiques, littéraires et industrielles de l'Auvergne, publiées par l'*Académie des Sciences, Belles-Lettres, et Arts de Clermont Ferrand,* sous la direction de *M. Lecoq.* Clermont-Ferrand. 8°. 1827–1844.

2. Congrès scientifique de France ; 6me session tenue à Clermont-Ferrand en 1838. Paris et Clermont, 1839, 1 vol. 8°.

DIJON.

1. Recueil de Mémoires, ou Collection de Pièces académiques, concernant la Médecine, l'Anatomie, la Chirurgie, la Chymie, la Physique expérimentale, la Botanique et l'Histoire naturelle, etc. Par *M. J. Berryat,* 16 vols. 4°. Dijon et Par. 1754–87, 5 vols. 4°.—Liège, 1781 (le 6e vol.).

2. Collection académique, composée de Mémoires, Actes, Journaux des plus célèbres Académies, concernant l'Histoire naturelle, la Botanique, la Physique expérimentale, la Chimie, la Médecine et l'Anatomie, etc. Dijon, Auxerre et Paris, 1755–79, 13 vols. 4°.

3. Mémoires de l'*Académie de Dijon.* 2 vols. 8°. Dijon, 1769–1774.

4. Nouveaux Mémoires de l'*Académie de Dijon.* 3 vols. 8°. Dijon, 1782–1784.

5. Notice publique de l'*Académie de Dijon.* An XI.

6. Séances publiques de l'*Académie des Sciences, Arts et Belles-Lettres de Dijon.* 1819.

7. Mémoires de l'*Académie des Sciences, Arts et Belles-Lettres de Dijon.* Nouv. Sér. 8°. Dijon, 1831–38.

Douai.

1. Mémoires de la *Société royale d'Agriculture, Sciences et Arts de Douai*. 1833, 8°.

2. Mémoires de la *Société royale et centrale d'Agriculture, Sciences et Arts du département du Nord*, séant à Douai. I, 1826–1828. Douai 8°.—Séances publiques, 1826, 8°.

3. Congrès scientifique de France ; 3me session tenue à Douai en 1835.

Draguignan, Dép. du Var.

1. Mémoires publiés par la *Société libre d'émulation du département du Var*. Draguignan, 1802, 8°.

2. Bulletin de la *Société d'Agriculture et du Commerce du département du Var*. Draguignan, 1820, 8°.

Dunkerque (Dunkirk).

1. Extrait du procès-verbal des séances de la *Société d'Agriculture de l'arrondissement de Dunkerque*. 1823.

Epinal, Dép. des Vosges.

1. Rapport sur les travaux de la *Société d'Agriculture du département des Vosges*. Epinal, 1822, 8°.

2. Journal de la *Société d'émulation du département des Vosges*. Epinal, 1823, 8°.

3. Annales de la *Société d'émulation du département des Vosges*. 8°. Epinal, 1837–1840.

Etampes.

1. Séance publique de la *Société d'Agriculture de l'arrondissement d'Etampes*.

Evreux, Dép. de l'Eure.

1. Séances de la Société d'Agriculture, Sciences, Arts et Belles-Lettres du Département de l'Eure. Evreux, 1819.

2. Bulletin d'Agriculture et Sciences accessoires de l'Eure. Evreux, 1822.

3. Journal d'Agriculture. Evreux, 1825.

4. Bulletin de l'*Académie Ebroïcienne du Département de l'Eure*. Evreux, 1833.

5. Recueil de la *Société libre d'Agriculture, Sciences, Arts et Belles-Lettres du Département de l'Eure*. 8°. 1839, 10 vols.

GAP, Dép. des Hautes Alpes.

1. Journal de la *Société d'émulation des Hautes Alpes*. Gap, 1806. 8º.

FOIX.

1. Journal de la *Société d'Agriculture de l'Arriège*. Foix, 1820–183 , 5 vols.

FALAISE.

1. Annales de la *Société académique agricole, industrielle et d'instruction de Falaise*. 8º. 1835.

LA ROCHELLE.

1. Annales de la *Société d'Agriculture, Arts et Commerce du département de la Charente*. 15 vols. 8º.

LE MANS, Dép. du Sarthe.

1. Analyse des travaux de la *Société royale du Mans*. Le Mans, 1820, 8º.

2. Procès-verbal de la séance publique de la *Société royale du Mans*. Le Mans, 1822, 8º.

3. Comptes-rendus des travaux de la *Société royale d'Agriculture, Sciences et Arts du Mans*. 1824, 8º.

4. Congrès scientifique de France; 7me session tenue au Mans en 1839.

LILLE (Insulæ, Lisle).

1. Séance publique de la *Société d'amateurs des Sciences de Lille*. 1816, 8º.

2. Mémoires de la *Société royale des Sciences, de l'Agriculture et des Arts de Lille*. 1819–1841. Lille, 8º.

3. Recueil des travaux de la *Société d'amateurs des Sciences, de l'Agriculture et des Arts à Lille*. 1819–1843, 15 vols. 8º.

4. Mémoires de la *Société d'amateurs des Sciences, de l'Agriculture et des Arts à Lille*. 1831, 6 vols.

LIMOGES, Dép. de la Haute Vienne.

1. Procès-verbal de la séance publique de la *Société d'Agriculture, de Limoges*. An XI. 8º.

2. *Société d'Agriculture, des Sciences et des Arts du département de la Haute Vienne*. 1809, 8º.

3. Bulletin de la *Société royale d'Agriculture de Limoges*. 1822.

LONS LE SAULNIER.

1. Séance publique de la *Société d'émulation du Jura*. Lons le Saulnier, 1823, 8°.

LYON (Lugdunum, Lyons).

(Vide etiam PARIS, 69.)

1. Histoire de la *Société royale des Sciences de Montpellier*, avec les Mémoires de Mathématique et Physique tirés des Registres de cette Société. Lyon, 4°. 1766.

2. Dictionnaire d'Histoire Naturelle, par *Bomare*. 15 vols. 8°. Lyons.

3. Mémoires de la *Société royale d'Agriculture, d'Histoire naturelle et Arts de Lyon*. 8°. Lyon, 1806.

4. Comptes-rendus des travaux de la *Société royale d'Agriculture, d'Histoire naturelle et Arts utiles de Lyon*. 8°. Lyon, 1812–1833.

5. Comptes-rendus des travaux de la *Société des Sciences, Arts et Belles-Lettres de Lyon*. 1813.

6. Comptes-rendus des travaux de l'*Académie de Lyon*. 1825.

7. Annales de la *Société Linnéenne de Lyon*. 8°. 1836.

8. Mémoires de la *Société royale d'Agriculture, Histoire naturelle et Arts utiles de Lyon*. 8°. Lyon, 1837.

9. Annales des Sciences physiques et naturelles d'Agriculture et d'Industrie, publiées par la *Société royale d'Agriculture etc. de Lyon*. 1838–1845. Lyon, 8 vols. 8°.

10. Mélanges d'Histoire Naturelle, par *M. Alléon Dulac*. Lyon.

11. Revue du Lyonnais. Lyon, 8°. tom. i.—xvii. 18 –18 .

12. Congrès scientifique de France ; 8me session tenue à Lyon en 1840.

MACON.

1. Rapport fait à l'Assemblée générale de l'*Académie de Mâcon*, par *M. Ch. Lacretelle*. Mâcon, 1836.

2. Comptes-rendus des travaux de la *Société des Sciences, Arts et Belles-Lettres de Mâcon*. 1807–1842. 8°.

MARSEILLES (Massilia, Marseille).

1. Mémoires publiés par l'*Académie de Marseille*. 1803–1813. 11 vols. 8°.

2. Comptes-rendus des travaux de la *Société de Statistique de Marseille*. 1828–1831–1832, 8°.

3. Annales des Sciences et de l'Industrie du Midi de la France, par la *Société de Statistique de Marseille.* Marseilles, 1832, 10 cah. 8°.

4. Répertoire des travaux de la *Société de Statistique de Marseille*, 1ʳᵉ année. Marseille, 1837–18

MENDE, Dép. de la Lozère.

1. Mémoires et Analyses des travaux de la *Société d'Agriculture, Commerce, Sciences et Arts de la ville de Mende*, département de la Lozère. 1826, 1841–1842, 1 vol. 8°.

METZ, Dép. de la Moselle (Metæ).

1. Mémoires de la *Société libre d'Agriculture de Metz.* An XI. 8°.

2. Procès-verbal de la séance publique de la *Société d'Agriculture du département de la Moselle.* Metz, 1806.

3. Mémoires de l'*Académie royale de Metz* (lettres, sciences, arts et agriculture). Metz, 1817–42, 15 vols. 8°.—*Société des Lettres, Sciences, Arts et Agriculture de Metz.* 1820, 10 vols. 8°.

4. Mémoires de la *Société d'Histoire naturelle du département de la Moselle.* 8°. Metz, 1843.

5. Bulletin de la *Société d'Histoire naturelle du département de la Moselle.* Metz, 2ᵐᵉ cahier, 1844, 8°.

MONTAUBAN.

1. Séances publiques de la *Société d'Agriculture, des Sciences et des Belles-Lettres du Tarn et Garonne.* Montauban, 1814, 8°.

2. Recueil agronomique de la *Société d'Agriculture, des Sciences et des Belles-Lettres du Tarn et Garonne.* Montauban, 1820, 8°.

MONT DE MARSAN, Dép. des Landes.

1. Mémoires de la *Société d'Agriculture, du Commerce et des Arts du département des Landes.* Mont de Marsan, 1806, 8°.

MONTPELLIER, Dép. de l'Hérault.

1. Recueil de Bulletins publiés par la *Société libre des Sciences et Belles-Lettres de Montpellier*, 1801–1813, 5 vols. 8°.

2. Bulletin de la *Société d'Agriculture de l'Hérault.* Montpellier, 1803–1836, 8°.

MOULINS.

1. Annales de la *Société d'Agriculture de l'Allier.* Moulins.

MULHOUSE.

1. Bulletin de la *Société industrielle de Mulhouse.* 14 vols. 1829.
2. Programmes des prix proposés par la *Société industrielle de Mulhouse.* Mulhouse, 1840–1843, 8°.

NANCY.

1. Précis analytique des travaux de la *Société académique des Sciences, Arts et Agriculture.* Nancy, 1811.
2. Précis des travaux de la *Société royale des Sciences, Lettres et Arts de Nancy*, 1829–1832. Nancy, 1833, 1 vol. 8°.
3. Mémoires de la *Société royale des Sciences, Lettres et Arts de Nancy.* Nancy, 1833–1842, 6 vols. 8°.

NANTES.

1. Séance publique de la *Société académique du département de la Loire inférieure.* Nantes, 1818.
2. Annales de la *Société royale académique de Nantes.* 8°. Nantes, 1834–1836.

NIORT, Dép. des Deux-Sèvres.

1. Séance publique de l'*Athénéum de Niort.* 1810, 8°.
2. Mémoires de l'*Athénéum de Niort du Société libre des Sciences et Arts des Deux Sèvres.* 8°.

NISMES.

1. Notice des travaux de l'*Académie du Gard.* 1807, 8°.

NOZAY, Dép. de la Seine Inf.

1. Revue trimestrielle de l'Agriculture de l'Ouest de la France; par *J. Rieffel.* Nozay, 1840.

ORLEANS (Aurelia).

1. Bulletin de la *Société des sciences physiques, médicales et d'agriculture d'Orleans.* 1810–1813, 7 vols.
2. Annales de la *Société des sciences physiques, médicales et d'agriculture d'Orleans.* 1818–1833, 13 vols.
3. Mémoires de la *Société des sciences physiques, médicales et d'agriculture d'Orleans.* 1841.
4. Annales de la *Société royale des Sciences, Belles-Lettres et Arts d'Orleans.* 1818.
5. Journal de la *Société d'Agriculture d'Orleans.* 182 , 8°.

GALLIA.

49

Paris (Lutetia Parisiorum, Parigi).

(Vide etiam Breslau, 6; Leipzig, 32, 33; Clermont, 2; Dijon, 1, 2.)

1. Histoire et Mémoires de l'*Académie Royale des Sciences*, depuis son établissement en 1666 jusqu'en 1790. Paris, 1701–1793, 164 vols. 4°. Amsterdam, 1706–1734, 12°.

2. Mémoires de l'*Institut National des Sciences*. Paris, 1798–1815, 14 vols. 4°.

3. Mémoires de l'*Académie Royale des Sciences de l'Institut de France*. Paris, 1818–1846, 4°.

4. Pièces, qui ont remporté le prix de l'*Académie Royale des Sciences de Paris*. 9 vols. 4°. Paris, 1727–1777.

5. Mémoires de Mathématique et Physique présentés à l'*Acad. Roy. des Sciences*, par divers savans, et lus dans ses Assemblées. Par. 1750–1786, 11 vols. 4°.

6. Mémoires présentés à l'*Institut des Sciences, Lettres et Arts* par divers savans, et lus dans ses Assemblées. Sciences Mathématiques et Physiques. 2 vols. 4°. Paris, 1806–1811.

7. Mémoires présentés par divers Savans à l'*Académie Royale des Sciences de l'Institut de France*. 1827–43, 8 vols. 4°.

8. Comptes-rendus Hebdomadaires des Séances de l'*Académie des Sciences*. Paris, 1835–1847, 19 vols. 4°.

9. Abrégé de l'Histoire et des Mémoires de l'*Académie Royale des Sciences à Paris*, par *Paul*. Paris, 1774, 4°. (formant le 5e vol. du Recueil de Mémoires, etc., par *Berryat*).

10. Table Alphabétique des Matières contenues dans l'Histoire et les Mémoires de l'*Acad. Roy. des Sciences*, publiée par son ordre et dressée par *M. Godin*, ann. 1666–1698. Paris, 1734, 4°. Ann. 1761–1770, Paris, 1774 (depuis le vol. v., par *Demours*). Table Générale des Matières, etc., depuis 1699–1734, inclus. Amst. 1741, 3 vols. 12°., depuis 1735–1751, inclus. ibid. 1760, vol. iv.

11. Nouvelle Table des Articles, contenus dans les volumes de l'*Académie Royale des Sciences de Paris*, depuis 1666 jusqu'en 1770, dans ceux des Arts et Métiers publiés par cette Académie, et dans la Collection Académique, par *M. l'Abbé Rozier*. Paris, 4°. 4 vols. 1775–76.

12. Recueil de Discours, Rapports et Pièces diverses lus dans les séances publiques et particulières de l'*Académie Française*. 4°. Paris, 1820–1839.

13. Recueil contenant les Délibérations de la *Société Royale d'Agriculture de la Généralité de Paris*, et les Mémoires publiés par son ordre. Paris, 1761, 8°.

VOL. I.   E

14. Bulletin des Sciences, par la *Société Philomatique*. Paris, 1791 –1805, 4°. 3 vols.

15. Nouveau Bulletin des Sciences, par la *Société Philomatique*. 8 vols. 4°. 1807–1826, Paris.

16. Mémoires de la *Société d'Histoire Naturelle de Paris*. 4°. Paris, 1792. Réimpression, 1799.

17. Actes de la *Société d'Histoire Naturelle de Paris*. 1 vol. fol. Paris, 1792.

18. Mémoires de la *Société d'Histoire Naturelle de Paris*. Paris, 1823–1834, 5 vols. 4°.

19. Mémoires de l'*Institut d'E'gypte*. Paris, 1800–1804, 8°. 4 vols.

20. Annales du *Muséum National d'Histoire Naturelle*, par les Professeurs de cet établissement. Paris, 1802–1813, 20 vols. 4°.

21. Mémoires du *Muséum d'Histoire Naturelle*, par les Professeurs de cet établissement. 1815–1832, 20 vols. 4°.

22. Nouvelles Annales du *Muséum d'Histoire Naturelle*, ou Recueil de Mémoires publiées par les Professeurs de cet établissement et par d'autres naturalistes, sur l'histoire naturelle, l'anatomie et la chimie. Paris, 1832–35, 4 vols. 4°.

23. Archives du *Muséum d'Histoire Naturelle*, par les Professeurs de cet établissement. Paris, 1839 et suiv. 4°.

24. Journal Asiatique. 11 vols. Paris, 1822–27, sér. 2, 1828.

25. Annales de la *Société Linnéenne de Paris*. 8°. 1822–1827, 6 vols.

26. Bulletin de la *Société Géologique de France*. Paris, 1830–1843, 8°. Deuxième série, 1844–   .

27. Mémoires de la *Société Géologique de France*. Paris, 1833–1842, 5 vols. 4°. Sér. 2, 1844–   .

28. Annales de la *Société Entomologique de France*. Paris, 1832–1842, 11 vols. 8°.—2ᵐᵉ série, 1843.

29. Mémoires de la *Société Entomologique*. 8°. Paris, 1845.

30. Bulletin de la *Société des Sciences Naturelles de France*. 1835, 4°. (1 cahier.)

31. Mémoires de la *Société Royale et Centrale d'Agriculture*. 8°. Paris, 18   .

32. Journal des Savans, par le sieur *Hedouville* (Denis Sallo), et continué par *J. Gallois, De la Roque, L. Cousin, Dupin*, etc. Paris, 1665–1792, 111 vols. 4°.—Tables jusqu'en 1750, par l'abbé *de Claustre*. Paris, 1753, 10 vols. 4°.

33. Journal des Observations physiques, mathématiques et bota-

niques, faites dans l'Amérique Méridionale. 4°. Paris, 1714–1725.

34. Observations curieuses sur toutes les parties de la Physique, extraites et recueillies des meilleurs Auteurs, par *Bougeant.* Paris, 1719, 1726, 1730, 3 vols. 12°.; et un 4me vol. en 1771.

35. Le Pour et le Contre, ouvrage périodique sur les sciences, les arts, les livres, les auteurs, etc. par les abbés *Prévost, Desfontaines* et *Le Fèvre de St. Marc.* Paris, 1733–1740, 20 vols. 32°.

36. Observations sur l'Histoire Naturelle, sur la Physique et sur la Peinture, par *Gautier.* Paris, 1751 et suiv. 6 vols. 4°.

37. Observations périodiques sur la Physique, l'Histoire Naturelle et les Beaux Arts; Paris, 1756, 4°. (Contin. par *Toussaint*), 1757 et 1758, 4 vols.

38. Bibliothèque de Physique et d'Histoire Naturelle; par *Lambert.* Paris, 1758, 5 vols. 12°.

39. Observations sur la Physique, sur l'Histoire Naturelle, et sur les Arts et Métiers, etc., par *l'Abbé Rozier, J. A. Mongez le jeune et J. de la Métherie,* Paris, 1771–73, 9 vols. 12°.—(Contin.) Tableau du Travail annuel de toutes les Académies de l'Europe, etc. 1773–81, 18 vols.

40. Introduction aux Observations sur la Physique, sur l'Histoire Naturelle et sur les Arts. Par *l'Abbé Rozier.* 2 vols. 4°. Paris, 1777.

41. Observations sur la Physique, sur l'Histoire Naturelle et sur les Arts; rédigées par *l'Abbé Rozier, J. A. Mongez et Jean de la Métherie.* 43 vols. Paris, 1784–1793, 4°.

42. Journal de Physique, de Chimie et d'Histoire Naturelle; par *J. C. de la Métherie* et *de Blainville.* 53 vols. 4°. Paris, 1794–1823.

43. La Nature considérée sous ses différens aspects, ou Lettres sur les Animaux, les Végétaux et les Minéraux. Ouvrage périodique par *Buchoz.* Paris, 1771–1783, 47 vols. 12°.

44. Journal d'Histoire Naturelle, par MM. *Lamarck, Bruguière, Olivier, Haüy* et *Lepelletier.* 2 vols. 4°. Paris, 1792.

45. Choix de Mémoires sur divers objets d'Histoire Naturelle, par *Lamarck,* 3 vols. 8°. Paris, 1792.

46. Magasin Encyclopédique, ou Journal des Sciences, des Lettres et des Arts, rédigé par *A. L. Millin.* Paris, 1795–1816, 122 vols. 8°.—Continuation sous le titre: Annales Encyclopédiques, 1817 et 1818, 12 vols.—Continuation: Revue Encyclopédique, par *M. Julien,* 1819–1834, 60 vols.—Table Générale des Matières, par ordre alphabétique, des 122 vols. qui composent le Magasin Encyclopédique, rédigée par *J. B. Lajou,* 4 vols. 8°.

47. Journal des Mines. Paris, 1795–1815, 38 vols. 8°.—Tabl. Analyt. 2 vols.

48. Annales des Mines, ou Recueil de Mémoires sur l'Exploitation des Mines et sur les Sciences et les Arts qui s'y rapportent. 13 vols. 8°. Paris, 1817–1826.—Sér. 2, 8 vols. 8°. Paris, 1826–1830.—Sér. 3, 20 vols. 8°. Paris, 1830–1841.

49. Journal de Physiologie expérimentale et pathologique, par *T. Magendie*. Paris, 1821–33, 13 vols. 8°.

50. Archives Générales de Médecine, par une Société de Médecins. Paris, 1823–1832; 30 vols. 8°.—2ᵉ série, 1833–1838, 8°. 3 vols. par an.

51. Bulletin Universel des Sciences et de l'Industrie; continuation du Bulletin général et universel des annonces et des nouvelles scientifiques; publié sous la direction de *M. de Férussac*. (Deuxième section) Bulletin des Sciences Naturelles et de Géologie. 27 vols. 8°. Paris, 1824–1831.

52. Annales des Sciences Naturelles, rédigées par MM. *Audouin, Brongniart et Dumas*. 8°. Paris, 1824–1833; avec la Revue bibliographique pour servir de complément, etc. 1824–1830, 30 vols. 8°.—(2ᵉ sér.), 1834–43, 20 vols. 8°.—(3ᵉ sér.), 1844–1846.

53. Annales des Sciences d'Observation, par *Saigey* et *Raspail*. 4 vols. 8°. Paris, 1829–1830.

54. Journal de Géologie; par *A. Boué, Jobert et Rozet*. 8°. Paris, 1830–1831.

55. Magasin de Zoologie, Journal destiné à établir une correspondance entre les Zoologistes de tous les pays, et à leur faciliter les moyens de publier les espèces nouvelles ou peu connues qu'ils possèdent. Par *F. E. Guérin-Méneville*. Paris, 1831–38, 8 vols. 8°.—Sér. 2, 1839–46.

56. Revue Zoologique, par la *Société Cuvierienne*, association universelle pour l'avancement de la Zoologie, de l'Anatomie comparée et de la Palæontologie, publiée sous la direction de *F. E. Guérin*. Paris, 1838–1846. 8°.

57. Journal de la *Société Phrénologique*. Paris, vol. 1, 8°. 1832.

58. Journal de l'Instruction publique et des Cours publics. Paris, 1833.

59. L'Institut, Journal général des Sociétés et travaux scientifiques de la France et de l'étranger. 1ᵉ section, 1833–1846. Paris, fol.

60. L'Echo du Monde Savant, Journal des nouvelles scientifiques. Paris, 1834–1843, fol.

61. Bulletin Zoologique. 1 vol. 8°. Paris, 1835.

62. Hermes, Journal des nouvelles scientifiques. Fol. Paris, 1836.

63. Annales Françaises et E'trangères d'Anatomie et de Physiologie appliquées à la Médecine et à l'Histoire Naturelle ; par MM. *Laurent* et *Bazin.* 3 vols. 8°. Paris, 1837–1839.

64. Annales des Sciences Géologiques, ou Archives de Géologie, de Minéralogie, de Paléontologie, et de toutes les parties de Géographie, d'Astronomie, de Météorologie, de Physique générale, etc., qui se rattachent directement à la Géologie pure et appliquée ; publiées par *A. Rivière.* 8°. Paris, 1842.

65. Mémorial Encyclopédique ; par *Bailly de Merlieux.*

66. Archives des Découvertes et Inventions nouvelles faites dans les Sciences, les Arts et les Manufactures, tant en France que dans les pays étrangers.

67. Archives d'Anatomie générale et de Physiologie. Par *L. Mandl.* 8°. Paris, 1846.

68. Encyclopédie, ou Dictionnaire Raisonné des Sciences, des Arts et des Métiers. Par *Diderot* et *D'Alembert.* 17 vols. fol. Paris, 1751.—Planches, 11 vols. fol. 1762–1772.—Supplément, 5 vols. fol. 1776–1777.—Table analytique, par *M. Mouchon.* 2 vols. fol. Paris et Amsterdam, 1780.

69. Dictionnaire Raisonné Universel d'Histoire Naturelle. Par *J. C. Valmont de Bomare.* 5 vols. 8°. et Supp. Paris, 1765–1768. —2ᵉ édition, 12 vols. 8°. Paris, 1768–1769.—3ᵉ édition, 6 vols. 4°. Paris, 1775.—4ᵉ édition, 15 vols. 8°. Lyon, 1791 ; Paris, 1800.—Supplément, par *P. R. Vicat.* 8°. Lausanne, 1778.

70. Encyclopédie Méthodique. 201 vols. 4°. Paris, 1782–1832.

71. Tableau Encyclopédique des trois Règnes de la Nature. Paris, 1788–1823, 4°.

72. Nouveau Dictionnaire d'Histoire Naturelle. 24 vols. 8°. Paris, 1802–1804.—2ᵉ édition, 36 vols. 8°. Paris, 1816–1819.

73. Dictionnaire des Sciences Naturelles, publié par les Professeurs du Jardin du Roi. Paris et Strasbourg, 1816–29, 60 vols. 8°.— Supplément, 1840–

74. Dictionnaire Classique d'Histoire Naturelle, par MM. *Audouin, Bourdon, Brongniart, Edwards, de Férussac, Drapiez, Flourens, Jussieu, Lucas, Richard, Bory de Saint-Vincent,* etc. Paris, 1824–1830, 17 vols. 8°.

75. Dictionnaire raisonné, étymologique, synonymique et polyglotte des Termes usités dans les Sciences Naturelles, par *A.-J.-L. Jourdan.* Paris, 1834, 2 vols. 8°.

76. Dictionnaire E'lémentaire d'Histoire Naturelle, publié sous la direction de *V. Meunier.* Paris, 2 vols. 8°. atl. 1842.

77. Dictionnaire Universel d'Histoire Naturelle, par une réunion de

savans. Dirigé par *Ch. d'Orbigny*. Paris, 1843-47, 16 vols. 8°. et atl.

78. Transactions Philosophiques des ann. 1731-1736, traduites par MM. *de Bremond* et *Demours*; 6 vols. 4°. Paris, 1741.—Table des Mémoires imprimés dans les Transactions Philosophiques de la *Société Royale de Londres*, de 1665 à 1735; par *Fr. de Bremond*. 4°. Paris, 1739.

79. Recueil des Mémoires les plus interessans de Chimie et d'Histoire Naturelle, contenus dans les Actes de l'*Académie d'Upsal* et dans ceux de l'*Académie Royale des Sciences de Stockholm*, publiés depuis 1720 jusqu'en 1760; trad. du Latin et de l'Allemand par M. le B. d'*O...* . Paris, 1764, 12°. 2 vols.

80. Mémoires de l'*Académie Royale de Prusse* concernant l'Anatomie, la Physiologie, la Physique, l'Histoire Naturelle, la Botanique, etc., extr. des 16 vols. in 4°. qui composent les Mémoires de la dite Académie, par M. *Paul*. Paris, 4°. et 12°. 7 vols.

81. Mémoires de l'*Académie des Sciences de l'Institut de Bologne*, concernant l'histoire naturelle, la physique, etc. Par *Paul*, 4°. Paris, 1773.

82. Mémoires de la *Société Linnéenne de Normandie*, publiés par *M. de Caumont*. 4 vols. 8°. et 1 vol. 4°. Caen, 1824-1828.—Sér. 2, 5 vols. Paris, 1835, 4°.

83. Séances publiques de la *Société Linnéenne de Normandie*. 8°. Paris, 1835-38.

84. Mémoires de la Société des Sciences Naturelles de Seine et Oise. Paris, 1835, 8°.

PAU.

1. Bulletin de la *Société des Sciences et Arts de Pau*. 1841-1844, 8°.

PÉRIGUEUX, Dép. de la Dordogne.

1. Annales Agricoles et Littéraires de la Dordogne. Périgueux, 1841.

PERPIGNAN.

1. *Société Agricole, Scientifique et Littéraire des Pyrénées Orientales*. 6 vols. 8°.

2. *Société Royale d'Agriculture, des Arts et du Commerce des Pyrénées Orientales*. Perpignan, 182 , 8°.

POITIERS, Dép. de Vienne.

1. Bulletin de la *Société d'Agriculture, Belles-Lettres, Sciences et Arts de Poitiers*. 8°. 1832.

2. Travaux de la *Société d'E'mulation de Poitiers.*

PUY.

1. Annales de la *Société d'Agriculture, des Sciences, des Arts, et du Commerce de Puy.* Puy, 1825, 8º.

PROVINS.

1. Séance publique de la *Société libre d'Agriculture, Sciences et Arts de Provins* (S. et M.), 1807.

RENNES.

1. Corps d'Observations de la *Société d'Agriculture, de Commerce et des Arts établie par les E'tats de Bretagne.* Rennes, 1760 et 1772, 2 vols. 8º.

2. Comptes-rendus des Travaux de la *Société des Sciences et Arts de Rennes.* Rennes, 1833–1836, 8º.

ROUEN.

1. Rapport sur les Travaux de la *Société d'E'mulation de Rouen.* An. IX. 8º.

2. Séance publique de la *Société libre d'E'mulation de Rouen.* 1805, 8º.

3. Précis Analytique des Travaux de l'*Académie Royale des Sciences, Belles-Lettres et Arts de Rouen.* 8º. 1814–1842.

4. Séance publique de la *Société centrale d'Agriculture de la Seine Inférieure.* Rouen, 1820.

5. Extrait des Travaux de la *Société centrale d'Agriculture de la Seine Inférieure.* Rouen, 1821.

6. Procès-verbal de la *Société libre d'E'mulation de Rouen.* 1823, 8º.—Bulletin, &c. 1837–1844.

7. Bulletin de la *Société centrale d'Agriculture de la Seine Inférieure.* Rouen, 1825.

SAINT-BRIEUC.

1. Annales de la *Société d'Agriculture de l'arrondissement de St.-Brieuc.* 1824, 8º.

SAINT-ETIENNE.

1. Bulletin de la *Société Industrielle de l'arrondissement de St.-E'tienne.* 19 vols. 8º.

2. Bulletin d'Industrie Agricole et Manufacturière de la *Société*

d'Agriculture, Arts et Commerce de la Loire. St.-E'tienne, 1822, 8°.

SAINT-QUENTIN.

1. Séance publique de la *Société des Sciences, Arts, Belles-Lettres, et Agriculture de St.-Quentin*, 1826, 8°.
2. Mémoires de la *Société Royale des Sciences, Arts, Belles-Lettres et Agriculture de St.-Quentin.* 1831, 8°.
3. Annales Agricoles du département de l'Aisne publiées par la *Société des Sciences, Arts, Belles-Lettres et Agriculture de St.-Quentin.* 1838, 8°.

SAINTES.

1. Procès-verbal de la Séance publique de la *Société d'Agriculture, des Arts et du Commerce de l'arrondissement de Saintes.* 1821.

SOISSONS.

1. *Société des Sciences, Arts et Belles-Lettres de Soissons.* 1807, 8°.

STRASBOURG (Argentoratum).

(Vide etiam PARIS, 73.)

1. Séance publique de la *Société d'Agriculture du Bas-Rhin.* Strasbourg, An. XII.
2. Tableaux Analytiques des Travaux de la *Société d'Agriculture du Bas-Rhin*; par *Hugot.* Strasbourg, 1809–1821, 8°.
3. Mémoires de la *Société des Sciences, Agriculture et Arts de Strasbourg.* 8°. Strasbourg, 1811 et 1823, 2 vols.
4. Comptes-rendus des Travaux de la *Société d'Agriculture du Bas-Rhin;* par Malte. Strasbourg, 1821–1833.
5. Journal de la *Société des Sciences, Agriculture et Arts du département du Bas-Rhin.* Strasbourg, 1824, 8°.
6. Compte-rendu des Travaux de la *Société des Sciences et Arts du Bas-Rhin*; par *Malle.* Strasbourg, 1833, 8°.
7. Mémoires de la *Société du Muséum d'Histoire Naturelle de Strasbourg.* 3 vols. 4°. Strasbourg, 183 –1846.
8. Revue Entomologique, par *H. G. Silbermann.* Strasbourg, 1833 et suiv.

TOULON.

1. Bulletin trimestriel de la *Société des Sciences, Belles-Lettres et*

Arts du département du Var, séant à Toulon. Toulon, 11ᵉ année. 8º. 1834.

TOULOUSE.

1. Histoire et Mémoires de l'*Académie Royale des Sciences, Inscriptions et Belles-Lettres de Toulouse*, 1782–92, 4 vols. 4º.

2. Histoire et Mémoires de l'*Académie Royale des Sciences, Inscriptions et Belles-Lettres de Toulouse*, depuis son rétablissement en 1807. Toulouse, 1827–1841, 4 vols. 8º.

3. Comptes-rendus des Travaux de l'*Académie Royale des Sciences, Inscriptions et Belles-Lettres de Toulouse*. 182 .

4. Journal des Propriétaires Ruraux pour le Midi de la France, par la *Société d'Agriculture de Toulouse*. 1805–1830, 26 vols. 8º.

5. Séances publiques de la *Société Royale d'Agriculture de la Haute Garonne*. Toulouse, 1825.

TOURS, Dép. de l'Indre et Loire.

1. Recueil des Délibérations et des Mémoires de la *Société Royale d'Agriculture de la Généralité de Tours*, pour l'année 1761. Tours, 1763.

2. Annales de la *Société d'Agriculture, Sciences et Belles-Lettres du département d'Indre et Loire*. Tours, 1821–1831, 12 vols. 8º.

3. Recueil des Travaux de la *Société Médicale d'Indre et Loire*. 1843, 8º.

TRÉVOUX.

1. Bulletin de la *Société d'Agriculture de l'arrondissement de Trévoux*.

TROYES.

1. Mémoires de la *Société d'Agriculture, Sciences, Arts et Belles-Lettres du département de l'Aube*. Troyes, 1822–1844, 8º. Nos. 1–86.

2. Bulletin de la *Société d'Agriculture, etc., de l'Aube*. Troyes, 18 –1839, Nos. 1–71, 8º.

VALENCIENNES.

1. Mémoires de la *Société d'Agriculture, des Sciences et des Arts de l'arrondissement de Valenciennes*. 8º. 1833.

2. Programme de la *Société d'Agriculture, des Sciences et des Arts de l'arrondissement de Valenciennes*. Valenc. 1837, 8º.

Vannes, Dép. du Morbihan.

1. Comptes-rendus des Travaux de la *Société Polymatique du Mor-bihan.* 8º. 1826–1829.

Verdun.

1. Mémoires de la *Société Philomatique de Verdun.* Tom. I.–II. 1840–1843, 2 vols. 8º.

Versailles.

1. Mémoires de la *Société des Sciences Naturelles de Seine et Oise.* 8º. Versailles, 1836–1842.

2. Mémoires de la *Société centrale d'Agriculture et des Arts du département de Seine et Oise.* 1800. Versailles.

Vésoul, Dép. de la Haute Saone.

1. Mémoires de la *Société d'Agriculture, des Sciences, du Commerce et des Arts de la Haute Saone.* Vésoul, 1806.

2. Recueil Agronomique publié par la *Société centrale d'Agriculture du département de la Haute Saone.* Vésoul, 1826, 8º.

HELVETIA.

Aarau.

1. Archiv für Thierheilkunde von der *Gesellschaft Schweizerischen Thierärtzte.* 1ter Jahrgang. Arau, 1816; 2ter Jahrg. Zug und Zürich. 1820–1839, 8º.

2. Verhandlungsblätter der *Gesellschaft für Vaterländische Cultur, im Kanton Aargau.* 1820, 8º.

3. Kurze Uebersicht der Verhandlungen der *Allgemeinen Schweizerischen Gesellschaft für die gesammten naturwissenschaften* in ihrer neunten Jahresversammlung zu Aarau, 1823, 8º.

4. Verhandlungen der *Schweizerischen Gesellschaft für die gesammten Naturwissenschaften* in ihrer 20te Versammlung zu Aarau, 1835. 8º. Aarau, 1836.

Altdorf.

1. Verhandlungen der *Schweizerischen Naturforschenden Gesellschaft* bei ihrer 27te Versammlung zu Altdorf. 8º. Altdorf, 1842.

Basel (Basilea, Basle).

1. Acta Helvetica physico-mathematico-botanico-medica. 8 vols. 4º. Basileæ, 1751–1777.

2. Nova Acta Helvetica physico-mathematico-anatomico-botanico-medica. Tabulis æneis illustrata, 1 vol. 4°. Basileæ, 1787.

3. Bericht über die Verhandlungen der *Naturforschenden Gesellschaft in Basel*. Basel, 8°. 1835–1843, 5 fasc.

4. Auserlesene Sammlung zum Vortheil der Staatswissenschaft, Naturforschung und Feldbaues; aus dem Schwed. übers. v. *G. Sig. Gruner*, mit *Haller's* Vorrede. Bas. 1762–69, 3 vols.

5. Eröffnungsrede der 7ten Jahres Versammlung der *allgem. Schweizerischen Gesellschaft für gesammte Naturwissenschaften* gehalten in Basel, 1821, 8°. Basel, 1821.

6. Verhandlungen der *Schweizerischen Naturforschenden Gesellschaft* bei ihrer Versammlung zu Basel, 1838. Basel, 8°.

BERN (Berna, Berne).

(Vide etiam BERLIN, 34.)

1. Mémoires et Observations recueillies par la *Société E'conomique de Berne*. 2 vols. Berne, 1772–1774.

2. Neue Sammlung physisch-ökonomischer Schriften, herausgegeben von der *Œkonomischen Gesellschaft in Bern*. 2 vols. 1779–1782.

3. Estratto della Litteratura Europæa. 18 vols. 8°. Bernæ, 1758–1762.

4. Excerptum totius Italicæ necnon Helveticæ Litteraturæ. 4 vols. Bernæ, 1759, 8°.

5. Bernisches Magazin der Natur, Kunst u. Wissenschaften; von *S. Wyttenbach*. Bern, 1775–1779, 8°. 5 Stücke.

6. Gemeinnützige Nachrichten. Bern, 8°. 1798.

7. Der Schweizerische Beobachter; herausg. von einer Gesellschaft Gelehrter. 2 vols. 8°. Bern, 1807–1809.

8. Museum der Naturgeschichte Helvetiens, v. *F. Meissner*. Bern, 1807–1820, 4°.

9. Naturwissenschaftliche Anzeiger der *allgem. Schweizerischen Gesellschaft der gesammten Naturwissenschaften*, v. *F. Meissner*, 5 vols. 8°. Bern, 1818–1823.

10. Annalen der *Schweizerischen Gesellschaft für die Naturwissenschaften*, v. *F. Meissner*. Bern, 1824–1825, 2 vols. 8°.

11. Eröffnungsrede der achten Jahres Versammlung der *Schw. Gesellsch. für die gesammten Wissenschaften* gehalten in Bern, 1822. Bern, 1823, 8°.

12. Verhandlungen der *Schweiz. Naturf. Gesellschaft* bei ihrer 24te Versammlung zu Bern, 1839. Bern, 8°.

13. Archiv für Thierheilkunde von der *Gesellschaft Schweizerischer Thierärzte.* 8°. Bern, 1829–1835.

14. Schweizerische Zeitschrift für Medicin, Chirurgie u. Geburtshülfe, v. *Lutki* u. *Bourgeois.* 8°. Bern, 1842.

15. Mittheilungen der *Naturforschenden Gesellschaft* in Bern. 8°. Bern, 1844.

CHUR (*Coire*).

1. Der Sammler, eine gemeinnützige Wochenschrift für Bundten, herausg. von der Œkonomischen Gesellschaft dasselbige, 1805–1812. Chur, Tubingen, Winterthur und Zürich, 8°.

2. Verhandlungen der *Allgemeinen Schweizerischen Gesellschaft für die gesammten Naturwissenschaften* in ihrer Jahresversammlung zu Chur, 1826 u. 1844, 8°.

FREIBURG (*Fribourg*).

1. Actes de la *Société Helvétique des Sciences Naturelles,* assemblée à Fribourg en 1840 (25ᵉ session). Fribourg, 8°. 1841.

2. Mémoires de la *Société E'conomique de Fribourg.* Fribourg, 1816, 8°.

GENF (Geneva, Genève).

1. Mémoires de la *Société établie à Genève pour l'encouragement des Arts et de l'Agriculture.* Genève, 1778, 4°.

2. Mémoires de la *Société de Physique et d'Histoire Naturelle de Genève.* Gen. 1821–1846, 10 vols. 4°.

3. Journal Historique du Commerce et des Arts et Manufactures. Genève, 1744, 4°.

4. Bibliothèque Britannique, rédigée par *Aug. Pictet* et *F. G. Maurice.* Genève, 1796–1815, 140 vols. 8°. Tables jusqu'en 1815, 4 vols.

5. Bibliothèque Universelle des Sciences, Belles-Lettres et Arts, faisant suite à la Bibliothèque Britannique ; rédigée à Genève, 1816–1838 ; 1ᵉ série, Littérature, 60 vols. ; 2ᵉ série, Sciences et Arts, 60 vols. ; 3ᵉ série, Agriculture, 14 vols. ; Nouv. série, 1836–1844, 4 vols.

6. Archives de l'Électricité. Supplément à la Bibliothèque de Genève, par *A. de la Rive.* Genève, 1841–1843, 3 vols. 8°.

7. Discours d'Ouverture de la Session de 1820 de la *Société Helvétique des Sciences Naturelles,* à Genève, 1820, 8°.

8. Actes de la *Société Helvétique des Sciences Naturelles.* Genève, 8°. 1832 u. 1845.

LAUSANNE.

(Vide etiam PARIS, 69.)

1. Mémoires de la *Société des Sciences Physiques de Lausanne*, 3 vols. 4°. Lausanne, 1784-1790.

2. Bibliothèque Medico-physique du Nord, ou recueil de ce qu'il y a d'essentiel dans les collections académiques et dans les autres ouvrages des savants du Nord. Par *V. R. Vicat*, 3 vols. 8°. Lausanne, 1783, 1784.

3. Feuille d'Agriculture, ou Feuille du Canton de Vaud. Lausanne, 1812-1846, 27 vols. 8°.

4. Discours prononcé à Lausanne le 27 Juillet 1828 en ouvrant la première séance de la réunion périodique de la *Société Helvétique des Sciences Naturelles;* par le président de la Société, *D. A. Chavannes.* Lausanne, 1818, 8°.

5. Actes de la *Société Helvétique des Sciences Naturelles.* Lausanne, 1829 et 1843.

6. Actes de la *Société Helvétique des Sciences Naturelles*, 15e réunion, à l'Hospice du Grand St. Bernhard, 1829. Lausanne, 1830, 8°.

7. Bulletin des Séances de la *Société Vaudoise des Sciences Naturelles.* Lausanne, 1842-44, 8°.

8. Journal de la *Société Vaudoise d'Utilité Publique*, faisant suite à la feuille du Canton de Vaud. 8°. Lausanne, 1832-1834. Par *M. Chavannes.*

9. Revue Suisse. Lausanne, 1838-1844, 7 vols. 8°.

LUGANO (Lucanum).

1. Atti della *Società Elvetica delle Scienze Naturali* in Lugano, 1833, sessione 18. Lugano, 1833, 8°.

2. Atti della *Società Ticinese d'Utilità Publica.* Lugano, 1835, 8°.

LUZERN (Lucerne).

1. Verhandlungen der *Landwirthschaftlichen Gesellschaft des Kantons Luzern,* von den Jahren 1824-1829, 8°.

2. Verhandlungen der *allgemeinen Schweizerischen Gesellschaft für die gesammten Naturwissenschaften* in ihrer 19te Versammlung zu Luzern, 1834. 8°. Luzern, 1835.

NEUCHATEL (Neocomium).

1. Mémoires de la *Société des Sciences Naturelles de Neuchâtel.* Neuch. 2 vols. 4°. 1836-1839.

2. Bulletin de la *Société des Sciences Naturelles de Neuchâtel.* Neuch. 1843–44, 8°. (No. 11.)

3. Actes de la *Société Helvétique des Sciences Naturelles,* réunie à Neuchâtel, 1837. Neuchâtel, 8°. 1837.

4. Neue Denkschriften der *Allgemeinen Schweizerischen Gesellschaft für die gesammten Naturwissenschaften.* (Nouveaux Mémoires de la *Société Helvétique des Sciences Naturelles.*) Neuchâtel, 1837–1842, 6 vols. 4°.

St. Gallen (St. Gall.).

(Vide etiam Berlin, 34.)

1. Eröffnungsrede der Jahresversammlung der *Allgemeinen Schweizerischen Gesellschaft für die gesammten Naturwissenschaften,* gehalten in St. Gallen, 1819. St. Gallen, 1819, 8°.

2. Verhandlungen der *Allgemeinen Schweizerischen Gesellschaft für die gesammten Naturwissenschaften* in ihrer 16te Jahresversammlung zu St. Gallen, 1830. St. Gallen, 1831, 8°.

Schaffhausen.

1. Kürze Uebersicht der Verhandlungen der *Allgemeinen Schweizerischen Gesellschaft für die gesammten Naturwissenschaften* in ihrer 10te Jahresversammlung zu Schaffhausen, 1824. Schaffhausen, 1824, 8°.

Solothurn (Soleure).

1. Verhandlungen der *Allgemeinen Schweizerischen Gesellschaft für die gesammten Naturwissenschaften* in ihrer Jahresversammlung zu Solothurn. Soloth. 1825 u. 1836, 8°.

2. Bericht der *Naturhistorischen Gesellschaft in Solothurn,* 1825–1827, 8°.

3. Schweizerisches ·Gewerbsblatt, herausg. v. *Bolley* u. *Möllinger.* Solothurn, 1840–1842, 8°.

Tubingen.

(Vide Chur, 1.)

Winterthur.

(Vide etiam Chur, 1 ; Zürich, 5.)

1. Helvetische Monatschrift, von *J. G. A. Höpfner.* 8 vols. 8°. Winterthur, 1799–1801.

2. Alpina, eine Schrift der genauern Kenntniss der Alpen gewidmet; v. *C. Ulr. Salis in Marschlin* u. *J. Rod. Steinmüller.* Winterthur, 1806–1809, 8°. 4 vols.

ITALIA.

AREZZO (Aretium).

(Vide etiam LEIPZIG, 38.)

1. Atti della *Imperiale e Reale Accademia Aretina di Scienze, Letteri ed Arti.* 8°. Arezzo, 1843.

BOLOGNA (Bononia).

(Vide etiam PARIS, 81.)

1. Bononiensis Instituti Commentaria, 1705. Vol. 1.

2. De *Bononiensi Scientiarum et Artium Instituto atque Academiâ* Commentarii. Bonon. 1731–1791, 10 vols. 4°.

3. Novi Commentarii *Academiæ Scientiarum Instituti Bononicnsis.* Bonon. 1834 et seq. 4°.

4. Notizie del nuovo *Istituto di Scienze ed Arti di Bologna.* Di *E. Corazzi.*

5. Dell'Origine e de' Progressi dell' *Istituto di Bologna* e di tutte le Accad. ad esso unite. Di *G. G. Bolletti.* Bol. 1769, 8°.

6. Rendiconto delle sessioni ordinarie dell' *Accademia delle Scienze dell' Istituto di Bologna.* 5 vols. 12ᵐᵒ. Bologna, 1833–1842.

7. Memorie dell' *Istituto Nazionale Italiano.* Classe di Fisica e Matematica. Bologna, 1806–10, 4°. 4 vols.

8. Memorie della *Società Medico-chirurgica di Bologna.* Bol. 1838 et seq. 4°.

9. Opuscoli Scientifici. Bologna, 1817–23, 4 vols. 4°.

10. Nuova Collezione di Opuscoli Scientifici. Bol. 1824–25, 2 vols. 4°.

11. Annali di Storia Naturale. Bologna, 1829, 4 vols. 8°.

12. Nuovi Annali di Storia Naturale. Bologna, 1838 et seq. 8°.

13. Nuovi Annali delle Scienze Naturali. 8°. Bologna, 1832–1842.

14. Bullettino delle Scienze Mediche, pubblicato per cura della *Società Medico-chirurgica di Bologna.* Bol. 1838 et seq. 8°.

BRESCIA (Brixia).

1. Commentarii dell'*Accademia di Scienze, Lettere, ed Arti dell' Ateneo di Brescia.* 8°. 1808 et seq.

CATANIA (Catana).

1. Atti dell'*Accademia Gioenia di Scienze Naturali di Catania,* 7 vols. 4°. Catan. 1825–1833.

2. Relazioni Accademiche dell'*Accademia Gioenia*, etc. 8°. Catan. 1834–1836.

3. Giornale del Gabinetto Letterario dell'*Accademia Gioenia di Catania*. Cat. 8°.

CHAMBERY.

1. Mémoires de la *Société Royale Académique de Savoie*. 6 vols. 8°. Chambery, 18 –1833.

FANO.

1. Giornale Letterario di Fano. Fano, 8°.
2. Raccoglitore Medico di Fano, 1827.

FERRARA.

1. Atti dell'Accademia Medico-chirurgica di Ferrara. 8°.

FIRENZE (Florentia, Florence).

1. Saggi di naturali Esperienze dell'*Accademia del Cimento*. Fol. Firenze, 1667–1691 ; Venez. 1711, 4°.; Firenze, 1841, 4°.

2. Atti e Memorie dell'*Accademia del Cimento*, da *Galileo Galilei* sino a *F. Redi* e *Viviani ;* pubblicate dal *Dr. Tozzetti Targioni.* Firenze, 4°. 1750, 4 vols.

3. Memorie di varia Erudizione della *Società Columbaria*, colla Notizia della origine di questa Accademia scritta da *Bindo Peruzzi*. Firenze, 1747, 4°.

4. Atti dell'*Accademia de Georgofili*. Firenze, 8°.

5. Atti dell'*Istituto di Firenze.*

6. Atti dell'*Accademia Italiana*. Firenze, 1808, 4°.

7. Annali del *Museo Imperiale di Fisica e Storia Naturale* di Firenze, 1808–12, 3 vols. 4°.

8. Atti della terza Riunione degli Scienziati Italiani tenuta in Firenze nel Settembre, 1841. Fir. 1842, 1 vol. 4°.

9. Giornale dei Letterati. Firenze, 1742–52, 22 vols. 12°.—Pisa, 1785–88, 12°.

10. Raccolta d'Opuscoli Fisico-medici, del Prof. *Targioni.* Firenze, 1775, 12°.

11. Collezione di Opuscoli Scientifici e Letterari. Firenze, 1807–15, 5 vols. 8°.

12. Giornale Enciclopedico di Firenze. Fir. 1809, 8°.

13. Giornale di Scienze ed Arti. Firenze, 1816–17, 5 vols. 4°.

14. Giornale Agrario Toscano, 1819–1821, 8º.

15. L'Antologia.   Firenze, 1822–32, 44 vols. 8º.

16. Giornale Toscano di Scienze Mediche, etc. 1840 et seq. 4º.

GENOVA (Genua, Genoa, Gènes).

1. Memorie della *Società Medica di Emulazione di Genova*, 1801.

2. Memorie dell'*Istituto Ligure.*   Genova, 1806–1809, 3 vols. 4º.

3. Memorie dell'*Accademia delle Scienze, Lettere ed Arti di Genova*, 1809–1814, 4º.

4. Giornale Ligustico di Scienze, Lettere ed Arti.   Genova, 1827–29, 8º.

5. Atti della ottava Riunione degli *Scienziati Italiani* tenuta in Genova, nel Settembre del 1846, 4º. Genoa, 1847.

LIVORNO (Leghorn).

1. Magazzino Toscano d'Instruzione e di Piacere.   Livorno, 1754, 2 vols. 4º.

2. Atti dell'*Accademia Italiana di Scienze, Lettere ed Arti.*   Livorno, 1810, 4º. 2 vols.

LUCCA.

1. Atti della *Accademia Lucchese di Scienze, Lettere ed Arti.*   Lucca, 1821–40, 10 vols. 8º.

2. Memorie sopra la Fisica e la Storia Naturale, di diversi valent' uomini.   Lucca, 1743–57, 4 vols. 8º.

3. Miscellanei di varia Letteratura, di *Seb. Donati* e *F. Cenami.*   Lucca, 1762–72, 8 vols. 12º.

4. Nuovi Miscellanei Lucchesi di *Seb. Donati.*   Lucca, 1775, 4º.

5. Atti della quinta Riunione degli Scienziati Italiani in Lucca.   Lucca, 1844, 4º.

MANTOVA (Mantua).

1. Memorie dell'*Accademia di Scienze e Belle Lettere ed Arti di Mantova*, 1795, 4º.

MESSINA (Messana).

1. Giornale del Gabinetto Letterario di Messina.   Messina, 8º.

MILANO (Mediolanum, Milan).

1. Atti della *Società Patriotica di Milano*, dalla istituzione, 1778, fino al 1783.   Milano, 1783–93, 3 vols. 4º.

2. Giornale della *Società d'Incoraggiamento di Milano.* 1808 et seq. 8°.

3. Memorie dell'*Imperiale Regio Istituto del Regno Lombardo-Veneto.* 5 vols. Milano, 1812–1838.

4. Giornale dell' *I. R. Istituto Lombardo-Veneto,* e Biblioteca Italiana. Milano, 1841 et seq. 8°.

5. Scelta d'Opuscoli interessanti (redatti da *C. Amozetti, Fortis, Fromond, Campi, Soave*). Milano, 1775–77, 36 vols. 12°.—(2ᵉ édit.) 3 vols. 4°. 1781–1784.

6. Opuscoli scelti sulle Scienze e sulle Arti. Milano, 1778–1804, 23 vols. 4°.

7. Nuova Scelta d'Opuscoli sulle Scienze e sulle Arti. Milano, 1804–1807, 2 vols. 4°.

8. Effemeridi Fisico-mediche. Milano, 1804–1807, 12 vols. 8°.

9. Annali di Agricoltura del Regno d'Italia, del Cav. *F. Ré.* Milano, 1809–14, 22 vols. 8°.

10. Annali di Scienze e Lettere. Milano, 1810–13, 2 vols. 8°.

11. Biblioteca Italiana ossia Giornale di Letteratura, Scienze ed Arti. Milano, 1816–39, 8°. 92 vols.

12. Annali Universali di Medicina istituiti dal Prof. *A. Omodei,* continuati dal Dr. *C. A. Calderini.* Milano, 1817 et seq. 8°.

13. Annali Universali di Statistica, Viaggi, etc. Milano, 1824 et seq. 8°.

14. Indicatore Lombardo. Milano, 1832.

15. Bibliografia Italiana, pubblicata dalla libreria *Stella.* Milano, 1835 et seq. 8°.

16. Il Politecnico Repertorio mensuale di Studii applicati alla Prosperita e Cultura sociale. Milano, 1839 et seq. 8°.

17. Giornale di Fisica, Chimica ed Arti del Prof. *Majocchi.* Milano, 1839 et seq. 8°.

18. Annali Universali di Agricoltura. Milano, 17 vols. 4°. Proseguiti col nome di Giornale Agrario Lombardo-Veneto. Milano, 1843, 17 vols. 8°.

19. Atti della sesta Riunione degli *Scienziati Italiani* in Milano. Milano, 1845, 4°.

20. Annali delle Scienze del Regno Lombardo-Veneto. Milano, 4°.

21. Il Raccoglitore, di *D. Bertolotti.* Milano, 8°.

22. Rivista Europea. Milano, 8°.

23. Giornale d'Agricoltura, Arti e Commercio. Milano, 8°.

24. Giornale di Farmacia, Chimica, e Scienze Accessorie, del Dott. *Cattaneo.* Milano, 8°.

F 2

25. Giornale Agrario del Regno Lombardo-Veneto. Milano, 8°.
26. Giornale di Chimica e Veterinaria, di L. Pozzi. Milano, 8°.

MODENA (Mutina).

1. Memorie di Matematica e Fisica della *Società Italiana delle Scienze in Modena*, 7 vols. 4°.—Verona, 1782–94, 9 vols. 4°.—Modena, 1799–1803.
2. Memorie di Fisica della *Società Italiana*, 4°. Modena, 1815–46.
3. Giornale Letterario-scientifico Modenese. Modena, 1839 et seq. 8°.
4. Saggi dell'*Accademia di Modena*.

NAPOLI (Neapolis, Naples).

1. Atti della *Reale Accademia delle Scienze e Belle Lettere di Napoli*, dalla fondazione sino al 1787, 1 vol. 4°. fig. Napoli, 1788.
2. Atti della *Reale Accademia delle Scienze*, sezione della *Società Reale Borbonica*. Napoli, 1819–1839, 4 vols. 4°.
3. Rendiconto delle adunanze della *Reale Accademia delle Scienze di Napoli*. 4°. Napoli, 1842.
4. Atti del *Reale Instituto d'Incoraggiamento alle Scienze Naturali di Napoli*. 4°. Napoli, 1820–1841.
5. Esercitazioni *Accademiche degli Aspiranti Naturalisti di Napoli*. Napoli, 1839–1842, 8°.
6. Bullettino dell'*Accademia degli Aspiranti Naturalisti di Napoli*. Napoli, 1842 et seq. 8°.
7. Annali dell'*Accademia degli Aspiranti Naturalisti de Giulio Avellino*. 8°. Napoli, 1843.
8. Saggio delle Transazioni filosofiche della *Società Regia*, compendiate, tradotte dall'Inglese nell' idioma Toscano. 4°. Napoli, 1729–1734.
9. Calendario Astronomico-geognostico-statistico del Regno di Napoli. Napoli, 1819–23, 12°.
10. Filiatre o Giornale delle Scienze Mediche. Napoli, 1830 et seq.
11. Lo Spettatore del Vesuvio; di *L. Pilla* e *F. Cassola*. Napoli, 8°. 1832.
12. Antologia di Scienze Naturali, pubblicata da *R. Pirta* ed *Arcang. Scacchi*. Napoli, 8°.
13. Il Progresso, Giornale delle Scienze, Lettere ed Arti. Napoli, 8°.

14. Giornale Enciclopedico di Napoli. Napoli, 8°.—L'Osservatore Medico. Napoli, 4°.

15. Atti della settima Riunione degli *Scienziati Italiani*, tenuta in Napoli nel Settembre del 1845, 4°. Napoli, 1846.

PADOVA (Patavium, Padua).

1. Saggi Scientifici e Letterari della *Accademia di Padova*, 4 vols. 4°. Padova, 1786–1794.

2. Memorie dell'*Accademia di Scienze, Lettere ed Arti di Padova*, 1809, 4°.

3. Nuovi Saggi della Imperiale Regia *Accademia de Scienze, Lettere ed Arti* di Padova, 4°. 1817–1840, 5 vols.

4. Relazione delle Memorie lette all'*Acc. di Sc. Lett. ed Arti* di Padova, nel 1840–41. Padova, 1842, 4°.

5. Atti della quarta Riunione degli *Scienziati Italiani* in Padova. Padova, 1843, 4°.

6. Giornale dell' Italiana Letteratura. Padova (di *Nicolò da Rio*), 1802 et seq. 8°.

7. Giornale di Padova, 1807.

8. Biblioteca Germanica di Lettere, Arti e Scienze. Padova, 1822–23, 6 vols. 8°.

9. Commentarii di Medicina del Dott. *Spongia*. Padova, 1836 et seq.

10. Annali delle Scienze del Regno Lombardo-Veneto, del Dott. *Fusinieri*. Padova, 183 et seq. 4°.

11. Giornale di Medicina pratica, di *V. L. Brera*. Padova.

PALERMO (Panormus, Palerme).

1. Specchio della Scienze, o Giornale Enciclopedico di Sicilia. Palermo, 1814, 8°.

2. Travaux de l'*Académie Royale des Sciences de Palerme*. Publiés par *A. Barbucci*, 1833, 8°.

3. Giornale di Scienze, Lettere ed Arti per la Sicilia, 8°. Palermo.

4. Effemeridi Scientifiche e Letterarie per la Sicilia. 8°. fig. 1832 et seq.

PARMA.

1. Giornale dei Letterati, di *B. Bacchini* e *G. Roberto*. Parma, 1686–92, 4°.; Modena, 1692–97.

2. Giornale della *Società Medica* di Parma. 1806–15, 15 vols. 4°.

PAVIA.

1. Biblioteca Fisica d'Europa, di *L. Brugnatelli.* Pavia, 1788–90, 18 vols. 8°.
2. Annali di Chimica e Storia Naturale, di *L. Brugnatelli.* Pavia, 1790–1802, 21 vols. 8°.
3. Giornale di Fisica, Chimica, Storia Naturale, etc. di *Brugnatelli, Brunacci* e *Configliacchi.* 4°. Pavia, 1815–1827.
4. Giornale delle Scienze Medico-chirurgiche di Pavia. Pavia, 1833–1842, 8°.

PERUGIA (Perusia).

1. L'Agricoltore. Perugia, 1784–86, 4°.
2. Giornale Scientifico e Letterario di Perugia.

PESARO.

1. Esercitazioni dell'*Accademia Agraria di Pesaro.* Pesaro, 8°.

PISA (Pisæ).

1. Atti della prima Riunione degli *Scienziati Italiani,* tenuta in Pisa nell' Ottobre di 1839. Pisa, 1840, 1 vol. 4°.
2. Giornale dei Letterati. Pisa, 1771–96, 102 vols. 12°.
3. Nuovo Giornale dei Letterati. Pisa, 1802–1808, 9 vols. 8°. 1822 et seq.
4. Memorie Scientifiche, di *P. Savi.* Dec. I. Pisa, 1828, 8°.
5. Miscellanea di Chimica, Fisica e Storia Naturale. Pisa, 1843, 8°.
6. Miscellanee Medico-chirurgico-farmaceutiche. Pisa, 1843, 8°.
7. Notizie Storiche dell'*Accademia Valdarnese del Poggio,* colle memorie concernenti le scienze naturali, raccolta da' due primi volumi delle Memorie Valdarnesi. Da *J. Corinaldi,* 8°. Pisa, 1839.
8. Giornale Toscano di Scienze Mediche, Fisiche e Naturali, dai Prof. *G. Amici, Bufalini, Georgini, G. Savi* e *P. Savi.* 8°. Pisa, 1843.

RAVENNA.

1. Memorie della *Società Ravennate.*

ROMA (Rome).

1. Il Giornale dei Letterati, di *Fr. Nazari.* Roma, 1675, 4°.—(di *Tinasso*), 1675–1681.

2. Saggi di Dissertazioni Accademiche dell'*Accademia di Cortona.* Roma, 1735.

3. Giornale Arcadico di Scienze, Lettere ed Arti. 8°. Roma, 1819–29, 44 vols. 8°.

4. Annali Medico-chirurgici. Roma, 1839 et seq.

5. Effemeridi Letterarie di Roma. Roma, 8°.

SIENA.

1. Atti dell'*Accademia delle Scienze di Siena,* detta dei *Fisiocritici,* 10 vols. Siena, 1761–1841, 4°.

TORINO (Taurinum, Turin).

1. Miscellanea Philosophico-physico-mathematica *Societatis privatæ Taurinensis;* Aug. Taur. 1759, 1 vol. 4°.

2. Mélanges de Philosophie et de Mathématiques de la *Société Royale de Turin.* Turin, 1760–73, 4 vols. 4°.

3. Memorie della *Reale Accademia delle Scienze di Torino.* Torino, 1784–1838, 35 vols. 4°.—(2de sér.) Torino, 1839 et seq. 4°.

4. Mémoires de l'*Académie Royale des Sciences de Turin,* avec planches. Années 1784–1810, 10 vols. Turin, 1786–1811, 4°

5. Memorie della *Reale Società Agraria di Torino,* dalla sua istituzione (1785) al 1790, 6 vols. 8°.—(Contin.) Memorie della *Società Centrale di Agricoltura di Torino,* etc.

6. Calendario Georgico della *Reale Società Agraria di Torino.* Torino, 1791–1806, 15 vols. 16°.; 1807–39, 26 vols. 8°.

7. Annali della *Reale Società Agraria di Torino.* Torino, 1840 et seq. 8°.

8. Atti della seconda Riunione degli *Scienziati Italiani,* tenuta in Torino nel Settembre, 1840. Torino, 1841, 1 vol. 4°.

9. Annales de l'Observatoire de Turin, avec des Notices statistiques concernant l'Agriculture et la Médecine. Turin, 1787–1818, 2 vols. 4°.

10. Giornale Scientifico e Letterario di una Società filosofica di Torino. Torino, 1789–90, 8 vols. 8°.

11. L'Eridano. Torino, 8°.

12. Biblioteca Oltramontana ad uso d'Italia. Torino, 1787–93, 28 vols. 8°.

13. Bibliothèque Italienne, ou Tableau des Progrès des Sciences et des Arts en Italie. Turin, 1802–1803, 5 vols. 8°.

14. L'Ape Subalpina. Torino, 1811, 1 vol. 12°.

15. Repertorio Medico-chirurgico. Torino, 1821 et seq. 8°.

16. L'Amico d'Italia, Miscellaneo di Letteratura, Scienze ed Arti. Torino, 1822–29, 16 vols. 8°.

17. Dizionario periodico di Medicina, dei Dott. *L. Rolando* e *L. Martini.* Torino, 1822–28, 8°.

18. Il Propagatore. Torino, 1824–29, 12 vols. 8°.

19. Annali di Veterinaria. Torino.

20. Annali Clinici o Repertorio generale delle dottrine relative alla medicina pratica. Torino, 1829–31, 3 vols. 8°.

21. Giornale delle Scienze Mediche di Torino. Torino, 1837 et seq. 8°.

22. Il Subalpino, Giornale di Scienze, Lettere ed Arti. Torino, 1836–39, 7 vols. 8°.

23. Repertorio d'Agricoltura Pratica ed' Economia Domestica, del Dott. *R. Ragazzoni.* Torino, 8°.

24. Repertorio di Medicina, di Chirurgia, etc. Torino, 4°.

TREVISO.

1. Memorie Scientifiche e Letterarie dell' *Ateneo di Treviso.* 1817, 1 vol. 4°.

VENEZIA (Venetiæ, Venice).

1. Raccolta di Memorie della *Accademia di Agricoltura, Arti e Commercio dello Stato Veneto,* sino al 1798, 18 vols.

2. Atti del *Veneto Ateneo.* Venezia, 4°.

3. Esercitazioni Scientifiche e Letterarie dell' *Ateneo di Venezia.* Venezia, 1827–1841, 4°.

4. Memorie dell' *I. R. instituto Veneto di Scienze, Lettere ed Arti,* 4°. Venezia, 1843.

5. Giornale dei Letterati d'Italia, di *Ap. Zeno.* Venezia, 1710 et seq. 12°. 43 vols. et 3 vols. Suppl.

6. Raccolta d'Opuscoli Scientifici e Filologici, da *A. Calogierà.* Venezia, 1728 et seq. 51 vols. 12°.

7. Nuova Raccolta d'Opuscoli Scientifici del *P. A. Calogierà.* Venezia, 1776 et seq. 30 vols.

8. Actorum Eruditorum Synopsis, id est Opuscula Actis Eruditorum inserta, quæ ad Mathesin, Physicam, Medicinam, Anat. Chirurg. et Philolog. pertinent. Venet. 1740–46, 7 vols. 4°.

9. Giornale d'Italia, spettante alla Scienza Naturale e principalmente all' Agricoltura, alle Arti ed al Commercio. Venezia, 1764–76, 12 vols. 4°.

10. Nuovo Giornale d'Italia, spettante alla Scienza Naturale, etc. 7 vols. 4°. Venezia, 1777-1782.

11. Osservazioni spettanti alla Fisica, alla Storia Naturale ed alla Arti. Venezia, 1776, et seq.

12. Antologia Medica, di *L. V. Brera.* Venezia, 1834 et seq. 2 vols. 8°.

13. Giornale di Patologia e di Terapeutica, dei Signori *Bufalini, Corneliani,* etc. Venezia, 8°.

14. Dizionario Classico di Storia Naturale, 16 vols. 8°. Venezia, 1835-1844.

VERONA.

1. Memorie di Matematica e Fisica della *Società Italiana,* 5 vols. 4°. Verona, 1782-1799.

2. Memorie dell' *Accademia di Agricoltura, Commercio ed Arti di Verona,* 1807 et seq. 8°.

3. Il Poligrafo. Verona, 8°.

VICENZA (Vincentia).

1. Annali delle Scienze del Regno Lomb.-Veneto, dal Prof. *A. Fusinieri.* Vicenza, 4°.

HISPANIA.

MADRID.

1. Memorias de la *Real Academia de la Historia.* Madrid, 1796-1801, et 1832, 7 vols. 4°.

2. Annales de Historia Natural. Madrid, 1799-1804, 7 vols. 8°.

3. Annales de Minas. Madrid, 1838- , 8°.

4. Revista de Ciencias Naturales de Madrid. 8°. Madrid, 1842.

LUSITANIA.

LISBOA (Ulyssipo, Lisbon).

1. Memorias da *Academia Real das Sciencias de Lisboa* (desde 1780 até 1788). Lisboa, 1797-1799, 4°. 3 vols.

2. Historia e Memorias da *Academia Real das Sciencias de Lisboa.* Lisboa, 1815-1837, 4°. 9 vols.

3. Memorias Economicas da *Academia Real das Sciencias de Lisboa,* para o adiuntamento de Agricultura, das Artes e da Industria em Portugal. Lisboa, 1789-1815, 5 vols. 4°.

BRITANNIA.

BATH (Bathonia).

1. Letters and Papers of the *Bath and West of England Society*, 14 vols. 8°. Bath, 1793–1816.

2. Bath and Bristol Magazine, 3 vols. 8°. Bath, 1832.

BERWICK-ON-TWEED.

1. Transactions of the *Berwickshire Naturalists' Club*, 8°. Berwick, 1834–46.

CAMBRIDGE (Cantabrigia).

1. Transactions of the *Cambridge Philosophical Society*. Cambridge, 1821–1844, 8 vols. 4°.

2. Cambridge Quarterly Magazine, 8°. 1833.

DUBLIN (Eblana).

1. Transactions of the *Royal Irish Academy*, 4°. Dublin, 1787–1846.

2. Proceedings of the *Royal Irish Academy*. Dublin, 1836–1843, 8°.

3. An Index to the Transactions of the *Royal Irish Academy*, from 1786 to the present time; by *N. Carlisle*. 4°. London, 1813.

4. Transactions of the *Dublin Society*. 6 vols. 8°. Dublin, 1800–1810.

5. Journal of the *Geological Society of Dublin*, 1833–1846.

6. Dublin and London Magazine. Dublin, 1833.

7. Dublin University Review, 1833.

8. Dublin Philosophical Journal and Scientific Review, 2 vols. 8°. Dublin, 1825–1826.

EDINBURGH (Edinbourg).

(Vide etiam ALTENBURG, 1; LONDON, 43.)

1. Essays and Observations, physical and litterary, read before a Society in Edinburgh, and published by them. Edinb. 1754, 1756, 1771, 3 vols. 8°. Trad. Franç. par *Demours*. Paris, 1760.

2. Transactions of the *Royal Society of Edinburgh*. Edinb. 4°. 15 vols. 1788–1844.

3. Proceedings of the *Royal Society of Edinburgh*. Edinburgh, 8°. 1832–1844.

4. Memoirs of the *Wernerian Natural History Society*. Edinburgh, 7 vols. 8°. 1811–1838.

5. Prize Essays and Transactions of the *Highland Society of Scotland*, to which is prefixed an Account of the institutions and principal proceedings of the Society, by *H. Mackenzie*. 6 vols. 8°. Edinburgh, 1812–1824.

6. The Quarterly Journal of Agriculture, and the Prize Essays and Transactions of the *Highland and Agricultural Society of Scotland*. 8°. Edinb. 1828–1846.

7. Statistical Account of Scotland, drawn up from the communications of the ministers of the different parishes. By *Sir J. Sinclair, Bart.* 21 vols. 8°. Edinburgh, 1791.

8. New Statistical Account of Scotland. 15 vols. 8°. Edinb. 1834.

9. Edinburgh Philosophical Journal. Edinb. 14 vols. 1819–1826, 8°. By *Brewster* and *Jameson*.

10. Edinburgh New Philosophical Journal, 8°. Edinb. 1826–1846. By *R. Jameson*.

11. Edinburgh Journal of Science, conducted by *D. Brewster*. Ser. 1, 10 vols. 1824–1829.—New Series, 6 vols. 8°. Edinburgh, 1829–1832.

12. Edinburgh Journal of Natural and Geographical Science, 3 vols. 8°. 1829–1830. By *W. Ainsworth* and *H. H. Cheek*.

13. Naturalist's Library, conducted by *Sir W. Jardine*. Edinb. 40 vols. 12°. 1833–1844.

14. Edinburgh Cabinet Library. Edinb. 1831, 12°.

15. Encyclopædia Britannica, 3 vols. 4°. Edinb. 1771.—4th edition, 20 vols. 4°. 1810.—Supplement, 6 vols. 4°. 1824.—7th edition, by *M. Napier*, 21 vols. 4°. Edinb. 18   –1844.

16. Edinburgh Cyclopædia, by *D. Brewster*, 18 vols. 4°. Edinb. 1810–1830.

17. *Johnstone's* Edinburgh Magazine, 1833.

18. Edinburgh Academic Annual, 1840.

FALMOUTH.

(Vide etiam LONDON, 77, 80.)

1. Reports of the *Royal Cornwall Polytechnic Society*, 8°. Falmouth, 1833–1846.

GLASGOW.

1. Proceedings of the Glasgow Philosophical Society, 1841–1842.

LEEDS.

(Vide etiam LONDON, 75.)

1. Reports of the Council of the Leeds *Philosophical and Literary Society*, 1843.

2. Proceedings of the *West Riding Geological and Polytechnic Society*, 8°. Leeds, 1839–1842.

LIVERPOOL.

1. First Report of the *Liverpool Natural History Society*, br. 8°. Liverpool, 1836.

LONDON (Londinum, Londres, Londra).

(Vide etiam AMSTERDAM, 12; LEIPZIG, 27, 28; WITTENBERG, 2; PARIS, 78; NAPOLI, 8.)

1. Philosophical Transactions, giving some account of the present undertakings, studies and labours of the Ingenious, &c. London, 1665–1846, 4°.

2. A general Index to the Philosophical Transactions from the first to the end of the seventieth volume, 1665–1780. By *P. H. Maty*, 4°. London, 1787.

3. A Continuation of the Alphabetical Index contained in the Philosophical Transactions, from 1821–1830, 4°. 1833.

4. An Index to the medical and physiological Papers contained in the Transactions, &c. 4°. 1814.

5. The Philosophical Transactions and collections to the end of the year 1700, abridged and disposed under general heads. By *John Lewthorp*, 3 vols. 4°. London, 1705.

6. The Philosophical Transactions from the year 1700 to 1720, abridged and disposed under general heads. By *Benjamin Motte*, 2 vols. 4°. London, 1721.

7. The Philosophical Transactions from 1700 to 1720. By *H. Jones*, 4°. London, 1721.

8. The Philosophical Transactions and collections, abridged and disposed under general heads. The fourth edition, 6 vols. London, 1732–33, 4°.

9. The Philosophical Transactions from 1719 to 1733, abridged and disposed under general heads. By *John Eames*, 4 pts. in 2 vols. 4°. London, 1734.

10. The Philosophical Transactions from 1743 to 1750, abridged and disposed under general heads, the Latin papers being translated into English. By *John Martyn*, 4°. London, 1756.

11. The Philosophical Transactions from the commencement in 1665 to 1800, abridged, with Notes and Illustrations by *C. Hutton, J. Shaw* and *R. Pearson*, 18 vols. 4°. London, 1809.

12. Memoirs of the *Royal Society*, being a new abridgement of the Philosophical Transactions, from 1665 to 1735, by *Mr. Baddam*, 10 vols. 8°. London, 1838–1841.

13. Abstracts of the Papers printed in the Philosophical Transactions of the *Royal Society of London,* from 1800–1837; 3 vols. 8º. London, 1832–1837.

14. Proceedings of the *Royal Society,* 8º. London, 1842–1846.

15. The History of the *Royal Society of London,* in which papers not published are inserted as a supplement to the Philosophical Transactions. By *Thomas Birch,* 4 vols. 4º. London, 1756–1757.

16. Transactions of the *Linnean Society of London,* 1791–1846, 19 vols. 4º.

17. Proceedings of the *Linnean Society of London,* 8º. 1838–1846.

18. Transactions of the *Geological Society of London,* 1811–1821, 6 vols. 4º. London.—2ᵈ series 1824–1847.

19. Proceedings of the *Geological Society of London,* 4 vols. 8º. London, 1834–1846.

20. Quarterly Journal of the *Geological Society of London,* 8º. London, 1845–1847.

21. Quarterly Journal of Science, Literature and the Arts, 29 vols. 8º. London, 1816–1830.

22. Journal of Sciences and the Arts, edited at the *Royal Institution of Great Britain,* 8º. London, 1831.

23. Transactions of the *Asiatic Society of Great Britain,* 4º. London, 1827.—Journal of the *Royal Asiatic Society of Great Britain and Ireland,* 8º. London, 1834.

24. Proceedings of the *Royal Asiatic Society,* 8º. London, 1832–1837.

25. Journal of the *Royal Geographical Society of London,* 8º. 1830–1847.

26. Transactions of the *Zoological Society of London,* 4º. 1833–1847.

27. Proceedings of the Committee of Science and Correspondence of the *Zoological Society of London,* 2 parts, 8º. 1830–1832.

28. Proceedings of the *Zoological Society of London,* 8º. London, 1833–47.

29. Reports of the Council and Auditors of the *Zoological Society of London,* read at the Annual General Meeting. London, 8º. 1830–1847.

30. Transactions of the *Entomological Society of London,* 1 vol. 8º. London, 1807–1812.

31. Transactions of the *Entomological Society,* 4 vols. 8º. London, 1834–47.

32. Journal of Proceedings of the *Entomological Society of London*, 8°. London, 1841.

33. Transactions of the *Microscopical Society of London*, 8°. 1842.

34. Reports of the Meetings of the *British Association for the Advancement of Science.* London, 1833–1847. 8°.

35. Journal of the *Royal Agricultural Society of England.* 8°. London, 1840–1847.

36. Memoirs for a Natural History of Animals, containing the anatomical description of several creatures dissected by the *Royal Academy of Sciences at Paris.* fol. London, 1701.

37. Gentleman's Magazine and Historical Chronicle ; by *Sylv. Urban.* 8°. London, 1730–1847.

38. The London Magazine, or Gentleman's Monthly Intelligences. London, 1732 et seq. 8°.

39. Miscellaneous Tracts, relating to Natural History, Husbandry and Physic. London, 1762, 8°.

40. The Wonderful Magazine, or Merveillous Chronicle. London, 8°. 1764.

41. The Philosophical Magazine, 68 vols. 8°. London, 1798–1826.

42. The Philosophical Magazine, or Annals of Chemistry, Mathematics, Astronomy, Natural History and General Sciences ; by *R. Taylor* and *R. Phillips.* London, 1827–32, 11 vols. 8°.

43. The London and Edinburgh Philosophical Magazine and Journal of Science ; third series. 8°. London, 1832–1847.

44. Annals of Philosophy, by *Th. Thomson.* 8°. London, series 1, 16 vols. 1813–20. Series 2, 12 vols. 1821–26.

45. Zoological Journal, conducted by *Th. Bell, J. G. Children, J. de C. Sowerby,* and *G. B. Sowerby.* 5 vols. London, 1824–34, 8°.

46. The Magazine of Natural History and Journal of Zoology, Botany, Geology and Mineralogy ; by *J. C. Loudon.* London, 1828–39, 9 vols. 8°.—New series, by *Ed. Charlesworth* 4 vols. London, 1837–1840.

47. Magazine of Zoology and Botany, conducted by *Sir W. Jardine, P. J. Selby* and *G. Johnston.* 2 vols. 1837–1838, 8°.

48. Annals of Natural History, or Magazine of Zoology, Botany and Geology ; by *W. Jardine, P. J. Selby, Dr. Johnston, Sir W. J. Hooker* and *R. Taylor.* London, 1838–1840, 5 vols. 8°. —(Contin.) The Annals and Magazine of Natural History, including Zoology, Botany and Geology ; by *Sir W. Jardine, P. J. Selby, Dr. Johnston, David Don* and *R. Taylor.* London, 1841–1847.

49. Field Naturalist's Magazine and Review; by *J. Rennie.* 2 vols. 8°. London, 1833–1834.

50. Zoological Magazine. 8°. London, 1833.

51. Entomological Magazine; by *Walker.* 5 vols. London, 8°. 1833–1837.

52. The Analyst. vols. 8°. London, 1834.

53. The Naturalist; by *Neville Wood,* &c. 8°. London, 1836–1839.

54. Records of General Science; by *R. D. Thompson* and *T. Thompson.* 4 vols. 8°. London, 1835–1836.

55. The Mining Review, or Journal of Geology, Mineralogy and Metallurgy. 8°. London, 1830–1836.—New series, 1837.

56. The Mining Journal. fol. London, 1835.

57. London Medical Gazette. fol.

58. Asiatic Journal and Monthly Register for British and Foreign India, China and Australasia, from the year 1815 to 1829. London, 28 vols. 8°.—New Series, 1830.

59. The Literary Gazette and Journal of Belles-Lettres, Arts, Sciences, &c. London, 1847, 4°.

60. The Athenæum Journal of Literature, Science, and the Fine Arts. London, 1828–1847, 4°.

61. The Microscopic Journal and Structural Record; by *D. Cooper* and *G. Busk.* 2 vols. 8°. London, 1841–1842.

62. Malacological and Conchological Magazine; by *G. B. Sowerby.* 8°. London, 1838.

63. The Geologist, a record of investigations in Geology, Mineralogy, &c.; by *Ch. Moxon.* London, 8°. 1842.

64. The Zoologist; by *E. Newman.* 8°. London, 1843–1846.

65. London Physiological Journal; by *S. J. Goodfellow* and *E. J. Queckett.* 8°. London, 1843–1844.

66. London Geological Journal and Record of Discoveries in British and Foreign Palæontology. 8°. London, 1846–47.

67. Scientific Memoirs; by *R. Taylor.* 4 vols. 8°. London, 1838–1846.

68. The New Cyclopædia, or Universal Dictionary of Arts and Sciences; by *A. Rees.* 39 vols. 4°. London, 1802–1819.

69. Encyclopædia Metropolitana. 26 vols. 4°. London, 1818–1845.

70. London Encyclopædia. 22 vols. 8°. London, 1826.

71. Penny Cyclopædia of the *Society for the Diffusion of Useful Knowledge.* London, 4°. 27 vols. 1833–1843.—Supplement, London, 1844–1846.

72. Cabinet Cyclopædia; by *D. Lardner.* 8°. London, 1829–1846.

73. British Cyclopædia of Natural History; by *C. F. Partington.*
3 vols. 8°. London, 1835–37.

74. Cyclopædia of Anatomy and Physiology; by *R. B. Todd.* 8°.
1835.

75. Transactions of the *Philosophical and Literary Society of Leeds.*
vol. i. 8°. London, 1837.

76. Transactions of the *Manchester Geological Society.* London,
8°. 1841.

77. Transactions of the *Geological Society of Cornwall.* 4 vols. 8°.
London, 1818–1832.

78. Transactions of the Literary Society of Bombay. 4°. London,
1819.

79. Transactions of the Literary Society of Madras. 4°. London,
1827.

80. Transactions of the Royal Cornwall Polytechnic Society, 1842,
No. 1.

81. The Quarterly Journal of Meteorology and Physical Science;
by *J. W. G. Gutch,* 1843, No. 1–6.

82. Memoirs of the Geological Survey of Great Britain and of the
Museum of Economic Geology in London. 8°. London, 1846.

MANCHESTER.

(Vide etiam LEIPZIG, 39, 40; LONDON, 76.)

1. Memoirs of the *Literary and Philosophical Society of Manchester.* 8°. Manchester, Series 1, 5 vols. 1785–1802.—Series 2, 1805–
1843.

2. Annals of Philosophical Discovery, and Monthly Reporter of the
Progress of Science and Art. 1 vol. 8°. Manchester, 1843.

NEWCASTLE-UPON-TYNE.

1. Reports of the *Natural History Society of the Counties of Northumberland, Durham and Newcastle-upon-Tyne.* Newcastle,
1831–1841, 8°.

2. Transactions of the *Natural History Society of Northumberland,
Durham and Newcastle-upon-Tyne.* 2 vols. 4°. Newcastle, 1831–
1838.

OXFORD (Oxonia).

1. Proceedings of the *Ashmolean Society of Oxford.* 8°. 1832–
1846.

PLYMOUTH.

1. Transactions of the *Plymouth Institution.* 8°. Plymouth, vol. 1.
1830.

York (Eboracum).

1. *Yorkshire Philosophical Society.*—Annual Report, 1827–1836, 5 vols. 8°. York, 1828–1837.

2. Proceedings of the *Yorkshire Philosophical Society.* 8°. 1847.

AFRICA.

Cape Town.

1. South African Quarterly Journal. 8°. Cape-town, 1830.

Mauritius.

1. Rapports Annuels sur les travaux de la *Société d'Histoire Naturelle de l'isle de Maurice.* 8°. Port Louis, 1830–1838.

ASIA.

Batavia.

(Vide etiam Leipzig, 36.)

1. Verhandelingen van het *Bataviaasche Genootschap van Konsten en Wetenschappen.* Batavia, 1779–1839, 17 vols. 8°.

Bombay.

(Vide etiam London, 78.)

1. Journal of the *Bombay Branch of the Royal Asiatic Society.* 8°. Bombay, 1841–1847.

2. Journal of the *Bombay Geographical Society.* 8°. Bombay, 1836.

Calcutta.

1. Asiatic Researches, or Transactions of the *Society instituted in Bengal for inquiring into the History and Antiquities, Arts, Sciences and Literature of Asia.* Calcutta, 1788–1839, 19 vols. 4°.

2. Transactions of the Physical Class of the *Asiatic Society of Bengal.* 4°. Calcutta, 1829.

3. Indostan Gleanings in Science. 3 vols. 8°. Calcutta, 1829–1831.

4. Journal of the *Asiatic Society of Bengal.* 8°. Calcutta, 1832–1846.

5. Calcutta Journal of Natural History, conducted by *John M'Clelland.* 8°. Calcutta, 1840–1846.

6. The Asiatic Miscellany, consisting of original Productions, Translations, Fugitive Pieces, &c. Calcutta, 1785–1786, 2 vols. 4°.

7. Quarterly Journal of the *Calcutta Medical and Physical Society.* 2 vols. 8°. Calcutta, 1837.

8. Bengal Sporting Magazine. Calcutta, 8°.

9. Indian Review and Journal of Foreign Science; by *F. Corbyn*. 2 vols. 8°. Calcutta, 1837–38.

10. Transactions of the *Agricultural and Horticultural Society of India*. Calcutta.

MADRAS.

(Vide etiam LONDON, 79.)

1. Madras Journal of Literature and Science. 8°. Madras, 1834–46.

AUSTRALIA.

HOBART-TOWN.

1. The Tasmanian Journal of Natural Science, Agriculture and Statistics. 8°. Hobart-town, 1842.

SYDNEY.

1. Australian Quarterly Journal of Theology, Literature and Science. By *C. P. N. Wilton*. 8°. Sydney, 1828.

AMERICA.

ALBANY.

1. Transactions of the *Albany Institute*. 2 vols. 8°. Albany, 1828–1839.

2. Albany Gazette, 1842, fol.

3. American Quarterly Journal of Agriculture and Science. By *E. Emmons* and *A. J. Prime*. 8°. Albany, 1845–1847.

BALTIMORE.

1. Transactions of the *Maryland Academy of Science and Literature*. Baltimore, 1 vol. 8°.

BOSTON, U.S.

1. Transactions of the *American Academy of Arts and Sciences*. 4 vols. 4°. Boston, 1780–1818.

2. Memoirs of the *American Academy of Arts and Sciences*. 4 vols. 4°. Boston, 1785–93.

3. The Boston Journal of Philosophy and Arts, conducted by *J. W. Webster, J. Ware* and *D. Treadwell*. Boston, 1824–27, 3 vols. 8°.

4. Boston Journal of Natural History, containing papers and communications read to the *Boston Society of Natural History*. 3 vols. 8°. Boston, 1837–1841.

5. Proceedings of the *Boston Society of Natural History*. 2 vols. 8°. Boston, 1841–1844.

6. Reports of the Meetings of the *Association of American Geologists and Naturalists*. 8°. Boston, 1843.

7. Medical Magazine. Edited by *J. B. Flint, E. Bartlett* and *A. A. Gould*. Boston, 8°.

8. Massachusetts Agricultural Repository and Journal. Boston, 1817, 8°. 4 vols.

CAMBRIDGE, U.S.

1. Memoirs of the *American Academy of Arts and Sciences of Cambridge;* new series, vol. i. 4°. Cambridge, 1833.

HANOVER.

1. Constitution and First Annual Report of the *Northern Academy of Arts and Sciences*. Hanover, New Hampshire, 1842.

HARTFORD.

1. Transactions of the *Natural History Society of Hartford*. Connect. 1836, 8°.

HAVANNAH.

1. Anales de Ciencias, Agricultura, Commercio y Arte. Havana. Por Don *R. de la Sagra*, 1827–1828, 8°.

JAMAICA.

1. Proceedings of the Society for the Encouragement of Horticulture and Agriculture in Jamaica. Jamaica, 1825, 4°.

2. Companion to the Jamaica Almanack, 1840.

LEXINGTON.

1. The Transylvania Journal of Medicine and the associate Sciences. By *J. E. Cooke* and *C. U. Short*. 4 vols. 8°. Lexington, 1828–1831.

LIMA.

1. Memorial de Ciencias Naturales. Da *M. Riviero* y *N. Pierola*. 8°. Lima, 1827–1828.

MONTREAL.

1. Transactions of the *Natural History Society of Montreal*, 1831.

NEW-HARMONY.

1. The New-Harmony Disseminator.

NEW HAVEN.

1. Memoirs of the *Connecticut Academy of Arts and Sciences.* 8°. New Haven, 1810.

2. Proceedings of the Sixth Annual Meeting of the *Association of American Geologists and Naturalists,* held in New Haven, April, 1845.

NEW YORK.

1. Introductory Discourse to the Literary and Philosophical Society of New York, May 4, 1814. By *Dewitt Clinton.*

2. Transactions of the *Literary and Philosophical Society of New York.* 1 vol. 4°. New York, 1815.

3. Transactions of the *Physico-Medical Society of New York.* 8°. New York, 1817.

4. Proceedings of the *New York Lyceum of Natural History.*

5. Annals of the *Lyceum of Natural History of New York,* 1824–1837, 4 vols. 8°.—Series 2, 1846.

6. Address to the *New York Lyceum,* by *J. Dekay.* 8°. 1826.

7. Transactions of the *American Geologists' and Naturalists' Association.* 8°. New York, 1840–42.

8. American Medical and Philosophical Register, by *D. Hosack* and *J. W. Francis.* 8°. New York, 1811.

9. The American Journal of Science, &c. conducted by *Benj. Silliman.* 8°. New York, 1818; New Haven, 1820–45, 50 vols.—Series 2, 1846.

10. Natural History of the State of New York. 4°. New York, 1840.

11. Report of the Geological Survey of the State of New York, 1841.

12. Documents relating to the Geological Survey of New York, 1841, 8°.

PHILADELPHIA.

1. Transactions of the *American Philosophical Society of Philadelphia.* Philad. 1771–1818, 6 vols. 4°.—New series, Philad. 1818–1844, 4°.

2. Proceedings of the *American Philosophical Society.* 8°. Philad. 1838.

3. Journal of the *Academy of Natural Sciences of Philadelphia.* 27 vols. 8°. Philad. 1818–1843.

4. Proceedings of the *Academy of Natural Sciences of Philadelphia.* 8°. Philad. 1841–1846.

5. Notice of the *Academy of Natural Sciences of Philadelphia.* Philad. 1837, 8°.

6. Journal of the *Franklin Institute.* 12 vols. Philadelphia.

7. Contributions of the *Maclurian Lyceum* to the Arts and Sciences. 3 Nos. Philad. 1829, 8°.

8. Transactions of the *Geological Society of Pennsylvania.* 8°. Philadelphia, 1834–35.

9. Literary Record and Journal of the *Linnæan Association of Pennsylvania College.* 8° Philadelphia, 1844.

10. Emporium of Arts and Sciences, by *Dr. Coxe.* 2 vols. 8°. Philadelphia, 1812.

11. Monthly American Journal of Geology and Natural Science, by *G. W. Featherstonhaugh.* Philad. 1831–32, 8°. '

12. Atlantic Journal and Friend of Knowledge, by *C. S. Rafinesque.* Philad. 1832, 8°.

13. The Advocate of Science and Annals of Natural History. By *W. P. Gibbons.* 8°. Philad. 1834–1835.

QUEBEC.

1. Transactions of the *Literary and Philosophical Society of Quebec.* 3 vols. 8°. Quebec, 1829–1833.

SALEM.

1. Journal of the *Essex County Natural History Society.* 8°. Salem, 1836.

WASHINGTON.

1. Bulletin of the Proceedings of the *National Institution for the Promotion of Science.* 8°. Washington, 1840–1845.

FINIS PARTIS I.

BIBLIOGRAPHIA

ZOOLOGIÆ ET GEOLOGIÆ.

~~~~~~~~~~~~~~~~~~~~~~~~~~~~~

PARS SECUNDA,

AUCTORUM NOMINA ALPHABETICE ORDINATA, ET SINGULORUM OPERA CONTINENS.

AALBORG ( ).

1. De Culturâ Apum.   Copenh. 1639.—*Eis.* p. 224.
2. Tractatus de Apibus.   Hafniæ, 1642.—*Eis.* 224.

AASKOW (Urb Bruun).

1. Tetrapodologia Danica, sistens recensionem Quadrupedum Daniæ et Norwegiæ, ad methodum Brissonianam.   Hafn. 1762, 1763, 8°.—*Böhm.* Bibl. II. 1, p. 270.

ABBATI (Bald. Angel.).

1. De admirabili Viperæ Naturâ, et de Mirificis ejus Facultatibus. Urb. 1587, 1589, 1591 et 1594, 4°. fig.—Noriberg. 1603, 4°. fig. —Hag. Com. 1660, 12°.—*Böhm.* Bibl. II. 2, p. 34.

ABBOT (John).

1. Observations on *Papilio Paniscus* (*Hesperia Paniscus*).—Tr. Lin. Soc. Lond. V. p. 276.
2. Natural History of the rarer Lepidopterous Insects of Georgia, 2 vols. fol.   London, 1797.

ABBOT (S. L.).

1. Report on some Birds received from Rev. J. H. Linsley, of Strat-
ford, Con.—Proc. Bost. Soc. N. H. 1842, p. 56.

2. Some remarks upon several Birds which had been recently pro-
cured and mounted for the Society.—Proc. Bost. Soc. N. H.
1842, p. 61.

3. Report on a Specimen of Syrnium cinereum—the great cinereous
Owl.—Proc. Bost. Soc. N. H. 1842, p. 57.

4. A Report on Specimens of Birds presented to the Society by the
Hon. Mr. Amos, of Bengal.—Proc. Bost. Soc. N. H. 1841, p. 16.

5. Report on the Surinam Birds presented by Dr. F. W. Cragin.—
Proc Bost. Soc. N. H. 1844, p. 171.

ABBOTT (K. E.).

1. Letter on various Zoological Subjects, accompanying a Collection of Birds, formed by him in the neighbourhood of Trebizond.
—Proc. Zool. Soc. Lond. II. p. 50, 133. III. p. 89. V. p. 70.

2. On Poisonous Honey of Trebizond, described by Xenophon.
Proc. Zool. Soc. Lond. II. p. 50.—*Wiegm.* Arch. 1835, II. p. 50.

ABEL (Cl.).

1. Narrative of a Journey in the interior of China.   Lond. 1814,
4°. fig.—*Fisch.* Mamm. p. 333.

2. Note sur le Chien Sauvage de l'Himalaya.—Quart. Oriental.
Magaz. Fév. et Mars, 1826, p. 153.—*Féruss.* Bull. 1826, IX. p. 213.

3. Sur le Crocodile du Gange.—*Brewst.* Edinb. Journ. of Science;
Avril, 1828, p. 339.—*Féruss.* Bull. 1829, XVII. p. 126.

4. Account of an Orang Outang, of remarkable height, from the
island of Sumatra, etc.—Am. J. XV. p. 1, 161.—Asiat. Research.
XV. p. 489.—Phil. Mag. ser. 2, I. p. 213.—Edinb. New Phil.
Journ. II. p. 371, III. p. 81.—*Féruss.* Bull. 1827, X. p. 285.
Brewst. Journ. Sc. ser. 1, IV. p. 193.

ABEL (J. C. A. M.).

1. Die Conchylien in dem Naturalien-Cabinet zu Mörsburg, nach
Martini u. *Chemnitz* systematisch eingetheilt.   Bregenz, 1787,
8°.—Cat. Bibl. Turic.

ABELIN (J. Ph. sub nomine *J. L. Gottfried*).

1. Neue Welt und Amerikanische Historie. Frankf. 1650 und 1655,
fol. fig.

ABERCROMBY (Dav.).

1. De Variatione Pulsus Observationes. Accessit ejusdem autoris nova medicinæ tum speculativæ, tum practicæ clavis. Lond. 1685, in 8º.—Act. Erudit. V. p. 215.

ABERNETHY (J.).

1. Anatomisch-physiologische Vorlesungen, gehalten am Bartholomäus-Spital in London.—The Lancet, 1827.—Isis, 1827, XI. p. 970.

2. An Enquiry into the probability and rationality of *Mr. Hunter's* Theory of Life. 8º. Lond. 1814.—Cat. R. Soc. Lond.

3. Physiological Lectures, exhibiting a general view of *Mr. Hunter's* Physiology, and of his Researches in comparative Anatomy. 8º. Lond. 1817–1822.—Cat. R. Soc. Lond.

4. On the Anatomy of a Whale.—Phil. Trans. LXXXVI. p. 27.

ABICH (H.).

1. Ueber die Geologische Natur des Armenischen Hochlandes. Dorp. 1843, 4º.

2. Vulkanische Forschungen in Italien.—*L. u. Br.* N. Jahrb. 1837, p. 439.

3. *Vues Illustratives de Phénomènes Géologiques observés sur le Vésuve et l'Etna pendant les années* 1833 et 1834—Par. 1836, fol. fig. oder: Erläuternde Abbildungen Geologischer Erscheinungen am Vesuv und Ætna, etc. Berl. 1837–1841, X. lith. Tafeln in fol.—*L. u. Br.* N. Jahrb. 1837, p. 320.

4. Ueber Erhebungs-Kratere u. das Band inneren Zusammenhanges, welches, in der Richtung bestimmter Linien, räumlich oft weit von einander getrennte vulkanische Erscheinungen u. Gebilde zu ausgedehnten Zügen unter einander vereinigt.—Ber. Vers. Naturf. in Prag. p. 140.—*L. u. Br.* N. Jahrb. 1839, p. 334.

5. Geologische Beobachtungen über die Vulkanischen Erscheinungen u. Bildungen in Unter- u. Mittel-Italien. I Bd. 1ᵉ Lief. Braunschw. 1841, 4º. atl. fol.

6. On the Paleozoic Rocks of Armenia. Bull. Soc. Geol. France, ser. 2, III. p. 138.—Journ. Geol. Soc. II. pt. 2, p. 93.

ABILGAARD (P. C.).

1. Beskrivelse af twende nye Insekter herhorende under den Linneische slaegt *Monoculus* og den Mullerske slaegt *Caligus*. Skrivt. af Naturhist. Selsk. III. pt. 2, p. 46.

2. Zoologia Danica, de Müller, 4ᵉ. cah. avec figures. Copenhague,
1806.—Dict. Sc. Nat. LX. p. *535.*

3. Kurze Anatomische Beschreibung des Saugers (Myxine Gluti-
nosa, Lin.). Schrift. der Gesellsch. Naturforsch. Freunde, Bnd.
10, 1792, p. 193, cum fig.

4. Beschreibung einer Holothuria Priapus, Lin. zweier Arten des
Terebella, Lin. und einer Sabella, Lin. Schrift. der Gesells. Naturforsch. Freunde, Bnd. 9, 1789, p. 133, cum fig.

ABILGAARD (Sören).

1. Beskrivelse over Stevens Klint.   Copenh. 1759, 4°. fig.—Beschreibung von Stevens Klint und dessen Merkwürdigkeiten, mit
mineralogischen und chemischen Betrachtungen erläutert, und
mit Kupferstichen versehen.   Aus dem Dänischen übersetzt.
Copenh. und Leipz. 1764, 8°. fig.—Bresl. 1769, 8°.—*Klein* Disp.
Echinod. 1778, p. VII.—*Böhm.* Bibl. I. 1, p. 609.—*Mod.* Bibl.
Helm.

2. Ueber Norwegische Titanerze u. eine neue Steinart aus Grönland, welche aus Fluss-spathsäure u. Alaunerde besteht.   Aus d.
Dän. Kopenh. 180!, 8°.

3. Physisk-mineralogisk Beskrivelse over Moens Klint.   Copenh.
1781, 8°. fig.—Physikal-mineralogische Beschreibung des Vor-
gebirgs auf der Insel Moen; aus dem Dänischen übers. v. *C. H.
Reichel.*   Leipz. 1783, 8°. fig.—*Böhm.* Bibl. I. 1, p. 610.—*Mod.*
B. Helm.

4. Preisschrift vom Torfe.   Aus dem Dän. vom Verfasser selbst
übersetzt u. vermehrt.   Kopenh. 1765, 8°.—*Böhm.* Bibl. IV. 1,
p. 200.

ABOVILLE (D'.).

1. Genauere Umstände der merkwurdigen Fortpflanzungswerke
der weiblichen Beutelratze (Didelphis Marsupialis).   Lichtenberg's und Voigt's Mag.   Bnd. 5, pt. 2, 1788, p. 29.

ABULFEDA.

1. Descriptio Ægypti, Arabicè et Latinè; edidit cum notis *J. D.
Michaelis.*   Gött. 1776, 8°.—Gött. gel. Anz. 1777, p. 209.—
Böhm. Bibl. I. 1, p. 709.

ACERBI (G.).

1. Voyage au Cap-Nord par la Suède.   Paris, 1804.—Travels in
Sweden, 2 vols. 4°. London, 1802.

2. Sur une espèce nouvelle de Procellaria observée dans l'Helles-pont, la mer de Marmara et le détroit de Constantinople.—Bibl. Ital. CXL. 1827, p. 294.—*Féruss.* Bull. 1829, XVI. p. 463.

ACHARD (J.).

1. Notice sur la Sangsue officinale, sa reproduction aux Antilles, etc. Saint-Pierre, 1823, 8°.—*Féruss.* Bull. 1824, II. p. 105.

2. Remarks on Swallows on the Rhine.—Phil. Trans. LIII. p. 101.

ACHARI (C. A.).

1. Beobachtung einer der Zauberkraft hoererer Thiere achnelnden Erscheinung bei Infusorien. *Nov. Act. Phys. Med. Acad. Cæs. Leop. Carol. Nat. Curios.* tom. 10, 1820, p. 127.—cum fig. sup. p. 711.

ACHARIUS ( ).

1. Beschreibung und Abbildung v. *Cynips inanita.*—Gotheb. We-tenek. 1778, I. p. 9.—*Eis.* p. 222.

2. Animadversiones quædam physico-medicæ de Tæniâ. Lund. 1782.—*Mod.* Bibl. Helm.

3. De Vermibus in Vesicâ natatoriâ Salmonis Eperlani repertis.— Vet. Acad. Handl. 1771, p. 52.

4. *Bulbocerus*, nouv. genre de Coléoptère.—Vet. Acad. N. Handl. 1781, p. 244 (Allem.), p. 243.

5. De Acanthro Sipunculoide (Echinorhynchi Spec.).—Vet. Acad. N. Handl. 1780, p. 49.

6. Anmärkningar vid Hr. *Martins* Rön, rörande en besynnerlig mask hos Norsen.—N. Act. Holm. I. p. 49, fig.—*Mod.* Bibl. Helm.

ACHILLES (Al.).

1. Traité sur les causes des tremblemens de terre et de l'agitation de la mer. (Allem.)—Biogr. un. I. p. 143.

2. Von den Ursachen der Erdbebung, wie auch der Erzte u. Mine-ralien in der Erde Beschaffenheit, u. daher entspringenden war-men Brunnen, etc. Frankf. 1666, 4°.—*Böhm.* Bibl. V. p. 49.

ACHILLINI (Al.).

1. Annotationes Anatomicæ. Bonon. 1520, 4°.—Venet. 1521, 8°. —Biogr. un. I. p. 145.

2. De Humani Corporis Anatomiâ. Venet. 1521, 4°.—Biogr. un. I. p. 145.

3. In Mundini Anatomiam Annotationes. Ven. 1522, fol.—Biogr. un. I. p. 145.

ACHRELIUS (Dan.).

1. Cetographia. Dissertatio de Cetis.   Aboœ, 1683, 8°.—*Böhm.*
Bibl. II. 1, p. 485.

2. Contemplationum Mundi Libri III.   Aboœ, 1682, 4°.

ACKERMANN (J. C. H.).

1. Ueber den Natzen der Eingeweidewürmer.   Leipz. 1790, 8°.

ACKERMANN (J. Fr.).

1. De Nervei Systematis Primordiis Commentatio.   Accedit de
Naturæ Humanæ Dignitate oratio Academica.   Mannh. 1813, 8°.
—Bibl. med.-ch. p. 4.

2. Versuch einer phys. Darstellung der Lebenskräfte organischer
Körper (2ᵉ éd.).   Jena, 1800, 8°. 2 vols.—Bibl. med.-ch. p. 4.

3. Infantis Androgyni Historia et Iconographia.   Jenæ, 1805, fol.
fig.—Bibl. med.-ch. p. 4.

4. Ueber die Körperl. Verschiedenheit des Mannes vom Weibe.
Aus dem Lat. v. *J. Wenzel.* Francf. 1788, 8°.—Bibl. med.-ch.
p. 4.

ACKERMANN (P.).

1. Note sur le *Coua, Famac-acora* des Malgaches, Hache-escargot
(traduct. litt.) ou casseur d'escargots.—Rev. Zool. 1841, p. 209.

2. Considérations Anatomico-physiologiques et Thérapeutiques sur
le Coipo du Chili.   Paris, 1844, 4°. fig.

ACOSTA (Jos. d').

1. Historia Natural y Moral de las Indias.   Séville, 1589, 4°.—
1591, 8°.—Madr. 1608 et 1610.—Ed. 6. 4°. Madrid, 1792. Trad.
Fr. par *Rob. Regnault,* 1598 et 1606, 8°.—(Gall. *R. Cauxois*):
Histoire Naturelle et Morale des Indes Orient. et Occident.   Paris,
1600, 8°.—Cat. Bibl. Turic.—Biogr. un. I. p. 159.—Trad. Angl.
par *E. G.* Lond. 1604, 4°. (*The Natural and Moral History of
the East and West Indies.*)

2. De Naturà Novi Orbis, libri duo.   Salam, 1589 et 1595, 8°.—
Cologne, 1596, 8°.—Biogr. un. I. p. 160.

3. De Procurandâ Indorum Salute.   Sevilla, 1592.

ACREL (J. G.).

1. Historia Vermium, Larvarum, necnon Insectorum variorum generum, per Biennium intra Corpus Humanum hospitantium.—
N. Act. Upsal. VI. p. 98.

ADAM ( ).

1. Ueber die Vertilgung der Maykäfer und ihrer Larven.—*Voigt's* Mag. IV. p. 71, 75.

ADAM (J.).

1. Geological Notices on the District between the Jumna and Nerbuddah.—Mem. Wern. Soc. IV. p. 24. *Féruss.* Bull. XVI. p. 379.

2. On Barren Island in the Bay of Bengal. J. A. S. B. I. 128.

3. On the Geology of the Banks of the Ganges, from Calcutta to Cawnpore.—Trans. Geol. Soc. ser. 1, vol. v. p. 346.—Isis, 1823, L. A. p. 180.

4. On the Ciconia Argala, or Adjutant Bird.—Trans. Med. et Phys. Soc. Calcutta, I. p. 240.—Edinb. New Phil. Journ. I. p. 327. Isis, 1832, VIII. p. 685.—*Féruss.* Bull. XV. p. 392.

ADAM (Walt.).

1. On the Osteological System of the Camel (*Camelus Bactrianus*).—Tr. Lin. Soc. Lond. XVI. p. 525.

2. On some Symmetrical Relations of the Bones of the Megatherium.—Rep Brit. Assoc. 1833, p. 437.

3. On the Osteology of the Hippopotamus.—Edinb. New Phil. Journ. XV. p. 361.

ADAM (Jos. Mar.).

1. De Renibus, Uretheribus et Vesicâ urinariâ.—De Motu duræ Meningis et Cerebri, etc. Taurini, 1761, 8°.

ADAM (J. St.).

1. Dissertatio de Osse Cordis Cervi. Giess. 1684, 4°. fig.—*Böhm.* Bibl. II. 1, p. 363.

ADAM (Paul).

1. Hydrographia Comitatûs Trenesiniensis. Viennæ, 1766, 8°.— *Böhm.* Bibl. V. p. 394.

ADAMS (B.).

1. On a great and extraordinary Cave in Indiana.—Edinb. Phil. Journ. VI p. 29.

ADAMS (C. B.).

1. First Annual Report on the Geology of Vermont. 8°. New York, 1846.

2. Mollusca. Fresh Water and Land Shells of Vermont, 1842, 1843, 8°. pam.

3. Descriptions of new Shells.—Bost. Soc. Febr. 1840.—*Sill.* Am. J. XXXIX. p. 373.

4. Catalogue of the Mollusca of Middleburg, Vt., and vicinity, with Observations —*Sill.* Am. J. XL. pp. 266, 408.

5. Note to *Mr. Lea's* paper on native Shells.—*Sill.* Am. J. XLII. p. 392.

6. Description of a new Species of *Thracia* (*Thr. inæqualis*).— *Sill.* Am. Journ. Jul. 1842, p. 145. An. & Mag. N. H. X. p. 238.

7. Remarks on some species of Shells found upon the S.E. shore of Massachusetts.—Bost. Soc. N. Hist. Aug. 1838.—*Sill.* Am. Journ. XXXVI. p 387.

8. Shells of Fresh Pond.—*Sill.* Am. Journ. XXXVI. p. 393.

9. *Delphinula minor.*—Bost. Soc. June 1839.—*Sill.* Am. Journ. XXXVIII. p. 193.

10. Shells obtained by dredging, and other new species.—Bost. Soc. Dec. 1839.—*Sill.* Am. Journ. XXXVIII. p. 396.

ADAMS (George).

1. Essays on the Microscope. Lond. 1787, 4°.—2^d ed. with additions by *F. Kanmacher.* Lond. 1798, 4°.—Trans. Linn. Soc. Lond. V. p. 277.

2. Lectures on Natural and Experimental Philosophy, with addit. by *W. Jones.* Lond. 1799, 8°. 5 vols.—Cat. Roy. Soc. Lond.

3. Micrographia illustrata, etc. Lond. 1743–46, 4°.

ADAMS (J.).

1. Descriptions of some minute British Shells.—Trans. Linn. Soc. Lond. V. p. 1, fig.

2. Description of minute Shells found on the Coast of Pembrokeshire.—Trans. Linn. Soc. III. p. 64, 252 ; V. p. 1 ;—VII. p. 234.

3. The specific Characters of some minute Shells discovered on the Coast of Pembrokeshire, with an Account of a new marine Animal (*Derris sanguinea*).—Trans. Linn. Soc. Lond. III. p. 64, fig.

4. Description of *Actinia crassicornis* and some British Shells.— Trans. Linn. Soc. Lond. III. p. 252.

5. Descriptions of some marine Animals found on the Coast of Wales.—Trans. Linn. Soc. Lond. V. p. 7, fig.

ADAMS (Jos.).

1. A Philosophical Treatise on the Hereditary Peculiarities of the Human Race, etc. Lond. 1815, 8°.—Cat. Roy. Soc. Lond.

ADAMS (Mich.).

1. Description de trois Coléoptères inconnus de la Sibérie orientale. —Mém. Natur. Mosc. III. p. 163.

2. Relation d'un Voyage à la Mer Glaciale, et Découverte des restes d'un Mammouth.—Journ. du Nord, 1807, XXXIII.—Phil. Mag. XXIX. p. 141.

3. Descriptio Insectorum novorum Imperii Rossici, imprimis Caucasi et Sibiriæ.—Mém. Natur. Mosc. V. p. 278.

ADAMSON (J.).

1. On Marine Deposits on the Margin of Loch Lomond.—Mem. Wern. Soc. IV. p. 334.—Féruss. Bull. 1824, I. p. 112.

2. Conchological Tables. 12°. Newcastle, 1823.

ADAMUCCI ( ).

1. Système Mécanique des Fonctions Nerveuses, 2 vols. 8°.

ADANSON (Mich.).

1. Histoire Naturelle du Sénégal avec la Relation abrégée d'un Voyage fait en ce pays en 1719-53.—Paris, 1757, 4°. fig.—Trans. Linn. Soc. Lond. VII. p. 229.—Deutsch. übers. v. Martini, Brand. 1773, 8°.; u. v. Schreber, Halle, 1773, 8°.—Mod. Bibl. Helm.— (Trad. Angl.), 8°. London, 1759.

2. Histoire Naturelle des Coquillages du Sénégal, 1775, 1 vol. 4°. —Cuv. R. an. III. p. 330.

3. Description d'une Nouvelle Espèce de Ver qui ronge les bois et les vaisseaux, observée au Sénégal.—Mém. Acad. des Sc. 1759, p. 249, fig:—Trans. Linn. Soc. Lond. VII. p. 222.—Mod. Bibl. Helm.

4. Cours d'Histoire Naturelle fait en 1772, publié sous les auspices de M. Adanson, son neveu, avec une introduction et des notes par M. L. P. Payer. 4 vols. 18°. Paris, 1844-45.

ADDISON (Jos.).

1. Remarks on several parts of Italy. Lond. 1705, 1718, 1736, 8°. —(Franc.). Paris, 1722, 12°.—(Deutsch.) Altenb. 1752, 8°.— Böhm. Bibl. I. 1, p. 553.

ADDISON (W.).

1. On the Sacculi of the Polygastrica.—An. & Mag. N. Hist. XII. p. 101.

2. On the ultimate Distribution of the Air-passages, and the formation of the Air-cells of the Lungs.—Phil. Trans. CXXXII. p. 157.

ADELON (N. P.).

1. Physiologie de l'Homme. Paris, 1829, 1831, 4 vols. 8°.—Rev. Bibl. Sc. Nat. p. 74. dans Ann. Sc. Nat. XVII. 1829.

ADELUNG (Johann Christopher).

1. Geschichte der Schiffarten und Versuche zur Entdeckung des Nordostlichen Weges nach Japan und China. Halle, 1768.

ADHÉMAR (J.).

1. Révolutions de la Mer. 8°. (Germ.) *Die Revolutionen des Meeres.* Leipz. 1843, 8°.

ADIE (A. J.).

1. On the Habits of a Mangouste. Mag. Nat. Hist. ser. 1, I. p. 20.

ADLER (C. F.).

1. Dissertatio de Noctilucâ Marinâ, præside Linnæo. Upsaliæ, 1752, 4°. cum fig.—Amœnit. Acad. vol. 3, p. 202, cum fig.

ADLERHEIM (   ).

1. Sur un moyen de détruire les Chenilles qui passent l'hiver sur les Arbres.—Vet. Acad. Handl. 1770, XXXII. p. 24.

ADMIRAL (Jac. l').

1. Insectes gravés en manière noire, avec l'explication des planches en Hollandais. fol. fig. 1740.—*Eis.* p. 136.

2. Naauwkeurige Waarnemingen van veele Gestalt verwisselnde gekorvene Dierkens. fol. Amst. 1746, 25 pl.—*Eis.* p. 136.

3. Naauwkerige Waarnemingen omtrent de Veranderingen van veele Insecten of gekorvene Diertjes, die in omtrent Vyftig Jaaren, zo in Vrankryk als in England en Holland, by een verzameld naar't Leven konstig aufgetekend en in't Koper gebragt zyn Door wylen.—Amsterdam, 1774; gr. fol. fig. Auch deutsch. übers.—*Eis.* p. 163.

ADOLPHI (Chr. Mich.).

1. Dissertatio de Aëre, Solo et Aquis Lipsiensibus. Lips. 1747.—*Böhm.* Bibl. I. 1, p. 603.

2. De Filis sic dictis meteoricis, seu filamentis Mariæ. Ephem. Ac. Nat. Cur. Cent. 8, p. 83.

ADRIEN ( ).

1. Gemmarum et Lapidum Historia   Lugd 1636, 8°.

AELIANUS (Claudius).

1. De Naturâ Animalium, Libri XVII., *Petro Gillio* et *Conr. Gesnero* interpretibus, græcè et latinè.   Tiguri, 1555, fol. (vel 1556, sec. *Gronovium*).—Lugd. 1562, 8°.; 1665, 8°.—Lugd. Bat. 1611, 12°.—Genevæ, 1616, 12°.—Colon. Allobr. 1616, 12°. (acc. Index locupletissimus).—Lond. 1744, 4°. 2 vols. (cur. *Gronovio*).—Basil. 1750, 4°.—Heilbr. 1765, 4°.—Tübing. 1768, 4°.—Lips. 1784, 8°. (cur. *J. G. Schneider*).—Jenæ, 1832, 2 vols. 8°. (illustr. *F. Jacob*).—*Böhm.* Bibl. I. 1, p. 9.—*Mod.* Bibl. helm.

AEMYLIANUS (J.).

1. Historia naturalis de Ruminantibus et Ruminatione.   Venet. 1584, 4°.—*Böhm.* Bibl. II. 1, p. 116.

AENETIUS (Theoph.).

1. De Mineralibus s. Corporibus subterraneis.   Jenæ, 1620, 4°.—*Böhm.* Bibl. IV. 1, p. 27.

AEPINUS (F. Ulr. Th.).

1. Cogitationes de distributione Caloris per Tellurem. 4°. Petrop. 1761.—Cat. R. Soc. Lond.

AFZELIUS (A.).

1. Observations on the genus *Pausus*, and Description of a new Species.—Tr. Lin. Soc. Lond. IV. p. 243, fig.   *Bemerkungen über das Geschlecht* Pausus *und eine neue Art* (sphærocerus).—*Wiedem.* Arch. I. p. 294.

2. *Afzelius (A.)* et *Brannius (F. W.)*, Achetæ Guineenses.   Upsal, 1804, 4°. fig.—*Eis.* p. 239.

AFZELIUS (J.).

1. Dissert. de Pissalitho Siberico.   Ups. 1800, 4°.

2. Dissert. de Baroselenite in Suecia reperto.   Ups. 1788, 4°.

AGAPITO (Girol.).

1. Delle Grotte di Trieste.   Vienna, 1823, 12°.

AGARD (Nic.).

1. Dissertatio super *Justini* sententiâ de Ignibus subterrancis. Soræ, 1653, 4°.—*Böhm.* Bibl. IV. 2, p. 371.

AGARDH (C. A.).

1. Ueber die Anatomie und den Kreislauf der Charen. fig.—Nov. Act. Nat. Cur. XIII. p. 113.

2. Beobachtung einer der Zauberkraft höherer Thiere ähnelnden Erscheinung bei Infusorien.—Leop. Ac. d. Naturf. 1820, II. p. 127. —Isis, 1821, X. p. 963.—*Féruss.* Bull. 1823, III. p. 627. (*Phé-nomène observé dans des Animaux infusoires, et analogue à la Fascination exercée par des Animaux d'un ordre plus élevé.*)

3. Nachtrag zur Abhandlung über die Zauberkraft der Infusorien. —Nov. Act. Nat. Cur. X. p. 711, App. XI. p. 721.

4. Observations relatives à la métamorphose et à l'animalité des Algues.—Physiogr. Selskap. Aarsber. Lund, 1825, p. 100.— *Féruss.* Bull. 1827, X. p. 427.

AGASSIZ (L.).

1. *Cynocephalus Wagleri.*—Isis, 1828, IX. p. 861. fig.—*Féruss.* Bull. XIX. p. 345. (*Déscription d'une nouv. espèce du genre Cy-nocéphale.*)

2. Selecta genera et species Piscium, quos in itinere per Brasiliam— jussu et auspiciis Max. Jos. J. B. R. A.—peracto collegit et pin-gendos curavit *Dr. G. de Spix*; digessit, descripsit et observa-tionibus anatomicis illustravit *Dr. L. Agassiz*; præfatus est et edidit itineris socius *Dr. de Martius.* Mon. et Lips. 1829, fol. fig. —Isis, 1829, VII. p. 715.

3. Beschreibung einer neuen Species aus dem Genus Cyprinus *Linn.*—Isis, 1828, X. p. 1046.

4. *Agassiz* et *Oken, Cyprinus uranoscopus,* nouvelle espèce trouvée par le premier à Munich, et présentée à la réunion des savans d'Allemagne à Berlin par ce dernier.—Isis, 1828, X. p. 1046; et 1829, III. et IV. p. 414.—*Féruss.* Bull. 1829, XIX. p. 117.

5. Recherches sur les Poissons de la Suisse, et particulièrement sur les Cyprins du Lac de Neuchâtel.—*Cuv. et Val.* Hist. n. des Poiss. XII. p. viii.

6. Ueber die natürlichen Verwandtschaften und die generische Eintheilung der Cyprinoiden. Vortrag in der Londner Zool. Ge-sellsch.—Proc. Zool. Soc. p. 150.— *Wiegm.* Arch. 1836, II. p. 240.

7. Ueber die lachsartigen Fische.—*James.* N. Philos. Journ. XVII. p. 380.— *Wiegm.* Arch. 1835, II. p. 265.

8. Histoire naturelle des Poissons d'eau douce de l'Europe centrale, ou Déscription anatomique et historique des Poissons qui habi-tent les lacs et les fleuves de la chaîne des Alpes et les rivières qu'ils reçoivent dans leur cours. Petit in-fol. avec pl. Munich, 1830. (Prospectus.)—*Féruss.* Bull. 1830, XXIII. p. 271.—Neu-

châtel, 1839.—*Sill.* Am. J. XXXIX. p. 390 (*Nat. Hist. of the Freshwater Fish of Central Europe*).

9. On the Anatomy of the Genus Lepisosteus, with Descriptions of two new Species.—Pr. Zool. Soc. Lond. IV. p. 119.

10. Algemeine Bemerkungen über fossile Fische.—*Leonh. u. Br.* N· Jahrb. 1834, p. 379.

11. Recherches sur les Poissons fossiles. Neuch. 1833–44, 5 vols. 4°. fig.—Isis, 1834, IV. p. 405 ; 1835, II. p. 135.—Amer. Journ. XXVIII. p. 193 ; XXX. p. 34.—*L. u. Br.* 1834, p. 242, 484 ; 1835, p. 595, etc. ; 1844, p. 250.

12. Sur les Belemnites (morceau communiqué à l'Acad. par *Férussac*).—C. 2, 1835, I. p. 341.—Instit. N° 132.—*Wiegm.* Arch. 1835, II. p. 244.—*L. u. Br.* 1835, p. 168 (*Ueber Belemniten*).

13. Ueber das Alter der Glarner Schieferformation nach ihren Fischresten.—*Leonh. u. Br.* N. Jahrb. 1834, III. p. 301.

14. On the Principles of Classification in the Animal Kingdom.— Rep. Brit. Assoc. 1835, Sect. p. 67.

15. Neue Entdeckungen über fossile Fische.—*L. u. Br.* N. Jahrb. 1833, p. 676.—Edinb. New Phil. Journ. XXXVII. p. 331.

16. Synoptische Uebersicht der fossilen Ganoiden.—Rech. Poiss. foss. II. p. I.—*L. u. Br.* N. Jahrb. 1833, p. 470.

17. Kritische Revision der in der *Ittiolitologia Veronese* abgebildeten fossilen Fische.—*L. u. Br.* N. Jahrb. 1835, p. 290.

18. Rapport sur les Poissons fossiles découverts en Angleterre. Neuch. 1835, 8°.—Feuill. add. p. 39.—*L. u. Br.* N. Jahrb. 1835, p. 491.

19. Opinions relative to Boulders quoted.—*Sill.* Am. Journ. XXXVI. p. 393.

20. Additions to Mr. *Wood's* Catalogue of Crag Radiaria.—Ann. a. Mag. N. Hist. VI. p. 343.

21. Notice sur les Poissons fossiles et l'Ostéologie du genre Brochet. Neuch. 1842, 4°.

22. Notice sur les Caractères zoologiques et anatomiques des Sauroïdes vivans et fossiles. Neuch. 1842, 4°.

23. Notice sur la succession des Poissons fossiles dans la série des Formations géologiques. Neuch. 1843, 4°.

24. Essai sur la Classification des Poissons. Neuch. 1844, 4°.

25. On a new Classification of Fishes, and on the Geological Distribution of Fossil Fishes.—Edinb. New Phil. Journ. XVIII. p. 175.

26. Ueber die äussere Organisation der Echinodermen.—Isis, 1834,

III. p. 254.—*Wiegm.* Arch. 1835, I. p. 36.—*Müll.* Arch. 1835, p. 87.

27. Prodrome d'une Monographie des Radiaires ou Echinodermes. —Mém. Neuch. I. p. 168.—Ann. Sc. n. 1837, VII. p. 257.— *Wiegm.* Arch. 1837, VI. p. 273.—*Sill.* Am. J. XXXIV. 1, p. 212 (*On the Echinodermata*).—*L. u. Br.* N. Jahrb. 1837, p. 223.— Ann. of Nat. Hist. I. p. 30, 297, 440 (*Prodromus of a Monograph of the Radiata a. Echinodermata*).

28. Monographies d'Echinodermes vivans et fossiles, avec fig. I^e livr. (Salénies) Neuch. 1838.—IIe Mon. (Scutelles) 1841.—IIIe Mon. (Galérites) 1842.—IVe Mon. (Dysaster) 1842.—*L. u. Br.* N. Jahrb. 1839, p. 486 ; 1841, p. 612 ; 1842, p. 485.—*Sill.* Am. J. XXXVII. p. 369 ; XLII. p. 378 ; XLIII. p. 390.

29. Embryologie des Salmones, par *C. Vogt.* Neuch. 1842, 8°. Atl. fol.—Ann. a. Mag N. Hist. IX. p. 49.

30. Déscription des Echinodermes fossiles de la Suisse. 2 vols. 4°. fig. Neuch. 1839, 1840.—N. Mém. Soc. Helv. Sc. n. III. et IV. —*L. u. Br.* N. Janrb. 1840, p. 502 ; 1842, p. 393.

31. Mémoire sur les Moules de Mollusques vivans et fossiles.— Mém. Soc. Neuch. II. 1839, 4°.

32. Catalogus systematicus Ectyporum Echinodermatum fossilium Musei Neocomensis, secundum ordinem zoologicum dispositus ; adjectis synonymis recentioribus, nec non stratis et locis in quibus reperiuntur. Sequuntur *Characteres diagnostici* generum novorum vel minus cognitorum. Neocomi Helvet. 1840.

33. Etudes sur les Glaciers. 8°. atl. in fol. Neuchâtel, 1840.— *L. u. Br.* N. Jahrb. 1841, p. 707.—Rep. Brit. Assoc. 1840, Sect. p. 113.

34. Untersuchungen über die Gletscher. 8°. mit Abbild. in folio. Solothurn, 1841.

35. Notice sur les Fossiles du terrain cretacé du Jura Neuchâtelois, avec fig.—Mém. Soc. Neuch. I. 1836, p. 126, fig.—*L. u. Br.* N. Jahrb. 1837, p. 102.

36. Notice sur quelques points de l'organisation des Euryales, accompagnée de la déscription détaillée de l'espèce de la Méditerranée, avec fig.—Mém. Soc. Neuch. II. 1839.

37. Notice sur la *Mya alba,* espèce nouvelle de Porto-Rico, avec fig.—Mém. Soc. Neuch. II. 1839.

38 Nomenclator Zoologicus, continens nomina systematica generum animalium tam viventium quam fossilium, secundum ordinem alphabeticum disposita, adjectis auctoribus, libris in quibus reperiuntur, anno editionis, etymologiâ et familiis ad quas pertinent in variis classibus. Soloduri, 1842, 4°.—*L. u. Br.* N. Jahrb. 1842, p. 496.—*Sill.* Am. J. XI. p. 57.

39. Remarques sur la Structure des Ecailles des Poissons.—Ann. Sc. n. (2ᵉ S.) XIII. p. 58.—C. R. X. p. 191.

40. Observations sur la structure et le mode d'accroissement des Ecailles des Poissons, et réfutation des objections de *Mr. Mandl.* —Ann. Sc. n. (2ᵉ S.) XIV. p. 97, fig.—Comptes Rendus, Feb. 3, 1840, p. 191.—Edinb. New Phil. Journ. XXVIII. p. 287.

41 Anatomie des Echinodermes. Iᵉ Monogr. (Echinus), par *G. Valentin.* 4°. Neuch. 1841.

42. Theorie der erratischen Blöcke in den Alpen.—*L. u. Br.* N. Jahrb. 1838, p. 304.

43. Künstliche Steinkerne von Konchylien;—Fische.—*L. u. Br.* N. Jahrb. 1838, p. 49.

44. Gletscherstudien mit *Studer.*—Färbende Infusorien im rothen Schnee.—*L. u. Br.* N. Jahrb. 1840, p. 92.—Rep. Brit. Assoc. 1840, Sect. p. 143.

45. Gegen *Wissmann's* Ansicht vom Ursprung erratischer Blöcke. —*L. u. Br.* N. Jahrb. 1840, p. 575.

46. Alte Morainen bei Baden-Baden.—*L. u. Br.* N. Jahrb. 1841, p. 566.

47. Genus Trigonia.—Gletscher-Theorie.—*L. u. Br.* N. Jahrb. 1841, p. 356.

48. Etudes critiques sur les Mollusques fossiles. Iᵉ livr. Genre Trigonie. Sol. 1841, 4°. fig.—*L. u Br.* N. Jahrb. 1841, p. 848.— IIᵉ livr. Myes du Jura et de la Craie Suisses. Neuch. 1842, 4°. —*L. u. Br.* op. cit. 1842, p. 862; 1843, p. 747.

49. Reise nach dem Aar-Gletscher.—*L. u. Br.* N. Jahrb. 1842, p. 313.

50. De la Succession et du Développement des Etres organisés à la surface du Globe terrestre dans les différens Ages de la nature. Neuch. 1841, 8°.—Edinb. New Phil. Journ. XXXIII. p. 388.— (Germ. N. *Gräger*): *Ueber die Aufeinanderfolge u. Entwickelung der organisirten Wesen auf der Oberfläche der Erde in den verschiedenen Zeitaltern.* Halle, 1843, 8°.

51. On the polished and striated surfaces of the rocks which form the beds of Glaciers in the Alps.—Proc. Geol. Soc. III. 321.— Phil. Mag. ser. 3. XVIII. p. 565.—Ann. and Mag. N. H. VI. p. 392.

52. Observations sur les progrès récens de l'Histoire nat. des Echinodermes. *Observations on the Progress recently made in the Nat. Hist. of the Echinodermata.* (Monogr. d'Echinod. No. II.) —Ann. a. Mag. N. H. IX. p. 189, 296.

53. Ueber die Famille der Karpfen.—Mém. Soc. Neuch. I. p. 33.— *Wiegm.* Arch. 1838, I. p. 73.

54. Remarks on Prof. Pictet's Treatise on Palæontology.—Bibl. Univ. de Genève, No. 112, 1845.—Edinb. New Phil. Journ. XXXIX. p. 295.

55. Rapport sur les Poissons Fossiles de l'Argile de Londres.—Rep. Brit. Assoc. 1844, p. 279.—Edinb. New Phil. Journ. XXXIX. p. 321 ; XL. p. 121.

56. Report on Fossil Fishes of the Devonian System or Old Red Sandstone.—Rep. Brit. Assoc. 1842, p. 80.—Bibl. Univ. 1843.— *L. u. Br.* N. Jahrb. 1843, p. 198, 750.

57. On different species of Salmon in Rivers and Lakes of Europe.— Rep. Brit. Assoc. 1834, p. 617.

58. On Fossil Fishes of Scotland.—Rep. Brit. Assoc. 1834, p. 646.

59. Synoptical Table of British Fishes, arranged in the order of the Geological Formations.—Rep. Brit. Assoc. 1843, p. 194.

60. On a new Classification of Fishes, and on the Geological Distribution of Fossil Fishes.—Proc. Geol. Soc. II. 97.—Phil. Mag. ser. 3. V. p. 459.

61. Déscription de quelques espèces de Cyprins du Lac de Neuchâtel, qui sont encore inconnues aux Naturalistes. Neuch. 1834, 4º.

62. Lettre sur les Ossemens fossiles de Stonesfield qu'on avait cru pouvoir rapporter à des Didelphes.—C. r. VII. p. 537.

63. Untersuchungen über die fossilen Süsswasser-Fische der Tertiaren- und Lias-Formation. 1832, 8º.

64. Notice sur les Dents et les Rayons des Placoïdes. Neuch. 1844, 4º.

65. Tableau général des Poissons fossiles rangés par Terrains. Neuch. 1844, 4º.

66. De la forme des Placoïdes.

67. La Théorie des Glaciers et ses progrès les plus récens. Genève, 1842, 8º.—Edinb. New Phil. Journ. XXIV. p. 364; XXXIII. p. 217.

68. Discours d'ouverture des Séances de la Soc. Helv. des Sciences natur. à Neuchâtel, le 24 Juillet 1837, 8º.

69. Eine Periode der Geschichte unseres Erdballes.   8º.—Edinb. New Phil. Journ. XXXV. p. 1.

70. Sur l'existence d'anciens Glaciers dans les Iles Britanniques. 8º.

71. On the Growth and on the bilateral Symmetry of Echinodermata.—Phil. Mag. ser. 3, V. p. 369.

72. On Glaciers, and the evidence of their having once existed in Scotland, Ireland, and England.—Proc. Geol. Soc. III. 327.— Phil. Mag. ser. 3, XVIII. p. 569.—Ann. and Mag. N. H. VI. p. 396.

73. Neue Beobachtungen auf den Gletschern.—*L. u. Br.* N. Jahrb. 1843, p. 84.

74. Struktur der Gletscher, &c.—*L. u. Br.* N. Jahrb. 1843, p. 86. —Bullet. Soc. Géol. France.—Edinb. New Phil. Journ. XXVII. p. 383.

75. Beobachtungen auf dem Aar-Gletscher im Sommer 1842.— Instit. X. p. 278, 305, 359.—*L. u. Br.* N. Jahrb. 1843, p. 364.

76. Monographie des Poissons fossiles du Vieux Grès-rouge ou Système Devonien. Sol. 1844, 4°. Livr. I. II.—Edinb. New Phil. Journ. XLI. p. 17.

77. Bewegung der Gletscher.—Bull. Soc. Sc. n. Neuch. 1843, Nov.—*L. u. Br.* N. Jahrb. 1844, p. 620.

78. Notice sur la succession des Poissons fossiles dans la série des Formations géologiqnes.—Ann. Sc. n. (3e S.) II. p. 251.

79. Sur les Blocs erratiques du Jura.—C. r. V. p. 506.—Edinb. New Phil. Journ. XXIV. p. 176.

80. Remarks on the observations of M. Durocher relative to the Erratic Phænomena of Scandinavia.—Comptes Rendus, Dec. 15, 1845.—Edinb. New Phil. Journ. XL. p. 237.

81. On the Fossil Fishes found by Mr. Gardner in the province of Ceará in the north of Brazil.—Edinb. New Phil. Journ. XXX. p. 82.

AGERIUS (Nic).

1. De Metallis. Argent 1634.—*Böhm.* Bibl. IV. 2, p. 105.

AGGERUP (Mich.).

1. Circa hieroglyphica quatuor Animalia. Hafn. 1731, 12°.—*Böhm.* Bibl. II. 1, p. 75.

AGNESIUS (J. Bapt.).

1. Apologia s. Liber de Avibus quæ in Albufera palude prope Valentiam degunt. (Oper. poetic. 1545, 8°.)—*Böhm.* Bibl. II. 1, p. 510.

AGRICOLA (G.).

1. De Ortu et Causis Subterraneorum; de naturâ eorum quæ effluunt e Terrà; de naturâ Fossilium; de veteribus et novis Metallis; Bermannus; recens. et Scholiis illustr. *T. Sigfrid.* Witteb. 1612, 8°.—Basil. 1546; etc.

2. Mineralogische Schriften, übers. u. mit erläuternden Anmerk. u. Excursionen begl. v. *E. Lehmann.* Th. I.-IV. Freiberg, 1806-12, 8°. fig.

3. De Animantibus subterraneis Liber. Basil. 1559, 8°. (*Mod.*),
1549, fol. Auch als Anhang zu seinem Hauptwerk : De Re metallicâ libri XII. etc. Basil. 1521, 1546, 1550, et 1556, fol.—Wittenb. 1614, 8°. Schweinf. 1607, 8°. Basil. 1621, fol.—*Böhm.*
Bibl. IV. 1, p. 16.

4. De naturâ Fossilium libri X. Basil. 1657, fol.—*Klein* Disp.
Echinod. p. vii.

5. De Succi nutritii per nervos transitu ; sub præsidio *J. G. Ber-
geri.* Viteberg. 1695, 4°.—Cat. Roy. Soc. Lond.

6. De Re metallicâ, libri XII. etc. Basil. 1556, 1561, 1657, fol.
fig.—*Mod.* Bibl. Helm.

7. Della Generazione delle cose che sotto la Terra sono, etc. 8°.
Venez. 1550.—Cat. R. S. L.

8. Cervi tum integri et vivi Natura et Proprietates, tum excoriati
et dissecti in Medicinâ usus. (German.) *Ausführliche Beschreibung des ganzen lebendigen Hirsches.* Amberg, 1603, 4°. 1617,
4°. fig —*Böhm.* Bibl. II. 1, p. 361.

9. Woher die warmen u. wilden Bäder, sonderlich auf dem
Schwarzwalde ihren Ursprung haben. Amberg, 1619, 4°.

AHLBERG (Dan.).

1. De Auro e Septentrione. 1748, 4°.—*Böhm.* Bibl. IV. 2, p. 210.

AHLERS (J. B. T.).

1. Erfahrung von einer Bienenkönigin.—Braunschw. Lüneb. Landw.
Ges. II. p. 578.

AHLMAN (J.).

1. Dissert. Fabricam, usum et differentias Palporum in Insectis exponens. Aboæ, 1792, 4°. (præside *Bonnsdorf*).

AHRENS (A.).

1. Uebersicht aller bis jetzt auf salzhaltigem Erdboden und in dessen
Gewässern entdeckten Käfer.—Isis, 1833, VII. p. 642.

2. Fauna Insectorum Europæ ; curâ *Germar et Kaulfuss.* Halæ,
1812 40, 18°., 20 fase —Isis, 1817, I. p. 113 ; 1822, XII. p. 1338.

3. Beiträge zu einer Monographie der Rohrkäfer. (*Donacia* Fabr.).
8°. Halle, 1817.—N. Schr. Naturf. Ges. Halle, I. 3, p. 1 ; 6, p. 25
(Nachtrag von *Germar*).

4. Beiträge zur Kenntniss deutscher Käfer.—N. Schr. Hall. Naturf.
Ges. II. 2, p. 1 ; II. 4, p. 60 (Bemerkungen von *Kunze*), fig. col.

5. Beschreibung der grossen Wasserkäferarten in der Gegend um

Halle im Sachsen (*Dytici*).—N. Schr. Hall. Naturf. Ges. I. 6, p. 47 ; II. 4, p. 58.

6. Würmer in einer Erdschnecke.—Berl. Mag. 1810, p. 292.—Isis, 1818, IX. p. 1467.

7. Déscription de la Larve de la *Pyrochroa coccinea*.—*Silb*. Rev. Ent. 1833, I. p. 247, fig.—Ann. Sc. Nat. (2ᵉ S.) XIII. p. 343.

8. Abhandlung über Wurmer, welche in einer Erdschnecke entdeckt worden sind.—Mag. Bul. Naturf. Fr. IV. p. 292, fig.

9. Monographie der deutschen Rohrkäferarten.—N. Schr. Nat. Ges. Halle, I.

AHRENS (G. Fr.).

1. Verzeichniss einiger Schmetterlinge, welche zu Schloss-Ballenstedt gefunden und beobachtet worden sind.   Halle, 1783.— *Fuessl.* N. Mag. 1785, p. 55, 64 ; mit Zusätzen v. F. L. *Brunn. Eis.* p. 203.

AHRENS (J. F. A.).

1. Beschreibung der bekannten deutschen Taumelkaefer Gyrini. Neue Schrift. der Naturforschend. Gesellsch. zu Halle.   Bnd. 2. Heft 2. 1812, p. 41.

AJASSON DE GRANDSAGNE (   )

1. Caii Plinii Secundi libri de Animalibus, cum notis variorum. Notas et excursus zoologici argumenti adjecit *G. Cuvier.* 8⁰. Vol. I. Paris, 1827.—*Féruss.* Bull. 1829, XVI. p. 392.

2. Résumé d'Ichthyologie ou d'Histoire Naturelle des Poissons. 2 vols. 32⁰.   Paris, 1829.—*Féruss.* Bull. 1829, XVII. p. 438.

3. *Ajasson de Grandsagne* et *Parisot,* Nouveau Discours sur les Révolutions du Globe, etc.   Paris, 1836, 2 vols. 18⁰.—*Quér.* Litt. Fr.

4. Notice sur la Vie et les Ouvrages de *Pline l'Ancien.*   Paris, 1829, 8⁰.—*Féruss.* Bull. XVI. p. 395.

5. Traité élémentaire d'Histoire Naturelle des Poissons, etc. un Appendice sur les Poissons Fossiles. (Encycl. port.)

AIGNER (Ant.).

1. Freymüthige Gedanken über einige in der Naturlehre noch unentschiedene Streitigkeiten.   Wien, 1782, 8⁰.—*Böhm.* Bibl. IV. 1, p. 239.

AIKIN (Arthur).

1. On the Geolog. Struct. of Cader Idris.—Geol. Trans. ser. 2, II.

p. 273. Phil. Mag. LXVIII. p. 58; ser. 2, II. p. 433.—*Féruss.* Bull. 1828, XV. p. 20.

2, Journal of a Tour through North Wales and part of Shropshire, with Observations in Mineralogy and other branches of Natural History. 8°. Lond. 1797.—Cat. R. Soc. Lond.

3. On the Wrekin, and on the great Coal-field of Shropshire.—Trans. Geol. Soc. ser. 1, I. p. 191.

4. *Aikin (A.)* and *Warburton (H.),* On the Vitreous Tubes found near to Drigg, in Cumberland.—Trans. Geol. Soc. ser. 1, II. p. 528.

5. On a Bed of Trap occurring in the Colliery of Birch Hill, near Walsall, in Staffordshire.—Trans. Geol. Soc. ser. 1, III. p. 251.

6. On the Shropshire Witherite.—Trans. Geol. Soc. ser. 1, IV. p. 438.

7. On a green waxy substance found in the alluvial soil near Stockport, in Cheshire.—Trans. Geol. Soc. ser. 1, IV. p. 445.

8. On some peculiarities observed in the Gravel of Litchfield.—Trans. Geol. Soc. ser. 1, IV. p. 426.

9. On a series of specimens (from Torre del Greco) presented to the Geological Society by H. G. Bennet.—Trans. Geol. Soc. ser. 1, V. p. 9.

10. On the Valleys and Watercourses of Shropshire and of parts of the adjacent Counties.—Trans. Geol. Soc. ser. 1, V. p. 73.

11. Calendar of Nature, or natural history of each month of the year. 8°. London, 1839.

AIKIN (W. E. A.).

1. Some Notices of the Geology of the Country between Baltimore and the Ohio River, with a Section illustrating the superposition of the rocks.—Am. J. XXVI. 2, p. 219.

AIMÉ ( ).

1. Sur du Corail fossile des environs d'Alger qui conserve encore une teinte rougeâtre.—C. R. VII. p. 903.

AIMEN ( ).

1. Sur la cause de la Nielle.—Acad. Sc. Par. 1760, fig.

2. Experimenta et Observationes de parvulis Insectis (*Podura*). 4°. fig.

AINSWORTH (W.).

1. Researches in Assyria, Babylonia and Chaldea. 8°. London, 1838.

2. Notice of Volcanic Island thrown up between Pantellaria and

Sciacca.—Mag. of Nat. Hist. ser. 1, IV. p. 545.—Amer. J. XXI. 2, p. 399.—*Féruss.* Bull. XXV. p. 231.

3. Sur l'âge des filons métallifères de Lead-hills, Wanlock-head et de Glendinning dans les comtés de Dumfries et de Lanark.—Edinb. Journ. Sept. 1830, p. 400.—*Féruss.* Bull. 1831, XXIV. p. 7.

4. Déscription des Landes.—Edinb. Journ. II. 1829, p. 99.—*Féruss.* Bull. 1830, XXII. p. 180.

5. Esquisse de la Gépgraphie physique des Monts Malvern-hills.— N. Édinb. Philos. Journ. Dec. 1827, p. 91.—*Féruss.* Bull. 1829, XVI. p. 366.

6. An Account of the Caves of Ballybunian, county of Kerry, with some mineralogical details. Dubl. 1834, 8º. fig.—Cat. Roy. Soc. Lond.—*L.* u. *Br.* N. Jahrb. 1834, p. 593.

7. Notes on the Pyrenees.—Mag. Nat. Hist. ser. 1, III. p. 496.

8. Observations on Mr. Rennie's Paper on the peculiar Habits of Cleanliness in some Animals.—Journ. Roy. Inst. ser. 2, I. p. 261. Am. Journ. XXI. 2, p. 382.

9. On the Caverns in the N. E. district of the High Peak of Derby-shire.—Edinb. Journ. Nat. and Geogr. Sc. II. p. 168.

10. On the Age of the Metalliferous Veins of Leadhills, &c. in Dumfries and Lanark.—Edinb. Journ. Nat. and Geogr. Sc. II. p. 400.

11. On Prof. Müller on the Structure of the Eyes in the Gastero-podous Mollusca.—Edinb. Journ. Nat. and Geogr. Sc. III. p. 282.

AINSWORTH (W.) et CHEEK (H. H.).—Edinburgh Journ. of Nat. Sc. Vide pars I. p. 75.

AIRY (G. B.).

1. On traces of Glaciers near Bantry Bay.—Rep. Brit. Assoc. 1843, Sect. p. 62.

AITKIN (J.).

1. An Essay on the Application of Natural History to Poetry. War-ringt. 1777, 8º.—*Böhm.* Bibl. I. 1, p. 179.

2. Relation of his Journey to Guinea, Brasil and the West Indies. Lond. 1723, 1735, 8º.—*Böhm.* Bibl. I. 1, p. 707.

3. Observations on the Coast of Guinea. Lond. 1758, 8º.—*Böhm.* Bibl. I. 1, p. 707.

AITKIN (Ph. J.).

1. Elements of Physiology, being an Account of the Laws and

Principles of the Animal Œconomy, especially in reference to the Constitution of Man. Edinb. 1838, 12°.

 AKERLEY (S.).

1. An Essay on the Geology of the Hudson River and the adjacent Regions, &c. N. York, 1820, 8°.

ALARDUS (Lamp.).

1. Jupiter pluvius, s. Dissertatio de Pluviâ. Lips. 1622, 4°.—*Böhm.* Bibl. V. p. 52.

ALBERS (J. A.).

1. Icones ad illustrandam Anatomen comparatam. Lips. 1818–22, fol. fig.—Isis, 1819, I. p. 130.

2. Bemerkungen über den Bau der Augen verschiedener Thiere. Denkschrift. der Academ. der Wissensch. zu Munchen 1808, p. 81.

3. Beiträge zur Anatomie u. Physiologie der Thiere. 4°. Bremen, 1802.

4. Amerikanische Annalen. Vide sup. pars I. p. 15.

ALBERS (J. Fr. H.).

1. Atlas der pathol. Anatomie für pract. Aerzte; nebst Erläuterungen. Bonn, 1832–1836. 8°. (Lief. 1–9.)—Bibl. med.-ch. p. 7.—(Lief. 11) Bonn, 1838. (Lief. 12 und 13) Bonn, 1839. (Lief. 14) 1840.

ALBERTI (F. v.).

1. Die Gebirge des Königreichs Würtemberg, in besonderer Beziehung auf Halurgie. Mit Anmerk. u. Beilagen v. *Prof. Schübler.* (*Les Formations du Würtemberg, par rapport aux dépôts salins. Avec des notes et des additions par* Schübler.) Stuttg. et Tübing. 1826, 8°.—*Féruss.* Bull. 1827, XII. p. 203.

2. Beitrag zu einer Monographie des Bunten Sandsteins, Muschelkalks und Keupers, und die Verbindung dieser Gebilde zu einer Formation. Stuttg. und Tübing. 1834, 8°.

3. Gyps des Muschelkalks bei Eisenach.—*L. u. Br.* N. Jahrb. 1837, p. 41, fig.

ALBERTI (Mich.).

1. Nova Paradoxa, d. i. Verhandlung von der Seele des Menschen, der Thiere und der Pflanzen. Halle, 1707, 8°.—*Böhm.* Bibl. II. 1, p. 92.

2. Dissertatio de Succino. Halæ, 1750, 4°.

ALBERTI (Val.).

1. De Figuris variarum rerum in Lapidibus et speciatim Fossilibus
Comitatûs Mansfeldici. Lips. 1675, 4°.—*Brückmann*, Epist. itin.,
Cent. II., Memorab. Mansfeld. p. 942.—*Böhm*. Bibl. II. 1, p.599 ;
IV. 1, p. 139.—*Mod*. Bibl. Helm.

ALBERTINUS (Ægid.).

1. Der Welt Tummel und Schauplatz, sammt der bitter-süssen
Wahrheit, in VIII. Theilen, darinnen von den grossen u. kleinen,
wilden u. zahmen, vierfüssigen u. kriechenden Thieren, von
allerhand Vögeln, Fischen u. Igl. *Münch*. 1613, 4°.—*Böhm*.
Bibl. I. 1, p. 238 ; II. 1, p. 26.

ALBERTUS MAGNUS.

1. De virtute Herbarum, Lapidum et Animalium. Bol. 1478.—
Venet. 1490, fol. Deutsch von *G. Apollinares*, etc. Strasb. 1541.
—*Eis*. p. 131.

2. De Animalibus, Libri XXVI. Romæ, 1478, fol.—Mantuæ, 1479,
fol.—Venet. 1495, 1498, 1519, fol.—Lugd. 1651, fol. (Oper. vol.
VI.).—German. titulo : *Thierbuch*, Frankf. 1545, fol.—Tr. Linn.
Soc. Lond. VII. p. 216.—*Böhm*. Bibl. II. 1, p. 15.

3. De Secretis Mulierum et Virorum. 4°. Lips. 1515.—Cat. Roy.
Soc.

4. Mineralium Libri V. Paduæ, 1476, fol.—Oppenh. 1518, 4°.—
Norib. 1518, 4°.—Aug. Vind. 1519, 4°.—Argent. 1541, 8°.—
Colon. 1568, 12°.—August. 1589, 4°.—Lugd. 1598, 12°.—Hamb.
1669, 4°.—*Böhm*. Bibl. IV. 1, p. 15.

5. Opusc. de Falconibus, Asturibus et Accipitribus. Aug. Vind.
1566, 8°.—*Böhm*. Bibl. II. 1, p. 545.

ALBIN (El.).

1. A Natural History of Birds, illustrated with 104 copper-plates.
Lond. 1731, 1734, 1738, 3 vols. 4°. fig.—(Franç.) *Histoire des
Oiseaux, avec les Notes de* DERHAM ; La Haye, 1749, 3 vols. 4°.
fig.—*Böhm*. Bibl. II. 1, p. 497.

2. A Natural History of English Insects, illustrated with 100 copper-plates, &c. Lond. 1720, 4°. (Ed. 2d.) with Notes and Observations by *W. Derham*. Lond. 1724, 4°. fig. 1749, *id*.—Ann.
Sc. n. II.—*Böhm*. Bibl. II. 2, p. 161.

3. A Natural History of Spiders and other curious Insects. Lond.
1736, 4°. fig.—*Cuv*. R. An. III. p. 330.—*Böhm*. Bibl. II. 2,
p. 349.

4. History of Esculent Fish. 4°. Lond. 1724.

5. Natural History of English Song-Birds, and such foreign ones as
   are esteemed for their Singing, &c.   Lond. 1737, 1741, 8°. fig.
   col.   3d ed. 1759.—*Böhm.* Bibl. II. 1, p. 497.

ALBINUS (Bernh. Sigfr.).

1. De Cantharibus Dissertatio.   Francf. 1687, 4°. ; 1694, 4°.—*Eis.*
   p. 197.

2. Icones Ossium Fœtus Humani.   Lugd. B. 1737, 4°. fig.—Bibl.
   med.-ch. p. 7.

3. Tabulæ Anatomicæ Sceleti et Musculorum Hominis ; cum expla-
   natione.   Lugd. B. 1747, fol.—Bibl. med.-ch. p. 8.

4. Tabulæ VII. Uteri gravidi in plano.   Lugd. B. 1747, 1750, fol.
   —Bibl. med.-ch. p. 8.

5. De Utero et Append.   Lugd. B. 1751, fol. fig.—Bibl. med.-ch.
   p. 8.

6. De Ossibus Corporis Humani. (Ed. 3d.)   Vindob. 1759, 8°.—
   Bibl. med.-ch. p. 7.

7. De Sceleto Humano.   Lugd. B. 1762, 4°.—Bibl. med.-ch. p. 7.

8. Explicatio Tabularum Anatomicarum *B. Eustachii.*   Lugd. B.
   1744 ; 1761, fol. fig. Francof. 1784.—Bibl. med.-ch. p. 7.

9. Historia Musculorum Hominis. Lugd. B. 1734, 4°. fig.—Francof.
   1784, *id.*   Ed. aucta.   Bamb. 1796, 4°. fig.—Bibl. med.-ch. p. 7.

10. De Naturâ Hominis et Supellectile Anatomico.   Lugd. B. 1786,
    8°. 2 vol.—Bibl. med.-ch. p. 7.

11. Diss. de Cervo, glande plumbeâ trajecto, et post tres horæ quad-
    rantes quatuor circiter passuum millia aufugiente. (*V.-Riedlinus,*
    De Ventre Cervi, &c., Dec. III. p. 78.)—*Böhm.* Bibl. II. 1,
    p. 363.

12. Diss. de Tarantismo.   Francf. s. O. 1691, 4°. fig.—*Böhm.* Bibl.
    II. 2, p. 357.

13. Oratio inaug. de Anatome Comparatâ.   Lugd. B. 1719, 4°.—
    Böhm. Bibl. II. 1, p. 83.

14. Academicarum Annotationum Libri VIII. continent Anatomica,
    Physiologica, Zoographica et Phytologica.   Lugd. B. 1755–61,
    2 vols. 4°. fig.—*Mod.* Bibl. Helm.

ALBINUS (P.).

1. Meisnische Land- und Berg-Chronica.   Wittenb. 1580, 4°.—
   Dresd. 1589, fol. fig —*Böhm.* Bibl. I. 1, p. 601.

ALBMAIER (Theod.).

1. I quatro Elementi Spiegati in 25 Discorsi, ne quale si ragiona delle cosi principali che nascano in essi, delle Pietre preziose, de Muschio, dell' Ambra, del Balsamo, del Zibetto, de Metalli, &c. Firenze, 1668, 4°.—*Böhm.* Bibl. IV. 1, p. 281.

ALBOSIUS (Joh.).

1. Fœtûs per annos 28 in Utero contenti et lapidefacti Historia. (V. *Rousset,* Fœtus vivi, &c.) 1591.—Cat. Roy. Soc. Lond.

2. Portentosum Lithopædion, s. Embryon petrefactum urbis Senonensis, unà cum *Sim. Provancherii* de eàdem re opinione et *Th. Bartholini* de eodem Fœtu judicio. Amst. 1662, 12°. (Gallicè) *Le prodigieux Enfant pétrifié de la ville de Sens,* trad. du Latin par *S. de Provenchère.* Sens, 8°.—*Böhm.* Bibl. IV. 2, p. 271.

ALBREOHT (J. F. E.).

1 Zootomische und physikalische Entdeckungen von der innern Einrichtung der Bienen, besonders von der Art ihrer Begattung. 8°. Gotha, 1775, fig.—*Böhm.* Bibl. II. 2, p. 305.

ALBRECHT (J. Fr.).

1. Versuch eines neuen Beweises, dass die Sündfluth allgemein gewesen sey. Nordh. 1752, 8°. fig.—*Böhm.* Bibl. IV. 2, p. 243.

ALBRECHT (J. Seb.).

1. Der Mensch und sein Geschlecht. Quedl.

2. De Lumbricite elegantissimo, Observatio.—Act. Nat. Cur. VI. p. 116, fig.—*Mod.* Bibl. Helm.

3. Dubitationes et Conjecturæ circa duo Petrefacta.—Act. Nat. Cur. X. p. 211.

4. Ducatûs Coburgensis agri cum vicinis Corporum petrificatorum, ex utroque regno, copià et veritate nullis secundi in Germaniâ.— Act. Nat. Cur. IX. p. 401.—*Mod.* Bibl. Helm.

5. De Insectorum Larvis in Puero.—Comm. Litter. Norimb. 1744, p. 102.

6. De Insectorum ovis sine prævià maris cum femellà conjunctione fœcundis.—Ephem. Nat. Cur. II. p. 26.—*Manget* Bibl. I. p. 93. —*Eis.* p. 162.

7. Spicilegium ad historiam naturalem Scarabæi maximi Platyceri, &c.—Ephem. Nat. Cur. VI. p. 404, fig.—*Eis.* p. 194.

8. Programma quo Fossilia figurata Diluvii universalis esse testimonia, adstruitur et affirmatur. Coburg, 1734, 4°. fig.—*Böhm.* Bibl. IV. 2, p. 236.—*Mod.* Bibl. Helm.

9. De ornatissimo, figuris hieroglyphicis quasi, Belemnite Trechhei-
mensi prope Coburgum, Observatio. Act. Nat. Cur. IV. p. 72,
fig.—*Mod.* Bibl. Helm.

ALBRINK (J. H.).

1. Dissertatio Caloris Animalis Theoriam sistens. 8°. Berlin, 1827.

ALDER (Josh.).

1. Note on *Euplocamus, Triopa* and *Idalia.*—Ann. and Mag. N.
Hist. XV. p. 262.

2. Notes on the Land and Freshwater Mollusca of Great Britain,
with a revised List of Species.—Mag. Zool. and Bot. 1838, II.
p. 101.

3. Remarks on *Lottia virginea.*—Ann. and Mag. N. Hist. VIII.
p. 404, fig.

4. Description of some new British species of *Rissoa* and *Odostomia.*
—Ann. and Mag. N. Hist. XIII. p. 323, fig.

5. A Catalogue of the Land and Freshwater Testaceous Mollusca
found in the vicinity of Newcastle-upon-Tyne.—Trans. Nat. Hist.
Soc. Newcastle, I. p. 26 ; II. p. 337.

6. Observations on the genus *Polycera* of Cuvier, with Descriptions
of two new British species.—Ann. and Mag. N. Hist. VI. p. 337,
fig.   Notices of *Eolis, Doris,* &c.—Rep. Brit. Assoc. 1843, Sec.
p. 69.

ALDER (Josh.) and HANCOCK (Alb.).

1. Remarks on the genus *Eolidina* of *M. de Quatrefages.*—Ann.
and Mag. N. Hist. XIV. p. 125.

2. Descriptions of *Pterochilus,* a new genus of Nudibranchiate Mollusca, and two new species of *Doris.*—Ann. and Mag. N. Hist.
XIV. p. 329.

3. Descriptions of several new species of Nudibranchous Mollusca
found on the coast of Northumberland.—Ann. and Mag. N. Hist.
IX. p. 31.

4. Notice of a British species of *Calliopæa,* d'Orb., and of four new
species of *Eolis,* with Observations on the Development and Structure of the Nudibranchiate Mollusca.—Ann. and Mag. N. Hist.
XII. p. 233.

5. Description of a new genus of Nudibranchiate Mollusca, with
some new species of *Eolis.*—Ann. and Mag. N. Hist. XIII. pp.
161, 407, fig.—Ann. Sc. n. (3ᵉ S.) I. p. 190 (*Déscript. d'un
genre nouveau de Mollusques Nudibranches, et de quelques espèces
nouvelles d'Eolides*).

6. On some Species of Mollusca Nudibranchiata, with Observations on the Structure and Development of the Animals of that order. —Rep. Brit. Assoc. 1843, Sect. p. 73.

7. Report on the British Nudibranchiate Mollusca. — Rep. Brit. Assoc. 1844, p. 24.

8. On a new Genus of Mollusca Nudibranchiata.—Rep. Brit. Assoc. 1845, Sect. p. 65 —Ann. and Mag. N. Hist. XVI. p. 311.

9. Monograph of the British Nudibranchiate Mollusca, with figures of all the species.—RAY SOCIETY, fol. London, 1845-1847.

10. Notices of some new and rare British species of Naked Mollusca. —Ann. and Mag. N. Hist. XVIII. 289. fig.

ALDERSON (J.).

1. On a Whale of the Spermaceti Tribe cast on shore on the Yorkshire Coast on the 28th April, 1825.—Trans. Cambr. Phil. Soc. II. p. 253.—*Féruss*. Bull. 1829, XVII. p. 282.—Isis, 1835, XII. p. 1006. (*Männlicher Walrath-Wal, an die Küste v. Yorksh. geworfen.*)

ALDES (Theod.).

1. Dissert. epistol. de Generatione Animalium contra *Harvaum*. Amst. 1666, 12°.—*Böhm*. Bibl. II. 1, p. 127.

2. Observationes naturales in Ovis factæ. Amst. 1673, 12°.—*Böhm*. Bibl. II. 1, p. 534.

ALDROVANDI (Ul.).

1. Historia naturalis. Bonon. 1599-1640, 14 vols. fol. fig.—Dict. Sc. nat. LX. p. 535.

2. Museum metallicum, in libros IV. distributum à *B. Ambrosino* compositum, edente *M. A. Bernia*. Bonon. 1648, fol. fig.— *Böhm*. Bibl. IV. 1, p. 30.—*Mod*. Bibl. Helm.

3. De Quadrupedibus solipedibus. *Uterverius* collegit et recensuit, *Tamburinus* in lucem edidit. Bonon. 1616, fol. fig.—Francf. 1623, fol. fig.—Bonon. 1639, fol. fig.—*Böhm*. Bibl. II. 1, p. 23. Bonon. 1647, fol.—*Fisch*. Cat.

4. De Quadrupedibus digitatis viviparis libri III., et de Quadruped. digit. oviparis libri II. *B. Ambrosinus* collegit. Bonon. 1621, fol. fig., 1637, 1645, 1665.—*Böhm*. Bibl. II. 1, p. 24.

5. Ornithologiæ libri XII. Bonon. 1599, 3 vols. fol.—Ornithologiæ, h. e. de Avibus Historiæ libri XII., in quibus Aves describuntur, descriptæ delineantur, natura earum, mores, proprietates ità declarantur, ut facilè quicquid de Avibus dici queat, hic peti possit.

(Vid. *Böhm.* Bibl. II. 1, p. 24.) 1646, 3 vols. fol. Francf. 1610, fol.—Trans. Linn. Soc. Lond. V. p. 277.

6. Serpentium et Draconum historiæ libri duo. Bonon. 1640, fol. fig. Cur. *Barth. Ambrosini.—Dum. et Bib.* I. p. 233.—*Böhm.* Bibl. II. 1, p. 25.

7. De Piscibus libri V., et de Cetis liber unus. Bonon. 1613, 1625, 1638, 1644, fol. fig.—Venet. 1616, fol. fig.—Francf. 1623, 1629, 1640, fol. fig.—Dict. Sc. n. XXII. p. 488.—*Böhm.* Bibl. II. 1, p. 24.

8. De Animalibus insectis libri VII. Bonon. 1602, fol. fig.— Francf. 1618, 1623, fol.—Bonon. 1620, 1638, fol. fig.—Tr. Linn. Soc. Lond. V. p. 277.—*Böhm.* Bibl. II. 1, p. 25.—*Mod.* Bibl. Helm.

9. De reliquis Animalibus exsanguibus, utpotè de Mollibus, Crustaceis, Testaceis et Zoophytis, libri IV. Bonon. 1606, 1642, fol. fig.—Francf. 1616, 1623, fol.—Tr. Linn. Soc. Lond. VII. p. 216. —*Wiegm.* Arch. 1837, I. p. 3.

10. Quadrupedum omnium bisulcorum Historia; *J. C. Uterverius* colligere incepit, *T. Dempsterus* perfectè absoluit, *M. A. Bernia* denuò in lucem edidit. Bonon. 1600, fol. fig., 1613, 1621, 1642. Francf. 1647, fol. fig.—*Böhm.* Bibl. II. 1, p. 23.

11. Monstrorum historia, cum Paralipomenis Historiæ omnium Animalium. *B. Ambrosinus* labore et studio volumen composuit. Bonon. 1642, 2 vols. fol.—Cat. R. Soc.

12. Opera omnia. Bonon. 1599, 1645, fol. 11 vols.—*Böhm.* Bibl. II. 1, p. 25.

13. Œconomia Formicarum (cum *Mich. Gehler*).—Amphith. Dorn. I.

ALESSANDRI (Aless. degli).

1. Dies geniales. 1840.

ALESSANDRINI (Ant.).

1. Apparatus Branchiarum Heterobranchi anguillaris.—N. Comm. Bonon. V. p. 149.

2. De Testudinum Linguâ atque Osso hyoideo.—N. Comm. Bonon. I. p. 53.

3. De Testudinis Casuaræ Larynge.—N. Comm. Bonon. I. p. 383.

4. Descriptio veri Pancreatis glandularis et parenchymatosi in Accipensere et in Esoce repertis.—N. Comm. Bonon. II. p. 335

5. An quidquam Nervi conferant ad evolutionem et incrementum Systematis Muscularis.—N. Comm. Bonon. III. p. 177.

6. De Piscium Apparatu Respirationis, tum speciatim Orthagorisci.
—N. Comm. Bonon. III. p. 359.

7. Observationes super intimâ Branchiarum Structurâ Piscium cartilagineorum.—N. Comm. Bonon. IV. p. 329.

8. Sull' Organo dell' Olfatto dei Cetacei in genere, ed in particolare sopra quello del Delfino volgare.—Ann. Sc. nat. Bologna, 1840.
—Nov. Act. Ac. Leop. Car. 1842.

9. Saggio di Osservazioni di Anatomia patologica comparata riguardanti il Sistema osseo.—Ann. Sc. nat. Bol. 1841.

10. Observations sur le Pancréas des Poissons.—Ann. Sc. nat. 1833, XXIX. p. 193.

11. Nervensystem der Scolopendra morsitans.—Ann. di St. nat. VIII. p. 190.—Isis, 1833, XI. p. 1041.

ALESSI (Gius.).

1. Sulla vera Origine del Succino.—Att. Soc. Gicen. Catan. VI. p. 17.

2. Introduzione alla Zoologia del Mare di Sicilia.—Att. Soc. Gicen. Catan. XI. p. 89.

3. Descrizione fisico-mineralogica di Castrogiovanni e del suo territorio.—Att. Soc. Gicen. di Catan. I. p. 99.

4. Abrégé de la Déscription physico-minéralogique de l'Etna.—Giorn. di Sc. etc. per la Sicilia, Nov. 1824, p. 187.—Féruss. Bull. 1829, XVI. p. 44.

5. Storia critica delle Eruzioni dell' Etna.—Att. Soc. Gicen. Catan. IV. p. 23 ; V. p. 43 ; VI. p. 85.

6. Sopra gli ossidi di Silicio ed i Silicati della Sicilia.—Att. Soc. Gicen. Catan. V. p. 95.

ALEXANDER ( ).

1. A Treatise of Vision. Lond. 1833.

ALEXANDER (Guil.).

1. De Cantharidum usu et historiâ. Edinb. 1769, 4º.—Böhm. Bibl. II. 2, p. 206.

ALEXANDER (Capt. H.).

1. On Mastodon Teeth from the Crag, and the occurrence of a particular bed containing Echini in the Coralline Crag at Sudbourne.—Proc. Geol. Soc. III. 10.—Phil. Mag. ser. 3, XIV. p. 146.

2. The Soils of East Suffolk considered Geologically. 8º. Woodbridge, 1840.

3. On the Annual Destruction of Land at Easton Bavent Cliff, near

Southwold.—Proc. Geol. Soc. III. 445.—Phil. Mag. ser. 3, XX. p. 57.

ALEXANDER (J. E.).

1. Expedition of Discovery into the Interior of Africa. 2 vols. 8°. London, 1838.

2. Notitz über einen Asphalt-oder Pech-See auf Trinidad.—Ed. N. Phil. J. XXVII. p. 94.—*L. u. Br.* N. Jahrb. 1833, p. 629.

ALFARIUS A CRUCE (Vinc.).

1. Vesuvius ardens, s. Exercitatio ad πιγοπυποετον, i. e. motum incendium Vesuvii in Campaniâ libr. III. comprehensa. Romæ, 1632, 4°.—*Böhm.* Bibl. IV. 2, p. 378.

ALGER (Fr.).

1. Notes on the Mineralogy of Nova Scotia.—*Sill.* Amer. Journ. XII. 2, p. 227.

2. Notice of Minerals from New Holland.—Bost. Soc. June 1840. —*Sill.* Am. J. XXXIX. p. 157.

ALGREN (M.).

1. Sur les soins à donner aux Abeilles.—Vet. Acad. Handl. 1776, p. 234; 1777, pp. 185, 328.

ALI KHAN (A.).

1. On the Baya, or Indian Grosbeak.—Asiat. Res. II. p. 109.

ALI-PONZONI (Gius.).

1. Metodo di preparare e conservare pe' Gabinetti di Storia naturale i Bruchi ed altri Insetti.—Opusc. Scelt. XII. p. 239.

ALISON (Edw.).

1. Relation d'une excursion à la cime du Pic de Ténériffe, en Févr. 1829.—Phil. Mag. a. Ann. of Philos. 1830, VIII. pp. 23, 140, 195, 248, 433.—*Féruss.* Bull. 1831, XXIV. p. 285.—*L. u. Br.* 1834, p. 572 (*Ersteigung des Gipfels v. Pico de Teyde*).

ALISON (W. P.).

1. Outlines of Physiology and Pathology. Lond. a. Edinb. 8°. 1833. —Supplement. 8°. Edinb. 1836.

2. On single and correct Vision by means of Double and Inverted Images on the Retina.—Tr. R. Soc. Edinb. XIII. p. 472.—*Müll.* Arch. 1840, p. v.

3. On the Theory which ascribes Secretion to the agency of Nerves.
—Quart. Journ. Sc. IX. p. 106.

4. On the Physiological principle of Sympathy.—Edinb. Med. Chirurg. Trans. II. p. 165.

5. Experiments on the vital power of Arteries, and on the immediate cause of death by Asphyxia.—Edinb. Med. and Surg. Journ. XLV. p. 98.

6. Articles—Asphyxia, Contractility, Exertion, and Instinct, in *Todd's* Cyclop. of Anat. and Physiol.

7. On the Principle of Vital Affinity.—Trans. Roy. Soc. Edinb. XVI.

ALISON (R. E.).

1. On the Earthquake of Chili of the 20th of Feb. 1835.—Proc. Geol. Soc. II. 209.—Phil. Mag. ser. 3, VIII. p. 74.—*L. u. Br. N.* Jahrb. 1836, p. 719.

ALLAMAND (jun.).

1. *Observ. sur les Mœurs des Animaux domest. (Ueber das Naturell unserer Hausthiere.)*—Mém. Soc. sc. nat. Neuch. I. p. 77.—*Wiegm.* Arch. 1837, V. p. 146.

2. Van de nitwerkzelen, welke een Americaans vis veroorzaakt op de geenen, die Nem aanraaken.—Verhand. Maatsch. te Haarl. II. p. 372.—Dict. Sc. nat. XXII. p. 546.

3. Histoire naturelle du Gnow, du grand Gerbu et de l'Hippopotame. Amst. 1776, 4°. fig.—*Böhm.* Bibl. I. 1, p. 282.

ALLAN (R.).

1. A Manual of Mineralogy, comprehending the more recent Discoveries in the Mineral Kingdom. Lond. 1834, 8°. fig.

ALLAN (T).

1. On the Formation of the Chalk Strata, and on the Structure of the Belemnite.—Trans. Roy. Soc. Edinb. IX. p. 393.—Phil. Mag. 1823, p. 391.—*Féruss.* Bull. 1824, I. p. 200.

2. On the Geology of the environs of Nice.—Trans. Roy. Soc. Edinb. VIII. p. 427.—Isis, 1818, p. 583.

3. On a mass of native Iron, etc. (*Sur une masse de Fer natif du désert d'Ataiama, au Pérou*).—Ann. Sc. n. 1829, XVII.; Rev. Bibl. p. 1.

4. On the Transition Rocks of Werner.—Trans. Roy. Soc. Edinb. VII. p. 109.—Phil. Mag. XLII. p. 15.

5. Geological account of the Lead Mine of Dufton, in Westmoreland.—Journ. Roy. Inst. II. 198.

6. On the Rocks in the vicinity of Edinburgh.—Trans. Roy. Soc. Edinb. VI. p. 405.

7. Mineralogical Synonymes and Analyses. Edinb. 1814, 8°.

8. On a Vegetable Impression found in the quarry of Craigleith.— Trans. Roy. Soc. Edinb. IX. p. 235.

ALLARDYCE (J.).

1. On the Granitic Formation of South India.—Madras Journ. Lit. IV. p. 327.

ALLEMAND (P. L.).

1. Zoologie générale d'après Cuvier. Une feuille. Paris, 1840.

ALLEN (Benj.).

1. Of the Manner of the Generation of Eels.—Phil. Trans. XIX. p. 664.—Dict. Sc. nat. XXII. p. 543.

2. Of the *Scarabæus pulsator galeatus.*—Phil. Trans. No. 245, p. 376. Budd. III. p. 392, fig.—*Eis.* p. 190.

3. Account of the Death-Watch and Gall-Bee.—Phil. Trans. XX. p. 376.

4. The Natural History of the Chalybeat and Purging Waters of England, with their particular essays and uses, etc. Lond. 1699, 8°.—*Böhm.* Bibl. V. p. 219.

5. The Natural History of the Mineral Waters of Great Britain, to which are added some Observations of the *Cicindela* or Glowworm. Lond. 1711, 8°.—*Böhm.* Bibl. V. p. 219.

ALLEN (J. A.).

1. On the Question, Whether there are any traces of a Volcano in the West-River Mountain.—Amer. Journ. III. 1, p. 73.

ALLEN (Joh.).

1. Beschreibung des Eppacher und Neuensteiner Heil- u. Gesundbrunnen. Oehring. 1725, 8°.—*Böhm.* Bibl. V. p. 305.

ALLEN (Paul).

1 History of the Expedition under the command of Captains Lewis and Clarke to the sources of the Missouri, and thence to the Pacific Ocean. London, 1814, 4°. Dublin, 1817, 8°. 2 vols.

ALLEN (Capt. R.N.).

1. On a Collection of Objects of Zoology, made in the Interior of Africa.—Proc. Zool. Soc. Lond. IV. p. 45.

2. On some Drawings of Fishes of the river Quorra.—Proc. Zool. Soc. Lond. IV. p. 147.

ALLEN (W.) and WOODS (S.).

1. Barometrical Measurements.—Trans. Geol. Soc. ser. 1, vol. iv. p. 434.

ALLEN (W.) and PEPYS (W. H.).

1. On the Respiration of Birds.—Phil. Trans. CXIX. p. 279.

ALLEON-DULAC (J. L.).

1. Melanges d'Histoire Naturelle, nouv. ed. 6 vols. 8°. Lyon, 1763-65.

2. Mémoires pour servir à l'Histoire Naturelle des provinces du Lyonnais, du Forez et du Beaujolais. 2 vols. 8°. Lyon, 1765.

ALLETR ( ).

1. Histoire des Singes et autres Animaux curieux. Paris, 1752, 12°.

ALLIES (Jabez).

1. Observations on certain curious Indentations in the Old Red Sandstone of Worcestershire and Herefordshire, considered as the Tracks of Antediluvian Animals, &c. Lond. 1835, 8°.

ALLIONE (Car.).

1. Manipulus Insectorum Taurinensium.—Mél. Soc. Roy. Turin, III. p. 158.—Eis. p. 154.

2. Oryctographiæ Pedemontanæ specimen, exhibens corpora fossilia terræ adventitia. Paris, 1752, 8°. fig.—Klein Disp. Echinod. p. VII.—Böhm. Bibl. IV. 1, p. 111.—Mod. Bibl. Helm.

3. Stirpium præcipuarum littoris et agri Nicæensis enumeratio, cum elencho aliquot animalium ejusdem maris. 8°. Paris, 1757.

4. De Firmitate, sivè Soliditate Corporis.—De Liene et Pancreate. —De Respiratione, etc. (Theses pro Aggreg.) Taurini, 1747, 8°.

ALLIS (T.).

1. The Sclerotic Bones of the Eyes of different Birds and Reptiles. —Brit. Assoc.—Amer. Journ. XXXIII. 2, p. 289.

2. On the Toes of the African Ostrich, and the number of Phalanges in the Toes of other Birds.—Rep. Brit. Assoc. 1838, Sect. p. 107.—Amer. Journ. XXXV. 2, p. 312.

3. On some Peculiarities in the Flight of Birds, especially as that is influenced in some species by the power they possess of decreasing and adjusting their own specific gravity.—Rep. Brit. Assoc. 1844, Sect. p. 72.

4. On the Windpipe of the Dun Diver.—Mag. Nat. Hist. ser. 1, V. p. 766.

5. On the Connection between the Furculum and Sternum of Birds. —Proc. Zool. Soc. 1835.

ALLIX (G. a. J.).

1. Teoria dell' Universo.   Milano, 1817, 12°.

ALLMAN (G. J.).

1. On the Stinging Property of the Lesser Weever-fish (*Trachinus Vipera*).—Ann. and Mag. N. H. VI. p. 161.

2. On the Occurrence of White's Thrush in Ireland.—Ann. and Mag. N. H. XI. p. 78.

3. On Plumatella repens.—Rep. Brit. Assoc. 1843, Sect. p. 74.

4. On a new Genus of Terrestrial Gasteropod.—Rep. Brit. Assoc. 1843, Sect. p. 77.—Ann. and Mag. N. H. XVII. p. 297.

5. Synopsis of the Genera and Species of Zoophytes inhabiting the Fresh Waters of Ireland.—Rep. Brit. Assoc. 1843, Sect. p. 77.— Ann. and Mag. N. Hist. XIII. p. 328.

6. On certain Peculiarities in the Arteries of the Six-banded Armadillo.—Rep. Brit. Assoc. 1843, Sect. p. 68.

7. On a new Genus of Nudibranchiate Mollusca.—Rep. Brit. Assoc. 1844, Sect. p. 65.—Ann. and Mag. N. H. XVII. p. 1.

8. On the Anatomy of Actæon viridis.—Rep. Brit. Assoc. 1844, Sect. p. 65.—Ann. and Mag. N. H. XVI. p. 145.

9. On a new Genus of Helianthoid Zoophytes.—Rep. Brit. Assoc. 1844, Sect. p. 66.—Ann. and Mag. N. H. XVII. p. 417.

10. On the Structure of the Lucernariæ.—Rep. Brit. Assoc. 1844, Sect. p. 66.

11. On the Muscular System of certain freshwater Ascidian Zoophytes.—Proc. R. Irish Acad. II. 319.

12. On a new genus of Hydraform Zoophytes.—Proc. R. Irish Acad. II. 395.

13. On certain Peculiarities in the Anatomy of Anthocephalus.— Proc. R. Irish Acad. II. 423.

14. On Fredericella Sultana.—Proc. R. Irish Acad. II. 545.

15. On Chelura terebrans.—Ann. and Mag. N. H. XIX. p. 361.

16. Description of a new genus of Entomostraca.—Ann. and Mag. N. H. XX. p. 1.

17. Description of a new genus of Tracheary Arachnidans.—Ann. and Mag. N. H. XX. p. 47.

18. Notice of a peculiar Tufa from a bog in the co. Tipperary.— Journal of the Geolog. Soc. of Dublin, III. p. 203.

ALLMER (C. H. E.).

1. Disquisitio anatomica de Pinguedine animali. Jenæ, 1823, 4°. fig.—Bibl. med.-ch. p. 9.

ALLOATTI ( ).

1. Sui Testacei fossili dell' Alto Monferrata.—Giorn. Sc. e Lett. Torin. I. 2, p. 124.

ALOPÆUS (S.).

1. Kurze Beschreibung der im Russ. K. Karelien befindlichen Marmor- u. anderen Steinbrüchen, Berg- und Steinarten. Petersb. 1787, 8°.

ALOS (Jo.).

1. Dissertatio de Viperis trochiscis. 1664, 4°.

ALPINO (Joh. Lud.).

1. Febris epidemicæ anno 1694 et 95, in Noricæ Ditionis oppido Herspruccensi et vicino tractu grassari deprehensæ, tandemque petechialis redditæ historica relatio, in observationum semicenturiam digesta. Norimb. 1697, fig.—Act. Erudit. XVI. p. 25.

ALPINO (Prosp.).

1. Historiæ Ægypti naturalis pars prima, quâ continentur rerum Ægyptiacarum libri quatuor. Leyde, 1735, 4°., 2 vols.—*Dum. et Bib.* I. p. 303.—*Böhm.* Bibl. I. 1, p. 710.   H

ALROT (Eric.).

1. Diss. I. et II. de Gestriciâ. Ups. 1722, 4°.—*Böhm.* Bibl. IV. 1, p. 156.

ALSARIUS A CRUCE (Vinc.).

1. De Verme admirando per nares egresso Comment. ad *Fulv. Angelinum*, cum hujusdem de eodem brevi Discursu. Ravenn. 1610, 4°.—*Böhm.* Bibl. II. 2, p. 389.

ALSOP (J.).

1. On the Toadstones of Derbyshire.—Rep. Brit. Assoc. 1844, Sect.
p. 51.

ALSTEDIUS (J. H.).

1. Compendium Lexici Philosophici. Hernborn, 1626, 8°.—*Böhm.*
Bibl. I. 1, p. 27.

ALSTRÖMER (Cl.).

1. Tal oun den fin-ullega fär Afveln, etc. (*De Ovium lanâ teneriori
vestitarum progenie*). Stockh. 1770, 8°.—*Böhm.* Bibl. II. 2, p. 388.

ALT (H. C.).

1. De Phthiriasi. 8°. Bonn, 1824.

ALTEN (J. W. v.).

1. Systematische Abhandlung über die Erd- und Fluss-Conchylien
um Augsburg. Augsb. 1812, 8°. fig.

ALTENA (H. L. van).

1. Commentatio ad quæstionem de speciebus Reptilium Batrachi-
orum. 1 vol. 4°. Leyden, 1829.

ALTH ( ).

1. Gebirgs-Profil und Hebungen in Ungarn u. Süd-Russland.—
L. u. Br. N. Jahrb. 1840, p. 347, fig.

ALTHAUS (C. L.).

1. Grundzüge zur gänzlichen Umgestaltung der bisherigen Geologie, oder Kurze Darstellung der Weltkörper- und Erdrinde-Bild-
ung. Koblenz, 1839, 12°.—*L. u. Br.* N. Jahrb. 1839, p. 444.

2. Ueber das Vorkommen der Sandstein-Spiegel in der Gegend von
Marburg.—*L. u. Br.* N. Jahrb. 1837, p. 536, fig.

3. Poröser Kieselschiefer am Heidekopf in Kurhessen, etc.—*L. u.
Br.* N. Jahrb. 1840, p. 83.

4. Mesotype vom Alpstein.—*L. u. Br.* N. Jahrb. 1842, p. 276.

5. Ueber *Alberti's* Monographie des Muschelkalks.—*L. u. Br.* N.
Jahrb. 1834, p. 406.

6. Schildkröten im Torf bei Dürrheim.—*L. u. Br.* N. Jahrb. 1834,
p. 537; 1835, p. 63.

ALTING (J. Chr.).

1. Dissert. de Lumbricis. Ultraj. 4°.

ALTMANN (J. G.).

1. Versuch einer historischen u. physischen Beschreibung der Helvetischen Eisberge. Zür. 1751, 1753, 8°.—*Böhm.* Bibl. I. 2, p. 547.

2. Exercit. de Tesseris Badæ Helveticorum erutis. Bernæ, 1750, 4°.—Mus. Helvet. XXVI. p. 311.—*Böhm.* Bibl. IV. 1, p. 226.

ALTMANN (L.).

1. Abriss der Entomologie. Leipz. 1837, 8°.

2. Die nützlichen u. schädlichen Forstkäfer für Forstbeamte. 8°. Dessau, 1844.

ALTMANN (M.).

1. De Cadaverum Conditurâ. Berol. 1841, 8°.

ALTON (E. d').

1. Naturgeschichte des Pferdes. 2 thle. fol. Weimar, 1810-16.

2. Beschreibung des Muskelsystems eines *Python bivittatus.*—*Müll.* Arch. 1834, pp. 346, 432, 528, fig.

3. De Pythonis ac Boarum Ossibus Commentatio. Hal. 1836, 4°.—*Val.* Repert. anat. u. phys. II. p. 22.—*Müll.* Arch. 1837, IV. p. LXV.

4. Descriptio Dentium Camelopardalis Giraffæ, quam loco appendicis ad *Bojani* de Merycotherio Sibirico commentationem proponit.—N. Act. Leop. Ac. XII. 1, p. 335, fig.—Isis, 1825, II. p. 203.

5. Zur vergleichenden Osteologie v. *Goethe,* mit Zusätzen und Bemerkungen.—N. Act. Leop. Ac. XII. 1, p. 323.—Isis, 1825, II. p. 203.

6. Das Riesen-Faulthier (*Bradypus giganteus*). Bonn, 1821, fol. fig.—Isis, 1821, IX. p. 862.

7. Sepia Octopus.—Isis, 1818, XI. p. 1930.

8. De Strigum Musculis Commentatio. Hal. 1837, 4°.—*Müll.* Arch. 1838, p. cxxv.

9. Ueber die von *Sellow* aus der Banda oriental mitgebrachten fossilen Panzer-Fragmente, etc.—Abh. Berl. Ak. (1833) 1835, p. 369, fig.—*L. u. Br.* N. Jahrb. 1837, p. 603.

10. *Alton* (*d'*) u. *Pander,* Skelete der straussartigen Vögel, abgebildet und beschrieben. Bonn, 1827, fol.

11. Die Skelete der Chiropteren und Insectivoren, abgebildet und beschrieben. Bonn, 1831, fol.—Cat. Bibl. Turic.

ALZATE Y RAMYRES ( ).

1. Sur les Poissons vivipares et quelques autres objets d'histoire naturelle. Rozier, Journ. de Phys. t. 1, 1773, p. 221.

AMAMA (Nic. ab).

1. Dissertationum marinarum Decas. Francof. 1651, 8°.—*Böhm.* Bibl. V. p. 81.

AMATO (P. Gaet. d').

1. Divisamento critico sulle correnti Opinioni intorno a Fenomeni del Vesuvio e degli altri Volcani. Nap. 1756, 8°.—*Böhm.* Bibl. IV. 2, p. 380.

AMBROISE (St.).

1. Divi Ambrosii Mediolanensis Episcopi, Hexaëmeron libri VI. Basileæ, 1566, fol.—Dict. Sc. nat. XXII. p. 486.
2. Oratio de Ingratitudine Cuculi. (In *Melanchtonis* Declamationibus, Argent. 1535, 8°. I. p. 643.)—*Böhm.* Bibl. II. 1, p. 559.

AMBRUN (Ant. d'Illy d').

1. Traité de l'âme et de la connaissance des Bêtes, suivant les principes de *Descartes.* La Haye, 1690, 12°. Amst. 1691, 12°. fig. —*Böhm.* Bibl. II. 2, p. 92.

AMELUNG (C. H.).

1. De Funiculi umbilicalis delapsu atque Umbilici formatione. Dorp. 1837, 8°.

AMELUNG (F. A.) u. BIRDTE (F.).

1. Beiträge zur Lehre von den Geisteskrankheiten. I. Bd. Darmst. und Leipz. 1832, 8°.—N. Act. Nat. Cur. XVI. p. xv.

AMICI (G.).

1. Giornale Toscano; vide sup. pars I. p. 70.

AMICI (J. B.).

1. Circulation des Saftes in der Chara.—Soc. Ital. in Modena, XVIII.—Isis, 1822, VI. p. 665.

AMICI (Vit.).

1. Sui Testacei della Sicilia.—Opusc. Sc. VIII.

Amio.

1. Lettre sur les Têtes des Caraibes.—Journ. de Phys. t. 39, 1791, p. 132.

Amman (Paul).

1. Diss. Σιδηροπεψια Struthionis. Lips. 1657, 4°. fig.—*Böhm*. Bibl. II. 2, p. 579.

Ammann (J.).

1. Descriptio Cameli Bactriani binis in dorso tuberibus, e scriptis G. *Messerschmidii* collecta.—Comm. Ac. Petr. 1738, X. p. 326.

Ammon (Fr. A. v.).

1. Die Entwickelungsgeschichte des menschl. Auges, nach eigenen Beobachtungen und Untersuchungen skizziert. 8°.—Isis, 1834, III. p. 326.

2. Eigenthümlichkeit der Choroidea im menschl. Fœtusauge.— Naturf. in Berl. 1828.—Isis, 1829, IV. p. 430.

3. De Physiologiâ tenotomiæ experimentis illustratâ. Dresd. fol.— *Val.* Repert. anat. u. phys. III. p. 30.

4. Die angebornen chirurgischen Krankheiten des Menschen in Abbildungen dargestellt u. durch erläuternden Text erklärt. Berl. 1839, fol.

5. Die angebornen Krankheiten des Auges und der Augenlieder. Berl. 1841, fol.

6. De genesi et usu Maculæ luteæ in Retinâ Oculi humani obviæ. Vinor. 1830, fig.

Amoretti (C.).

1. Viaggio da Milano ai tre Laghi, Maggiore, di Lugano e di Como. Mil. 1801, 8°; 1806, 8°.

2. Su un Dente e parte di Mandibola di un Mastodonte.

3. Osse fossili del Dip^{to} del Panaro.—Att. Istit. d' Ital. Il.

4. Sur le Trap du Mont Simmolo près d'.... sur le Lac Majeur.— Mem. Accad. Tor. XI. pp. 89, 268.

5. Nuova Dottrina della Vitalità. 8°.

Amoreux (N.).

1. Notice sur les Insectes de la France réputés vénimeux. Paris, 1789, 8°. fig.—*Cuv*. R. an. III. p. 331.

2. Déscription méthodique d'une espèce de Scorpion commune à Souvignargues en Languedoc.—Journ. de Phys. XXXV. p. 9.

3. Observations sur une Tortue à cuir, prise dans les parages du port de Cette.—Journ. de Phys. de *Rozier*, II. p. 65.

4. Observations sur des Fossiles marins des environs d'Aubai en Languedoc.—J. de Phys. XXIII. p. 350.—*Mod.* Bibl. Helm.

AMOUREUX (P. J.).

1. Dissertatio de Noxiâ Animalium. Avon. 1762, 4°.—*Böhm.* Bibl. II. 1, p. 147.

2. Revue de l'Histoire de Licorne, par un Naturaliste de Montpellier. 8°. Paris, 1818.

AMOZETTI (C.).

1. Scelta d' Opusc. interessanti ; vide sup. pars I. p. 67.

AMPÈRE ( ).

1. Rapport sur un Mémoire de *M. de Meyraux*, relatif aux Mollusques Céphalopodes. Instit. 1833, I. No. 21.

2. Essai sur la Philosophie des Sciences, ou Exposition analytique d'une Classification naturelle de toutes les Connaissances humaines. Par. 1834, 8°.—Naturliches System aller Naturwissenschaften, aus dem Franz. v. *G. Widenmann.* 8°. Stuttgart, 1844. —Bull. Soc. Géol. VI. p. 62.

AMSTEIN (J. G.).

1. Geschichte des Fichtenspinners.—*Fuessl.* Mag. II. 1779, p. 232. —(Nachtrag) N. Mag. I. p. 44.

2. Spielarten des rothen Augenspiegels (*Pap. Apollo* L.)—*Fuessl.* N. Mag. I. p. 183.

AMUSSAT (J. Z.).

1. Recherches sur l'introduction accidentelle de l'Air dans les Veines, et particulièrement sur cette question : L'air, en s'introduisant spontanément dans une veine blessée par une opération chirurgicale, peut-il causer subitement la mort ? Par. 1839, 8°.

AMYOT (C. J. B.).

1. Moluris Pierreti.—*Guér.* Magas. de Zool. IX. p. 129.

2. *Amuet (C. J. B.)* et *Audinet Serville.*—Histoire naturelle des Insectes Hémiptères. Paris, 1843, 8°. fig.

3. Considérations sur un *Nouveau Système de Nomenclature.*—Rev. Zool. 1838, p. 133.

ANATOLIUS ( ).

1. Fragmentum insigne ineditum περὶ συμπαθειῶν καὶ ἀντιπαθειῶν unà cum Apospasmatio ejusdem Argumenti, sub *Democriti* nomine, quod nunc primùm vulgatur, versione et notis illustr. a *J. Rendtorfio.—Fabr.* Bibl. Græc. IV. c. 29, p. 295.—*Böhm.* Bibl. I. 1, p. 343.

ANCHIETA (Jos. de).

1. Epistola quamplurimarum rerum naturalium quæ St. Vincentii (nunc St. Pauli), provinciam incolunt, sistens descriptionem.— Coll. de Not. para a hist. et geogr. das nacoes ultramarinas, etc. I. p. 127.—*Féruss.* Bull. 1828, XIV. p. 206.

ANCORA (Gaet. d').

1. Ricerche filosofico-critiche sopra alcuni Fossili metallici della Calabria. Livorno, 1791, 8°.—Tr. Linn. Soc. Lond. VII. p. 311.

ANDERMANN (A. A. J.).

1. De Pulmonum formæ necnon voluminis Aberrationibus. Vratisl. 1838, 4°.

ANDERSCH (C. Sam.).

1. Tractatio anatomico-physiologica de Nervis humani corporis aliquibus. Regiom. 1797, 8°. 2 P. fig.—Bibl. med.-ch. p. 11.

ANDERSECK (F. J. Ph.).

1. Exercitatio anatomica circa Monstra duo humana Spinâ bifidâ singulari affectâ. Vratisl. 1842, 4°.— *Val.* Repert. VIII. pp. 18, 42.

ANDERSEN (Joh.).

1. Nachrichten von Island, Grönland und der Strasse Davis.— Hamb. 1746, 8°. fig. Frankf. u. Leipz. 1747, 8°. fig.—Trad. Franç. Paris, 1750, 2 vols. 8°; 1754, 1764, 12°. (*Hist. natur. de l'Islande, du Grœnland, etc.*)—(Dan.) Copenh. 1748, 8°.— (Angl.) Lond. 1758, fol.—(Belg.) Amst. 1756.—*Cuv.* R. an. III. p. 331.—*Böhm.* Bibl. I. 1, p. 769.

ANDERSEN (Jürg.) u. VOLQUARD YVERSEN.

1. Orientalische Reise-beschreibung, mit Erklärung v. *G. Olearius.* Schlesw. 1669, fol. fig.—*Böhm.* Bibl. I. 1, p. 659.

ANDERSON (Ben. Quist).

1. Tal innehällande nägra Anmärkningar öfver Metall- och Mine-

ralvaror, samt deras Affättning. Stockh. 1776, 8°.—*Böhm.* Bibl.
IV. 2, p. 133.

ANDERSON (G.).

1. On the small District of Primitive Rocks near Stromness, in the
Orkney Islands.—Mem. Wern. Soc. IV. p. 173.—*Féruss.* Bull.
1828, XIV. p. 408.

2. Guide to the Highlands and Islands of Scotland.   12°. London,
1834, and Edinb. 1836.

3. Geognostical Sketch of part of the Great Glen of Scotland.—
Mem. Wern. Soc. IV. p. 190.—*Féruss.* Bull. 1828, XIV. p. 408.

4. On the Quartz District in the neighbourhood of Inverness.—
Brewst. Journ. Science, ser. 1, III. p. 212 ; IV. p. 91.—*Féruss.*
Bull. 1826, VIII. p. 322 ; IX. p. 12.

5. On the Bituminous Rock which occurs in Ross-shire and the
neighbourhood of Inverness.—Brewst. Journ. Science, ser. 1, IV.
p. 93.—*Féruss.* Bull. 1826, VIII. p. 322.

ANDERSON (Jam.).

1. Recreations in Agriculture, Nat. History and Arts.   6 vols. 8°.
1799–1802.

2. A Practical Treatise on Peat Moss.   Lond. 1794, 1797, 8°.

3. Letters on Cochineal Insects discovered at Madras.   Madr.
1788–90.—Tr. Linn. Soc. Lond. V. p. 277.—*Licht. u. Voigt* Mag.
VI. 1, p. 24 (*Ueber einige zu Madras entdeckte Cochenille-Arten*).

ANDERSON (John).

1. Sketch of the Comparative Anatomy of the Nervous System ;
with Remarks on its Development in the Human Embryo. Lond.
1837, 4°. fig.— *Val.* Repert. anat. u. phys. III. p. 14.

ANDERSON (J.).

1. Organic Remains in the Old Red Sandstone of Fife.—Edinb.
New Phil. Journ. XXIII. p. 137.

2. On the Strata at the Diamond Mines of Mallivully.—Edinb.
Phil. Journ. III. p. 72.

3. Account of Morne Garou, a mountain in the Island of St. Vincent, with a Description of the Volcano on its Summit.—Phil.
Trans. LXXV. p. 16.

ANDERSON (W.).

1. On some Poisonous Fish in the South Seas.—Phil. Trans. LXVI.
p. 544.—*Cuv. et Val.* Poiss. I. p. 169.

ANDRAL (M. G.).

1. Précis d'Anatomie pathologique. Par. 1829, 3 vols. 8º. (*Grund-riss der pathologischen Anatomie.* Aus dem Franz. v. *F. W. Becker.*) Leipz. 1829, 2 vols. 8º.—Bibl. med.-ch. p. 11.

2. Die Krankheiten des Gehirns. Aus dem Franz. von *Kahler.* 8º. I. u. II. B.— *Val.* Repert. anat. u. phys. III. p. 19.

ANDRAL (M. G.) et GAVARET (   ).

1. Recherches sur les Modifications de Proportion de quelques Principes du Sang dans les Maladies. (Extrait.) Par. 1841, 8º.— Ann. Sc. n. (2ᶜ S.) XIV. p. 361 ; XV. p. 129.

2. Réponse aux principales objections dirigées contre les procédés suivis dans les Analyses du Sang et contre l'exactitude de leurs résultats. Paris, 1842, 8º.

3. Recherches sur la Quantité d'Acide Carbonique exhalé par le Poumon dans l'Espèce humaine. Paris, 1843, 8º. 1 pl. 4º.

ANDRAL, GAVARET et DELAFOND.

1. Recherches sur la Composition du Sang de quelques Animaux domestiques. Par. 1842, 8º.—Ann. Sc. n. (2ᶜ S.) XVIII. p. 213.

ANDRAL (fils).

1. Note sur des Acéphalocystes contenues dans les Veines pulmo-naires.—Bull. Soc. Phil. 1823, p. 15.

ANDRÉ (Ch. K.).

1. Anleitung zum Studium der Mineralogie für Anfänger. Wien, 1804, 8º. fig.

2. Der Mineraloge, oder kompendiöse Bibliothek alles Wissens-würdigen aus dem Gebiete der Mineralogie. Gotha, 1792–94, 2 H. 8º.

3. Deutsch, französisch, technologisches u. naturhistorisches Hand-wörterbuch. 8º. Halle, 1797–1800.

4. Hesperus ; *vide sup.* Pars I. p. 32.

ANDRÉ (L.).

1. Sur un Granit secondaire.—Edinb. Phil. Journ. April 1823, p. 403.—*Féruss.* Bull. p. 6.

ANDRÉ (M.).

1. Sur la Silice pure nouvellement découverte aux environs de Vierzon (Dépt. du Cher).—Bull. Soc. Phil. 1823, p. 19.

ANDRÉ (Noël).

1. Théorie de la Surface actuelle de la Terre.   Par. 1806, 8°.

ANDRÉ (Will.).

1. A Description of the Teeth of the *Anarrhicas lupus*, L., and of those of the *Chætodon nigricans* ; with an Attempt to prove that the Teeth of Cartilaginous Fishes are perpetually renewed.—Phil. Trans. LXXIV. p. 274.—Dict. Sc. Nat. XXII. p. 541.

2. Microscopic description of the Eye of *Monoculus polyphemus*, Lin.—Phil. Trans. LXXII. p. 440.

ANDREA (de).

1. Eruzione del Vesuvio.   Vue de l'Éruption du Vésuve du 22 Octobre 1822. Milan.—*Féruss.* Bull. 1826, VII. p. 15.

ANDREÆ (J. G. R.).

1. Briefe aus der Schweiz nach Hannover geschrieben in dem Jahre 1763.   Zür. u. Winterth. 1776, 4°. fig. — *Klein*, Disp. Echinod. p. VII.—*Böhm.* Bibl. I. 1, p. 543.—Hann. Mag. 1774-76.—*Mod.* Bibl. Helm.

ANDREÆ (Tob.).

1. Bilanx exacta Bilsianæ et Clauderianæ Balsamationis ; quâ ostenditur, *D. Clauderi* inventam Balsamationem non minùs ac veterum longè à Bilsiana differre, etc.   Amstel. 1682, 12°.—Act. Erudit. II. p. 270.

ANDREJEWSKY (Er. St.).

1. De Thermis Aponensibus Commentatio Physiographica.   Berol. 1831, 4°.

ANDRÉOSSY (Comte).

1. Untersuchungen über den See Merzalch, über den Thal der Natronsee, u. über den See Möxis, etc.   Aus d. Franz.   Leipz. u. Gera, 1801, 8°.

2. Mémoire sur les Dépressions de la Surface du Globe.   8°. 1826.

3. Mémoire du Bosphore.   8°. Paris.

ANDRIA (Nic.).

1. Trattato delle Acque Minerali.   Nap. 1775, 8°; 1783, 8°. 2 vols. —*Böhm.* Bibl. V. p. 137.

ANDRY (Nic.).

1. Examen de différens points d'Anatomie, de Chirurgie, de Physique et de Médecine. Paris, 1723, 8°.—Biogr. Un. II. p. 154.

2. Traité sur la Génération des Vers dans le corps de l'Homme, de la nature et des espèces de cette maladie, etc., avec trois Lettres écrites à l'Auteur sur le sujet des Vers, les deux premières par *N. Hartsœcker*, et l'autre par *G. Bagliv.* Par. 1700, 12°. fig.— Amst. 1701, 8°.—Par. 1741, 12°. 2 vols.—*Böhm.* Bibl. II. 2, p. 376. —Act. Erudit. XIX. p. 519. (*Tractatus de generatione Vermium in corpore humano.*)

3. Eclaircissemens sur le Livre de la Génération des Vers, etc. Paris, 1704, 8°. Amst. 1705, 8°.—Biogr. Un. II. p. 154.

4. Epistola ad *Baglivium*, hujusque Responsio de Lumbricis latis (in *Baglivii* Oper. ed. 4ª, p. 687).—*Böhm.* Bibl. II. 2, p. 407.— *Mod.* Bibl. Helm.

5. Gründlicher Unterricht von Erzeugung der Würmer im menschl. Leibe. Leipz. 1716, 8°.—*Mod.* Bibl. Helm.

6. Vers solitaires et autres, dont il est traité dans son Livre de la Génération des Vers, représentés en plusieurs planches. Paris, 1718, 4°.—*Mod.* Bibl. Helm.

ANDRZEJOWSKI (A.).

1. Notice sur quelques Coquilles fossiles de Volhynie et de Podolie. fig.—Bull. Soc. Natur. Mosc. 1830, p. 90.—*Féruss.* Bull. 1831, XXIV. p. 78.—Bull. Soc. Géol. Fr. I. p. 63.

ANFORNI (J. Th.).

1. De Fontium perennium Origine.—De Intestinorum potissimùm tenuium Fabricâ.—De Motu peristaltico Intestinorum, etc. (Theses pro Aggreg.) Taurini, 1767, 8°.

ANGELINI (Bern.).

1. Ascalafi Italiani, con nuova specie. Milano, 1827, 8°.—Bibl. Ital. XLVII.

2. Dei Danni principalmente causati nel 1826 dalla *Noctua Gamma* alla Provincia Veronese.—Bibl. Ital. XLV.

ANGELINUS (Fulv.).

1. De Verme admirando per Nares egresso. Ravennæ, 1610, 4°.

ANGELY (J. L.).

1. De Oculo Organisque Lachrymalibus. 8°. Erlang. 1803.

ANGILERIUS (Bonavent.).

1. Lux magica physica, etc., cœlestium, terrestrium, et inferiorum origo, ordo et subordinatio; XXIV. vol. divisa. Pars I. Venet. 1686, 4°. Pars II. ibid. 1687, 4°.—*Böhm.* Bibl. I. 1, p. *253.*

ANGLADA (J.).

1. Traité des Eaux minérales et des Etablissemens thermaux du Dép^t. des Pyrénées-Orientales. Paris, 1833, 8°. 2 vols. fig.

2. Mémoires pour servir à l'histoire générale des Eaux minérales, sulfureuses, et des Eaux thermales. Paris, 1827–28, 2 vols. 8°.

ANGLICUS (Barth. de Glanvilla).

1. De Proprietatibus Apum. 1573.—*Eiselt.* p. 220.—*Böhm.* Bibl. II. 2, p. 281.

2. De Rerum Proprietatibus libr. XIX.; de Avibus, Aquâ, Terrâ, Lapidibus pretiosis, Arboribus, Herbis. Colon. 1470, 1481, 1482, 1483, fol.—(Holl.) *Van de Proprietäten der Dingen.* Harl. 1484, 1485, fol.—(Angl.) Lond. 1535, fol. etc. etc. (Vid. *Böhm.* Bibl. I. 1, pp. 224, 225.)

ANGUISOLA (Ant.).

1. Historia Unicornu, annexo ejus Compendio Simplicium. Placent. 1578, 4° —*Böhm.* Bibl. II. 1, p. 481.

ANHALT (H.).

1. Ambra à Philosophorum cunis ad aërem et meteora usque velut in exilium relegata, ad Mineralia revocata. Neo-Rupp. 1707, 4°. fig.—*Böhm.* Bibl. IV. 1, p. 463.

2. Ambram à Sordibus Nauci alicujus Philosophi, ficto sibi cognomento, in Cunis, modestiùs repurgatam et Mineralibus restitutam denuò sistit. 1710, 4°.—*Böhm.* Bibl. IV. 1, p. 464.

ANINO (Casim.).

1. Osservazione di due Lacertole acquatiche o Salamandre, uscite del basso ventre di un fanciullo di Tortona. 4°. Torino, 1762.

ANKARCRONA (Theod.).

1. Beskrifning ofver famfingers fisken.—Vet. Acad. Handling. 1740, p. 457.—Dict. Sc. nat. XXII. p. 537.

2. Blennius sinensis descriptus.—Anat. Transalp. I. p. 103.—Dict. Sc. nat. XXII. p. 537.

ANKE ( ).

1. Von der Blutbewegung in den Venen, dem Venenpulse und der Abdominalpulsation. Moskwa, 1835, 8°.—*Müll.* Arch. 1836, p. cxxxv.

ANKER (M. J.).

1. Kurze Darstellung der mineral-geognostischen Gebirgsverhält-nisse der Steyermark. Grätz, 1835, 8°.—Isis, 1835, II. p. 103.— *L.* u. *Br.* 1835, p. 702.

2. Geognostische Charte der Steyermark. 1832, fol.—Isis, 1834, V. p. 527.

3. Kurze Darstellung einer Mineralogie v. Steyermark, oder syste-matische Aufzählung Steyermärkischer Fossilien mit Angabe ihrer Fundörter u. ihrer technologisch-ökonomischer Nutzbarkeit. Grätz, 1809–10, 2 vols. 8°.

4. On the occurrence of Bones in a Coal-Mine near Gratz in Sty-ria.—Trans. Geol. Soc. ser. 2, III. 488.—Proc. Geol. Soc. I. 467.

5. Fossile Knochen in Braunkohle v. Schönegg.—*Leonh.* u. *Br.* Jahrb. 1833, p. 61.

6. Geognosie von Grätz.—Steyerm. Zeitschr. IX.—*L.* u. *Br.* N. Jahrb. 1833, p. 507.

7. Fossile Reste um Grätz.—*L.* u. *Br.* N. Jahrb. 1835, p. 524.

ANNING (Miss Mary).

1. Découverte géologique.—Lond. a. Par. Observ. 1829, p. 26.— *Féruss.* Bull. 1830, XXI. p. 33.

ANNONE (J. J. d').

1. De Balanis fossilibus, præsertim Agri Basileensis.—Act. Helvet. II. p. 242, fig.—(Gall.) *Sur les Glands de mer fossils, et princi-palement sur ceux du territoire de Bâle.*—J. de Phys. I. p. 209.

2. De Petrificatis quibusdam minùs cognitis.—Act. Helvet. IV. p. 275, fig.—Miner. Belustig. V. p. 161, fig.—*Mod.* Bibl. Helm.

3. Epistola de Pisciculis et Vermibus 1758 circa Birsam repertis. —Act. Helvet. IV. p. 301.—*Mod.* Bibl. H.

4. De Piscibus, in quorum Abdomine Vermes inclusi erant.—Act. Helvet. IV. p. 301.

5. De Cancris lapidefactis Musei sui.—Act. Helvet. III. p. 265.

ANOISOF (G.).

1. Observations géognostiques sur les Monts Ourals situés dans le district de Zlatooustof Gornoy.—Bull. Natur. Mosc. 1826, p. 3.

(Russ.)—Journ. d. Mines, 1826, V. p. 3.—*Féruss.* Bull. 1828, XIII. p. 301.

ANONYMOUS.

1. Descriptiones quorundam Capensium Insectorum. Erl. 1787, 4°. fig. col.
2. Monita quædam de speciebus nigris Ichneumonum. Wratisl. 1829, 4°.
3. Guide du Naturaliste. Brux. 1792, 8°.
4. Tentyriæ et Opatra Collectionis Stevenianæ. 4°.
5. Calendario Entomologico, ossia osservaz. sulle stagione proprie all' insetti nel clima piemontese. Torino, 1791, 12°.
6. Mémoire sur la chasse aux Coléoptères et sur la manière de les conserver. Par. 1833, 8°. fig.

ANSCHEL ( ).

1. Anfangsgründe der Naturwissenschaft. Mainz, 1801.

ANSCHÜTZ (J. M.).

1. Ueber die Gebirgs- und Steinarten des Chursächs. Hennebergs, nebst einer allgemeinen Uebersicht aller bis jetzt bekannten Mineralien dieses Landes u. einem Anhange vom Schneekopfe u. Rupberg. Leipz. 1788, 8°.
2. Berichtigungen u. Zusätze zu der Schrift über die Gebirgs- u. Steinarten, etc., nebst einem neuen nach *Werner*'schen Systeme geordneten Verzeichnisse der Mineralien dieses Landes. Leipz. 1798, 8°.

ANSLIJN (N.).

1. Systematische Beschryving der voor ons meest belangryke Voortbrengselen uit de drie Rijken der Natuur. (*Déscription systématique des Productions indigènes tirées des trois Règnes de la Nature. Mammifères.*) 13 livr. 8°. Leyd. 1829.—*Féruss.* Bull. 1826, VII. p. 24.
2. Catalogue des Insectes des Pays-Bas qui se trouvent la plupart dans les environs de Harlem.— Nat. Verh. Maatsch. d. Vet. Harl. XVI. 1, 1828, p. 125.—*Féruss.* Bull. 1828, XV. p. 189.
3. Liste supplémentaire des Insectes des Pays-Bas, trouvés pour la plupart dans les environs de Harlem.—Nat. Verh. d. Vet. XVII. 2.—*Féruss.* Bull. 1830, XXI. p. 168.
4. Afbeeldingen van Nederlandsche Dieren. 8°. Leyden, 1820–1830.

ANSON (G.).

1. A Voyage round the World in the years 1740–1744. Lond. 1748, 4º. fig.; *ibid.* 8º.—Edinb. 1776, 8º. 2 vols.—(Deutsch.) Gött. 1749, 4º; 1763, 8º.—(Franç.) Amst. 1749, 4º; Par. 1750, 12º. 3 vols.—Genève, 1750, 4º.—(Belg.) *Reize-ronds om de Wereldt*; Amst.—*Echt Verhaal. der Reistogt ronds om den Aardkloot*; Delft.—*Böhm.* Bibl. I. 1, p. 464.

ANSPACH (C. A.).

1. Geschichte und Beschreibung von Neufoundland und der Küste Labrador. Aus d. Engl. Weimar, 1822, 8º.—Cat. Bibl. Turic.

ANSPACH (Margrave of).

1. On some remarkable Caves in the Principality of Bayreuth, and on the Fossil Bones found therein.—Phil. Trans. LXXXIV. p. 402.

ANSTED (D. T.).

1. On the Carboniferous and Transition Rocks of Bohemia.—Proc. Geol. Soc. III. 167.—Phil. Mag. ser. 3, XVII. p. 226.

2. On the Zoological Condition of Chalk Flints, and the probable causes of the Deposit of Flinty Strata alternating with the Upper Beds of the Cretaceous Formation.—Ann. and Mag. N. Hist. XIII. p. 241.—*L. u. Br.* N. Jahrb. 1844, p. 617. (*Ueber das Zoologische Verhältniss der Kreide-Feuersteine, etc.*)

3. Geology, introductory, descriptive and practical, with numerous Illustrations, comprising Diagrams, Fossils and Geological Localities. Lond. 1844, 8º. 2 vols. fig.—*L. u. Br.* N. Jahrb. 1844, p. 868.

4. Geologists' Text Book. 12º. London, 1845.

5. On a new Genus of Fossil Multilocular Shells found in the Slate Rocks of Cornwall.—Trans. Cambr. Phil. Soc. VI. p. 415.

6. On a Portion of the Tertiary Formations of Switzerland.—Trans. Cambr. Phil. Soc. VI. p. 141.

7. The Ancient World, or Picturesque Sketches of Creation. 8º. London, 1847.

ANSTICE (R.).

1. On the Discovery of some Fossil Bones.—Trans. Geol. Soc. ser. 1, V. p. 611.

2. Ein Stück Arragonit von den Quantock-Hügeln.—Geol. Soc.— Isis, 1818, IV. p. 594.

ANTELME (Adr.).

1. Galérie zoologique, ou Exposé analytique et synthétique de l'Histoire des Animaux. Sous la direction de *Geoff. St. Hilaire.* 2 vols. 12º. Paris, 1837-38.

2. Hist. nat. des Insectes et des Mollusques. 2 vols. 8º. Paris, 1841– 2

ANTHOINE (D').

1. Cynipédologie du Chene Roure (*Quercus Robur*).—Journ. de Phys. XLIV. pp. 34, 391.

ANTHONY (J. G.).

1. Angeborner Bruch der Baucheingeweide durch das Zwergfell. —Journ. Hebd. 1835, No. 8.—*Müll.* Arch. 1836, VI. p. cxciii.

2. Fossil Encrinite.—Amer. Journ. XXXV. 2, p. 359.

3. On the Byssus of *Unio* ; with Notes by *J. E. Gray.*—Ann. and Mag. N. H. VI. p. 77.

4. Description of a new Fossil (*Calymene Bucklandii*).—West. Acad. Jan. 1839.—*Sill.* Am. Journ. XXXVI. p. 106, fig.

ANTIO (D').

1. Moyen simple de dessécher les Larves pour les conserver dans des Collections entomologiques à côté des Insectes qu'elles produisent.—*Rozier.* Journ. de Phys. XXVI. 1785, p. 241.—*Lichtenb.* Mag. III. 2, p. 81. (*Ueber ein einfaches Mittel die Larven der Insekten zu trocknen, um sie in entomologischen Sammlungen aufzubewahren.*)

ANTOMARCHI (Fr.).

1. Sur la Communication des Vaisseaux Lymphatiques avec les Veines.—Ann. Sc. N. 1829, XVIII. ; Rev. Bibl. pp. 108, 125.

2. Prodroma della grande Anatomia di *Mascagni.* 1819.—Isis, 1821, V., Litt. A. p. 268.

3. Mémoire sur un cas de Monstruosité produit par l'espèce Brebis et du genre *Synotus* ; monstre à deux corps portant une seule face et quatre oreilles.—Ann. Sc. Nat. 1828, XIV. p. 395.

4. Planches Anatomiques du Corps Humain, exécutées d'après les dimensions naturelles. Paris, 1823, fol.

ANTON (H. E.).

1. Diagnosen einiger neuen Conchylien-Arten.—*Wiegm.* Arch. 1837, I. p. 281.

2. Verzeichniss der Konchylien, welche sich in seiner Sammlung befinden. Halle, 1839, 4°.—*L.* u. *Br.* N. Jahrb. 1839, p. 234.

ANVITY ( ).

1. Observation sur des Vers sortis par le Canal de l'Urèthre.—J. de Phys. XIII. p. 379.—*Mod.* Bibl. Helm.

ANZOUX (L.).

1. Lecons élémentaires d'Anatomie et de Physiologie, ou Déscription succincte des Phénomènes physiques de la Vie dans l'Homme et les différentes classes d'Animaux, à l'aide de l'Anatomie classique. Par. 1839, 8°.

APEL (G.).

1. Leitfaden zum Unterricht in der Naturgeschichte für Schullehrer-Seminarien, etc. Magdeb. 1836, 8°.

AQUÆUS (Steph.).

1. In omnes *Plinii* nat. hist. argutissimi Scriptoris libros Commentaria. Par. 1533, fol.—*Böhm.* Bibl. I. 1, p. 207.

AQUINO (Th. de).

1. Tractatus Sextus de Esse et Essentiâ Mineralium.—Colon. 1592, 8°.

ARAGO ( ).

1. Instructions concernant la Physique du Globe, redigées pour le Voyage de la Bonite.—Compt. Rend. 1835, I. p. 380.

2. Instructions concernant la Météorologie et la Physique du Globe, pour la Commission chargé de l'Exploration scientifique de l'Algérie.—Compt. Rend. VII. p. 206.

3. Liste des Volcans actuellement enflammés.—Ann. Bur. des Longit. 1824, p. 108.—*Féruss.* Bull. 1827, X. p. 47.

4. Recherches à entreprendre pour découvrir la Cause de la Chaleur des Sources thermales de Sextius, à Aix en Provence.—Compt. Rend. 1835, I. p. 445.

5. Unterhaltungen aus dem Gebiete der Naturkunde. Aus d. Franz. v. *K. v. Remy.* 8°. Stuttgart, 1837-44.

6. Ueber den thermometrischen Zustand der Erdkugel.—Ed. N.

Phil. J. 1834, XVI. pp. xxxii, 205.—*L.* u. *Br.* N. Jahrb. 1835, p. 564.

7. On Springs, Artesian Wells and Spouting Fountains.—Edinb. New Phil. Journ. XVIII. p. 205.—*L.* u. *Br.* N. Jahrb. 1836, p. 90.

ARAGO et DE BLAINVILLE.

1. Rapport sur les Résultats scientifiques du Voyage autour du Monde de la Frégate *la Venus.*—Comptes Rendus, XI. p. 298.— Rev. Zool. 1840, p. 253.

ARAGONA (

1. De quibusdam Coleopteris Italiæ Tentamen. Tic. Reg. 1830, 8º.

ARANJO Y ASCARRAGA (B. L.).

1. Dissertacion zoologica sobre la Existencia del Hippopotamo. 4º. Madrid, 1749.

ARANTIUS (J. Cæs.).

1. Anatomicarum Observationum liber (avec 2 autres ouvr. de Médecine). Lugd. 1580, 8º.—Lugd. Bat. 1639, 1641, 12º.—Biogr. Un. II. p. 355.
2. De Humano Fœtu liber. Venet. 1571, 8º ; 1587 et 1595, 4º.— Basil. 1579, 8º.—Lugd. Bat. 1664, 12º.—Biogr. Un. II. p. 355.

ARBUTHNOT ( ).

1. Natural History of those Fishes, etc. that are indigenous to, or occasionally frequent the Coasts of Buchan, etc. Aberdeen, 1815, 8º.

ARCANGELI (C.).

1. Saggio di una Introduzione alla Fisica del Corpo umano. Firenze, 1839, 8º.

ARCÈRE (L. E. d').

1. Histoire naturalle du Pays d'Aunis, de ses cotes et des provinces limitrophes. Paris, 1757, 4º.—*Böhm.* Bibl. I. 1, p. 509.

ARCET (J. d').

1. Notice ou Précis sur la Mine de Sel gemme de Vic, dép. de la

Meurthe, et sur les principales Mines de Sel d'Europe, suivi du
Rapport fait à l'Académie Roy. des Sciences, au nom d'une Commission.   Paris, 1824, 8°.—*Féruss.* Bull. 1824, II. p. 115.

2. Discours sur l'état actuel des Montagnes des Pyrénées et sur les
causes de leur dégradation, etc. Par. 1776, 8°.—(Allem.) *Abhandlung über die Pyrenäischen Gebirge*; Berl. 1779, 8°.—*Böhm.*
Bibl. I. 1, p. 504.

ARCHIAC (d').

1. Versuch über die Coordination der Tertiär-Gebirge von Nord-
Frankreich, Belgien u. England.—*L.* u. *Br.* N. Jahrb. 1839,
p. 631, fig.

2. Discours sur l'ensemble des Phénomènes qui se sont manifestés
à la Surface du Globe depuis son origine jusqu'à l'époque actuelle.
Par. 1840, 4°.

3. Notice sur le genre *Murchisonia.*—Bull. Soc. Géol. XII.—Ann.
and Mag. N. H. IX. p. 278.

4. Mémoire sur la Formation Crétacée du Sud-ouest de la France,
4°. Paris, 1843.

5. Études sur la Formation Crétacée des Versans S.O. et N.O. du
Plateau central de la France.   Paris, 1843, 8°. I.

ARCHIAC (d') et VERNEUIL (Ed. de).

1. Memoir on the Fossils of the Older Deposits in the Rhenish Provinces, etc. Par. 1842, 4°. fig.—Tr. Geol. S. Lond. VI. p. 303,
fig.—*L.* u. *Br.* N. Jahrb. 1843, p. 624 (*Die Fossil-Reste der ältern
Ablagerungen in den Rhein-Provinzen,* etc.).

2. On the Fossils of the Older Deposits in the Rhenish Provinces;
preceded by a General Survey of the Fauna of the Palæozoic
Rocks, and followed by a Tabular List of the Organic Remains
of the Devonian System in Europe.—Trans. Geol. Soc. ser. 2,
VI. p. 303.

ARCONS (Cés. d').

1. Sur les Mines métalliques de France.—(Du Flux et du Reflux
de la Mer, 1667, 4°.)—*Böhm.* Bibl. IV. 1, p. 97.

ARDERON (W.).

1. On the Precipices or Cliffs on the North-East Sea-Coast of the
County of Norfolk.—Phil. Trans. XLIV. p. 275.

2. On the Perpendicular Ascent of Eels.—Phil. Trans. XLIV. p. 395.

3. On the Bansticle or Prickle-back, and also on Fish in general.—
Phil. Trans. XLIV. p. 424.

4. On the Formation of Pebbles.—Phil. Trans. XLIV. p. 467.

5. On the Hearing of Fish.—Phil. Trans. XLV. p. 149.

6. A Supposition how the White Matter is produced that floats about the Air in Autumn.—Phil. Trans. 1747, p. 428.

ARDIZZONI (Fabr.).

1. Discorse sopra l'essenza, cosa e effetti delle Acque minerali, singolarmente del Monte di Corsema. Genova, 1680, 4°.—*Böhm.* Bibl. V. p. 126.

ARDUINI (J.).

1. Ossa Fossili di Coccodrillo.—Giorn. di Grisell. I. p. 204.

2. Saggio fisico-mineralogico di Lithogonia ed Orognosia. Pad. 1774, 4°.—*Böhm.* Bibl. IV. 1, p. 77.

3. Raccolta di Memorie chimico-mineralogiche, metallurgiche e orittografiche, tratte dal Giorn. d'Italia. Venez. 1775, 8°.— (Germ. *A. C. v. Ferber.*) Dresd. 1778, 8°. fig.—*Böhm.* Bibl. IV. 1, p. 77.

4. Notizie sopra una Sorgente di Acqua acidula medicinale scoperta nei Monti di Arzignoro. Pad. 1775, 12°.—*Böhm.* Bibl. V. p. 244.

5. Memoria epistolare sopra varie produzioni vulcaniche minerali e fossili. 12°. Venezia, 1782.

ARENDES (Chr. L.).

1. De Dracone et Basilisco. Helmst. 1661, 4°. fig.—Halberst. 1670, 4°.—*Böhm.* Bibl. II. 2, p. 532.

ARENDS (Isid.).

1. Ostfriesland in geognostischer u. besonders landwirthschaftlicher Hinsicht. 1818, 8°.—*Ballenst.* Arch. I. 9, p. 216.

ARENDT (Ed.).

1. De capitis ossei *Esocis Lucii* structurâ regulari. Regiom. 1822, 4°. fig.

ARENS (J.).

1. Beskrivelse over Stavanger-Amt i Norge. Copenh. 1779, 8°.— *Böhm.* Bibl. I. 1, p. 615.

ARENSWALD (C. F. v.).

1. Galanterie-Mineralogie, und Vorschläge zur Naturwissenschaft für die Damen in sieben Unterhaltungen. Halle, 1780, 8°.— *Böhm.* Bibl. IV. 1, p. 82.

2. Geschichte der Pommerischen u. Mecklenburgischen Versteinerungen.—Naturf. V. p. 145; VIII. p. 224.

ARETÆUS ( ).

1. Zoologia, seu Animalium contemplatio physica. 8°. Tiguri, 1709.

ARETIUS (Ben.).

1. Stockhornii et Nessi in Bernatium Helveticorum ditione Montium, et nascentium in iis Stirpium Descriptio. Tig. 1561, fol.—
Böhm. Bibl. I. 1, p. 548.

ARGENVILLE (A. J. Desallier d').

1. Histoire Naturelle éclaircie dans une de ses principales parties,
la Conchyliologie. Paris, 1742, 4°; 2e éd. augm. de la Zoomorphose, 1757; 3e éd., augm. par *Favanne,* 1772–1780, 3 vols.
(Germ. *Martini*): *Natürliche Geschichte der Conchylien, vermehrt.*
Nürnb. 1767, 4°.—*Mod.* Bibl. Helm.

2. L'Histoire Naturelle éclaircie dans deux de ses parties principales,
la Lithologie et la Conchyliologie. Par. 1742, 4°. fig.; 2e éd.
(augmentée de la Zoomorphose), 1757, 4°; 3e éd. (par *MM. de
Favanne,* père et fils), La Conchyliologie, ou Histoire Naturelle
des Coquilles de mer, d'eau douce, terrestres et fossiles. Paris,
1780, 2 vols. 4°.—(Germ.) *Conchyliologie, oder Abhandl. v.
Schnecken, Muscheln u. dgl. nebst der Zoomorphose,* etc. Wien,
1772, fol. 2 vols.—*Böhm.* Bibl. II. 2, p. 441.

3. L'Histoire Naturelle éclaircie dans une de ses parties principales,
l'Oryctologie, qui traite des Pierres, Métaux, Minéraux et autres
Fossiles, etc. Par. 1755, 4°. fig.—*Böhm.* Bibl. IV. 1, p. 48.—
Mod. Bibl. Helm.

4. Enumerationis Fossilium quæ in omnibus Galliæ Provinciis reperiuntur, Tentamina. Par. 1751, 8°. fig.—*Böhm.* Bibl. IV. 1,
p. 98.

ARGILLANDER ( ).

1. Fécondation du Brochet.—Mém. Stockh. 1753.—*Cuv.* et *Val.*
Poiss. I. p. 168.

ARIÆMONTANUS (Ben.).

1. Naturæ Historia. Antwerp, 1601, 4°. fig.—*Böhm.* Bibl. I. 1,
p. 237.

ARINGHI (Paul.).

1. Roma Subterranea. Lut. Par. 1659, fol. 2 vols.; *ibid.* 1651, fol.
—*Böhm.* Bibl. I. 1, p. 568; IV. 1, p. 111.

ARIOSTUS (Fr.).

1. De Oleo Montis Zibinii, s. Petroleo Agri Mutinensis. Hafn.
1690, 8°.—Tr. Linn. Soc. Lond. V. p. 277.—*Böhm.* Bibl. IV. 1,
p. 479.

ARISTOTELES.

1. Περι ζωων ιστοριας (Test. το Δ Κεφ. δ.).—Tr. Linn. Soc. Lond.
VII. p. 216.

2. Historia Animalium, lib. X. Paris, 1533, fol.—Francf. 1587, 4°.
—Tolosæ, 1619, fol.

3. Histoire des Animaux, éd. de *Scaliger,* Toulouse, 1619.—Éd. de
Schneider, Leipz. 1811, 4 vols. 8°.—Éd. de *Camus,* Paris, 1783,
2 vols. 4°, avec la trad. Fr.—Trad. All. de *Strack,* Francf. 1816, 8°.
—*Cuv.* et *Val.* Poiss. I. p. 18.

4. De mirabilibus Auscultationibus; ed. de *Beckmann.* Gœtting.
1786, 4°.—*Cuv.* et *Val.* Poiss. I. p. 18.

5. Opera omnia quæ extant, Græcè et Latinè, auct. *G. Duval.* Lut.
Par. 1619, fol. 2 vols.—Cat. Bibl. Turic.

6. Ueber die wissenschaftliche Behandlungsart der Naturkunde
uberhaupt, vorzüglich aber der Thierkunde; Griech. u. Deutsch
mit Anmerk. v. *Titze.* Leipz. 1823, 8°.

7. Works translated by *T. Taylor,* 10 vols. 4°. London, 1812.

ARMSTRONG (John).

1. The History of the Island of Minorca. Lond. 1752, 8°. ed. 2,
1756.—(Franç.) *Histoire Naturelle et Civile de l'Ile de Minorque,
trad. de l'Angl.* Par. 1769, 12°. fig.—(Allem.) Ein Auszug mit
Zusätzen; Hamb. 1771, 8°.—*Böhm.* Bibl. I. 1, p. 503.

ARNAUD DE NOBLEVILLE (L. D.).

1. Ædonologie, ou Traité du Rossignol. 12°. Paris, 1751; éd. 2,
1773.

2. Histoire Naturelle des Animaux, pour servir de continuation à
la Matière médicale de Geoffroy. Paris, 1756, 1757, 6 vols. in
12°.—*Dum.* et *Bib.* I. p. 304.

ARNAULD (G. d').

1. Anatomische u. chirurgische Abhandlung v. Hermaphroditen.
Aus dem Franz. Strasb. 1777, 4°. fig.—Bibl. Méd.-ch. p. 17.

ARNBERG (J. A.).

1. Diss. Entomologica de Hemipteris maxillosis capensibus. 4°.
Upsal, 1822.

ARND (J.).

1. Von Erzeugung der Metalle in libro Naturæ. Francf. 1723, 8°. —*Böhm.* Bibl. IV. 1, p. 165.

ARNEMANN (J.).

1. Versuche über die Regeneration an lebenden Thieren. Gött. 1787, 2 Th. fig. 8°.
2. Versuche über das Gehirn und Rückenmark. Gött. 1787, 8°. fig.
3. Ueber die Reproduktion der Nerven. Gött. 1786, 8°.

ARNETH (H.).

1. Die menschliche Stimme und der Einfluss des Gesanges auf die Athmungsorgane, etc. Wien, 1842, 12°.

ARNOLD (A. A.).

1. Einige Beobachtungen ueber den Winterschlaf der Thiere. *Müll.* Arch. 1837, I. p. 63.

ARNOLD (Fr.).

1. Lehrbuch der Physiologie des Menschen. ¸Th. I.–III. Zürich, 1836–1842, 8°. 2 vols.—*Val.* Repert. Anat. u. Phys. II. p. 19; VIII. pp. 21, 44.
2. Der Kopftheil des vegetativen Nervensystems beim Menschen, in anatomischer und physiologischer Hinsicht. Heidelb. 1831, 4°. fig.—Isis, 1832, II. p. 211.—*Müll.* Arch. 1837, II. p. 275.
3. Die Gelenke und Bänder des menschlichen Körpers. Berl. 1844, fol.—(Lat.) *Icones Articulorum et Ligamentorum Corporis humani.*
4. Mémoire sur le Ganglion optique, publié par *G. Breschet.* Paris, 1830, 4°.
5. Ueber den Ohrknoten. Heidelb. 1828, 4°. fig.
6. Anatom. u. physiol. Untersuchungen über das Auge des Menschen. Heidelb. 1832, 4°. fig.
7. Icones Nervorum Capitis. Heidelb. 1834, fol. fig.
8. Ueber die Knoten der hintern Wurzel der letzten Spinalnerven. —*Tiedem.* Zeitschr. V. 2, 1835, p. 176.—*Müll.* Arch. 1837, II. p. 291.
9. Ueber den Fontanaschen Canal im Menschenauge. *Tiedem.* Zeitschr. V. 2, p. 181.—*Müll.* Arch. 1837, II. p. 291.
10. Die Erscheinungen u. Gesetze des lebenden menschl. Körpers

im gesunden und kranken Zustande.   Zürich, 1836; 1841, 8º.
2 vols. fig.

11. Tabulæ Anatomicæ, quas ad naturam accuratè descriptas in
lucem edidit.   Turici, fol. 1838, 1839.   Fasc. I. & II.—*Val.*
Repert. V. p. 16.

12. Annotationes Anatomicæ de Velamentis Cerebri et Medullæ
Spinalis.   Turici, 1838, 4º. fig.—*Val.* Repert. IV. p. 17.

13. Bemerkungen über den Bau des Hirns und Rückenmarks, nebst
Beiträgen zur Physiologie des zehnten und eilften Hirnnerven,
mehreren kritischen Mittheilungen, so wie verschiedenen pathologischen und anatomischen Beobachtungen.   Zürich, 1838, 8º.
fig.—*Val.* Repert. IV. p. 17.

ARNOLD (Gabr.).

1. Dissertatio de Zoophytis.   Lips. 1670, 4º. fig.—*Böhm.* Bibl. II.
2, p. 492.

ARNOLD (J. B.).

1. Letters on the Naturalization of Sea-Fishes in a Lake chiefly
supplied with Fresh Water.—Proc. Com. Zool. Soc. Lond. I. p. 126.

ARNOLD (J. W.).

1. Lehrbuch der pathologischen Physiologie des Menschen.   Bd. I.
Zürich, 1836, 8º.—II. Bd. Zür. 1839.

ARNOLD (W.).

1. Die Lehre von der Reflex-Function für Physiologen und Aerzte.
Heidelb. 1842, 8º.

ARON (J. Leo).

1. Dissertatio de Lumbricis.   Lugd. Bat. 1728, 4º.—*Böhm.* Bibl.
II. 2, p. 379.

ARPHEY VILLAFANE (J. de).

1. Quilatador de Oro, Plata y Piedras conforme allas Leyes reales.
Valladol. 1572, 4º.—Madr. 1598, 12º.

ARRAGONA (L.).

1. De quibusdam Insectis Italiæ novis aut rarioribus.   Ticini Reg.
1830, 8º.

ARRAGOSIUS (Guil.).

1. De naturâ et viribus Hydrargyri Epistola (in *Th. Zwingeri*
Fascic. Disputationum).   Basil. 1710.—*Böhm.* Bibl. IV. 2, p. 15.

ARRHENIUS (J. E.).

1. Monographia Cantharidum et Malachiorum Sueciæ. Lund.
1807.

ARROWSMITH (A.).

1. Geological Map of Scotland, constructed from original materials.
fol. Text 8°. Lond. 1840.

ARSAKY (L.).

1. Commentatio de Piscium Cerebro et Medullâ Spinali, scripta
auspiciis et ductu *J. F. Meckelii.* 4°. Halæ. 1813. Denuò edita,
fragmentis de eâdem re additis à *G. G. Mintero.* Lips. 1836, 4°.
fig.—*Val.* Repert. Anat. u. Phys. II. p. 21.—*Müll.* Arch. 1835,
p. 247.

ARTAUD ( ).

1. Notice pour servir à l'Histoire Naturelle du Goramy (*Osphro-
menus olfax, Commers.*), et Rapport sur cette notice par *Duvernoy.*—J. Sc. Agr. et Arts B.-Rhin, 1828, I. p. 117.

2. Organizazione e Funzioni della Mitra.—Ann. un. Medic. V.
p. 134.

3. Manuel d'Histoire Naturelle. Metz, 1803, 2 vols. fig.

ARTEDI (P.).

1. Genera Piscium. Lugd. Bat. 1738, 8°.—Ed. *J. J. Walbaum.* 4°.
Grypheswald, 1792.—*Fisch.* Cat.

2. Descriptiones Specierum Piscium. Lugd. Bat. 1738, 8°.—*Fisch.*
Cat.

3. Ichthyologia, s. opera omnia de Piscibus, edidit *C. Linnæus.*
Lugd. Bat. 1738, 8°. (emend. à *J. J. Walbaum*). Gryph. 1788–
89.—Trans. Linn. Soc. Lond. V. p. 277.—*Böhm.* Bibl. II. 2, p. 55.

4. Synonymia Nominum Piscium. Lugd. Bat. 1738, 8°.—Gryph.
1793.

5. Bibliotheca et Philosophia ichthyologica. Lugd. Bat. 1738, 8°.
—Grypheswald. 1789, 8°. cur. *J. J. Walbaum.*

6. Synonymia Piscium græca et latina, sivè Historia Piscium naturalis et litteraria. Acced. Disputatio de veterum Scriptorum
Hippopotamo. Auct. *Schneider.* Lips. 1789, 4°. fig.

7. Bibliotheca ichthyologica, seu Historia litteraria Ichthyologiæ.
Ed. *C. Linnæo.* 8°. Lugd. Bat. 1738.—Ed. *J. J. Walbaum.* 4°.
Gryph. 1788.

ARTHAUD ( ).

1. De la Bête à mille pieds à St. Domingue.—Journ. de Phys. XXX. 1787, p. 427.

2. Sur les effets de la figure de l'Araignée-Crabe des Antilles.— Journ. de Phys. XXX. 1787, p. 422.

3. Sur la Conformation de la Tête des Caraibes, et sur quelques usages bizarres attribués à des nations sauvages.—Journ. de Phys. XXXIV. 1789, p. 250.

ARTHUR (J. F.).

1. Théorie élémentaire de la Capillarité, suivie de ses principales applications à la Physique, à la Chimie, et aux Corps organisés. Paris, 1842, 8°.

ARTIS (Ed. Tyrell).

1. Antediluvian Phytology, illustrated by a Collection of the Fossil remains of Plants peculiar to the Coal Formations of Great Britain. Lond. 1825, 4°; 1838, 4°. fig.—*Féruss.* Bull. 1829, XVII. p. 189.

ARTUR (F.).

1. Thèse sur la Loi rélative à la Densité des Couches intérieures de la Terre et sur son Applatissement. Par. 1834, 4°.

2. Observations sur l'espèce de Ver nommé Macaque (*Œstrus*). —Mém. Acad. Sc. Par. 1753, Hist. p. 72.

ARVIEUX (L. d').

1. Mémoires concernant ses Voyages à Constantinople, dans l'Asie, la Syrie, la Palestine, l'Egypte et la Barbarie, la déscription et l'hist. nat. de ces pays, etc., recueillis et mis en ordre par *J. B. Labat.* Par. 1735, 6 vols. 12°.—(Trad. Allem.) Kopenh. et Leipz. 1753– 56, 8°. 6 vols.—(Dan.) Copenh. 1759 *et seq.*, 6 vols. 8°.—*Böhm.* Bibl. I. 1, p. 460.

ASCANIUS (P.).

1. *Philine quadripartita* et förut obekant Sjökräk, etc.—Act. Holm. XXXIII. p. 329, fig.—*Mod.* Bibl. Helm.

2. Sur la *Doris frondosa.*—Act. Nidr. V. p. 155, fig. (Beskrivelse over Norsk Södyr *Doris frondosa.*)

3. Descriptio *Aphidis Tremulæ.*—Prodr. Act. Hafn. p. 127.

4. *Icones rerum naturalium,* ou Figures enluminées d'Histoire natu-

relle du Nord. Copenhagen, 1767–1805, 5 cah. fol.—*Cuv.* R. An. III. p. 332.

5. Beskrivelse over en Norsk Sneppe og et Södyr (*Doris frondosa*). —Trondh. Selsk.V. 5, p. 153.—*Reuss*, Repert. Comment. II. p. 84.

6. Of a mountain of Iron Ore at Taberg in Sweden.—Phil. Trans. 1755.

ASCH (G. Th.).

1. Dissertatio inauguralis de primo pare Nervorum Medullæ spinalis. 4°. Gott. 1750.—Cat. R. Soc. L.

ASCH (P. Ern.).

1. Dissertatio de naturâ Spermatis, observationibus microscopicis indagatâ. Gött. 1756, 4°. fig.—*Böhm.* Bibl. II. 1, p. 138.

ASCHERSON (F. M.).

1. Ueber die Hautdrüsen der Frösche.—*Müll.* Arch. 1840, p. 15, fig.

2. Ueber den physiologischen Nutzen der Fettstoffe u. über eine neue, auf deren Mitwirkung begründete und durch mehrere neue Thatsachen unterstützte Theorie der Zellenbildung.—*Müll.* Arch. 1840, p. 44.

3. Ueber die relative Bewegung der Blut u. Lymphkörnchen in den Blutgefässen der Frösche.—*Müll.* Arch. 1837, p. 452.

ASELLI (Gasp.).

1. De Venis lacteis, cum figuris elegantissimis. Milan, 1627, 4°.— Bâle, 1628, 4°.—Leyde, 1640, 4°. etc.—Biogr. Un. II. p. 572.

ASH (St. Georg.).

1. De Puellâ quâdam Hibernicâ, e cujus corpore varia excrevere cornua.—Phil. Trans. Nov. 1685, No. 176, p. 1202.—Act. Erudit. V. p. 487.

ASHBURNER ( ).

1. On Dentition and some coincident disorders. Lond. 1834, 18°.

ASHE ( ).

1. Memoirs of Mammoth and other Bones from Ohio. 8°. Liverpool, 1806.

ASHTON (R. I.).

1. On some Peculiarities observable in the Cornea of the Eyes of certain Insects.—Trans. Entom. Soc. II. p. 253.

2. On the Wings of the Hemiptera.—Trans. Entom. Soc. III. p. 95.

3. On the Metamorphosis of Caterpillars.—Trans. Entom. Soc. III. p. 157.

ASMUS (Herm.).

1. Monstrositates Coleopterorum Rigæ 1835, 8°. fig.—*Val*. Repert. nat. u. phys. I. p. 12.—Rev. Zool. 1840, p. 54.

2. Notice sur les *Apate elongata* et *substriata* Payk.—Ann. Soc. Ent. Fr. V. p. 625.—*Wiegm*. Arch. 1837, p. 304.

3. Ueber die Knochen- u. Schuppen-Reste im Boden Lieflands.— Bull. Ac. Pétersb. 1839, VI. p. 220.—*L*. u. *Br*. N. Jahrb. 1840, p. 738.

ASQUINO (Fab.).

1. Discorso sopra la scoperta della Torba in mancanza de Boschi e del Legname. Udine, 1770, 8°.—*Böhm*. Bibl. IV. 1, p. 200.

ASSMANN (C. G.).

1. Progr. de Geologiæ et Anthropologiæ nexu. Witteb. 1807, 4°.

2. Reise im Riesengebirge, ein geologischer Versuch. Leipz. 1798, 8°. fig.

3. De Fossilibus volutatis, et præcipuè de iis quæ in Wittebergensi Regione inveniuntur. Witteb. 1795, 4°.—Comm. II. ibid. 1801, 4°.

ASSMANN (Fr. Guil.).

1. Prodromus observationum circa ganglion Arnoldi oticum in homine variisque animalibus. 4°. Lips. 1832.

ASSO (I. de)

1. Introductio in Oryctographiam et Zoologiam Aragoniæ. 1 vol. 8°. (Madrid?) 1784.

2. Abhandlung von den Heuschrecken und ihren Vertilgungsmitteln. Aus dem Span. von *O. G. Tychsen*. 8°. Rostoch. 1787.

AST (J. Chr.).

1. De Corporum dispositione ad Morbos. 4°. Hal. 1715.—Cat. R. Soc. L.

ASTERIOS ( ).

1. Ansichten über die neuere Geogenie u. Geognosie.—Isis, 1831, XI. p. 1178.

ASTI (Fel.).

1. Compendio di Notizie circa il Veneno di rabiosi Animali. Mantua, 1778, 4°.—*Böhm.* Bibl. II. 1, p. 312.

ASTRUC (Jean).

1. Mémoires pour servir à l'Histoire naturelle du Languedoc. Paris, 1737; 1749, 4°. fig.—*Dum.* et *Bib.* I. p. 304.—*Böhm.* Bibl. I. 1, p. 510.

2. Mémoire sur les Pétrifications de Boutonnet, près de Montpellier. Montp. 1708.—Comment. Soc. Reg. Monsp. I. p. 48.— (Deutsch übers. in den Mineral. Belustig. II. p. 460.)—*Böhm.* Bibl. IV. 2, p. 258.

ATHENÆUS.

1. Δειπνοσοφισται, sivè Deipnosophistarum libri XV.—Venise, 1514, fol.; *ibid.* (Latinè) 1556.—Bâle, 1535, fol.—Heidelb. 1597, fol.— Lyon, 1583, 1612, 1657, fol.—Genèv. 1597, fol.—Strasb. 1801 à 1807, 4 vols. 8°.—Paris, 1789, 4°, 5 vols. (Trad. Fr. de *Lefebvre.*) —*Böhm.* Bibl. I. 1, p. 216.

ATKINSON (G. C.).

1. On Tortuous Casts of Vermiform Bodies in Sandstone.—Proc. Geol. Soc. III. 126.—Phil. Mag. ser. 3, XV. p. 406.

2. On the Island of St. Kilda, on the north-west coast of Scotland. —Trans. Nat. Hist. Soc. Newcastle, II. p. 215.

ATKINSON (J.).

1. Compendium of the Ornithology of Great Britain. 1 vol. 8°. London, 1820.

ATKINSON (J.) et SANDERSON (Edw.).

1. Déscription d'un Animal fossile non décrit, trouvé dans le Terrain houiller de Yorkshire.—Geol. Soc. Lond.—*Féruss.* Bull. 1826, VIII. p. 23

ATKINSON (St.).

1. Relatio de Metallis in Scotiâ et eorum locis natalibus et fossoribus. (In *Sibbaldi* Nunt. Scot. Brit. p. 10.)—*Böhm.* Bibl. IV. 1, p. 106.

2. The Discovery and History, etc. (*La Découverte et l'Histoire des Mines d'Or d'Ecosse*; MS. de l'an 1619, publié par *G. L. Meason.*)—Ed. J. of Sc. July 1827, p. 174.—*Féruss.* Bull. 1828, XIV. p. 190.

ATTENHOFER (H. L. v.).

1. Lymphatologia, oder Abhandlung über das lymphatische System u. dessen Leiden. Wien, 1808, 8°.—Bibl. Med.-ch. p. 20.

ATTUMONELLI (Mich.).

1. Della Eruzione del Vesuvio nel anno 1779. 8°. Napoli, 1779.

ATWATER (Caleb).

1. Facts and Remarks relating to the Climate Diseases, Geology and Organized Remains of parts of the State of Ohio, etc.— Amer. Journ. XI. 2, p. 224.

2. On the Prairies and Barrens of the West.—Amer. Journ. I. 2, p. 116.

3. Notice of the Scenery, Geology, Mineralogy, etc. of Belmont County, Ohio.—Amer. Journ. I. 3, p. 226.

4. Facts relating to certain parts of the State of Ohio.—*Sill.* Amer. Journ. X. p. 1.—*Féruss.* Bull. VII. p. 183 (*Faits relatifs à une partie de l'Etat de l'Ohio*).

5. Account of ancient Bones and of some fossil Shells found in Ohio.—Amer. Journ. II. 2, p. 242.

AUBÉ (Ch.).

1. Note sur la famille des Psélaphiens.—Ann. Soc. Ent. Fr. II. p. 502.

2. *Lucanus capreolus.*—Ann. Soc. Ent. Fr. p. LXXXI.—Ann. Sc. n. IV. p. 377.

3. Note sur les premiers états de l'*Agrilus viridis.*—Ann. Soc. Ent. Fr. VI. p. 189.

4. Déscription de deux Coléoptères nouveaux, des genres *Ptilium* et *Hister.*—Ann. Soc. Ent. Fr. II. p. 94.

5. Essai sur le genre *Monotoma.*—Ann. Soc. Ent. Fr. VI. p. 453.

6. Considérations sur la Gale et l'Insecte qui la produit. Paris, 1836, 4°.

7. Spécies général des Coléoptères de la collection de M. le Comte Dejean : *Hydrocanthares* et *Gyrinites.* Paris, 1838, 1 vol. 8°. —Rev. Zool. 1838, p. 268.

8. Iconographie des Coléoptères d'Europe (Tome V. de *Dejean* et *Boisduval*). Paris, 1836, 8°. fig. col.

9. Monographia Pselaphiorum, cum Synonymiâ extricatâ. Paris, 1834, 8°. fig.—*Guér.* Mag. Entom. II.

Aubenton (L. J. M. d').—Vide Daubenton.

Aubert (R. P.).

1. Observations sur des Coquillages.—Journ. d. Sav. 1713, p. 543. —*Mod.* Bibl. Helm.

Aubertus (Jac.).

1. De ortu et causis Metallorum contra Chemicos brevis et dilucida Explicatio. Lugd. 1575, 8°.—*Böhm.* Bibl. IV. 1, p. 163.

Auboin (Steph.).

1. Ornithologie, ou Traité des Oiseaux. 18°. Paris, 1831.

2. Mammalogie, ou Traité des Mammifères. 18°. Paris, 1831.

3. Erpetologie et Ichthyologie, ou Traité des Reptiles et des Poissons. 18° Paris, 1831.

4. Entomologie, ou Traité des Insectes. 2 vols. 18°. Paris, 1831.

Aubrey (John).

1. The Natural History and Antiquities of the County of Surrey. 8°. 5 vols. Lond. 1719.—Cat. R. Soc. L.

Aubriet (Cl.).

1. Deux suites de Papillons, d'Oiseaux et de Poissons.—Bibl. de Paris.—Biogr. Un. III. p. 16.

2. Recueil de Coquillages et de Poissons.—Bibl. de Paris.—Biogr. Un. III. p. 16.

Aubuisson de Voisins (J. Fr. d').

1. Traité de Géognosie, ou Exposé des Connaissances actuelles sur la Constitution physique et minérale du Globe terrestre. Strasb. 1819, 2 vols. 8°. fig.—(2ᵉ éd. revue et continuée par *Burat.*) Par. 1828–35 3. vols. 8°. fig.—*Féruss.* Bull. 1828, XV. p. 321 ; 1829, XVI. p. 1.

2. Mémoire sur les Basaltes de la Saxe, accompagné d'observations sur l'Origine des Basaltes en général. Par. 1803, 8°.

Audebert (Germ.).

1. Voyage d'Italie.   Par. 1656, 8°.—*Böhm.* Bibl. I. 1, p. 552.

Audebert (J. B.).

1. Histoire naturelle des Singes et des Makis, peints d'après nature. Paris, 1800, fol. fig.—*Cuv.* R. An. III. p. 332.

2. Hist. nat. et générale des Colibris, Oiseaux-mouches, Jacamars et Promerops. Paris, 1802, 2 vols. 4°. fig. (Continués par *Vieillot.*) —*Cuv.* R. An. III. p. 332.

Audinet Serville (J. G.)

1. Tableau méthodique des Insectes de l'ordre des Orthoptères. 8°. Paris, 1831.

2. Insectes Coléoptères.   8°. Paris, 1831.

3. Déscription du genre *Peirate,* de l'ordre des Hemiptères.— Ann. Sc. Nat. 1831.

4. Sur une Lettre de *Westermann,* sur les Mœurs d'Insectes des Indes orientales et du Cap de Bonne Espérance.   8°. Paris.

5. Revue méthodique des Insectes de l'ordre des Orthoptères.— Ann. Sc. Nat. XXII.

6. Hist. nat. des Insectes Orthopteres.   8°. Paris, 1839.

Audoin de Chaignebrun (H.).

1. Cartes microcosmographiques, ou Déscription du Corps humain. Paris, 1770, 4°.—Biogr. Un. III. p. 27.

Audouin (J. V.).

1. Explication sommaire des Planches du grand Ouvrage d'Égypte relatives aux Anim. sans vert.—Déscript. des Mammifères, conjointément avec *Geoffroy St. Hilaire.*—*Cuv.* R. An. III. p. 333.

2. Explication sommaire des Planches de Reptiles (Supplément) publiées par *J. C. Savigny* dans le grand Ouvrage sur l'Égypte. Paris, fol. et 8°.—*Dum.* et *Bib.* I. p. 305.

3. Recherches sur les rapports naturels qui existent entre les Trilobites et les Animaux articulés.—Ann. Gén. Sc. Phys. VIII. p. 233, fig.—Isis, 1822, I. p. 87.

4. Mémoire sur les rapports naturels qui existent entre les appendices masticateurs et loco-moteurs des Crustacés, et ceux de même nature chez les Insectes apodes et les Arachnides.—*Cuv.* Anal. des Trav. de l'Acad. pendant 1820.—Ann. Sc. n. (2e S.) XVI. p. 372.

AUD

5. Observations sur les phénomènes qui précèdent souvent la Reproduction des Pattes chez certains Crustacés.—Ann. Sc. Ent. Fr. I. p. 238.—Ann. Sc. n. (2e S.) XVI. p. 375.

6. Examen des Crustacés qui habitent les Salines de Marignane.—Compt. Rend. III. p. 545.—Ann. Sc. n. 1836, VI. p. 226.

7. Observations sur un genre nouveau d'Entomostracé bivalve, remarquable par son volume.—Ann. Soc. Ent. Fr. VI. p. ix.—Ann. Sc. n. (2e S.) XVI. p. 376.

8. Mémoire sur l'Achlysie, nouv. genre d'Arachnides trachéennes. Paris, 1823, 4o. fig.—Mém. Soc. d'Hist. Nat. I. p. 98, fig.—Bull. Soc. Philom. 1822, p. 12.—Isis, 1827, IX. p. 751.—*Féruss.* Bull. 1823, IV. p. 46.

9. Observations sur la Structure du Nid de l'Araignée pionnière.—Ann. Soc. Ent. Fr. II. p. 69, fig.

10. Instructions relatives aux Animaux sans Vertèbres, faisant partie du rapport de la Commission chargée de rédiger les instructions pour un voyage de *M. Lefebvre* en Abyssinie et dans les contrées qui avoisinent la Mer Rouge.—Rev. Zool. 1839, p. 35.—Compt. Rend. VIII. p. 160.—Ann. Sc. Nat. sér. 2, XVI. p. 375.

11. Remarques sur la Phosphorescence de quelques Animaux articulés, à l'occasion d'une Lettre de *Mr. Forester* sur la Phosphorescence des Lombrics terrestres.—Ann. Sc. n. (2e S.) XV. p. 253. —Compt. Rend. XI. p. 747.

12. Observations sur les Organes copulateurs mâles des Bourdons. —Ann. génér. Sc. ph. VIII. p. 285.

13. Observations sur le Nid d'une Araignée, construit en terre et remarquable par une grande perfection de travail.—Ann. Soc. Ent. Fr. I.

14. Mémoire sur une Larve de Taupin (*Elater segetis*) qui exerce de grands ravages dans les champs d'Avoine.—Soc. Ent. Fr. 3 Juin 1839. (inédit.)

15. Note sur la Demeure d'une Araignée maçonne originaire de l'Amérique du Sud.—Compt. R. IV. p. 853.—Ann. Sc. n. VII. p. 227.

16. Note sur une nouvelle espèce d'Achlysie.—Ann. Sc. nat. II. p. 497.

17. Recherches sur quelques Araignées parasites des genres Ptéropte, Caris, Argas et Ixode. (Lettre à *M. L. Dufour.*)—Ann. Sc. n. 1832, XXV. p. 401, fig.

18. Observations sur les Podures (*Podura nivalis*, L.), observés à la surface de la neige dans les Alpes.—Ann. Soc. Entom. Fr. V. p. xi.—Ann. Sc. n. (2e Sér.) XVI. p. 376.

19. Note sur un Insecte fossile, découvert dans le terrain houillier.
—Ann. Soc. Entom. Fr. II. p. 17.—Feuillet. du *Temps*, 27 févr.
1833.—Ann. Sc. n. (2ᵉ Sér.) XVI. p. 375.

20. Histoire des Insectes nuisibles à la Vigne. 4°. Paris, 1840–42.

21. Quelques remarques sur le développement excessif de la Lèvre
inférieure dans les Stènes.—Ann. Soc. Entom. Fr. IV. p. 166.—
Ann. Sc. n. (2ᵉ Sér.) XVI. p. 375.

22. Anatomie comparative des parties solides des Insectes.—Ann.
gén. Sc. phys. VII. p. 396.—Isis, 1822, I. p. 80; 1832, I. p. 89.

23. Recherches anatomiques sur le Thorax des Animaux articulés et
celui des Insectes hexapodes en particulier.—Ann. Sc. n. I. p. 97,
416, fig.—*Féruss.* Bull 1824, II. p. 215.—Bull. Soc. Philom. 1820,
p. 72.—Isis, 1824, XI. p. 1140. (*Ueber die Brust der Kerfe.*)

24. Anatomie der festen Theile der Kerfe.—Ann. gén. Sc. phys.
VII. p. 396.—Isis, 1832, I. p. 89.

25. Observations sur le mode singulier d'Accouplement des Cé-
brions.—Ann. Soc. Entom. Fr. II.—Ann. Sc. n. (2ᵉ Sér.) XVI.
p. 375.

26. Lettre à *M. Arago* sur la Génération des Insectes.—Ann. Sc.
nat. II. p. 281.—*Müll.* Arch. 1837, p. 392.—Isis, 1830, VIII.
p. 782. (*Ueber die Zeugung der Insecten.*)

27. Observations sur l'Accouplement entre des individus d'espèces
différentes du genre Coccinelle.—Ann. Soc. Entom. I. p. 232.
Ann. Sc. n. (2ᵉ Sér.) XVI. p. 374.

28. Observations sur les Ecailles des Ailes de la Pyrale de la Vigne
et sur la structure de la Verge de cet insecte.—Ann. Soc. Entom.
Fr. VIII. p. 111.—Ann. Sc. n. (2ᵉ Sér.) XVI. p. 377.

29. Recherches anatomiques sur la femelle du Drile jaunâtre et sur
le mâle de cette espèce.—Ann. Sc. nat. II. p. 443, fig.

30. Observations sur la faculté que possèdent les Callidies de ronger
des corps très-durs.—Ann. Soc. Entom. Fr. II. p. 76.—Ann. Sc.
n. (2ᵉ Sér.) XVI. p. 375.

31. Observations sur la manière de vivre des Larves du *Sitaris
humeralis.*—Ann. Soc. Entom. Fr. IV.—Ann. Sc. n. (2ᵉ Sér.)
XVI. p. 375.

32. Observations sur les Coques construites par divers Insectes qui
subissent leurs métamorphoses dans la terre.—Ann. Soc. Entom.
II. p. 71.—Ann. Sc. n. (2ᵉ Sér.) XVI. p. 375.

33. Observ. sur les métamorphoses d'une Chenille du genre *Dosi-
thea*, et sur les habitudes d'une Larve d'Ichneumon qui it à ses
dépens.—Ann. Soc. Entom. III. p. 417, fig.—*Wiegm.* Arch. 1835,
II. p. 60.

84. Observations sur l'*Artemia salina.*—Compt. Rend. IX. p. 570.
—Ann. Sc. n. (2ᵉ Sér.) XVI. p. 377.

35. Observations sur le Vol des Cétoines.—Ann. Soc. Entom. VIII.
p. xlviii.—Ann. Sc. n. (2ᵉ Sér.) XVI. p. 377.

36. Observations sur les mœurs des Odynères. (Lettre à *M. Léon
Dufour.*)—Ann. Sc. n. (2ᵉ Sér.) XI. p. 104, fig.; XVI. p. 377.

37. Observations sur un Insecte coléoptère qui passe une grande
partie de sa vie sous la mer.—N. Ann. Mus. III. p. 117.— *Wiegm.*
Arch. 1835, II. p. 25 (*Blemus fulvescens*).

38. Note additionelle au Mémoire sur le *Blemus.*—Ann. Sc. n.
1835, (2ᵉ Sér.) III. p. 33.

39. *Meloë collegialis.*— *Guér.* Mag. Zool. 1836.—*Erichs.* in *Wiegm.*
Arch. 1837, VI. p. 302.

40. Prodrome d'une Histoire naturelle chimique, etc. des Cantharides. Paris, 1826, 4º.—*Cuv.* R. An. III. p. 333.

41. Recherches pour servir à l'Histoire naturelle des Cantharides.
—Ann. Sc. nat. IX. p. 31, fig.—*Féruss.* Bull. 1827, XII. p. 175.

42. Quelques observations sur le Parasitisme des Insectes.—Congr.
Sc. de Pise, 7 Oct. 1839.—Rev. Zool. 1840, p. 28.

43. Lettre sur la Génération des Insectes. Paris, 1824, 4º. et 8º.
—Ann. Sc. n. 1824.

44. Note sur la distribution qui a été faite, à divers Éducateurs, de
la variété de Ver-à-soie dite *Trevoltini,* envoyée à l'Acad. par
M. Bonafous.—C. R. VIII. p. 953.

45. Calculs trouvés dans les Canaux biliaires d'un Cerf-volant femelle (*Lucanus capreolus*).—Ann. Sc. n. 1836.

46. *Cicindela quadrimaculata.*— *Guér.* Mag. Zool. fig.

47. Remarques sur la Phosphorescence de quelques Animaux articulés.—Compt. Rend. II. p. 747.—Ann. Sc. n. (2ᵉ Sér.) XVI.
p. 378.—Rev. Zool. 1840, p. 345.

48. Analyse de deux Calculs d'acide urique, trouvés dans les canaux
biliaires des Insectes.—Ann. Sc. n. (2ᵉ Sér.) V. p. 129.—*Müll.*
Arch. 1836, p. cxxxi.

49. Sur une Éducation, faite à Paris, d'un Ver-à-soie de la Louisiane
(le *Bombyx Cecropia*).—Compt. Rend. II. p. 96.—Ann. Sc. n.
(2ᵉ Sér.) XVI. p. 378.

50. Recherches anatomiques et physiologiques sur la Maladie contagieuse qui attaque les Vers-a soie, et qu'on désigne sous le nom
de *Muscardine.*—Ann. Sc. n. 1837, VIII. p. 229.—Compt. Rend.
III. p. 82.

51. Remarques sur la Cochenille du Nopal.—Compt. Rend. IX.

p. 69.—Ann. Soc. Entom. VIII. p. xlvi.—Ann. Sc. n. (2e Sér.)
XVI. p. 377.

52. Exposé sommaire de diverses observations recueillies pendant
plusieurs années sur les Insectes nuisibles à l'Agriculture.—Ann.
Sc. n. (2e Sér.) IX. p. 54 ; XVI. p. 377.—Compt. Rend. VI.
p. 138.

53. Observations sur certains Insectes qui attaquent les Bois employés dans les Constructions.—Ann. Sc. n. (2e Sér.) XIV. p. 39 ;
XVI. p. 378.—Rev. Zool. 1840, p. 150.—Compt. Rend. X. p. 689.

54. Observations sur les Dégâts occasionnés par le *Ptinus fur* dans
les Farines conservées en magasin.—Ann. Soc. Entom. (Bull.)
VI. p. ix.—*Erichs.* in *Wiegm.* Arch. 1837, VI. p. 295 (*Ueber den
Ptinus fur, der in den Mehlmagazinen zu Versailles in grosser
Menge vorgekommen*).

55. Recherches sur la Cause de certaines Fissures, qu'on remarque
sur les tiges des Poiriers et qu'on attribue à la gelée.—Ann. Soc.
Entom. Fr. V. p. 68.—Ann. Sc. n. (2e Sér.) XVI. p. 374.

56. Remarques sur les Dégâts occasionnés aux Ormes de nos routes
par les Insectes.—*Loud.* Arboretum, etc. p. 1387.—Ann. Sc. n.
(2e Sér.) XVI. p. 377.

57. Observations sur la manière dont les Scolytes nuisent aux
arbres forestiers.—Ann. Soc. Entom. VI. p. ii.—Ann. Sc. n. (2e
Sér.) XVI. p. 377.

58. Histoire des Insectes nuisibles à la Vigne, et particulierement de
la Pyrale.   Paris, 1840–42, 4°. atlas.—Ann. Sc. n. (2e Ser.) XVI.
p. 278.

59. Notice sur les Ravages causés dans quelques cantons du Mâcon-
nais par la Pyrale de la Vigne, et sur les moyens qui ont été jugés
les plus convenables pour arrêter ce fléau.—Ann. Sc. n. 1837,
VIII. p. 5.—Paris, 1837, 8°.—C. R. V. p. 384.

60. Considérations nouvelles sur les Dégâts occasionnés par la Pyrale
de la Vigne, particulièrement dans la commune d'Argenteui.—
Ann. Sc. n. 1837, VIII. p. 65.—Paris, 1837, 8°.—C. R. V. p. 471.

61. Observations sur les altérations que produit le Puceron lanigère
sur les Pommiers.—Ann. Soc. Entom. Fr. V. p. 9.—Ann. Sc. n.
(2e Sér.) XVI. p. 376.

62. Observations sur les Insectes qui, depuis plusieurs années, dé-
vastent le bois de Vincennes.—Ann. Soc. Entom. Fr. V. p. xv.
—Ann. Sc. n. (2e Sér.) XVI. p. 376.

63. Observations sur le dépérissement de plusieurs Chênes, qui a
eu pour cause la piqûre faite à l'écorce par des milliers d'insectes
du genre *Coccus*.—Ann. Soc. Entom. Fr. V. p. xxix.—Ann. Sc.
n. (2e Sér.) p. 376.

64. Observations relatives à plusieurs Mollusques inconnus, ou jusqu'ici incomplètement décrits.—*Féruss.* Bull. 1829, XVIII. p. 292.

65. Observations pour servir à l'histoire de la formation des Perles. —Mém. Mus. XVII. p. 176, fig.—*Féruss.* Bull. 1829, XVIII. p. 297.

66. Monographie des Térébratules.—Ann. Sc. n. 1829, XVI.; Rev. Bibl. p. 47.

67. Mémoire sur l'animal de la Glycimère et sur l'anatomie de ce Mollusque.—Ann. Sc. n. XXVIII. p. 331, fig.—*Féruss.* Bull. 1829, XVII. p. 136.

68. Observations sur l'animal de la Siliquaire.—Ann. Sc. n. 1829, XVI.; Rev. Bibl. p. 13.—*Féruss.* Bull. XVII. p. 310.

69. Observations sur un Mollusque de la Méditerranée, qui se rapproche beaucoup des Clavagelles.—Ann. Sc. n. 1829, XVII.; Rev. Bibl. p. 78.

70. Emarginula u. Siphonaria. (*Sav.* Égypte, XXII.)—Isis, 1832, VI. p. 670, fig.

71. Nouvelles expériences sur l'origine et le mode de propagation de la Muscardine.—C. R. V. p. 712.

72. Sur des Œufs de Vers-à-soie rapportés de l'Inde par *M. Gaudichaud.*—C. R. VI. p. 19.

73. Quelques remarques sur la contagion de la Muscardine.—C. R. VIII. p. 622.

AUDOUIN (J. V.) et LACHAT ( ).

1. Anatomie d'une Larve apode, trouvée dans le Bourdon des pierres.—J. de Phys. LXXXVIII.—Mém. Soc. d'H. n. I. p. 329, fig.—Ann. Sc. n. 1841, XVI. p. 372.

AUDOUIN (J. V.) et MILNE EDWARDS (H.).

1. Voyage sur les côtes de la France.—Ann. Sc. n. XVIII.; Rev. Bibl. p. 149.

2. Recherches pour servir à l'histoire naturelle du littoral de la France, ou recueil de mémoires sur l'anatomie, la physiologie et les mœurs des animaux de nos côtes. Paris, 1832, 2 vols. 8°. fig. —Ann. Sc. n. (2e Sér.) XVI. p. 374.

3. Recherches anatomiques sur le Système nerveux des Crustacés. —Ann. Sc. n. 1828, XIV. p. 75, fig.—Isis, 1834, X. p. 1023.

4. Note sur le Système nerveux des Crustacés.—Ann. Sc. n. 1830, XX. p. 181.—*Féruss.* Bull. 1830, XXII. p. 344.—Isis, 1834, XI. p. 1111.

5. De la respiration aërienne des Crustacés, et des modifications

que l'appareil branchial présente dans les Crabes terrestres.— Ann. Sc. n. XV. p. 85.—*Féruss.* Bull. 1829, XIV. p. 308.

6. Recherches anatomiques et physiologiques sur la Circulation dans les Crustacés. Paris, 1827, 4°. fig.—Ann. Sc. n. XI. p. 283, 352, fig.; XIV. p. 77.—Isis, 1834, IX. p. 936 (*Ueber den Kreislauf der Crustaceen*).—*Heus.* Zeitschr. I. p. 732 (*Anatomisch-physiologische Untersuchungen über den Kreislauf der Crustaceen*).— *Féruss.* Bull. XIV. p. 380.

7. Déscription des Crustacés nouveaux ou peu connus, et remarquables par leur organisation, conservés dans la collection du Muséum. I^r Art. genre Sérole; II^r Art. Écrevisse de Madagascar.—Arch. Mus. II. p. 5.—Ann. Sc. n. (2e Sér.) XVI. p. 378.

8. Résumé d'Entomologie, ou Histoire naturelle des Animaux articulés, &c. Par. 1825–28, 4 vols. 32°.—Ann. Sc. n. 1829, XVI.; Rev. Bibl. p. 29.

9. Résumé des recherches sur les Animaux sans vertèbres, faites aux îles Chausey.—Ann. Sc. n. XV. p. 5.—*Féruss.* Bull. 1829, XVII. p. 127.—Isis, 1834, X. p. 1029 (*Untersuch. über die wirbellosen Thiere, angestellt auf den Chausey-Inseln*).

10. Histoire naturelle des Annélides, Crustacés et Arachnides, formant le T. I^r de l'Hist. nat. des Animaux articulés, et faisant partie de l'Encyclopédie portative. Paris, 1829, 18°. fig.—*Féruss.* Bull. 1829, XIX. p. 124.

11. Mémoire sur la Nicothoë, genre nouv. de Crustacé qui suce le sang du Homard.—Ann. Sc. n. IX. p. 345, fig.—*Féruss.* Bull. 1827, X. p. 183.—Isis, 1831, XI. p. 1228.

12. Appendice au mémoire sur la Nicothoë, à l'occasion d'un petit Crustacé isopode qui vit sous le test de la Callianasse.—Ann. Sc. n. IX. p. 359, fig.

13. Déscription de l'Hipponoë, nouveau genre d'Annélides.—Ann. Sc. n. XX. p. 156, fig.—*Féruss.* Bull. 1830, XXII. p. 334.—Isis, 1831, I. p. 100.

14. Classification des Annélides, et déscription de celles qui habitent les côtes de la France.—Ann. Sc. n. XXVII. p. 337; XXVIII. p. 187; XXIX. p. 195, 388; XXX. p. 411, fig.—Isis, 1835, XII. p. 1061, fig. (*Classification der Anneliden*).

15. Déscription et Classification des Annélides de France.—Ann. Sc. n. XXI. (2e Sér.) XVI. p. 374.—Isis, 1835, VII. p. 637; VIII. p. 678; IX. p. 768.

16. Des Poils des Annélides, considérés comme moyen de défense. —Ann. Sc. n. XXI. p. 317.

17. Observations sur divers faits relatifs à l'anatomie des Crustacés et à la découverte de plusieurs Mollusques nouveaux.—Ann. Sc. n. XXI. p. 317; (2e S.) XVI. p. 374.

18. Précis-d'Entomologie. (Encycl. portat.)   Paris, 1829, 8°. fig.

19. Mémoires pour servir à l'Histoire naturelle. des Crustacés. (Fasc. I.)   Paris, 1829, 8°. fig.—Ann. Sc. n.

20. Traité élémentaire d'Histoire naturelle des Insectes. (Encycl. port.)

21. Traité élémentaire d'Hist. nat. des Annélides, Crustacés et Arachnides. (Encycl. port.)

22. Iconographie des Insectes, ou collection de figures représentant les Insectes qui peuvent servir de types pour chaque famille. 32°. Paris, 1828.

AUDOUIN (J. V.) et BRULLÉ (   ).

1. Histoire naturelle des Insectes, traitant de leur Organisation, de leurs Mœurs en général, &c.   Par. 1834–36, 4 vols. 8°. fig. col.

2. Déscription des Espèces nouvelles ou peu connues de la famille des Cicindélètes faisant partie de la Coll. du Muséum.   1839, 4°. fig. col.

AUDOUIN (J. V.) et PAYEN (   ).

1. Note sur des Animaux qui colorent en rouge les Marais salans. —Ann. Sc. n. (2e Sér.) 1836.

AUDOUIN (J. V.), BRONGNIART (A.) et DUMAS (   ).

1. Annales des Sciences ; *vide sup.* Pars I. p. 52.

AUDOUIN (J. V.), etc.

1. Dict. Classique d'Hist. Nat. ; *vide sup.* Pars I. p. 53.

AUDUBON (J. J.).

1. *Faits et Observations relatives à la résidence permanente des Hirondelles dans les Etats-Unis.*   Ann. Lyc. New York, 1824, I. 6, p. 166.—*Féruss.* Bull. 1826, VII. p. 109.

2. The Birds of America, from drawings made in the United States and their territories.   4 vols. fol. Lond. 1828–1840.—*Sill.* Am. Journ. XXXIX. p. 343 ; XLII. p. 130.—*Féruss.* Bull. 1828, XV. p. 301.—Lyc. of New York, May 1829.—Amer. Journ. XVI. 2. p. 353.—Mag. Nat. Hist. ser. 1, I. p. 43.—Second Ed. 7 vols. 8°. New York, 1839–1844.

3. Ornithological Biography, or an Account of the Habits of the Birds of the United States of America.   5 vols. 8°. Edinb. 1831–1839.—*Wiegm.* Arch. 1836, II. p. 261.

4. A Synopsis of the Birds of North America.   8°. Edinb. 1839.— *Sill.* Am. Journ. XXXIX. p. 343.

5. On the Natural History of the Alligator. Edinb. New Phil. Journ. II. p. 270.—Isis, 1832, VII. p. 687.—*Féruss.* Bull. XV. p. 162.

6. On the Habits of the Turkey Buzzard.—Ed. New Phil. Journ.— Mag. Nat. Hist. ser. 1, VI. pp. 83, 163; VII. pp. 164, 276.— *Féruss.* Bull. 1828, XIII. p. 239.

7. On the Bird of Washington (*Falco Washingtoni*) or Great American Sea Eagle.—Mag. Nat. Hist. ser. 1, I. p. 115.

8. On the Rattlesnake (*Crotalus horridus*).—Edinb. New Phil. Journ. III. p. 21.—*Féruss.* Bull. 1828, XIII. p. 357.—Isis, 1832, VII. p. 690.

9. On the Habits of the Wild Pigeon of America (*Columba migratoria*).—Brewst. Journ. Science, ser. 1, VI. p. 257.—*Féruss.* Bull. 1827, XII. p. 125.

10. On the Habits of *Falco palumbarius.*—Edinb. Journ. Nat. and Geogr. Sc. III. p. 145.

11. On the Carrion Crow, or *Vultur atratus.*—Brewst. Journ. Sc. ser. 1, VI. p. 156.—*Féruss.* Bull. 1828, XIII. p. 239.

AUDUBON (J. J.) and BACHMAN (J.).

1. Descriptions of some new species of Quadrupeds inhabiting North America. Philad. 1841, 8°. pam.

2. Viviparous Quadrupeds of North America. 8°. and fol. Philadelphia, 1843–47, fig.

AUFSCHLÄGER (J. F.).

1. Déscription historique et topographique des deux Départemens du Rhin. 8°. Strasb. 1828.

AUGENIUS (Hor.).

1. Quod Homini certum non sit nascendi tempus; adjectum Embryon petrefactum urbis Senonensis, cum Exercitatione de hujus indurationis causis. Venet. 1595, 8°.—Francof. 1597, fol.— *Böhm.* Bibl. IV. 2, p. 271.

AUGUSTIN (F. L.).

1. Lehrbuch der Physiologie des Menschen. 8°. Berl. 1809, 1810.

AULAGNIER (Alph.).

1. Apercu sur la Géologie et l'Agriculture du Dép. de la Haute-Loire et pays limitrophes, précédé de notes historiques sur l'an-

cien état du Velay, et suivi d'un itinéraire pour faciliter les recherches des amateurs en histoire naturelle.—Le Puy, 1823, 8°. —*Féruss.* Bull. 1824, II. p. 325 ; 1828, XIII. p. 380.

AULD (W.).

1. Notes upon Nelson River, near York Fort, Hudson's Bay.— Trans. Geol. Soc. ser. 1, V. p. 599.

AULDJO (J.).

1. Sketches of Vesuvius, with short Accounts of its principal Eruptions. 8°. Lond. 1832, fig.—Cat. R. Soc. Lond.
2. Narrative of an Ascent to the summit of Mont Blanc. 4°. London, 1828.
3. Spouting Fountain of Mineral Water, discovered in 1832 near Cape Uncino, kingdom of Naples.—Bibl. Univ. 1833, Mars.— Amer. J. XXV. 1, p. 194.

AUMÜLLER ( ).

1. De Glandulæ lacrymalis Fungo medullari. Berol. 1833.—*Müll.* Arch. 1835, p. 242.

AUNANT (J.).

1. L'Art de planter et de cultiver les Meuriers blancs, d'élever les Vers-à-soye ; pour servir d'Instruction aux provinces d'Allemagne. Hanau, 1744, 8°. fig.—(Allem.) *Anweisung zum Seidenbau u. dazu gehörigen Maulberplantagen* ; 2te verm. Aufl. Leipz. 1754, 8°.—*Böhm.* Bibl. II. 2, p. 256.

AURIFABER (A.).

1. Succini Historia, oder Bericht woher der Agt-oder Börnstein ursprünglich komme. Königsb. 1551, 4°.—*Ibid.* 1557 et 1572.

AUSONIUS.

1. Mosella. Burdig. 1580, fol. (cum notis *P. Duezii* et *M. Freteri*) 1619, fol.—*Böhm.* Bibl. I. 1, p. 514 ; II. 2, p. 60.

AUST (H. G. A.).

1. De sectione Tendinum. Berol. 8°.—*Val.* Repert. anat. u. phys. III. p. 30.

AUSTEN (R. A. C.).

1. Considerations on Geological Evidence and Inferences.—Rep. Brit. Assoc. 1838, Sect. p. 93.

2. Note on the Organic Remains of the Limestones and Slates of South Devon.—Rep. Brit. Assoc. 1839, Sect. p. 69.

3. On the raised Beach near Hope's Nose in Devonshire, and other recent disturbances in that neighbourhood.—Proc. Geol. Soc. II. 102.—Phil. Mag. ser. 3, VI. p. 63.

4. On the part of Devonshire between the Ex and Berry Head, and the Coast and Dartmoor.—Proc. Geol. Soc. II. 414.—Phil. Mag. ser. 3, IX. p. 495.

5. On the Geology of the South-east of Devonshire.—Trans. Geol. Soc. ser. 2, VI. p. 433.—Proc. Geol. Soc. II. 584 ; III. 123.— Phil. Mag. ser. 3, XII. p. 564 ; XIV. p. 404.

6. On the Origin of the Limestones of Devonshire.—Proc. Geol. Soc. II. 669.—Phil. Mag. ser. 3, XIII. p. 228.

7. On Orthoceras, Ammonites, and other cognate genera; and on the position they occupy in the animal kingdom.—Proc. Geol. Soc. III. 179.

8. On the Bone Caves of Devonshire.—Proc. Geol. Soc. III. 286.— Phil. Mag. ser. 3, XVIII. p. 228.

9. On the Geology of the South-east of Surrey.—Proc. Geol. Soc. IV. pp. 167, 196.—Phil. Mag. ser. 3, XXIV. p. 65.

10. On the Coal Beds of Lower Normandy.—Journ. Geol. Soc. II. p. 1.

Austin (R.).

1. On Coralloidal Arragonite.—Trans. Geol. Soc. ser. 1, V. p. 613.

Austin (T.).

1. On the Elevation of Land on the shores of Waterford Haven during the Human Period, and on the Geological Structure of the District.—Proc. Geol. Soc. III. p. 360.—Phil. Mag. ser. 3, XIX. p. 318.

2. On the Geology around the shores of Waterford Haven.—Proc. Geol. Soc. III. 154.—Phil. Mag. ser. 3, XVII. p. 68.

3. Descriptions of several new Genera and Species of *Crinoidea*.— Ann. and Mag. N. Hist. XI. p. 195.

4. On *Sterna arctica*.—Ann. and Mag. N. Hist. IX. p. 434 ; X. p. 75.

5. Proposed Arrangement of the *Echinodermata*, particularly as regards the *Crinoidea*, and a subdivision of the class *Adelostella* (*Echinidæ*).—Ann. and Mag. N. Hist. X. p. 106.

6. Remarks on the Habits of Birds which are natives of the British Islands.—Ann. and Mag. N. Hist. XIII. p. 92.

7. On the Habits of the Godwit.—Ann. and Mag. N. Hist. XIV. p. 382.

8. Note on *Bowerbank's* Paper on the genus *Dunstervillia*, with Remarks on the *Ischadites Kœnigii*, the *Tentaculites*, and the *Cornularia.*—Ann. and Mag. N. Hist. XV. p. 406.

9. Monograph on Recent and Fossil *Crinoidea.*—4°. Bristol, 1845.

AUTENRIETH (J. H. F.).

1. Ansichten über Natur und Seelenleben. Stuttg. 1836, 8°.—*Val.* Repert. II. p. 18.

2. Handbuch der empirischen menschlichen Physiologie. Tüb. 1801, 8°. 3 Bde.—Bibl. med.-ch. p. 21.

3. Supplementa ad historiam Embryonis Humani, quibus accedunt observata quædam circa palatum fissum. Tüb. 1797, 4°.—Bibl. med.-ch. p. 21.

4. Ueber die Asymmetrie in dem Lobus olfactorius.—*Wiedem.* Arch. 1800, I. 2, p. 63.—*Müll.* Arch. 1835, p. 476.

5. Kleine Naturhistorische Bemerkungen aus dem Thierreiche.— *Voigt's* Mag. X. 1805, p. 41.—Ad Œconomiam vertebrarum animalium spectant.

6. Bemerkungen über die verschiedenen Menschenracen und ihren gemeinschaftlichen Ursprung.— *Voigt's* Mag. V. 1803, p. 420.

7. Bemerk. über die psychologische Gleichheit des ganzen Thierreichs.— *Wiedem.* Arch. II. 2, p. 225.

8. Bemerkungen über den Bau der Scholle (*Pleuronectes Platessa*) insbesondere, u. den Bau der Fische im Allgemeinen.—*Wiedem.* Arch. I. 2, p. 47.

9. Ueber das Gift der Fische. 8°. Tübingen, 1833.

AUTENRIETH (J. H. F.) et BOHNENBERGER (J. G. F. v.).

1. Tübingen Blätter für Naturwissensch.; *vide sup.* Pars I. p. 34.

AUTEROCHE (Chappe d').

1. Relation d'un Voyage en Sibérie fait par ordre du Roi en 1761. Par. 1768, 4°. 3 vols. fig.—Amst. 1769-70, 8°. 4 vols.—(Angl.) Lond. 1770, 4°.—*Böhm.* Bibl. I. 1, p. 638.

2. Voyage en Californie pour l'observation du passage de Vénus sur le disque du Soleil, etc., publié par *M. de Cassini,* fils. Par. 1772, 4°. fig.—*Böhm.* Bibl. I. 1, p. 742.

AUTOPTA ( ).

1. Beschreibung des Gesundbrunnen so unweit Dölitsch, nahe bey

einem Dorfe, Drossig genannt, entsprungen. Leipz. 1704, 8°.—
Böhm. Bibl. V. p. 370.

AUVITY (M.).

1. Observ. sur des Vers sortis par le canal de l'Urètre.—*Roz.* Journ.
de Phys. XIII. p. 379.

AUZOUX (L.).

1. Lecons élémentaires d'Anatomie et de Physiologie. 8°. Paris,
1829.

AVELIN (G. E.).

1. Dissert. de Miraculis Insectorum, præside Linnæo. Upsaliæ,
1752, 4°.—Am. Acad. III. p. 313.

AVELLINO (G.).

1. Ann. Acad. Aspir. Naturalisti ; *vide sup.* Pars I. p. 68.

AVERTRANI ( ).

1. Considerazioni sopra le Dottrine fisiologiche e patologiche del
Prof. *Medici.* Perug. 1842.

AVICENNA (Abou Ali Alhussein ben Abdoullah).

1. De Animalibus *Aristotelis* Libri XIX., par *Mich. Schot* ; ex Arabico in Latin. translatum. fol.—*Böhm.* Bibl. II. 1, p. 8.
2. Tractatus de Mineralibus.—*Vid. Böhm.* Bibl. IV. 1, p. 14.

AVRIL (Phil.).

1. Voyage en divers états d'Europe et d'Asie, avec une Description
de la Grande Tartarie. Par. 1691, 1692, 4° ; 1693, 12°. fig.—
(Allem.) *Curieuse Reisen, mit Anmerk. v. L. Fr. Vischer* ; Hamb.
1705, 8°.—*Böhm.* Bibl. I. 1, p. 450.

AVRILLY (Chanoine d').

1. Du Myzoxile, Puceron lanigere. Louviers, 1834, 8°. fig.

AXFORD (Th.).

1. Ueber das Schnabelthier.—Ed. n. Phil. J. VI. p. 399.—Isis, 1832,
VIII. p. 806.

AXTELMEIER (St. R.).

1. Ebenbild der Natur in dem Entwurfe der Gewächse, Unge-
ziefer u. anderer Thiere. Augsb. 1699, 8° ; 1715, 8°.—*Böhm.*
Bibl. I. 1, p. 256.

2. Naturlicht vom Bergwesen, Schatzgraben, u. dgl. m. Augsb.
1706, 4°.—*Böhm.* Bibl. IV. 1, p. 42.

AYCKE (J. C.).

1. Fragmente zur Naturgeschichte des Bernsteines. Danzig,
1835, 8°.

AYRER (J. G.).

1. De Vermibus Intestinorum. Jena, 1670, 4°.

AYRES (Will. O.).

1. Enumeration of the Fishes of Brookhaven, L. I., with remarks
upon the species observed.—Proc. Bost. Soc. N. H. 1842, p. 58.

2. Descriptions of Four Fishes taken at Brookhaven, L. I.—Proc.
Bost. Soc. N. H. 1842, p. 64.

3. Descriptions of four species of Fishes from Brookhaven, L. I,
two of which are believed to be new.—Proc. Bost. Soc. N. H.
1842, p. 67.

4. General description of a new Species of *Leuciscus.*—Proc. Bost.
Soc. N. H. 1843, p. 130.

AZAIS (H.).

1. Précis du Systéme Universel. Paris, 1826, 8°.—*Féruss.* Bull.
VIII. p. 25.

AZARA (F. de).

1. Apuntamientos para la Historia natural de los Paxaros del Paraguay. 3 vols. 8°. Madrid, 1802.

2. Voyage dans l'Amérique méridionale, de 1781 à 1801 ; trad.
par *Walkenaer* et *Sonnini.* Paris, 1809, 4 vols. 8°.—(Germ. *W.
Lindau*) : *Reisen im Südamerika,* etc. Leipz. 1810, 8°.—Cat.
Bibl. Turic. —Dass. von *C. Weyland.* 8°. Berlin, 1810.

3. Apuntamientos para la Historia natural de los Quadrupedos del
Paraguay, 2 vols. 8°. Madr. 1802. (*Essai sur l'Histoire natu-
relle des Quadrupèdes du Paraguay* ; trad. par *Moreau de St.
Méry.*) Paris, 1801, 2 vols. 8°.—*Wied.* Arch. IV. pp. 190, 237.

—*Quadrupeds of Paraguay*, tr. by *W. P. Hunter.* 8°. London, 1838.

4. On the Wild Horses in Spanish America.—Phil. Mag. V. p. 330.

AZEMA (Aug.).

1. Nouveau Gisement d'Os fossiles de grands Mammifères.—C. R. IV. p. 978.

AZES (Ch. G.).

1. Naturlehre für Frauenzimmer. Bressl. 1781, 8°.—*Böhm.* Bibl. I. I, p. 337.

AZUNI (Alb.).

1. Histoire géographique, politique et naturelle de la Sardaigne. Paris, 1802, 2 vols. 8°.—(Germ. v. *K. M. Brede*) 8°. Leipz. 1803, et 8°. Hamb. 1803.

AZZOGUIDI ( ).

1. Compendio dei Discorsi che si tengono nella R. Univers. di Bologna dalla cattedra di Fisiologia e di Notomia comparata. Bol. 1808, 8°.

AZZOGUIDI, PALETTA et BRUGNONI.

1. Observationes ad Uteri constructionem pertinentes ; nova Gubernaculi Testis Hunteriani et Tunicæ vaginalis, etc., ed. *E. Sandifort.* Lugd. Bat. 1780, 8°.

B.

BAADER (F. M. v.).

1. Ueber einige Neuerungen in der Naturgeschichte. 4°. Münch. 1790.

BABBAGE (C.).

1. On the Temple of Serapis at Pozzuoli near Naples, with Remarks on certain Causes which may produce Geological Cycles of great extent.—Proc. Geol. Soc. II. 72.—Phil. Mag. ser. 3, V. p. 213.—Amer. J. XXVII. 2, p. 408.—*L. u. Br.* N. Jahrb. 1835, p. 539.

2. On Impressions in Sandstone resembling those of horses' hoofs. —Proc. Geol. Soc. II. 439.—Phil. Mag. ser. 3, X. p. 474.

3. Note sur un Système descriptif des Mammifères.—Ed. J. of Sc. 1829, I. p. 187.—*Féruss.* Bull. XXV. p. 296.

4. The Ninth Bridgewater Treatise. 8°. London, 1838.

BABINGTON (C. C.).

1. On *Arvicola pratensis.*—Mag. Zool. a. Bot. 1838, II. p. 92.

2. On *Malachius bipunctatus.*—Mag. Nat. Hist. ser. 1. V. p. 329; VII. pp. 378, 524.

3. On certain Species of the Genus *Dromus.*—Trans. Entom. Soc. I. p. 80.—*Wiegm.* Arch. 1836, II. p. 305.

4. On *Haliplus ferrugineus* of Authors, being an attempt at its sub-division into several species.—Trans. Entom. Soc. I. p. 175.

5. Distinction of Sex in *Papilio Machaon.*—Mag. Nat. Hist. ser. 1, II. p. 67.

6. Additions to the list of British Insects.—Mag. Nat. Hist. ser. 1, V. p. 327.

7. Dytiscidæ Darwinianæ; or, Descriptions of the Species of Dytiscidæ collected by C. Darwin, Esq. in South America and Australia, during his voyage in H.M.S. Beagle.—Trans. Entom. Soc. III. p. 1.

8. On *Malachius ruficollis* (Panz.) and *M. bipunctatus* (Bab.).— Entom. Mag. IV. p. 365.

9. On *Macroplea Zosteræ.*—Entom. Mag. IV. p. 438.

10. On a new British *Colymbetes.*—Ann. Nat. Hist. VI. p. 53.

BABINGTON (S.).

1. On the Island of Salsette.—Trans. Geol. Soc. ser. 1, vol. V. p. 1.

2. On the Geology of the Country between Tellicherry and Madras. —Trans. Geol. Soc. ser. 1, vol. V. p. 328.—Isis, 1823, IV. L. A. p. 179.

BABINGTON (W.).

1. A systematic Arrangement of Minerals, founded on the joint considerations of their chemical, physical, and external characters, etc. Lond. 1795, 4°.

2. A new System of Mineralogy, in the form of Catalogue, after

the manner of *Born's* Systematic Catalogue of the Collection of Fossils of *Mlle. El. de Rabb.* 4°. Lond. 1799.—Cat. R. Soc. Lond.

BACCHINI (B.) et ROBERTO (G.).

1. Giornale dei Letterati; *vide sup.* Pars I. p. 69.

BACCI (Andr.).

1. De Gemmis et Lapidibus pretiosis in S. Scripturâ relatis. 4°. Rome, 1577, 1587.

2. Tractus de Monocerote, s. Unicornu, ejusque admirandis viribus et usu; Italicâ linguâ conscriptus et ab *A. Marino* in Latin. conversus et editus. Venet. 1566, 4°.—nunc a *W. Gabelchover*, etc. Stuttg. 1598, 8°.—*Böhm.* Bibl. II. 1, p. 480.

3. De Thermis, Libri VII. opus locupletissimum, in quo agitur de universâ Aquarum naturâ, deque earum differentiis omnibus ac mistionibus cum terris, cum ignibus, cum metallis; de Fontibus, Fluminibus, Lacubus, de Balneis totius Orbis, etc. Venet. 1571 et 1578, fol. fig.; ibid. ab Auct. recognitum et locupletatum, 1588, fol. fig.—Romæ, 1622, fol. (sine fig.)—Lugd. Bat. 1699, fol.—Patav. 1711, fol. (cum libro VIII°. de novâ Methodo Thermarum explorandarum, etc.)—*Böhm.* Bibl. V. p. 120.

4. Discorso della gran Bestia, giunt al suo Discorso dell Pietre pretiose. Roma, 1587, 4°.—*Böhm.* Bibl. II. 1, p. 370.—(Latin.) *De Magnâ Bestiâ, s. Alce, ejusque proprietatibus,* etc., interprete *W. Gabelchover*; annex. ejusd. Monocerologiâ. Stuttg. 1598, 8°.

BACH (Ch. E.).

1. Annotationes anatomicæ de Nervis Hypoglosso et Laryngeis. Turici, 1834, 4°. fig.—Isis, 1835, I. p. 96.—*Müll.* Arch. 1835, p. 239; 1836, p. 22.

2. Leçons sur l'Embryologie, faites en 1839 à la Faculté de Médecine de Strasbourg. Strasb. 1840, 8°.

BACH (W. E.).

1. Stirpes et genera *Pselaphorum.*—Zool. Journ. VIII. 1826.—Isis, 1830, X. p. 1054.—*Féruss.* Bull. Sept. 1827.

2. Monographie des Cébrionides.—Zool. Journ. 1824, p. 33.—*Féruss.* Bull. Juin 1824.

3. Caractère d'un nouveau genre d'Insecte coléoptère de la famille des Byrrhidies.—Trans Lin. Soc. XIII. 1, p. 41.—*Féruss.* Bull. Juin 1824.

BACHELEY ( ).

1. Nouvelles observations lithologiques sur la formation du Silex, &c.—Journ. de Phys. XXI. 1782. Suppl. p. 81.—*Mod.* Bibl. Helm.

BACHELIER (J. B.).

1. De Vermibus intestinalibus. Monspel. 1777, 4°.

BACHELTEL (J. Chr.).

1. Beschreibung des Fichtelberges.   Leipz. 1716, 4°. fig.—*Böhm.* Bibl. I. 1, p. 587.

BACHEM (Guill.).

1. De Liene. Berol. 1839, 8°.

BACHIENE (M. Alb.).

1. Historische u. geographische Beschreibung von Palästina, nebst Landcharte; aus dem Holländ. ubers. u. mit Anmerk. begleitet v. *G. A. Maas.* Clev. u. Leipz. 1768–75, 4 vols. 8°.—*Böhm.* Bibl. I. 1, p. 683.

BACHMAN (J.).

1. Observations on the genus *Scalops,* with Descriptions of the species found in North America.—Proc. Bost. Soc. N. H. 1841, p. 40.

2. Quelques remarques sur le genre *Sorex* (Musaraigne) et Mono-graphie des espèces Nord-Américaines qui s'y rapportent.—Journ. Ac. Philad. 1837, VII. p. 362.—Rev. Zool. 1838, p. 290.

3. Observations on the Changes of Colour in Birds and Quadrupeds. —Trans. Am. Phil. Soc. ser. 2, VI. p. 197.

4. Monograph of the Genus *Sciurus,* with Descriptions of new Species and their Varieties, as existing in North America.—Mag. Nat. Hist. ser. 2, III. pp. 113, 154, 220, 330, 378.—*Sill.* Am. J. XXXVII. p. 290.

5. On the Migration of the Birds of North America.—Lit. a. Phil. Soc. of Charlest. 1833.—Amer. J. XXX. p. 81.— *Wiegm.* Arch. 1837, V. p. 198 (*Ueber die Wanderungen der nordamerikan. Vögel*).

6. Description d'une nouvelle espèce de Lapin, trouvée dans la Caroline du Sud.—Journ. Ac. Philad. 1837, VII. p. 194.—Rev. Zool. 1838, p. 291.

BACHMANN (Fr.).

1. *N. G. Geven's* Conchylien-Cabinet.—Lüneb. 1830–33, 1 vol. 4º.
fig.—*Féruss.* Bull. XXV. p. 125.

BACHSTRÖM (J. Fr.).

1. Nova Æstûs Marini Theoria. Lugd. Bat. 1734, 8º.—*Böhm.*
Bibl. V. p. 83.

BACK (Abr.).

1. De Cornu Piscis planè singulari carinæ navis impacto.—Act.
Acad. Nat. Cur. VIII. p. 199.—Dict. Sc. Nat. XXII. p. 536.

BACK (Alb.).

1. Animalia composita.—Diss. Acad. 4º. Upsal. 1759.

BACK (G.).

1. Narrative of the Arctic Land Expedition. Lond. 1836, 8º.—
Wiegm. Arch. 1837, II. p. 136.

BACK (Jac. de).

1. Dissertatio de Corde. 12º. Roterd. 1640.—Cat. R. Soc. Lond.

BACKHOUSE (E.).

1. Notice of the Annual Appearance on the Durham Coast of some
of the *Lestris* tribe.—Rep. Brit. Assoc. 1838, Sect. p. 108.—Rev.
Zool. 1839, p. 155.

BACKWELL (Jos.).

1. Introduction to the Study of Mineralogy. 12º.   Halifax, 1826.

BACMEISTER (J.).

1. Essai sur la Bibliothèque et le Cabinet des Curiosités et d'Histoire
Naturelle de l'Académie des Sciences de St. Pétersbourg. Pé-
tersb. 1776, 8º.—(Allem.) Pet. u. Leipz. 1777, 8º.—*Böhm.* Bibl.
I. p. 394.

BACO DE VERULAMIO (Fr.).

1. Ten Centuries of Natural History. Lond. 1621–1627, 4º.; ibid.
1639–1670, fol.—(Latinè) Lond. 1622, 4º.—(Gallicè) Par. 1631,
8º.—Lugd. Bat. 1648, 12º. (tit. *Sylva Sylvarum, s. Historia
naturalis*), à *J. Grutero.*—Amst. 1661, 12º.—*Böhm.* Bibl. I. 1,
p. 239.

Bacounin ( ).

1. Mémoire sur le *Gordius* d'eau douce des environs de Turin.—Mem. Acad. Tor. IX. p. 23.—Journ. de Phys. XXXIX. p. 204.

Badariotti (J. Aut.).

1. De Lacteorum Vasorum fabricâ et positionibus.—De Liquoribus Salivariis.—De Lacte, etc. (Theses pro aggreg.) Taurini, 1743, 8°.

Baddam ( ).

1. Philos. Transactions; *vide* sup. Pars I. p. 76.

Baddeley (P. F. H.).

1. On a species of Gall Insect inhabiting the leaves of the Ficus racemosa.—Ind. Rev. II. p. 34.

2. On a Dipterous Fly, the lava of which produces a kind of Gall on the leaves of the Ficus racemosa.—Ind. Rev. I. 274.

3. On a species of Ichneumon, inhabiting the interior of the Gall by an Insect on the leaves of the Ficus racemosa.—Ind. Rev. I. 329.

Baddely ( ).

1. Geognosie der Magdalenen-Inseln im Lorenz-Golfe.—Tr. Liter. a. Hist. Soc. Queb. 1833, III. 2, p. 147.—*L. u. Br.* N. Jahrb. 1835, p. 718.

2. Miner. Examination of the Sulphate of Strontian, from Kingston (U. C.), with miscellan. Notices of the Geology of the vicinity.—Amer. J. XVIII. 1, p. 104.

Badenach (J.).

1. On an uncommon Bird from Malacca.—Phil. Trans. LXII. p. 1. Journ. de Phys. 1774, p. 444.

Badham (David).

1. The Question concerning the Sensibility, etc.; Questions sur la Sensibilité, l'Intelligence, et l'Instinct des Insectes. Paris, 1837, 8°.—Rev. Zool. 1838, p. 60.

Badier (de).

1. Observations sur les Œufs du Mabouya-Colant.—*Rozier*, Journ. de Phys. X, 1777, p. 414.

2. Lettre sur le Scolopendre-polype.—Journ. de Phys. XXXIV. 1789, p. 55.

3. Observations sur la Reproduction des Pattes de Crabes.—*Rozier*, Journ. de Phys. XI. 1778, p. 33.

4. Observations sur la Nourriture des Colibris.—*Rozier*, Journ. de Phys. XI. 1778, p. 38.

5. Oiseaux mouches.—*Rozier*, Journ. de Phys. 1778, XIV. p. 32.

BÆCHTOLD (J. J.).

1. Unters. über die Giftwerkzeuge der Schlangen. 4°. Tübing. 1843.

BÆCK (Abr.).

1. Oratio de Memorabilibus Insectis.—Biogr. un. III. p. 205.

2. Insectes nuisibles aux Graminées.—Vetensk. Ac. Handl. 1742, p. 40.—Analecta Transalp. I. p. 200.—Bibl. Ent. I. p. 12.

3. Berättelse om Watten Polypen, i anledning af dem som äro fundne omkring Stockholm.—Act. Holm. VII.—*Mod.* Bibl. Helm.

4. Animalia Composita, dissert. sub præsid. Linnæi. Upsaliæ, 1759, 4°.—Amœnit. Academic. V. p. 343.

BÆCKMAN (A. P.).

1. Fundamenta Ornithologiæ, dissert. præsid. Linnæo. Upsaliæ, 1765, 4°. cum fig.—Amœnitat. Acad. VII. p. 109.

BÆCKNER (M. Andr.).

1. De Noxâ Insectorum. Ups. 1752, 4°.—Amœn. Ac. III. p. 305. —Bibl. Ent. I. p. 13.

BAER (K. E. v.).

1. Noch ein Wort über den After der Distomen.—*Heusing.* Zeitschr. f. die organ. Physik, II. 1828. p. 197.—*Sieb.* in *Wiegm.* Arch. 1835, I. p. 59.

2. Zurechtweisung wegen *Aspidogaster.*—Isis, 1828, VII. p. 671.— *Féruss.* Bull. XVII. p. 155 (*Réclamation contre M. Raspail*).

3. Ueber eine Süsswasser-Miessmuschel (*Mytilus Hagenii*).—Isis, 1826, V. p. 525.—*Féruss.* Bull. 1826, VIII. p. 141.

4. De fossilibus Mammalium reliquiis in Prussiâ adjacentibusque regionibus repertis Dissertatio. 4°. Kœnigsb. 1823, fig.—*Féruss.* Bull. 1826, IX. p. 153.

5. Sur un Mammouth semblable à l'Éléphant actuel d'Afrique.—Mém. Acad. Pétersb. 1830, I. 2, p. 16 du Bull. Scient.—*Féruss.* Bull. 1830, XXII. p. 320.

6. Sur les Entozoaires, ou Vèrs intestinaux.—*Féruss.* Bull. 1826, IX. p. 123.

7. Recherches sur plusieurs Animaux inférieurs.—N. Act. Nat. Cur. XIII. 2, p. 523.—*Féruss.* Bull. XVI. p. 291.

8. Die Metamorphose des Eies der Batrachier vor der Erscheinung des Embryo, u. Folgerungen aus ihr für die Theorie der Erzeugung.—*Müll.* Arch. 1834, p. 481.

9. Beitrag zu der Entwickelungsgeschichte der Schildkröten.—*Müll.* Arch. 1834, p. 544, fig.

10. Ueber das Gefäss-System des Braunfisches.—N. Act. Nat. Cur. XVII. p. 393.

11. Die Nase der Cetaceen, erläutert durch Untersuchung der Nase des Braunfisches.—Isis, 1826, VIII. p. 811.—*Féruss.* Bull. XIII. p. 117 (*Sur le Nez des Cétacés*, etc.).

12. Nachträgliche Bemerkungen über die Riechnerven des Braunfisches.—Isis, 1826, IX. p. 944.

13. Doppelter Muttermund des einfachen Fruchthälters vom Ameisenfresser.—*Müll.* Arch. 1836, V. p. 384.

14. Observations sur la Génération des Moules, et sur un système de Vaisseaux hydrofères dans ces animaux.—*Fror.* Notiz. 1826, No. 265, p. 1.—*Féruss.* Bull. 1826, IX. p. 369.

15. Beschreibung u. Abbildung der Geflechte der Armarterie beim Braunfisch- u. Manatifœtus, und der Verzweigung der Armarterie beim jungen Wallross.—Mém. Acad. Pét. 1833.—*Müll.* Arch. 1835, p. 44.

16. Depth of the Frozen Ground in Siberia.—Ed. N. Phil. Journ. XXIV. p. 435.—*Sill.* Am. Journ. XXXVI. p. 210.

17. Ueber die sogenannte Erneuerung der Magens der Krebse u. die Bedeutung der Krebssteine.—*Müll.* Arch. 1834, p. 510.

18. Wanderung eines sehr grossen Granit-Blockes über den Finnischen Meerbusen.—Bull. Ac. Pét. V. p. 154.—*L. u. Br.* N. Jahrb. 1841, p. 599.

19. Untersuchungen über die ehemalige Verbreitung und die gänzliche Vertilgung der von *Steller* beobachteten nordischen Seekuh (*Rytina* Ill.). Petersb. 1838, 4°.

20. Nochmalige Untersuchungung der Frage: Ob in Europa in historischer Zeit zwei Arten von wilden Stieren lebten?—Bull. Acad. Pétersb. IV. No. 8.—*Wiegm.* Arch. 1839, I. p. 62.

21. Berichte über die Zoographia Rosso-Asiatica von *Pallas*, ab-

gestattet an die K. Acad. der Wiss. zu St. Petersb. Königsb. 1831, 4°.

22. Zwei Worte über den jetzigen Zustand der Naturgeschichte. Königsb. 1821, 4°.

23. Begleiter durch das königl. zoologische Museum zu Königsberg. 8°. Königsb. 1842.

24. Ueber die Geflechte in welche sich einige grössere Schlagadern der Saügethiere fruh auflösen.—Mém. prés. Ac. Pétersb. 1835.

25. Ueber doppelleibige Missgeburten in Wirbelthieren. 4°. Petersb. 1845.

26. Schilderung des thierischen Lebens auf Novaïa Zemlia.—Bull. Pétersb. III. 22.—*Wiegm.* Arch. 1839, I. p. 160.—An. of Nat. H. III. p. 145.—Edinb. New Phil. Journ. XXVIII. p. 93 (*Description of Animal Life in Nova Zembla*).

27. Schädel- und Kopfmangel an Embryonen von Schweinen, aus der frühesten Zeit der Entwickelung beobachtet. fig.—Nov. Act. Nat. Cur. XIV. p. 287.

28. Untersuchungen über die Gefässverbindung zwischen Mutter und Frucht in den Säugethieren. Ein Glückswunsch zur Jubelfeier *Sam. Thom. v. Sœmmerring's.* Leipzig, 1828, fol. fig.—N. Act. Nat. Cur. XIV. p. xiii.

29. Ueber die Entstehungsweise der Schwimmblase ohne Ausführungsgang.—Bull. Pétersb. I. p. 15.—*Wiegm.* Arch. 1837, I. p. 248.

30. Untersuchungen ueber die Entwickelungsgeschichte der Fische, nebst einem Anhange über die Schwimmblase. Leipz. 1835, 4°. —*Müll.* Arch. 1836, p. lxxvi. clxii.

31. Vorträge aus dem Gebiete der Naturwissenschaften und der Oeconomie. Königsb. 8°.—*Müll.* Arch. 1835, p. 237.

32. Ueber den Weg, den die Eier unserer Süsswassermuscheln nehmen, um in die Kiemen zu gelangen.—*Meck.* Arch. 1830, p. 320.—*Müll.* Arch. 1837, IV. p. 392.

33. Observations sur l'exfoliation de l'épiderme de l'embryon des Mammifères, appliquées à la connaissance des métamorphoses des Insectes. (Publ. par *Breschet.*)—Ann. Sc. Nat. XXVIII. 1833, p. 5.—Isis, 1835, XI. p. 973. (*Abschuppung der Oberhaut des Embryo's b. d. Säugethieren.*)

34. De Ovi Mammalium et Hominis Genesi. Königsb. 1827, 4°. fig. Lips. 1828, id.—Bibl. med.-ch. p. 26.—Trad. Fr. 4°. Par. 1829.

35. Des Branchies et des Vaisseaux branchiaux dans les embryons des animaux vertébrés.—Ann. Sc. N. 1828, XV. p. 266, 280.

36. De Ovo Mammalium.—*Wiegm.* Arch. 1836, II. p. 255.

37. Beiträge zur Kenntniss der niedern Thiere.—N. Acta Ac. Nat. Cur. XIII. p. 523, fig.—Isis, 1829, II. p. 133.—*Féruss.* Bull. 1829, XVI. p. 291. (*Recherches sur plus. Animaux inférieurs.*)

38. Ueber die Entwickelungsgeschichte der Thiere. Königsb. 1828–1837, 2 vols. 4°. fig. col.—Isis, 1829, II. p. 206.—*Féruss.* Bull. 1829, XVIII. p. 95. (*Hist. du développement des Animaux,* etc. trad. par *Breschet.* Paris, 1836, 4°.)

39. Beobachtungen über die Planarien.—N. Act. Nat. Cur. XIII. p. 2.—Isis, 1830, II. p. 183.—Ann. Sc. N. XV. 1828, p. 139, fig. (*Observations sur les Planaires,* etc.)

40. Ueber den Bau v. *Medusa aurita,* in Bezug auf *Rosenthal's* Darstellung derselben.—Isis, 1826, VIII. p. 847.—*Féruss.* Bull. 1828, XIII. p. 262. (*Sur la structure de la Medusa aurita.*)

41. Ueber die Spermatozoen.—N. Act. Nat. Cur. XIII. 2, p. 595. —*Burdach's* Physiol. I. p. 93. (2te Aufl. p. 116.)—*Fror.* Notiz. XIII. No. 265, p. 3.—*Müll.* Arch. 1837, IV. p. 382.

42. Schmarotzer Thiere der *Paludina vivipara.*—N. Act. Nat. Cur. XIII. 2, p. 610.—*Müll.* Arch. 1835, III. IV. p. 254.

43. Ueber den Braunfisch (*Delphinus Phocæna*).—Isis, 1826, VIII. p. 807.—*Féruss.* Bull. 1828, XIII. p. 117. (*Sur l'anatomie du Marsouin.*)—Bull. Ac. Pétersb. p. 26.—*Wiegm.* Arch. 1837, II. p. 193.

44. Selbstbefruchtung an einer hermaphroditischen Schnecke beobachtet.—*Müll.* Arch. 1835, p. 224.

45. Vorlesungen über Anthropologie für den Selbstunterricht. Königsb. 1824, 8°. fig. fol.—Isis, 1826, IX. p. 937.

46. Ueber den Zubr oder Auerochsen des Kaukasus.—Bull. Sc. Pétersb. I. No. 20, p. 153.—*Wiegm.* Arch. 1837, I. p. 268.

47. Noch ein Wort über das Blasen der Cetaceen.—Isis, 1828, IX. p. 927.

48. Häutung der Krebse.—*Müll.* Arch. 1834, p. 510.

49. Von der Verwandtschaft der Spermatozoen u. Cercarien in Lebensverhältnissen, Bildungsstätte und Form.—*Burd.* Physiol. I. p. 95.—*Müll.* Arch. 1835, p. 220.

50. Ueber *Linné's* im Wasser gefundene Bandwürmer.—Verhandl. der Ges. Naturforsch. Fr. in Berlin, 1829, I. 6, p. 388.—*Féruss* Bull. 1829, XVIII. p. 313.

BAER (K. E. v.) et HELMERSEN (G. v.).

1. Beitr. zur Kenntniss des Russ. Reiches ; vide sup. pars I. p. 7.

BAERENS ( ).

1. Lentis crystallinæ Monographia physiologico-pathologica. Tub.
1819, 4°.

BAERIUS (Nic.).

1. Arctophonia, h. e. Ursi laus et fraus, virtus et virus, rythmis La-
tino-germanicis 160 strophis modulata.   4°.—*Böhm.* Bibl. II. 1,
p. 331.

2. Phalainodia et Crocodilophonia.   Wallfisch u. Crocodillgedichte
aus heil. Schrift u. Weltgeschichte.   Brem. 1702, 4°.—*Böhm.*
Bibl. II. 1, p. 485.

3. Korakophonia, h. e. Corvi laus et vituperium, virtus et vitium,
versibus 672 rythmiticis leoninis et 78 stroph. Germ.   Bremæ,
1700, 4°.—*Böhm.* Bibl. II. 1, p. 556.

4. Ornithophonia, sive Harmonia melicarium avium.   4°.   1695.

BAGARD (Ch.).

1. Diss. sur la cause physique des Tremblemens de terre, et les épi-
démies qu'ils occasionnent.   Par. 1763, 4°.—Biogr. un. III. p. 210.

BAGET (J.).

1. Osteologie ; premier Traité dans lequel on considère chaque Os
par rapport aux parties qui le composent, aux cavités qui s'y
trouvent, et à ses jonctions avec les autres Os.   12°.   Paris, 1731.
—Cat. R. Soc. L.

BAGGE (H.).

1. Diss. inaug. de evolutione Strongyli auricularis et Ascaridis acu-
minatæ Viviparorum.   Erlang. 1841, 4°. fig.

BAGGESEN ( ).

1. Hebung Dänemarks.—*L. u. Br.* N. Jahrb. 1843, p. 107.

BAGLIVI (G.).

1. Opera omnia med.-practica et anatomica.   4°.   Lyon, 1704, 1710,
1715, 1745 ; Paris, 1711 ; Anvers, 1715 ; Bâle, 1737 ; Vénise,
1754, etc.—1788, 2 v. 8°. éd. de *Pinel.* 1827 et 1828, 2 vols.
8 pl. ed. *C. G. Kühn.*—Biogr. un. III. p. 216.

2. De circulatione sanguinis in Testudine experimenta, cum ejus
animalis cordis anatome. 1700. *Dum. et Bib.* I. p. 437.

3. De Praxi medicâ ad priscam observandi rationem revocandâ libri
duo.   Accedunt dissertationes novæ, de Anatome, morsu et effec-

tibus Tarantulæ, ubi obiter de Ovis Ostrearum detectis, de naturâ Lapidis serpentini, vulgò *Cobra de Capelo.*—Romæ, 1696, 8°. fig. —Act. Erudit. XVII. p. 33.

4. Specimen quatuor librorum de Fibrâ motrice et morbosâ, cui annexæ sunt quatuor Dissertationes. Romæ, 1702; Ultrajecti, 1703 ; Basileæ, 1703, 8°. fig. 19.—Act. Erudit. XXII. p. 330.

5. De Vegetatione Lapidum. 1703.—Opera, ed. 1710, p. 497.

BAHRING (D. Eb.).

1. Beschreibung der Saala.   Lemgo, 1744.—*Böhm.* Bibl. I. 1, p. 596.

BAIER (Chr. W.).

1. De generatione Insectorum in Corpore humano.   Altd. 1740, 4°. fig.—*Böhm.* Bibl. II. 2, p. 380.

BAIER (F. Jac.).

1. Epistola itineraria ad *Trew* (recensens memorabilia in Franconiâ observata). Norib. 1765, 4°. fig.—Act. Nat. Cur. IV. App. p. 1. —(German. à *Krünitz*) N. Hamb. Mag. LVIII. p. 313.—*Böhm.* Bibl. I. 1, p. 587.

BAIER (J. J.).

1. Monumenta Rerum petrificatarum præcipua, Oryctographiæ Noricæ Supplementi loco jungenda, edita per filium *F. J. Baierum.* Norib. 1757, fol. fig.—*Böhm.* Bibl. IV. 1, p. 124.

2. Epistolæ ad Viros eruditos, eorumque Responsiones, Historiam litterariam et Physicam specialem explanantes, cur. filio *F. Jac. Baiero.* Francof. et Lips. 1760, 4°. fig.—*Böhm.* Bibl. I. 1, p. 99.

3. Sciagraphia Musei sui ; acc. Supplementa Oryctographiæ Noricæ. Norib. 1730, 4°. fig.—*Böhm.* Bibl. I. 1, p. 390.

4. Oryctographia Norica, sive Rerum fossilium et ad regnum minerale pertinentium, in territorio Norimbergensi ejusque viciniâ observatarum succincta descriptio; supplementis A. 1780 editis. Norimb. 1708, 4°. fig.—*Mod.* Bibl. Helm.—*Böhm.* Bibl. IV. p. 123.—Cum Supplementis, Nor. 1758, fol. fig.

5. De Ephemeri Vitâ. (Adagium medicinalium Centuria, p. 54. Tref. et Leipz. 1718, 4°.)—Bibl. Entom. I. p. 13.

BAIER (J. W.).

1. Specimen Physicæ conciliatricis variarum de Origine Fontium Sententiarum.   Altd. 1719, 4°.—*Böhm.* Bibl. V. p. 65.

2. Diss. de Fontibus annonæ difficultatem portendentibus, vulgò *Hungerbrunnen.* Altd. 1709, 4°.—*Böhm.* Bibl. V. p. 68.

3. Diss. de Aquilâ et Muscâ ferreâ. Altd. 1707, 4°.—*Böhm.* Bibl. II. 1, p. 552.

4. Fossilia Diluvii universalis Monumenta. Altd. 1712, 4°.—*Böhm.* Bibl. IV. 2, p. 234.

5. Diss. de Behemoth et Leviathan, i. e. Elephante et Balænâ. Altd. 1708, 4°.—*Böhm.* Bibl. II. 1, p. 78.

6. Diss. de Sensu Brutorum. 4°.—*Böhm.* Bibl. II. 1, p. 102.

BAILEY (J. W.).

1. On the existence of Siliceous Spiculæ in the exterior Rays of Actiniæ.—Ann. and Mag. N. Hist. XII. p. 38.—Proc. Bost. Soc. N. H. 1842, p. 77.

2. A Sketch of the Infusoria of the family *Bacillaria,* with some account of the most interesting species which have been found in a recent or fossil state in the United States.—*Sill.* Am. J. XLI. p. 284; XLII. p. 88; XLIII. p. 321.

3. On spirally-dotted or scalariform Ducts in Anthracite of Pennsylvania.—*Sillim.* Journ. ser. 2, I. p. 407.—Journ. Geol. Soc. II. pt. 2, p. 94.

4. Account of an Excursion to Mount Katahdin, in Maine.—Amer. J. XXXII. 1, p. 20.

5. Fossil *Foraminifera* in the Greensand of New Jersey.—*Sill.* Am. J. XLI. p. 213.

6. American Polythalamia from the Upper Mississippi and from the Cretaceous formation on the Upper Mississippi.—*Sill.* Am. Journ. XLI. p. 400.

7. On some Infusorial Deposits in America.—Ann. and Mag. N. Hist. XV. p. 214.

BAILEY (S.).

1. Review of Berkeley's Theory of Vision. Lond. 1842, 8°.

BAILEY (T.).

1. On the Gravel Deposits in the neighbourhood of Basford.—Proc. Geol. Soc. III. 411.—Phil. Mag. ser. 3, XIX. p. 525.

BAILLIE (J.).

1. Mémoire contenant des observations géologiques recueillies dans un voyage en Perse, à Bushire, dans le Golfe Persique, à Téhéran.—*Féruss.* Bull. p. 117.

BAILLIE (Matth.).

1. The Morbid Anatomy of some of the most important parts of the Human Body. 8°. Lond. 1793 ; 1797.—(Trad. Fr. par *Guerbois*) *Anatomie pathologique des Organes les plus importans du Corps humain ; ouvr. enrichi de Notes et de Planches.* 8°. Par. 1815.—(Germ.) *Anatomie des Krankhaften Baues von einigen der wichtigsten Theile des menschlichen Körpers.* Berl. 1794.

BAILLIF (Roch. le).

1. Petict Traité de l'Antiquité et Singularité de la Bretagne Armorique. 1577, 8°.—*Böhm.* Bibl. IV. 1, p. 101 ; V. p. 174.

BAILLON (Em.).

1. Sur les Sables mouvans qui couvrent les côtes du Dépt. du Pas-de-Calais, et les moyens de s'opposer à leur invasion. Par. 1791. —Biogr. Un. III. p. 235.

BAILLON (L. A. J.).

1. Observations sur le Cygne de Bewick. (Lettre à *M. de Blainville.*)—C. R. VII. p. 1021.

2. Catalogue des Mammifères, Oiseaux, Reptiles, Poissons et Mollusques testacés marins observés dans l'Arrondissement d'Abbeville. Abbev. (sans date).

BAILLY (E. M.).

1. Considérations sur l'influence des circonstances extérieures dans les conceptions et les naissances masculines et féminines.—Ann. Sc. n. 1825, V. p. 47.

2. Mémoire sur l'usage des Cornes dans quelques animaux et particulièrement dans le Buffle.—Ann. Sc. nat. 1824, II. p. 369, fig.

BAILLY (Jos.).

1. Essai géologique et physique sur la possibilité d'obtenir des eaux jaillissantes dans le Dép^t du Doubs, au moyen des Puits Artésiens. Besanç. 1830, 8°.

BAILLY (M.).

1. On the Angling Filaments of the *Lophius piscatorius*, or Baudroie of the French, and Angler, Frog-fish, and Sea Devil of the English Naturalists.—Ann. des Sciences nat. II. 1824, p. 523 — Dublin Phil. Journ. I. p. 94.

BAILLY ( ).

1. Geognostische Bemerkungen über Isle-de-France.—(*Milb.* Voy. pittor. à l'I. de Fr. etc.)—*Leonh.* Zeitschr. 1825, I. p. 136.

BAILLY DE MERLIEUX ( ).

1. Mémorial Encyclopédique ; *vide sup.* Pars I. p. 53.

BAIN (A. G.).

1. On the Discovery of the Fossil Remains of Bidental and other Reptiles in South Africa.—Trans. Geol. Soc. ser. 2, VII. p. 53.—Edinb. New Phil. Journ. XXXIX. p. 333.—Journ. Geol. Soc. I. p. 317.

2. On the Discovery of the Piths and portions of the Head of an Ox in the alluvial banks of the Modder, South Africa.—Proc. Geol. Soc. III. 152.

3. On the Geology of the S.E. extremity of Africa.—Geol. Soc. Lond. 1845, Jan.—Ann. and Mag. N. Hist. XV. p. 138.

BAINBRIDGE (W.).

1. Descriptions of new species of *Cetoniadæ*, etc., with Observations on the genus *Osmoderma.*—Trans. Entom. Soc. III. p. 214.—Ann. and Mag. N. H. VI. p. 481.

2. On several species of *Bolboceras*, Kirby, from New Holland, in the Collection of the Rev. F. W. Hope.—Trans. Entom. Soc. III. p. 79.

BAIRD (J.).

1. On the Geology of the Rock of Gibraltar and the adjacent country.—Edinb. Phil. Journ. VII. p. 75.

2. On the Rocks in the neighbourhood of St. John's, Newfoundland.—Mem. Wern. Soc. IV. p. 151.—Phil. Mag. LX. p. 206.—*Féruss.* Bull. 1829, XVI. p. 386.

BAIRD (W.).

1. On the Luminousness of the Sea.—Mag. Nat. Hist. ser. 1, III. p. 308 ; IV. pp. 284, 510.

2. On a specimen of *Lemur tardigradus*, Lin., kept alive at Edinburgh.—Mag. Nat. Hist. ser. 1, I. p. 208.—Ed. New Phil. Journ. III. p. 195.—Isis, 1832, VII. p. 692.

3. On the Portuguese Man-of-War.—Mag. Nat. Hist. ser. 1, IV. p. 475.

4. On the Natural History of the British Entomostraca.—Mag. Zool. and Bot. I. pp. 35, 309, 514 ; II. pp. 132, 400.—*Wiegm.*

Arch. 1837, II. p. 256.—Ann. and Mag. of N. Hist. I. p. 245 ; XI. p. 81.

BAJON ( ).

1. Mémoires pour servir à l'Hist. de Cayenne, etc.   Paris, 1777, 2 vols. 8°.—*Cuv.* R. An. III. p. 334.—(Germ.) Erfurt, 1780, 2 vols. 8°.—*Böhm.* Bibl. I. 1, p. 752.

2. Sur un Poisson électrique, connu à Cayenne sous le nom d'*Anguille tremblante.*—Journ. de Phys. III. p. 47.—Mém. Hist. Cay. II. p. 287.

3. Observation sur les Corps Lumineux qui brillent dans l'obscurité sur la mer.—*Rozier,* Journ. de Phys. III. 1774, p. 106.

BAKER (D. Ersk.).

1. A Letter to *M. Folkes,* containing Considerations on two extraordinary Belemnitæ.—Phil. Trans. XLV. p. 598, fig.—*Mod.* Bibl. Helm.

2. A Letter concerning the property of Water Efts in slipping off their skins as serpents do.—Phil. Trans. XLIV. N° 483, p. 529. —Dict. Sc. nat. XV. p. 254.

BAKER (H.).

1. Of Microscopes and the discoveries made thereby. Lond. 1745, 8°. 2 vols.—ed. 2, 1785.—Bibl. Ent. I. p. 13.

2. Experimenta et Observationes de Scarabæo qui tres annos sine alimento vixit.—Phil. Trans. N° 457, p. 441.—Bibl. Ent. I. p. 13.

3. Essays on the Natural History of the Polypes. Lond. 1743, 8°. fig.—(Gall. *Demours*) *Essai sur l'H. n. des Polypes.* Paris, 1744, 8°. fig.—*Böhm.* Bibl. II. 2, p. 513.

4. On the Grubs destroying the Grass in Norfolk.—Phil. Trans. XLIV. p. 576.—Bibl. Ent. I. p. 13.—*Mod.* Bibl. Helm.

5. On a new discovered Sea-Insect, which he calls the Eye-sucker. —Phil. Trans. XLIII. p. 35.

6. The Microscope made easy. Lond. 1743, 8°; 1769, 8°.—(Gall.) *Le Microscope à la portée de tout le monde.*   Paris, 1754, 8°.— (Germ.) *Zum Gebrauch leicht gemachtes Microscopium.*   Zür. 1756, 8°.—*Mod.* Bibl. Helm.

7. Employment for the Microscope, in two parts : 1. An examination of Salts and saline substances, their amazing configurations and crystals ; 2. An account of various Animalcules, with observations and remarks. Lond. 1753, 8°. fig. ; 1764, 8°.—(Holl.) *Nuttig Gebruyk,* etc. Harl. 1754 ; Amst. 1756, 8°. fig—*Böhm.* Bibl. I. 1, p. 359.—(Germ.) *Beyträge zum nützlichen u. vergnä-*

genden Gebrauch u. Verbesserung des Microscopii. Augsb. 1754, 8°. fig.

8. Microscopical Observations. Lond. 1768.—*Böhm.* Bibl. I. 1, p. 360.

9. Observations on a Polype dried.—Phil. Trans. XLII. No. 471, p. 616.—*Böhm.* Bibl. II. 2, p. 514.—*Mod.* Bibl. Helm.

10. A Description of a curious Echinites.—Phil. Trans. XLIV. p. 432, fig.—*Mod.* Bibl. Helm.—*Klein,* Disp. Echin. p. VIII.

11. A Letter concerning some Vertebræ of Ammonitæ or Cornu Ammonis.—Phil. Trans. XLVI. p. 37, fig.—*Mod.*

12. An Account of the Sea Polypus (*Sepia*).—Phil.Trans. L. p. 777, fig.—*Mod.* Bibl. Helm.

13. An Account of some uncommon Fossil Bodies.—Phil. Trans. XLVIII. p. 117.

14. A Letter concerning an extraordinary large Fossil Tooth of an Elephant, found in Norfolk.—Phil. Trans. XLIII. p. 331.— (Germ.) *Von einem in der Erde gelegenen ausserordentlich grossen Elephanten-Zahne.*—Hamb. Mag. p. 453.

15. On an extraordinary Fish, called in Russia Quab; and on the Stones called Crab's-eyes.—Phil. Trans. XLV. p. 174.

BAKER (T. B. L.).

1. Ornithological Index, arranged according to the Synopsis Avium of *Mr. Vigors.* Lond. 1835, 8°.— *Wiegm.* Arch. 1836, II. p. 263.

BAKER (W. E.).

1. On a Fossil Elephant's Tooth from Somrotee near Nahun.—J. A. S. B. III. p. 638.

2. On the Fossil Elk of the Himalaya.—J. A. S. B. IV. p. 506.

3. Selected Specimens of the Sub-Himalayan Fossils in the Dadupur collection.—J. A. S. B. IV. p. 565.

4. On the Fossil Camel of the Sub-Himalayas.—J. A. S. B. IV. p. 694.—Ann. Sc. nat. 1837, VII. p. 62.

BAKER (W. E.) and DURAND (H. M.).

1. Fossil Remains of the smaller Carnivora from the Sub-Himalayas. —J. A. S. B. V. p. 579.

2. On the Fossil Jaw of a Gigantic Quadrumanous animal allied to the genera Semnopithecus and Cynocephalus.—Journ. Asiat. Soc. Bengal, V. p. 739.—Phil. Mag. ser. 3, XI. p. 33.—Ann. Sc. nat. 1837, VII. p. 370.

3. Table of Sub-Himalayan Fossil Genera in the Dádúpur collection.—J. A. S. B. V. pp. 291, 486, 661, 739.

BAKEWELL (R.).

1. On the Geology of Northumberland and Durham; and Remarks on Mr. Westgarth Forster's Section of the Strata, with a Sketch of the Physical Structure of that part of England from the German Ocean to the Irish Channel.—Phil. Mag. XLV. p. 81.

2. On the Great Derbyshire Fault.—Phil. Mag. XLII. p. 121.

3. On the Thermal Waters of the Alps.—Phil. Mag. ser. 2, III. p. 14.—*Féruss.* Bull. XV. p. 343.

4. On the Theory of the Progressive Development of Organic Life. —Phil. Mag. ser. 2, IX. p. 33.—Ann. Sc. nat. XXII.

5. On the Coal-field at Bradford near Manchester.—Trans. Geol. Soc. ser. 1, II. p. 281.

6. An Introduction to Geology, illustrative of the general surface of the Earth, etc. 8°. Lond. 1813; 1828; 1833, 8°. fig.—*Féruss.* Bull. 1828, p. 321.—(Germ. *K. H. Müller*) *Einleitung in die Geologie,* etc. Freyb. 1819, fig.—*Leonh.* Tasch. 1822, I. p. 158.—(Amer. edit. with an Appendix by *Prof. Silliman.*) N. Hav. 1839, 8°.—*Sill.* Am. Journ. XXXVI. p. 201.

7. Élémens de Géognosie, traduits d'après la 1ʳᵉ édition, avec planche. Berlin, 1830.—*Féruss.* Bull. 1830, XXI. p. 336.

8. Voyage dans la Tarentaise. 2 vols. in 8°. Partie géologique. —*Féruss.* Bull. 1826, VII. p. 183.

BAKEWELL (R., Jun.).

1. Visit to the Mantellian Museum at Lewes.—Mag. Nat. Hist. ser. 1, III. p. 9.—*Féruss.* Bull. XX. p. 376.

2. On the Falls of Niagara, and the physical structure of the adjacent country.—Mag. Nat. Hist. ser. 1, III. p. 117.—*Féruss.* Bull. XXIV. p. 141.

3. On the Fossil Remains of Elephants and other large Mammalia in Norfolk.

4. On the recent discovery of Gold Mines in the United States.— Mag. Nat. Hist. ser. 1, V. p. 434.

BAKKER (Gerb.).

1. Icones pelvis feminæ catagraphicè sectæ, etc.; addito schemate duplici capitis et trunci infantilis. 4°.—Isis, 1817, I. p. 116.

2. De Naturâ Hominis liber elementarius. Gron. 1827, 2 vols. 8°. —Biogr. Un. Suppl. LVII. p. 78.

3. Osteographia Piscium, Gadi præsertim æglefini, comparati cum

Lampride guttato, specie rariori.   Gron. 1822, 4°. fig.—N. Act. Nat. Cur. XI. p. lxvii.—Isis, 1829, I. p. 102.

BALBI (Adr.)

1. Série chronologique des plus importans envahissemens connus, faits par la mer, depuis le VIIIe siècle jusqu'à nos jours.   Tableau rédigé par cet auteur, pour le Traité élémentaire de Géographie par *Malte-Brun.—Féruss.* Bull. 1830, XX. p. 14.

BALBINUS (B.).

1. Curiosa Naturæ Arcana inclyti Regni Bohemiæ, etc.   Pragæ, 1724, fol.

2. Miscellanea historica Regni Bohemiæ.   Pragæ, 1679 et 1682, 2 vols. fol.—*Böhm.* Bibl. I. 1, p. 577.

BALBINUS (Bah. Al.).

1. Examen melissæum, id est novarum Apicularum Colonia.   Vien. 1670, 12°.—*Böhm.* Bibl. II. 2, p. 282.

BALBO (Prosp.).

1. Sur le Sable aurifère de l'Orco et de ses environs.—Mém. Acad. Tur. VII. p. 401.

2. Extrait des Mémoires de *M. Belly* sur la Minéralogie de la Sardaigne.—Mém. Ac. Tur. IX. p. 145.

BALD (R.).

1. On the Coal Formation of Clackmannanshire.—Mem. Wern. Soc. I. p. 479; III. p. 123.

2. On the Coal-field and accompanying Strata in the vicinity of Dalkeith, Mid-Lothian.—Edinb. New Phil. Journ. IV. p. 115.— *Féruss.* Bull. 1828, XV. p. 343.

3. On the Fossil Elephant of Scotland.—Mem. Wern. Soc. IV. p. 58.—*Féruss.* Bull. 1829, XVII. p. 187.

4. On the discovery of the Skeleton of a Whale on the estate of Airthrey, near Stirling.—Edinb. Phil. Journ. I. p. 393.

5. " Coal," and other articles in Edinb. Encyclopædia.

6. On the Mushet Band, commonly called the Black-band Ironstone of Scotland.—Rep. Brit. Assoc. 1846, Sect. p. 62.

BALDACCONI (Fr.).

1. Esposizione di un Metodo per ottenere la perfetta ridozione delle

Sostanze animali a solidità lapidea.—Att. Accad. Fisiocrit. di Siena, X.

BALDÆUS (Phil.).

1. Beschryving deer Ost-Indische Küsten Malabar, Coromandel, Ceylon. Amst. 1672, fol. fig.—(Deutsch) Amst. 1672, fol. fig. —*Böhm.* Bibl. I. 1, p. 685.—(Angl.) *Churchill,* Voy. III. p. 561.

BALDANUS (Ant.).

1. Locustæ majores, quibus Johannes in deserto vitam tolerare dicitur.—Comm. Bonon. V. p. 53.—Bibl. Entom. I. p. 14.

BALDASSARI (Gius.).

1. Osserv. sopra il Sale della Creta, sopra un Saggio di Produzioni naturali dello Stato Senese. Siena, 1750, 8°.—(Deutsch) Hamb. Mag. X. pp. 227, 339.—*Böhm.* Bibl. IV. 1, p. 191.

2. Delle Acque minerali di Chianciano nel Senese. Siena, 1756, 4°. —*Böhm.* Bibl. V. p. 247.

3. Tentamen observationum super res quasdam naturales in pratis et aliis locis maritimis Maremmæ ditionis Senensis.—Atti Acad. Sc. Sien. II. p. 1.—*Böhm.* Bibl. I. 1, p. 567.

4. Dissertationes; 1ª de omnium hominum primo; 2ª de virtute Arboris Vitæ; 3ª de antediluvianorum hominum cibo; 4ª de universitate Diluvii. Venet. 1757, 4°.

5. Osserv. geologiche e paleontologiche. Atti di Siena, III. p. 243.

BALDASSINI (Fr.).

1. Sull' emissione di un Liquido colorante per parte dei Molluschi, e sulla causa della simmetrica ed uniforme sua distribuzione alla superficie delle Conchiglie.—Mem. Accad. Torin. 1842.

2. Considerazioni sul modo con cui si suppone che i Molluschi litofagi perforino le Roccie. Bol. 1830, 8°.—Ann. di St. nat. Fasc. X. p. 47.—Isis, 1833, XI. p. 1099.—(*Ueber die Art, wie die steinfressenden Weichthiere die Felsen durchbohren sollen.*)

3. Sopra le Conchiglie considerate come parti integranti del corpo dei Molluschi, etc.—Ann. R. Lomb.-Ven. 1835, V. p. 225; VI. p. 3.

4. Dell' anteriorità del *Marsigli* sul *Reaumur* nella Teoria della formazione ed accrescimento delle Conchilie.

5. Considerazioni intorno all' *Argonauta.*—Ann. R. Lomb.-Ven. 1836, VI. p. 209.

6. Cenni ulteriori intorno alle Conchiglie considerate come parti integranti dell' organizzazione dei Molluschi.—Ann. med.-chir. di Roma, 1839, I. 2.

7. Nota intorno al genere *Bullea* di *Lamarck*—Ann. Sc. Regn. Lomb.-Ven. 1833, V. p. 225.

8. Della distribuzione dei Colori sulla superficie delle Conchilie, etc. —Mem. Accad. Tor. V. Ser. 2ª.

BALDEN (J.).

1. On Insects injurious to Forest Trees.—Prize Essays, Highland Soc. XIV. p. 114.

BALDENSTEIN (H. v.).

1. Beiträge zur Naturgeschichte des Bartgeyers.—Denkschr. Allg. Schweiz. Ges. I. 1, p. 86.

BALDINGER (E. G.).

1. Sur l'âge d'un Brochet.—Med. Journ. V.—*Cuv.* et *Val.* Hist. nat. des Poiss. J. p. 169.

2. Neues med. u. phys. Journal ; *vide sup.* Pars I. p. 29.

BALDIUS (Baldus).

1. De admirabili Viperæ naturâ. Hag. 1660, 12°.

2. Dissertatio in Hippocratem de Aëre, Aquis et Locis, in quâ et de Calculorum causis disseritur. Romæ, 1637, 4°.—*Böhm.* Bibl. IV. 2, p. 332 ; V. p. 11.

3. De Magnetis Antipathiâ. Romæ, 1612, 8°.—*Böhm.* Bibl. IV. 2, p. 169.

BALDRACCO ( ).

1. Ueber einige Gold-Gänge in den Ligurischer Apenninen.—Isis, 1841, p. 559.—*L.* u. *Br.* N. Jahrb. 1843, p. 361.

BALINGHEM (Ant. de).

1. Zoopædia, seu Morum a Brutis petita Institutio. Andomari, 1621. 12°.—*Böhm.* Bibl. II. 1, p. 102.

BALK (Laur.).

1. Museum Adolpho-Fridericianum, dissert. præs. *C. Linnæo.* Holmiæ, 1746, in 4°. fig.—Amœn. Acad. I. pp. 278, 556.

BALL (J.).

1. Remarks upon the Geology and Physical Features of the country west of the Rocky Mountains, with miscell. facts.—Am. J. XXVIII. 1, p. 1.—*L.* u. *Br.* N. Jahrb. 1837, p. 688.

BALL (R.).

1. On the Species of Seals (Phocidæ) inhabiting the Irish Seas.—Proc. R. Irish Acad. 2, p. 17.—Trans. R. Irish Acad. XVIII. p. 89. —*Wiegm.* Arch. 1837, V. p. 190.

2. Occurrence of the *Cuculus glandarius* in Britain.—Ann. and Mag. N. Hist. XII. p. 149.

3. On Noises produced by one of the Notonectidæ.—Rep. Brit. Assoc. 1845, Sect. p. 64.—Ann. Nat. Hist. XIV. p. 129.

4. Description of the *Cydippe pomiformis*, Patters. (*Beroë ovata*, Flem.), with Notice of an apparently undescribed species of *Bolina*, also found on the Coast of Ireland.—Ann. of Nat. H. III. p. 60.

5. On the Remains of Oxen found in the Bogs of Ireland.—Ann. of Nat. H. III. p. 270.—Proc. R. Irish Acad. 15, p. 253.

6. On a species of *Loligo* found on the Shore of Dublin Bay (*L. Eblanæ*, B.).—Proc. R. Irish Acad. 19, p. 362.—Ann. of Nat. H. V. p. 68.

7. On the *Cephalopoda* of the Irish Seas.—Proc. R. Irish Acad. 32, p. 192.—Ann. and Mag. N. H. IX. p. 348.

8. Description of the *Felis melanura*.—Proc. Zool. Soc. 1844, p. 128. —Ann. and Mag. N. Hist. XV. p. 286.

9. On a new British Cuckoo (*Coccyzus americanus*).—Field Naturalist's Mag. I. p. 6.

10. On the Pectinated Claws of Birds.—Field Naturalist's Mag. I. p. 372.

11. On a new Sturgeon (*Acipenser Thompsoni*).—Proc. Roy. Irish Acad. No. 25, p. 21.

BALLENSTEDT (J. G. J.).

1. Archiv für Entdeckungen, etc.; *vide sup.* Pars I. p. 33.

2. Entdeckung von Insecten-Nestern der Urwelt in Bernstein.— *Ball.* Arch. V. p. 28.

3. Ueber die neuesten Entdeckungen aus der Urwelt in Obersachsen. —*Ball.* Arch. V. p. 53.

4. Neuer Beweis des Daseyns von Riesenmenschen der Urwelt.— *Ball.* Arch. I. p. 48.

5. Ueber die Anthropolithen gegen *Ludwig.*—*Ball.* Arch. I. p. 181.

6. Zersetzung von urweltlichen Menschen in der Lehmgrube v. Pabstorf.—*Ball.* Arch. II. p. 95.

7. Neue Gründe für das Vorkommen v. menschlichen Ueberresten aus der Urwelt.—*Ball.* Arch. III. p. 247.

8. Die Urwelt, oder Beweis von dem Daseyn und Untergange v. mehr als einer Vorwelt. Quedl. u. Leipz. 1818–19, 2 vols. 8°. (3te Aufl.)—Cat. Bibl. Turic.

9. Wie konnten die tropischen Thiere der Urwelt in nördlichen Clima leben?—*Ball.* Arch. II. p. 75.

10. Ueber die in dunkler Vorzeit stattgefundenen Wasserbedeckungen der Festländer.—*Ball.* Arch. VI. p. 40.

11. Fernere Schicksale der Urwelt in Holland.—*Ball.* Arch. III. p. 407.

12. Die Versteinerungen des Elmgebirges.—*Ball.* Arch. IV. p. 44.

13. Die neuesten Entdeckungen aus der Vorwelt am Elmwalde.— *Ball.* Arch. V. p. 282.

14. Merkwürdige Ausgrubungen in Böhmen u. Mähren im 17^{den} Jahrhunderte.—*Ball.* Arch. V. p. 68.

BALLENSTEDT (L. G. S.).

1. (*Kleine Schriften*, etc.)—Petits écrits géologiques, historiques, topographiques, étymologiques, concernant les Antiquités. 2 vols. Nordhausen, 1826.—*Féruss.* Bull. 1828, XIV. p. 171 ; 1829, XVIII. p. 321.

BALLING (Fr. A.).

1. Kissingen, ses Eaux minérales et ses Bains.   Francf. 12°.

BALLING (J. G.).

1. System der Naturphilosophie.   8° Würzburg, 1828.

BALLUS (P. v.).

1. Beschr. der den Obstbaümen schädlichsten Raupenarten. 8°. Presburg, 1829.

BALOGH DE F. ALMAS (P.).

1. De Evolutione et Vitâ Encephali.   Pest. 1823, 8°.—Isis, 1826, III. p. 327.

BALSAMO-CRIVELLI (Gius.).

1. Descrizione d' un nuovo Rettile fossile, della famiglia dei *Plesiosauri*, e di due Pesci fossili, trovati nel Calcareo nero, sopra Varenna sul Lago di Como, dal nob. Sign. *L. Trotti*, etc.—Polit. di Milano, 1839, fig.

2. Mem. per servire all' illust. dei grandi Mammiferi fossili nel gabinetto di Santa Theresa.   8°. Milan, 1842.

BALTHASAR (Theod.).

1. Nachricht von einem Gesundbrunnen, welcher ohnweit Erlangen jüngsten erfunden worden. Erlang. 1709, 4°.—*Böhm*. Bibl. V. p. 305.

BAMBERG (C. T.).

1. De Avium Nervis Rostri atque Lingua. Halis, 1842, 8°.—*Val*. Repert. VIII. pp. 7, 34.

BANAU ( ).

1. Histoire naturelle de la Peau, et de ses rapports avec la santé et la beauté du corps. Paris, 1802, 8°.

BANCROFT (Ed.).

1. Experimental Researches concerning the Philosophy of Permanent Colours. Lond. 8°. 1794.—Bibl. Ent, I. p. 14.

2. An Essay on the Natural History of Guyana in South America, etc. Lond. 1769, 8°.—(Deutsch) *Naturgeschichte von Gujana, aus dem Engl*. Frankf. u. Leipz. 1769, 8°.—*Eis*. p. 149.

BANCROFT (E. N.).

1. Remarks on some Animals sent from Jamaica.—Zool. Journ. 1829, No. 17, p. 80.—Isis, 1831, VII. p. 726.

2. On the Fish known in Jamaica as the *Sea-Devil*.—Zool. J. IV. 1829, p. 444.—Isis, 1831, XII. p. 1363.

3. On several Fishes of Jamaica.—Pr. Com. Zool. Soc. Lond. I. p. 134.

BANISTER (John).

1. Histoire de l'Homme, extraite de la quintessence des meilleurs anatomistes de son temps. Lond. 1578, fol.—Biogr. Un. III. p. 314.

BANISTER (John II.).

1. Mollusques de l'Amérique sept.—Phil. Trans. XVII. p. 671. (1693.)—Trans. Linn. Soc. Lond. VII. p. 227.

2. Some observations concerning Insects, made in Virginia ann. 1680, with remarks on them, by *J. Petiver*.—Phil. Trans. 1701, No. 270, pp. 795, 807.—Badd. 4, p. 15.—Bibl. Ent. I. p. 14.

BANKIER (R. A.).

1. A new species of the Australian genus *Alcyone* (*A. ruficollaris*). —Ann. and Mag. N. H. VI. p. 394.

Banks (J.) and Solander (Dr.).

1. Letters from Iceland, containing Observations on the Civil and Natural History, Antiquities, Volcanos, Basalts, &c. Lond. 1780, 8°. fig.

Banks (Jos.).

1. Observations on the Nature and Formation of the Stone incrusting the Skeletons which have been found in the Island of Guadeloupe, with some Account of the Origin of those Skeletons.— Trans. Linn. Soc. Lond. XII. 1, p. 53.

2. Short Account of the Cause of the Disease in Corn called Blight, Mildew and Rust. Lond. 1805, 4°.—Trans. Linn. Soc. Lond. VIII. p. 360.

3. On the *Musca Pumilionis.*—Ann. of Agric. XVI.

4. On the first appearance of *Aphis lanigera* in this country.— Trans. Hort. Soc. II. p. 162.

Bar (E.).

1. Albert, ou le petit Naturaliste. Histoire des Animaux apprivoisés. Paris, 1837, 18°.

Baratte (   ).

1. De Vermibus in Sanguine.—*Vanderm.* Rec. VI. p. 306.—*Mod.* Bibl. Helm.

Barba (A.).

1. Osservazioni microscopiche sul Cervello e sue parti adjacenti. (2ᵉ ed.) Nap. 1819, 8°. fig.—(Germ. *v. Schönberg*): *Mikroskopische Beobachtungen über das Gehirn und die damit Zusammenhängenden Theile.* Würzb. 1829, fig.

Barbaleni (   ).

1. Sul modo di moltiplicare le Api.—Ann. Agric. Regno d'Ital. XXV.

Barbançois (De).

1. Observations sur la Filiation des Animaux depuis le Polype jusqu'au Singe.—Journ. de Phys. LXXXII. 1816, p. 444.

2. Observations pour servir à une nouvelle Classification des Animaux.—J. de Phys. LXXXIII. p. 67.—(Rapport de *Latreille.*) Acad. Sc. 1816.—Isis, 1817, VI. p. 709.

BARBARUS (Herm.).

1. Compendium Scientiæ Naturalis, ex Aristotele. Ven. 1545.

2. Castigationes in *Caji Plinii* libros XXXVII. Romæ, 1492 et 1493.—Mediol. 1494, fol.—Crem. 1495.—Veneti, 1497, fol.— Basil, 1534.—Tr. Linn. Soc. Lond. V. p. 277.—*Böhm.* Bibl. I. 1, p. 205.

BARBAULT (Ant. Fr.).

1. Splanchnologie, suivie de l'Angiologie et de la Névrologie. 1793, 12º.—Biogr. Un. III. p. 332.

BARBECK (F. G.).

1. Dissert. de Generatione Animalium. Duisb. 1693, 4º.—*Böhm.* Bibl. II. 1, p. 129.

BARBETTE (Paul).

1. Anatomie pratique. Amst. 1659, 8º.—Biogr. Un. III. p. 344.

BARBEU-DUBOURG (Jac.).

1. Recherches sur la durée de la Grossesse et le terme de l'Ac- couchement. 1765, 8º.—Biogr. Un. III. p. 445.

BARBIERI (Lud.).

1. Nuovo Systema intorno l' Anima delle Bestie. Venez. 8º. 1754. —*Böhm.* Bibl. II. 1, p. 95.

2. Trattato della Origine delle Sorgenti e dei Fiumi. Venez. 1750, 8º.—*Böhm.* Bibl. V. p. 66.

3. Storia del Mare, e Confutazione della favola dove scopronsi in- signi errori di vari Scrittori, e specialmente del *Sign. de Buffon.* Venez. 1781, 8º.—*Böhm.* Bibl. V. p. 82.

4. Tractatus de Corporum Principiis. Patav. 1744, 8º.

BARBOT (J.).

1. A Description of the Coasts of N. and S. Guinea, and of Ethiopia infer., vulgarly Angola.—*Church.* Coll. II. p. 1.—*Fisch.* Cat.

BARBOTEAU ( ).

1. Description d'une Mouche maconne.—Acad. Sc. 1776, Hist. p. 19.—Bibl. Ent. I. p. 14.

2. Essais sur la Fourmi.—J. de Phys. VIII. pp. 384 et 444 ; IX. pp. 21 et 88.—Bibl. Ent. I. p. 14.

BARBUCCI (A.).

1. Travaux de l'Acad. de Palerme ; *vide sup.* Pars I. p. 69.

BARBUT (Jam.).

1. The Genera Insectorum of *Linnæus* exemplified by various specimens of English Insects. (Les genres des Insectes de Linné constatés par divers échantillons d'Ins. d'Angleterre.)—Angl. et Fr. Lond. 1781, 4°. fig. col. et n.—Bibl. Ent. I. p. 14.—*Böhm.* Bibl. II. 2, p. 162.

2. The Genera Vermium exemplified by various specimens of the animals contained in the orders of the Intestina and Mollusca *Linnæi*, drawn from nature. Lond. 1783, fol. fig. (Angl. et Fr.) —*Böhm.* Bibl. II. 2, p. 373.—(Helminthologia) Portug. tr. por *F. J. Velloso.* 4°. Lisbon, 1799.

3. Testacea. 4°. Lond. 1788.

BARCHEWITZ (E. Chr.).

1. Ostindianische Reisebeschreibung. Chemni, 1730, 8°. fig.— Erfurt, 1751, 8°. fig.—*Ibid.* 1762, 8°. fig.—*Böhm.* Bibl. I. 1, p. 669.

BARCLAY (A.).

1. Ueber die Landkrabben auf Jamaica.—Ed. N. Phil. J. VII. p. 280. —Isis, 1832, VIII. p. 817.

BARCLAY (A. W.).

1. Inaugural Dissertation on Temperament. Berl. 1840, 8°.

BARCLAY (J.).

1. Series of Engravings, representing the Bones of the Human Skeleton; with the Skeletons of some of the lower Animals. Edinb. 1819, 4°.

2. On some parts of the Animal that was cast ashore on the Island of Stronsa, Sept. 1808.—Mem. Wern. Soc. I. p. 418.—Isis, 1818, p. 2096.

3. On the Structure of the Cells in the Combs of Bees and Wasps. —Mem. Wern. Soc. II. p. 259.

4. On the Causes of Organization.—Mem. Wern. Soc. II. p. 537.

5. An Inquiry into the Opinions, ancient and modern, concerning Life and Organization. Edinb. 1822, 8°.—Bibl. Ent. I. p. 14.

BARCLAY (J.) and NEILL (P.).

1. Account of a Beluga, or White Whale, killed in the Frith of Forth.—Mem. Wern. Soc. III. p. 371.

BARDENAT (J. Ph.).

1. Les Recherches physiologiques de *X. Bichat* sur la Vie et la Mort, réfutées dans leurs doctrines. 8°. Par. 1824.—Cat. Roy. Soc. L.

BARDIES ( ).

1. Discours de la Connoissance des Bestes. Par. 1672, 8°.—*Böhm.* Bibl. II. 1, p. 33.

BARELLI (Vinc.).

1. Cenni di Statistica mineralogica degli Stati di S. M. il Re di Sardegna, overo Catalogo raggionato della Racolta formatasi pressa l'Azienda gener. dell' Interno. Torino, 1835, 8°.—*L.* u. *Br.* N. Jahrb. 1835, p. 683.

BÄRENSPRUNG (Fel. v.).

1. *Excerpta Zoologica,* or Abridged Extracts from Foreign Journals.—Ann. and Mag. N. H. X. p. 47 ; XII. p. 427.

BARES (H.).

1. Notes sur quelques Coquilles.—Ann. Lyc. N. York, Jan. 1826, p. 383.—*Féruss.* Bull. 1827, XI. p. 445.

BARETTI (Jos.).

1. A Journey from London to Genoa, through England, Portugal, Spain and France. Lond. 1770, 8°. 4 vols.—(German.) *Reisen von London nach Genua, etc. aus dem Engl.* Leipz. 1772, 8°. 2 Thle.—*Böhm.* Bibl. I. 1, p. 476.

BARFEKNECHT (O. C.).

1. Dissertatio, an omne vivens ex Ovo. Par. 1733, 4°.—*Böhm.* Bibl. II. 1, p. 131.

BARHAM (H.).

1. An Essay upon the Silk-worm.   Lond. 1719, 8°.—Bibl. Ent. I.
   p. 15.—*Böhm.* Bibl. II. 2, p. 255.

2. Experiments and Observations on the production of Silk-worms,
   and of their Silk in England.—Phil. Tr. XXX. No. 362, p. 1036.
   —Bibl. Ent. I. p. 15.

BARICELLI (J. C.).

1. Hortulus genialis.   Bol. 1617, 12° ; 1621, 12°.   Genève, 1623,
   12°.—Bibl. Ent. I. p. 15.

BARING (D. E.).

1. Museographia Brunsvico-Luneburgica, oder Nachricht von Kunst-
   u. Raritäten Kammern, die curieuse Herren in den Braunschw.
   Lüneburgischen Landen gesammlet haben.   Lemgo, 1744, 4°.
   —*Böhm.* Bibl. I. 1, p. 394.

BARKAM (C. F.).

1. Quelques observations sur la Température des Mines.—Tr. Geol.
   Soc. Cornw. III. p. 150.—*Féruss.* Bull. 1829, XVI. p. 174.

BARKER (R.).

1. On a Stag's Head and Horns found at Alport, in the Parish of
   Youlgreave in the County of Derby.—Phil. Trans. LXXV. p. 353.

BARKOW (H.).

1. Syndesmologie, oder die Lehre von den Bändern, durch welche
   die Knochen des menschlichen Körpers zum Gerippe vereint
   werden.   Bresl. 1841, 8°.

BARKOW (J. C. L.).

1. Monstra Animalium duplicia per anatomen indagata, et descr. et
   icon. illustrata.   Lips. 1830–36, 2 vols. 4°. 15 pl.—Ann. Sc. nat.
   XVII. 1839.—*Val.* Repert. II. p. 24.

2. Commentatio anat.-physiolog. de Monstris duplicibus verticibus
   inter se junctis.   Lips. 1821, 4°. fig.—Bibl. med.-chir. p. 26.

3. Ueber angebornen Mangel des Unterkiefers bei Säugethieren.—
   N. Act. Nat. Cur. XV. 2, p. 289, fig.

4. Disquisitiones circa originem et decursum Arteriarum Mammalium.   Lips. 1829, 4°. fig.—Ann. Sc. nat. XVII. 1829.

5. Ueber den Verlauf der Schlagadern am Kopfe des Schafes. Ein Beitrag zur vergleichenden Gefässlehre.—N. Act. Nat. Cur. XIII. p. 359.

6. Recherches anatomico-physiologiques sur le Système artériel et quelques autres parties des Oiseaux.   fig.—*Meck.* Arch. f. Anat. u. Physiol. 1829, p. 305.—*Féruss.* Bull. 1830, XXII. p. 102.

7. Disquisitiones nonnullæ angiologicæ.   Vratisl. 1830, 4°.—Bibl. med.-chir. p. 26.

8. Disquisitiones nevrologicæ.   Vratisl. 1836, 4°. fig.—Bibl. med.-chir. p. 26.

9. Bemerkungen über Nervenanschwellungen.—N. Act. Nat. Cur. XIV. p. 515, fig.

10. Der Winterschlaf nach seinen Erscheinungen im Thierreich. 8°. Berl. 1846.

11. Entwickelung des Fetts beim Guckguck.—Isis, 1834, VII. p. 696.

BARKOW (L.).

1. Disquisitiones recentiores de Arteriis Mammalium et Avium.— N. Act. Nat. Cur. 1844, XX. 2, p. 607.

BARLÆUS (Casp.)

1. Rerum per octennium in Brasiliâ et alibi gestarum Historia. Edit. 2ª, cui accedit *G. Pisonis* Tract. de Aëre, Aquis et Locis in Brasiliâ, etc.   Cliv. 1660, 8°. fig.—*Böhm.* Bibl. I. 1, p. 761.

BARLES (L.).

1. Nouvelles Découvertes sur les Organes des Hommes servant à la Génération.   Lyon, 1675, 12°, et 1680, 4°.—Biogr. Un. III. p. 383.

2. Nouvelles Découvertes sur les Organes des Femmes servant à la Génération. Lyon, 1674, 12°, et 1680, 4°.—Biogr. Un. III. p. 383.

BARLOW (F.).

1. Diversæ Avium species studiosissimè delineatæ.   fol. Lond. 1655.

BARLOW (T. W.).

1. Chart of British Ornithology.   London, 1847.

BARLOW (Will.).

1. A paper concerning the *Mola salv.* or Sun-fish, and a Glue made of it.—Phil. Trans. XLI. No. 456, p. 343.—Dict. Sc. nat. XXII. p. 528.

BARNES (Archdeacon).

1. On the Carnelians of Cambay.—Trans. Geol. Soc. ser. 1, IV. p. 447.

BARNES (D. W.).

1. Notice of several species of Shells.—Annals of the Lyceum of New York, I. 1824, pp. 131, 383.—*Féruss.* Bull. 1827. XI. p. 445. (*Note sur quelques Coquilles.*)

2. Description of five species of Chiton.—Amer. Journ. VII. 1, p. 69.

3. On the genera *Unio* and *Alasmodonta,* with introductory remarks. —Amer. Journ. VI. pp. 107, 258; XIII. p. 358.—*Féruss.* Bull. 1828, V. p. 178. (*Remarques sur les espèces du genre Unio,* etc.) —Isis, 1832, X. p. 1055. (*Aelterrecht auf Gattungen v. Unio.*)

BARNES (H.).

1. An Arrangement of the genera of Batrachian animals, with a description of the more remarkable species ; including a Monography of the doubtful Reptiles.—Amer. Journ. XI. p. 268.—Isis, 1832, X. p. 1051 (*Anordnung der Frösche*).

2. Notes on the doubtful Reptiles.—Amer. Journ. XIII. p. 66.— *Féruss.* Bull. 1828, XV. p. 397 (*Notes sur quelques Reptiles douteux*).—Isis, 1832, X. p. 1054 (*Ueber zweifelhafte Lurche*).

3. A Geological Section of the Canaan Mountain, with observations on the soil and productions of the neighbouring region.—*Sill.* Amer. Journ. V. 1, pp. 8, 235.—*Féruss.* Bull. II. p. 51 (*Coupe géologique du Mont Canaan, avec des observations sur le sol,* etc.).

4. Sur le *Murex Corona,* Gmel. Dillw. (*Fusus Corona,* Lamk.).— Ann. Lyc. N. York, 1827, p. 291.—*Féruss.* Bull. 1828, XV. p. 175.

BARNES (Mrs.).

1. Note on the Rearing of a species of Humming-bird.—Pr. Zool. Soc. Lond. IV. p. 33.

BAROLES (W.).

1. Geschichte der Spanischen Heuschrecken. Madrid, 1781.

BAROLO (Ferd.).

1. De Glandulis ;—Glandulorum usus, etc. (Theses pro aggreg.) Taurini, 1785, 8°.

BARON (A.).

1. Recherches sur les Sauterelles et sur les moyens de les détruire. —Journ. de Phys. XXIX. p. 321.—Bibl. Ent. I. p. 15.
2. Album du Jardin des Plantes de Paris. 4°. Paris, 1837.

BARON (J.).

1. Delineations of the origin and progress of various changes of Structure which occur in Man and some of the inferior Animals. 4°. Lond. 1828.—Cat. R. Soc. L.

BARRAL (De).

1. Mémoire sur l'Histoire naturelle de l'Isle de Corse, avec un Catalogue lithologique de cette Isle. Lond. et Par. 1783, 8°. fig.— (Deutsch) 8°. Francf. 1789.

BARRAS (Mar. Thér.).

1. Mémoire sur l'Éducation des Abeilles. Paris, 1800, 8°.

BARRELIER (Jac.).

1. Specimen Insectorum quorundam marinorum, in Libro de Plantis per Galliam, Hispaniam et Italiam observatis (*A. de Jussieu*). Paris, 1714, fol. fig.—Tr. Linn. Soc. Lond. VII. p. 228.
2. Icones Plantarum rariorum per Galliam, Hispaniam et Italiam observat. ad vivum exhibitarum. Opus posthumum editum curâ et stud. *A. de Jussieu.* Paris, 1714, fol. fig.—*Lamx.* Polyp. Corall. p. 520.—*Mod.* Bibl. Helm.

BARRÈRE (Pierre).

1. Essai sur l'Histoire naturelle de la France équinoxiale. Paris, 1741, 12°; 1749, 8°.—*Dum.* et *Bib.* I. p. 426.
2. Ornithologiæ specimen novum, s. Series Avium in Ruscinone, Pyrenæis montibus, atque in Galliâ æquinoctiali observatarum, in classes, genera et species, novâ methodo digesta. Perpignan, 1745, 4°. fig.—*Cuv.* R. An. III. p. 334.—*Böhm.* Bibl. II. 1, p. 510.

3. Observations sur l'origine et la formation des Pierres figurées, et sur celles qui, tant extérieurement qu'intérieurement, ont une figure régulière et déterminée. Par. 1746, 8°.—*Böhm.* Bibl. IV. 2, p. 239.

4. Sur la Cause physique de la Couleur des Négres. 8°. Paris, 1741.

BARRINGTON (Daines).

1. On some particular Fish found in Wales.—Phil. Trans. LVII. p. 204.

2. On a Mole from North America.—Phil. Trans. LXI. p. 292.

3. On the Specific Characters which distinguish the Rabbit from the Hare.—Phil. Trans. LXII. p. 4.—Journ. de Phys. 1778, Supp. p. 255.

4. On the Periodical Appearing and Disappearing of certain Birds at different times of the year.—Phil. Trans. LXII. p. 265.

5. On a Fossil lately found near Christ-church in Hampshire.— Phil. Trans. LXIII. p. 171.

6. On the Lagopus or Ptarmigan.—Phil. Trans. LXIII. p. 224.

7. On the Singing of Birds.—Phil. Trans. LXIII. p. 249.—Journ. de Phys. 1774, p. 393.

8. On the Gillaroo Trout.—Phil. Trans. LXIV. p. 116.—*Cuv.* et *Val.* Poiss. I. p. 138.

9. Miscellanies. 4°. Lond. 1781.

10. Naturalist's Calendar. Lond. 1767 ; 4°. 1818.

BARROW (E. T.).

1. Description of *Mus castorides,* a new species.—Trans. of the Linn. Soc. XI. 1815, p. 169.

BARROW (J.).

1. An Account of Travels into the Interior of Southern Africa, in the years 1797, 1798, etc. Lond. 1801, 4°.—Trad. Franç. par *Grandpré* (Voyage dans la partie mérid. de l'Afrique). Paris, 1801, 8°.—*Lamx.* Pol. Corall. p. 520.—(Germ. *M. C. Sprengel*) *Reisen durch die innern Gegenden des Südl. Afrika,* etc. Weim. 1801 ; 1806, 8°.—(Germ. *J. A. Bergk*) Leipz. 1801–5, 2 vols. 8°.

2. Die Heuschreckenzüge in mittäglichen Africa.—*Barr.* Reise, etc. *Migrations de Sauterelles* (Extr. de son Voyage).—*Ill.* Mag. IV. p. 220.—Bibl. Ent. I. p. 15.

BARROWSKY (G. H.).

1. Gemeinnützige Naturgeschichte des Thierreichs. Berl. 1780–

1789, 8°. 10 vols. fig. col. (les 5 dern. cah. par *Herbst*).—Bibl. Ent. I. p. 16.

BARRUEL (G.).

1. Traité élémentaire de Géologie, Minéralogie et Géognosie, suivi d'une Statistique minéralogique des Départemens par ordre alphab. Par. 1835, 8°. fig.
2. Histoire naturelle inorganique: Géologie, Minéralogie et Géognosie. Par. 1835, 8°. fig.

BARRUEL (J. P.).

1. Notice sur le Fossile humain trouvé près de Moret. Paris, 1824, 8°.—Dublin Phil. Journ. I. p. 161.
2. Réponse aux principaux Écrits qui ont paru sur le Fossile humain trouvé près de Moret. Paris, 1824, 8°.

BARRY (De).

1. Mémoire sur les Fourmis des Cannes a Sucre. Paris, 1783, 4°.

BARRY (Edm.).

1. Eine Bemerkung eines blutigen Urins von einem Wurm in der Blase.—Edinb. Versuch. u. Bemerk. V. p. 388.

BARRY (M.).

1. Researches in Embryology.—Phil. Trans. CXXVIII. p. 301; CXXIX. p. 307; CXXX. p. 529; CXXXI. p. 193.
2. On the Corpuscles of the Blood.—Phil. Trans. CXXX. p. 595; CXXXI. pp. 201, 217.
3. On the Chorda Dorsalis.—Phil. Trans. CXXXI. p. 195.
4. On Fibre.—Phil. Trans. CXXXII. p. 89.—Ann. and Mag. N. H. VIII. pp. 502, 545; IX. p. 258.
5. Spermatozoa observed within the Mammiferous Ovum.—Phil. Trans. CXXXIII. p. 33.
6. The Cells in the Ovum compared with Corpuscles of the Blood. —On the difference in size of the Blood-Corpuscles in different animals.—Phil. Mag. ser. 3, June 1843.—Edinb. New Phil. Journ. XXXV. p. 320.
7. On the first changes consequent on Fecundation in the Mammiferous Ovum.—Rep. Brit. Assoc. 1840, Sect. p. 129.
8. On Fissiparous Generation.—Proc. Roy. Soc.—Edinb. New Phil. Journ. XXXV. p. 205.

9. On the Unity of Structure in the Animal Kingdom.—Edinb. New Phil. Journ. XXII. pp. 116, 345.—*Wiegm.* Arch. 1838, II. p. 310.

10. Observations in reply to *T. Wharton Jones's* Strictures. Lond. 1839, 8°.

11. Mémoire sur l'application du Baromètre à l'Étude de la Circulation du Sang et de la Respiration chez les Animaux vertébrés.— Ann. Sc. nat. 1827, X. p. 415.

12. Recherches sur le Passage du Sang à travers le Cœur.—Ann. Sc. n. XI. p. 113.—Isis, 1834, X. p. 929 (*Ueber den Durchgang des Bluts durch das Herz*).

13. Ueber die Absorption.—Ann. Sc. n. VIII. p. 315.—Isis, 1834, IX. p. 890.

14. Ascent to the Summit of Mont Blanc in 1834. Lond. and Edinb. 1836, 8°.—*Val.* Repert. III. p. 28.

Barse (J.).

1. Châtel-Guyon et ses Eaux minérales. Riom, 1840, 8°.

Bartalini (B.).

1. Catalogo dei Corpi marini dei contorni di Siena. Siena, 1776, 4°.

2. Osservazioni di Storia naturale fatte in alcuni luoghi dello Stato di Siena ed attorno di' Lagoni di Castelnuovo di Valdicecina presso Volterra.—N. Miscell. Lucch.

Bartels (A. Chr.).

1. De Janis inversis ac de Duplicitate generatim. Dissert. inaug. anat.-physiol. Berol. 1830, 4°. fig.—Bibl. med.-chir. p. 27.

Bartels (C. M. N.).

1. Beiträge zur Physiologie des Gesichtssinns. Berl. 1834, 4°. fig. —*Müll.* Arch. 1835, p. 145.—Isis, 1835, I. p. 95.

Bartels (Er. Dan. A.).

1. Anfangsgründe der Naturwissenschaft. Leipz. 1821, 2 vols. 8°. —*Féruss.* Bull. 1824, I. p. 333.

2. Die Respiration, als vom Gehirn abhängige Bewegung u. als chem. Process, nebst ihren physiol. u. pathol. Abweichungen. Bresl. 1813, 8°.—Bibl. med.-ch. p. 27.

3. Physiologie der menschlichen Lebensthätigkeit. Freiberg, 1810. —Bibl. med.-ch. p. 27.

4. Anthropologische Bemerkungen über das Gehirn u. den Schädel

des Menschen; mit beständiger Beziehung auf die Gall'schen Entdeckungen. Berl. 1806.—Bibl. med.-ch. p. 27.

5. Ueber innere und äussere Bewegung im Pflanzenreiche und Thierreiche, und insbesondere über Ersatz der äusseren durch innere oder chemische; mit Rücksicht auf Gestaltungsverschiedenheit. Marburg, 1828, 8°.

6. Systematischer Entwurf einer allgem. Biologie. 8°. Frankf. 1808.

7. Nachricht von einigen Versuchen u. Enthaupteten die Irritabilitäts-verhältnisse betreffend.—Schriften der Gesellschaft zu Marburg. Bnd. 1, 1823, p. 108.

BARTELS (H. W.).

1. Diss. inaug. de usu, quem præbet Agnus cyclops monstrosus in explicatione visûs simplicis ope binorum oculorum. Marb.1840, 8°.

BARTELS (J. H.) et FRICKE (J. C. G.).

1. Bericht deutsch-Naturf. zu Hamburg; *vide sup.* Pars I. p. 24.

BARTH (Dr.).

1. Obs. sur les Dents fossiles d'Éléphans qui se trouvent en Toscane. 8°. Florence.

BARTH (A.).

1. Dissertatio anatom.-physiologica de Retibus mirabilibus. Berol. 1837, 4°. fig.—*Val.* Repert. III. p. 14.—*Müll.* 1838, p. cxxvii.

2. Muskellehre. Wien, 1819, fol. fig.

BARTH (J. M.).

1. Schreiben, darinnen von dem Rhinoceros Nachricht gegeben und untersucht wird, ob dieses Thier der von Hiob beschriebene Behemot sey. Regensb. 1747, 4°.—*Böhm.* Bibl. II. 1, p. 479.

2. Dissertatio de Culice. Ratisb. 1737, 4°. fig.—Bibl. Ent. I. p. 16. H —*Böhm.* Bibl. II. 2, p. 341.

BARTHÉLEMY

1. Deux Cicindèles nouvelles.—Ann. Soc. Ent. de Fr. IV. p. 597, fig.—*Burm.* in *Wiegm.* Arch. 1836, II. p. 306 (*Cicindela Audouini et Rouxii*).

2. Observations sur le genre *Plochionus*, Dej.—Ann. Soc. Ent. Fr. III. p. 429.—*Wiegm.* Arch. 1835, II. p. 24.

3. Description d'une nouvelle espèce de Procruste.—Ann. Soc. Ent. Fr. VI. p. 245.

4. Description du Ricin de l'Houbara.—Ann. Soc. Ent. Fr. V. p. 689, fig.

5. Note sur la Foulque caronculée.—Rev. Zool. 1841, p. 307.

BARTHELSEN (Gottfr.).

1. Verzeichniss des von *Dr. J. Ph. Breyne* nachgelassenen Natura-lienkabinets. Danzig, 1765, 8°.—*Böhm.* Bibl. I. 1, p. 399.

BARTHES ( ).

1. Mémoire sur les Abeilles et la manière de faire le Miel. (In ejus Mémoires d'Agricult. et de Méchanique. Par. 1763, 8°.)—*Böhm.* Bibl. II. 2, p. 290.

BARTHEZ (F.).

1. Des Propriétés électives des Vaisseaux absorbantes chez l'Homme et les Animaux. 8°. Paris, 1843.

BARTHEZ (P. Jos.).

1. Nouvelle Mécanique des Mouvemens de l'Homme et des Ani-maux. Carcass. 1798, 4°.—(Germ.) *Neue Mechanik der will-kührlichen Bewegungen der Menschen und Thiere.* Aus dem Franz. v. *Sprengel.* Halle, 1800, 8°.—Bibl. med.-ch. p. 28.

2. Dissertatio de Aëris naturâ. Accedit Corollarium de aëre, aquis et locis Foruialiensibus. Monsp. 1767, 4°.—*Böhm.* Bibl. I. 1, p. 511 ; V. p. 185.

3. Nouveaux Élémens de la Science de l'Homme. Paris, 1806, 2 vols. 8°.

BARTHOLIN (Gasp.).

1. Opuscula IV. de Unicornu ejusque adfinibus et succedaneis, La-pide nephritico, etc. Hafn. 1628 ; 1630, 8°.—*Böhm.* Bibl. II. 1, p. 481.

2. Dissertatio Tetras Quæstionum physicarum. Lips. 1647, 4°.— *Böhm.* Bibl. II. 2, p. 78.

3. De Aquis Libri II., naturam et accidentia Aquæ, etc. etc., suc-cinctè explicantes, etc. Hafn. 8° ; Rost. 1620, 12°.—*Böhm.* Bibl. V. p. 13.

4. De Naturæ Mirabilibus. 4°. Hafn. 1674.

5. Exercitationes miscellaneæ varii argumenti, imprimis anatomici. Leyde, 1675, 8°.—Biogr. Un. III. p. 453.

6. Institutiones anatomicæ. Albi, 1611, 8°.—Lugd. B. 1645, 8°.— Trad. Fr. par *Duprat,* Paris, 1647, 4°.—*Cuv.*et *Val.* Poiss. I. p. 67.

BARTHOLIN (Gasp. II.).

1. Specimen historiæ anatomicæ partium Corporis humani, ad recentiorum mentem accommodatæ, novisque observationibus illustratæ. Hafn. 1701, 4°. fig.—Act. Erudit. XXI. p. 56.

2. De Ductu salivali hactenùs non descripto observatio anatomica. Hafniæ, 1684, 4°. fig.—Act. Erudit. IV. p. 30.

3. Diaphragmatis structura nova. Paris, 1676, 8°.—Biogr. Un. III. p. 453.

4. De Respiratione Animalium. Hafn. 1700, 4°.—Biogr. Un. III. p. 453.

5. De Glossopetris. Hafn. 1704, 4° ; 1706, 12°.—*Cuv. et Val.* Poiss. I. p. 68.

BARTHOLIN (Th.).

1. Act. med. et phil. Hafn.; *vide sup.* Pars I. p. 2.—*Cuv. et Val.* Poiss. I. p. 71.

2. De Luce Animalium Libri III. Leyde, 1643 ; 1647, 8°.—Copenh. 1693, tit. De Luce Hominum et Brutorum ; 1669.—*Cuv. et Val.* Poiss. I. p. 67.—*Böhm.* Bibl. II. 1, p. 104.

3. Anatomia, ex *Gasparis* parentis institutionibus, omniumque recentiorum et propriis observationibus locupletata. Leyde, 1641, 8°.—Biogr. Un. III. p. 452.

4. Historiarum anatomicarum et medicarum Centuriæ IV. Copenh. 1654–1661, 5 vols. 4°.—*Cuv. et Val.* Poiss. I. p. 67.

5. De Lacteis thoracicis in Homine Brutisque nuperrimè observatis, Historia anatoïica. Copenh. 1652, 4°.—Biogr. Un. III. p. 452.

6. Vasa lymphatica nuper Hafniæ in Animantibus inventa et in Homine, et hepatis exequiæ. Copenh. 1653, 4°.—Biogr. Un. III. p. 452.

7. De Cygni anatome ejusque cantu. Copenh. 1650, 4° ; 1668, 8°. —Biogr. Un. III. p. 452.

8. Diss. de Confectione Alchermes, quam Hafniæ *J. G. Becker* dispensare constituit. Hafn. 1672, 4°.—Bibl. Ent. I. p. 16.

9. De Unicornu Observationes novæ. Patav. 1641 ; 1645, 8°.— (Edit. 2ª à filio *Casp.*) Amst. 1678, 12°. fig.—*Böhm.* Bibl. II. 1, p. 481.

10. De Unicornu Africano.—Hist. anatom. Cent. II. hist. 61, p. 278. —*Böhm.* Bibl. II. 1, p. 481.

11. De raris Naturæ in Insulâ Melitâ observatis.—Epist. Cent. I. 53, p. 223.—*Böhm.* Bibl. I. 1, p. 576.

12. De Cygno nigro.—Orat. var. Argum. 52, p. 371.—*Böhm.* Bibl. II. 1, p. 563.

13. Neuverbesserte Künstliche Zerlegung des menschlichen Leibes. Nürnb. 1677.

BARTHOLIN (Th. II.).

1. De Vermibus in aceto et semine. Copenh. 1671, 12°.—Biogr. Un. III. p. 453.

BARTLETT ( ).

1. On the Post-tertiary Formations of Cornwall and Devon.—Rep. Brit. Assoc. 1841, Sect. p. 61.—*L.* u. *Br.* N. Jahrb. 1844, p. 105. —Instit. IX. p. 421.

2. Index geologicus. Lond. 1841.—(Deutsch übers. von *Ebenau* u. *Thomä*) Stuttg. 1841.

BARTLETT (A. D.).

1. On a new British species of the genus *Anser*, with remarks on the nearly-allied species.—Proc. Zool. Soc. Jan. 1839.—Ann. of Nat. H. IV. p. 64.

BARTOLI (Seb.).

1. Thermologia Aragonia, s. Historia naturalis Thermarum in occidentali Campaniæ orâ inter Pausilippum et Misenum scatentium, jam ævi injuriâ deperditarum, et *P. Ant. ab Aragoniâ* studio restauratarum, etc. Neapol. 1679, 2 vols. 8°. fig.—*Böhm.* Bibl. V. p. 255.

BARTOLINI (B.).

1. Catalogo delle Piante che nascono spontaneamente intorno alla città di Siena, coll' aggiunta d' altro Catalogo dei Corpi marini fossili, che si trovano in detto luogo. Siena, 1776, 4°.—*Mod.* Bibl. Helm.

BARTON (B. Sm.).

1. Papers relative to certain American Antiquities. Philad. 1796, 4°.—Tr. Linn. Soc. Lond. V. p. 278.

2. Discourse on some of the principal Desiderata in Natural History. Philad. 1807, 8°.—Tr. Linn. Soc. Lond. IX. p. 326.

3. Collections for an Essay towards a Materia Medica of the United States. Philad. 1798, 8°.—Tr. Linn. Soc. Lond. V. p. 278.

4. *Archæologiæ Americanæ Telluris Collectanea et Specimina,* or Collections, with Specimens, for a series of Memoirs on certain extinct Animals and Vegetables of North America, etc. 8°. Part I. Philad. 1814.—Cat. R. Soc. Lond.

5. Fragments of the Natural History of Pennsylvania. Philad. 1799, fol.—Cat. R. Soc. L.

6. On the *Tantalus ephouscyca*, a rare American Bird.—Tr. Linn. Soc. Lond. XII. p. 24.

7. On the Natural History of North America.—Phil. Mag. XXII. p. 97.

8. On a new species of N. American Lizard.—Trans. Am. Phil. Soc. VI. p. 108.

9. Memoir concerning the Fascinating Faculty ascribed to the Rattlesnake. Philad. 1796, 8°.—Trans. Am. Phil. Soc. IV. p. 74. —Phil. Mag. XV. pp. 193, 294.—Tr. Linn. Soc. Lond. V. p. 277. —(Deutsche Uebers. v. *Zimmermann*) *Abhandlung über die vermeinte Zauberkraft der Klapperschlange, etc.* Leipz. 1798, 8°.— *Licht.* u. *Voigt's* Mag. XI. 4, p. 21.—Bibl. med.-chir. p. 28.

10. Supplement to a Memoir concerning the Fascinating Faculty which has been ascribed to the Rattlesnake, etc. Philad. 1800, 8°.—Cat. Roy. Soc. Lond.

11. On the Torpidity of Animals.—Phil. Mag. XXXV. p. 241.

12. Facts, Observations and Conjectures relative to the Generation of the *Opossum* of N. America. Philad. 1806, 8°.—Tr. Linn. Soc. Lond. XI. 2, p. 422.—Additional Facts, &c. 8°. Philad. 1813.

13. A Memoir concerning the *Alligator*. (*Mémoire sur un Reptile nommé aux Etats-Unis* Alligator, ou Hellbender.) (Angl.) Philad.1812, 8°. fig.—*Cuv.* R. An. III. p. 334.—Phil. Mag. XXIII. p. 143.

14. Some account of the *Siren lacertina* and other species of the same genus of Amphibious Animals. Philad. 1808, 8°. fig.— *Voigt's* Mag. XII. p. 486.

15. An Inquiry into the question, whether the *Apis mellifica*, or true Honey-bee, is a native of America?—Tr. Am. Phil. Soc. III. p. 241.—Bibl. Ent. I. p. 16.

16. On an American Species of *Dipus* or Jerboa.—Trans. Am. Phil. Soc. IV. p. 114; VI. p. 143.

17. On the poisonous and injurious Honey of N. America.—Trans. Am. Phil. Soc. V. p. 51.

BARTON (D. W.).

1. Notice of the Geology of the Catskills.—*Sill.* Amer. Journ. IV. 2, p. 249.

BARTON (Dr.).

1. On Indian Dogs.—Phil. Mag. XV. pp. 1, 136.

BARTON (Rich.)

1. Dialogue concerning some things of importance to Ireland, particularly to the County of Armagh, being part of a design to write the Natural, Civil, and Ecclesiastical History of that County. Dubl. 1751. 4°. fig.—*Böhm.* Bibl. I. 1, p. 531.

2. Some Remarks towards a full Description of Upper and Lower Lough Lene near Killarny in the county of Kerry. Dubl. 1751, 4°. fig.—*Böhm.* Bibl. I. 1, p. 531.

3. Lectures in Natural Philosophy, designed to be a Foundation for reasoning pertinently upon the Petrifications, Gems, Crystals and Sanative quality of Lough Neagh in Ireland, etc. 4°. Dubl. 1751, fig.--Cat. R. Soc. Lond.—*Böhm.* Bibl. I. 1, p. 531.—*Mod.*

BARTRAM (J.).

1. A Letter to *Mr. P. Collinson,* containing some observations concerning the Salt-marsh Muscle, the Oyster-banks, and the Fresh-water Muscle of Pennsylvania.—Phil. Trans. 1744, XLIII. p. 157, fig.—Tr. Linn. Soc. Lond. XII. p. 227.

2. Observations on Inhabitants, Climate, Soil, Rivers, Productions, Animals and other matters worthy of notice, made in his Travels from Pennsylvania and Canada, etc. Lond. 1751, 8°. fig.—*Böhm.* Bibl. I. 1, p. 746.

3. An Account of some very curious Wasp-nests made of clay in Pennsylvania.—Phil. Tr. XLIII. p. 363, fig.—Bibl. Ent. I. p. 17.

4. Description of the Great Black Wasp from Pennsylvania.—Phil. Tr. XLVI. p. 278.—Bibl. Ent. I. p. 17.

5. Observations on the Dragon-Fly, or Libella, of Pennsylvania.—Phil. Tr. XLVI. pp. 323, 400.—Oekon. physik. Abhandl. p. 224 (Allem.).—Bibl. Ent. I. p. 17.

6. Observations on the Yellowish Wasp of Pennsylvania.—Phil. Tr. LIII. p. 37.—Bibl. Ent. I. p. 17.

7. A Letter concerning a cluster of small Teeth observed at the root of each Fang in the Head of a Rattlesnake.—Phil. Tr. XLI. p. 358.—Dict. Sc. nat. XV. p. 254.

8. A Description of East Florida, with a Journal, etc. upon a Journey from St. Augustine up the River St. John's as far as the Lakes, etc. Lond. 1769, 4°.—*Böhm.* Bibl. I. 1, p. 745.

BARTRAM (Mos.).

1. On the native Silk-worms of N. America.—Tr. Amer. Phil. Soc. I. p. 294.—J. de Phys. II. p. 51.—Bibl. Ent. I. p. 16.

BARTRAM (W.).

1. Travels through N. and S. Carolina, Georgia, E. and W. Florida,
   etc. Philad. 1791, 8°.—(Germ. *Zimmermann*): *Reisen durch
   N.- u. S.-Carolina, Georgien, O.- u. W.-Florida, etc.* 1792. 8°.
   —Bibl. Ent. I. p. 16.

2. Voyage dans les parties du sud de l'Amérique septentrionale,
   traduit de l'Anglais. 1779, 2 vols. 8°. Philad. 1784, fig.—*Dum.*
   et *Bib.* I. p. 424.

BASOCHE (De).

1. Mémoire sur un Mollusque fossile, inédit et remarquable, du
   Terrain secondaire de l'Arrond. de Falaise. fig.—Mem. Soc.
   Linn. Calv. p. 210.

2. Sur quelques Fossiles du Terrain intermédiaire des environs de
   Falaise.—Ann. Sc. nat. 1825, p. 472.—*Féruss.* Bull. 1826, VII.
   p. 390.

BASSETT (W.).

1. On Lizards in Chalk.—*Sill.* Am. J. XXXVII. p. 402.

BASSI (Agost.).

1. Del Mal del Segno, calcinario, o Moscardino, che afflige i Bachi
   de Seta. Lodi, 1835-36, 8°.—Compt. Rend. 1836, II. p. 434.

2. Rapport sur l'examen fait par *MM. Audouin* et *Savi* des dépouilles d'une larve d'Oryctés.—Congr. sc. de Pise, 11 Oct.
   1839.—Rev. Zool. 1840, p. 29.

BASSI (C.).

1. Notice sur le genre *Cardiomera*, nouv. genre de Coléoptères de
   la famille des Carabiques.—Ann. Soc. Ent. Fr. III. p. 319, fig.—
   Wiegm. Arch. 1835, II. p. 24.

2. Notice sur une monstruosité du *Rhizotrochus castaneus.* fig.—
   Ann. Soc. Ent. Fr. III. p. 373.

3. Description de quelques nouvelles espèces de Coléoptères de
   l'Italie. fig.—Ann. Soc. Ent. Fr. III. p. 463.—*Wiegm.* Arch. 1835,
   II. p. 43.

4. Missbildungen von Insecten.—Instit. 1834, No. 45.—*Müll.* Arch.
   1836, VI. p. cxciii.

5. Description du genre *Malacogaster.*—*Guér.* Mag. de Zool. 1833,
   No. 99.—Bibl. Ent. I. p. 17.

BASSI (Ferd.).

1. De quibusdam exiguis Madreporis.—Comment. Bonon. IV. p. 398, fig.—*Mod.* Bibl. Helm.

2. De Bononiensi Phytotypolito.—Comment. Bonon. V. p. 141, fig. —*Mod.* Bibl. Helm.

3. De Petrefactis Agri Bononiensis.—Comment. Bonon. V. p. 33.

BASTER (Job).

1. Over het, etc. (Observations sur l'usage des Antennes des Insectes.)—Verh. Maatsch. te Haarl. XII. p. 147.—Bibl. Ent. I. p. 17.

2. Observationes de Corallinis iisque insidentibus Polypis aliisque animalculis marinis.—Phil. Trans. L. 1, p. 258, fig.—*Mod.* Bibl. Helm.—(Belg.) *Moll.* Mag. III. p. 95.

3. Opuscula subseciva, observationes miscellaneas de Animalculis et Plantis quibusdam marinis eorumque ovariis et seminibus continentia. Harl. 1759, 1765, 4°. fig. 2 vols.—*Böhm.* Bibl. I. 1, p. 290.—*Mod.* Bibl. Helm.

4. Dissertation on the Worms which destroy the Piles on the coast of Holland and Zealand.—Phil. Trans. XLI. p. 276, fig.—Tr. Linn. Soc. Lond. VII. p. 222.

5. Natuurkundige Uitspanningen. Haarl. 1766, 4°. 2 vols. fig.— *Mod.* Bibl. Helm.

6. Over de Voorteeling en Eyernesten van sommige Hoorns en Zee-Insecten.—Act. Harl. IV. p. 473, fig.—*Mod.* Bibl. Helm.

7. Dissert. de Zoophytis.—Phil. Trans. LII. p. 108.—*Mod.* Bibl. Helm.

BASTEROT (B. de).

1. On the Strata in the vicinity of Folkstone.—Trans. Geol. Soc. ser. 2, II. 334.—Phil. Mag. ser. 2, I. p. 69.—Ann. Sc. nat. XVI.

2. On Fossil Shells.—Mém. de la Soc. d'Hist. nat. de Paris, II.— Phil. Mag. ser. 2, II. p. 102.

3. Mémoire géologique sur les environs de Bordeaux. Paris, 1825, fig.—*Féruss.* Bull. 1826, VII. p. 129.

4. Description du Bassin tertiaire du Sud-Ouest de la France. 4°. fig. (*Geologische Beschreibung des tertiären Beckens v. südlichen Frankreich.*)—Isis, 1832, V. p. 457.

5. Notice sur le Gisement des Ossemens fossiles des environs d'Argenton.—Mém. Soc. d'Hist. nat. I. 2, p. 233.—*Féruss.* Bull. 1823, I. p. 220; 1824, III. p. 166.—Bull. Soc. Philom. p. 188.

6. Sur le Webstérite, ou Alumine sous-sulfatée. (Extrait.)—Bull. Soc. Phil. 1822, p. 19.

BASTIANI (A.).

1. Animale bipede evacuato per secesso, etc.—Atti. Accad. Sien. VI. p. 241.
2. Delle Acque minerali di S. Casciano ai Bagni. Firenze, 1770, 8°.

BATARRA ( ).

1. Reproduction des Raies.—Mém. Sienn. IX. p. 353.—*Cuv.* et *Val.* I. p. 169.

BATHYANI (V. von).

1. Reise durch einen Theil Ungarns, Siebenbürgens, der Moldau u. Buckovina. Leipz. 1812, 8°.

BATSCH (A. J. G. K.).

1. Nachricht einer naturf. Gesellsch. zu Jena ; *vide sup.* Pars I. p. 25.
2. Taschenbuch für topograph. u. mineralog. Excursionen in die Gegend von Jena. 2 vols. 12°. Weimar, 1800–1802.
3. Versuch einer Anleitung zur Kenntniss und Geschichte der Thiere u. Mineralien, für akademische Vorlesungen entworfen. Jena, 1788–89, 2 vols. 8°.
4. Beiträge und Entwürfe zur pragmatischen Geschichte der drei Naturreiche nach ihren Verwandtschaften. Weimar, 1800, 4°.
5. Versuch einer Mineralogie für Vorlesungen u. für anfangende Sammler von Mineralien entworfen. Jena u. Leipz. 1796, 8°.
6. Umriss der gesammten Naturgeschichte. Jena, 1796, 8°.
7. Analytische Tabellen über die Art der Mineralien, ein Versuch zu genauerer Bestimmung u. zu eigener Auffindung. Jena, 1799, 4°.
8. Sechs Kupfertafeln mit Conchylien des Seesandes gezeichnet u. gestochen. Jena, 1791, 4°.
9. Naturgeschichte der Bandwurmgattungen, etc. Halle, 1786, 8°. fig.
10. Verzeichn. der Reuss-Plauischen Naturaliensammlung zu Kostritz. 8°. Jena, 1786.
11. Einleitung zum Studium der Naturgeschichte. Weimar, 1805, 1806, 3 vols. 8°. fig.

12. Grundzüge der allgemeinen Naturgeschichte nach den drei Reichen, oder Handbuch für Lehrer der Naturgesch. Weimar, 1801, 8°. fig.

13. Uebersicht der Kennzeichen zur Bestimmung der Mineralien, u. kurze Darstellung der Geologie für seine Vorlesungen entworfen. Jena, 1796, 8°.

BATTARA (J. Ant.).

1. Epistolæ selectæ de Re naturali observationes complect. Arim. 1774, 4°. fig.—Biogr. Un. III. p. 523.

2. Rerum Naturalium Historia, nempè Quadrupedum, Insectorum, etc. existentium in Museo Kircheriano, novâ methodo distributa. Romæ, 1773, fol.—Bibl. Ent. I. p. 18.

BATTEN (E.).

1. On the Explanation of certain Geological Phænomena by the agency of Glaciers.—Rep. Brit. Assoc. 1844, Sect. p. 57.

BATTEN (J. H.).

1. Note of a Visit to the Niti Pass of the grand Himalaya Chain. —J. A. S. B. VII. p. 310.

BATTINI (Const.).

1. Cosmogonia Mosaica. Florent. 1817, 8°.

BAUDELOT (   ).

1. Lettre sur une grosse Pierre trouvée dans le corps d'un Cheval. Par. 1708, 8°.—*Böhm.* Bibl. IV. 2, p. 337.

BAUDER (J. Fr.).

1. Nachricht von denen, seit einigen Jahren bey Altdorf von ihm entdeckten versteinerten Körpern. Jena, 1772, 8°.—*Böhm.* Bibl. IV. 2, p. 248.—(Gallicè) *Relation des Fossiles,* etc. Altd. 1772, 8°.—*Böhm.* l. c. p. 265.—*Mod.* B. Helm.

2. Beschreibung des Kostbaren Altdorfischen Ammoniten- u. Belemnitenmarmors, wie solche 1754 (in den Fränk. Samml. I. p. 298) zum ersten male gemacht, etc.; mit einem Anhange, der die neuesten Entdeckungen des 1770 u. 1771sten Jahres von Encriniten, Astroiten, Nautiliten u. andern Versteinerungen beschreibt, wieder herausgegeben. Altd. 1771, 4°.—*Böhm.* Bibl. IV. 2, p. 315.

3. Anmerkungen über Ammoniten, Nautiliten, Belemniten u. vornehmlich über das Crocodilskelet.—*Schröt.* Journ. VI. p. 516.

4. Nachricht von einem in Nürnbergischem Gebiete entdeckten Muschelsande.—Hamb. Mag. XII. p. 639.

BAUDET-LAFARGE (M. J.).

1. Monographie des Carabiques du Puy-de-Dôme. Clermont. Ferr. 1836, 8°.
2. Essai sur l'Entomologie du Dép. du Puy-de-Dôme. Clerm. 1809. 8°.—Bibl. Ent. I. p. 18.

BAUDIER (H. Ph.).

1. Description d'une espèce de *Lema* nouvelle pour la Faune française.—Ann. Soc. Linn. Par. 1825, p. 839, fig.—*Eis.* p. 192.

BAUDOUIN (Jul.).

1. Notice géologique sur une Caverne à Ossemens des environs de Châtillon. Châtill. 1843, 8°.

BAUDRIMONT (A.).

1. Traité élémentaire d'Histoire naturelle. Paris, 1840, 8°.

BAUDRY DES LOZIÈRES.

1. On Animal Cotton and the Insect which produces it.—Phil. Mag. XIX. p. 120.—Transact. Americ. V.—*Voigt's* Magaz. Bnd. 7, 1804, p. 146.

BAUER (Fr.).

1. Disquisitiones circa nonnullarum Avium systema arteriosum. Berlin, 1825.—*Féruss.* Bull. 1826, VIII. p. 394.
2. Microscopical Observations on the Suspension of the Muscular Motions of *Vibrio Tritici (Microscopische Beobachtungen über die Aufhebungen der Muscularbewegungen in dem* Vibrio Tritici). Phil. Trans. 1823, p. 1.—Isis, 1825, VII. L. A. p. 37; 1830, VIII. p. 775.—*Féruss.* Bull. 1824, I. p. 301. (*Observations microscopiques sur la Suspension des Mouvemens musculaires du Vibrion du blé.*)—Ann. Sc. n. 1830, II. p. 154.
3. Note sur la Structure intime du Corps caverneux.—Bull. Soc. Phil. 1823, p. 79.—*Féruss.* Bull. 1823, III. p. 285.

BAUER (J. L.).

1. Versuch eines Unterrichts für den Forstmann zur Verhütung der Waldverheerungen durch Insecten. Erlang. 1801, 8°. fig.

P 2

BAUER (J. Phil.).

1. Der Mensch in Bezug auf sein Geschlecht; oder Aufsatz über Zeugung, etc. Nach den neuesten Werken der franz. Aerzte. 3e Aufl. Leipz. 1834, 8º.—Bibl. med.-ch. p. 29.

BAUER (J. Val.).

1. Beschreibung des Biberacher Heilbrunnens, genannt der Jordan, etc. Biber. 1710, 8º.—*Böhm.* Bibl. V. p. 313.

BAUER (S.).

1. Landwirthschaftliche und technische Naturgeschichte, oder die Naturg. in Anwendung auf Gewerbe, Land- u. Forst-Wirthschaft, etc. Amberg, 1839, I. Bd.

BAUGIER et SAUZE.

1. Notice sur quelques Coquilles de la famille des Ammonidées, recueillies dans le Terrain jurassique des Deux-Sèvres. Niort, 1844, 8º.

BAUHIN (Gasp.).

1. Anatomes Liber primus (II.-IV.). Basil. 1590, 12º.—Cat. R. Soc. L.

2. De Lapidis Bezoaris Orientalis et Occidentalis, cervini item et germanici ortu, naturâ, differentiis et usu liber. Basil. 1613, 8º. fig.—Ibid. 1625, 8º. fig.—*Böhm.* Bibl. IV. 2, p. 340.

BAUHIN (J.).

1. De Aquis medicatis nova methodus. Montisbel. 1605, 1607, 1612, 4º. fig.—Bibl. Ent. I. p. 18.—(Germ.) Stuttg. 1603, 4º. fig.

2. Historia novi et admirabilis fontis balneique Bollensis in ducato Wirtemb. ad acid. Gœpingenses cum plurimis figuris variorum Fossil. Stirp. et Insector. quæ in et circa hunc Fontem reperiuntur. Montisbel. 1598–1600, 4º. fig.—(Deutsche Uebersetz.) Stuttg. 1602.—Bibl. Ent. I. p. 18.—*Böhm.* Bibl. I. 1, p. 589.—*Mod.* B. Helm.

3. Ein neu Badbuch, und historische Beschreibung fast aller heilsamen Bäder und Sauerbrunnen, insonderheit von dem Wunderbrunnen zu Boll; übers. durch *M. D. Förster.* Stuttg. 1602, 4º. fig.—*Böhm.* Bibl. V. p. 313.

4. Traité des Animaux ayant ailes, qui nuisent par leurs piqueures ou morsures. Montbel. 1593, 8º.—*Böhm.* Bibl. II. 1, p. 321.

5. De Lapidibus Metallisque miro naturæ artificio in ipsis Terræ

visceribus figuratis. Montisbel. 1598 et 1600, 4°. fig. (Edit. cum ejus Hist. Fontis Boll.)—*Böhm.* Bibl. IV. 2, p. 229.

6. De Luporum Rabie memorabili ex Vermibus Ascaridibus.— *Mod.* Bibl. Helm. p. 19.

BAUMANN (J.).

1. Naturgeschichte für das Volk. Luzern, 1837, 8°. (2te Aufl.) Luz. 1840, 8°. fig.—Verz. Bibl. d. Schweiz. Naturf. Ges.

2. Naturgeschichte für Volksschulen. Luzern, 1838, 8°. fig.

BAUMEISTER (W.).

1. Abbild. der Rindvich- Schaf- u. Schweine-Racen. fol. Stuttgart, 1840.

BAUMER (J. G.).

1. Anthropologia anatomico-physica. Francf. 1784, 8°.—Bibl. med. chir. p. 29.

BAUMER (J. Phil.).

1. Diss. de Apum culturà, imprimis in Thuringià. Erf. 1770, 4°. —Bibl. Ent. I. p. 18.—(Germ.) *Oekonomisch-physische Abhandlung über die Bienenpflege, besonders in Thüringen ; mit gemeinnützigen Anmerkungen u. einem Anhang begleitet,* etc. Ansp. 1774, 8°.—*Böhm.* Bibl. II. 2, p. 296.

BAUMER (J. W.).

1. Naturgeschichte des Mineralreichs mit besonderer anwendung auf Thüringen. 2 vols. 8°. Gotha, 1763–64 ; 1767, 8°.—*Klein,* Disp. Echinod. p. VII.—*Böhm.* Bibl. IV. 1, p. 64.

2. Historia naturalis Lapidum pretiosorum omnium, necnon Terrarum et Lapidum hactenùs in usum medicum vocatorum ; additis Observationibus Mineralogiam generatim illustrantibus. Francf. 1771, 8°.—(Germ.) übers. v. *Meidinger,* Wien, 1774, 8°. —*Böhm.* Bibl. IV. 1, p. 177.

3. Dissertatio de Calce vivâ. Giess. 1776, 4°.—*Pfingst.* Mag. II. p. 181 (Deutsch).—*Böhm.* Bibl. IV. 1, p. 241.

4. Dissert. de Montibus argillaceo-calcareis et argillaceo-gypseis. Erf. 1761, 8°.

5. Fundamenta Geographiæ et Hydrographiæ subterraneæ. Giess. 1779, 8°.—*Böhm.* Bibl. V. p. 19 ; IV. 1, p. 65.

6. Observationes ad Geographiam subterraneam pertinentes.—Act. Acad. Mogunt. I. p. 127.—*Böhm.* Bibl. IV. 1, p. 65.

7. Historia naturalis Regni mineralis. Francf. 1780, 8°. fig.—*Böhm.* Bibl. IV. 1, p. 65.

8. Dissertatio de Mineralogiâ Territorii Erfurtensis. Erf. 1759, 4°. —*Böhm.* Bibl. IV. 1, p. 137.

BAUMER (K. von).

1. Geognostische Fragmente, mit einer Charte. Nürnb. 1811.

BAUMES ( ).

1. Lettre sur le Tænia.—Journ. de Méd. LVI. p. 406.

BAUMGARTEN-CRUSIUS (Aug. Maur.).

1. Periodologie, oder die Lehre v. den periodischen Veränderungen im Leben des gesunden u. kranken Menschen. Halle, 1836, 8°. —Bibl. med.-chir. p. 29.

2. Ueber den Mechanismus, durch welchen die venösen Herzklappen geschlossen werden.—*Müll.* Arch. 1843, p. 463.

BAUMGARTEN (M.).

1. Das Schielen und dessen operative Behandlung, etc. Leipz. 1841, 8°.

BAUMGARTEN (Mart. à).

1. Peregrinatio in Ægyptum, Arabiam, Palæstinam, etc. Norib. 1594, 4°.—*Böhm.* Bibl. I. 1, p. 710.

BAUMGÄRTNER (K. H.).

1. Beiträge zur Entwickelungsgeschichte.—*Müll.* Arch. 1835, p. 563.

2. Einfluss der Nerven auf die Blutbewegung.—Isis, 1830, VI. p. 595.

3. Beiträge zur Physiologie und Anatomie. Stuttg. 1842, 8°.

BAUMGÄRTNER (M. à) et ETTINGHAUSEN ( ).

1. Zeitschr. für Physik; *vide sup.* Pars I. p. 35.

BAUMHAUER ( ).

1. Nouvelle Classification des Mouches à deux ailes. Paris, 1800, 8°.—Bibl. Ent. I. p. 18.

BAUNIER ( ).

1. Traité pratique sur l'Éducation des Abeilles. Vendôme, 1806, 8°. fig.—Bibl. Ent. I. p. 18.

BAUR (Chr. J.).

1. Tractatus de Nervis anterioris superficiei trunci humani, thoracis præsertim abdominisque. Tub. 1818.—Bibl. med.-chir. p. 29.

2. Anatomische Abhandlung über das Bauchfell des Menschen. Stuttg. 1835, 8°.—*Müll.* Arch. 1836, p. XLI.—*Val.* Repert. I. p. 11.

BAUSCH (J. Laur.).

1. Schediasma de Unicornu fossili. Jenæ, 1666, 8°. fig.—*Böhm.* Bibl. IV. 2, p. 282.

2. Tractatus de Lapide Hæmatite et Aëtite. Vratisl. 1664, 8°. fig. —Lips. 1665, 8°. fig.—*Böhm.* Bibl. IV. 2, p. 350.

BAUSSARD ( ).

1. Mémoire sur deux Cétacés échoués vèrs Honfleur le 19 Sept. 1783.—Journ. de Phys. t. XXXIV. 1789, p. 201.

BAUTMANN (J. Chr.).

1. De Verme in Lapide reperto.—Misc. Nat. Cur. Dec. III. 1699–1700, p. 51, Obs. 28.—*Mod.* B. Helm.

BAVIER ( ).

1. Mémoire sur le *Cancer gamarellus pulex.*—Bull. Ac. Brux. 1837, IV. p. 76.—Rev. Zool. 1838; p. 143.

BAXER ( ).

1. De Cutis mutatione in Locustâ aquaticâ.—Phil. Tr. No. 483.— Bibl. Ent. I. p. 18.

BAYARD ( ).

1. Examen microscopique du Sperme desséché sur le linge ou sur les tissus de nature et de coloration diverse. Paris, 1839, 8°

BAYER (J. J.). Vide BAIER.

BAYER (Theoph.).

1. Denkwürd. Erzählungen aus dem Thierreiche. 8°. Jena, 1825.

BAYFIELD (Capt.).

1. On the Geology of the North Coast of the St. Lawrence.—Trans. Geol. Soc. ser. 2, V. p. 89.—Proc. Geol. Soc. II. 4.—Phil. Mag. ser. 3, IV. p. 51.—*L. u. Br.* N. Jahrb. 1834, p. 443.

2. On the Transportation of Rocks by Ice.—Proc. Geol. Soc. II. 223.—Phil. Mag. ser. 3, VIII. p. 558.

3. On the Junction of the Transition and Primary Rocks of Canada and Labrador.—Journ. Geol. Soc. I. 450.

BAYLE (A. L. J.).

1. Petit Manuel d'Anatomie descriptive. Paris, 1823, 18°.—*Féruss.* Bull. 1823, III. p. 81.

2. Traité élémentaire d'Anatomie descriptive. 4ᵉ éd. Par. 1833, 18°; 5ᵉ éd. Paris, 1843, 18°.

3. Atlas élémentaire d'Anatomie descriptive, etc. Paris, 1839, 4°.

BAYLE (Fr.)..

1. Dissertationes physicæ, in quibus principia proprietatum in mixtis, Œconomia corporùm in Plantis et Animalibus demonstrantur. Tolosæ, 1677 et 1681, 12°; 1701, 4°.—*Böhm.* Bibl. II. 1, p. 82.

2. Dissertationes tres : 1° De causis Fluxûs menstrui. 2° De Sympathiâ Uteri. 3° De usu Lactis ad Tabidos reficiendos. Tolosæ, 1670, 4°.—Hag. Com. 1678, 12°.—Brugis, 1678, 12°.—*Böhm.* Bibl. II. 1, p. 280.

3. Opera omnia. Toul. 1701, 4 vols. 4°.—Act. Erudit. XXII. p. 76. —Biogr. Un. III. p. 606.

4. Institutiones physicæ ad usum Scholarum. Toul. 1700, 4°. 3 vols. fig. —Act. Erudit. XXI. pp. 387, 452.

5. Histoire anatomique d'une Grossesse de 25 ans. Toul. 1678, 12°. Paris, 1679, 12°.—Biogr. Un. III. p. 605.

BAYLE (   ).

1. Traité des Maladies cancéreuses. Paris, 1834, 8°.—*Müll.* Arch. 1836, VI. p. ccxiii.

BAYLE-BARELLE (   ).

1. Saggio intorno agli Insetti nocivi ai vegetabili economici, etc. Milan, 1809, 8°. fig. col.—Bibl. Ent. I. p. 19.

2. Degli Insetti nocivi all' Uomo, alle Bestie e all' Agricoltura. 12°. Milan, 1824.

BAYRHOFFER (C. Th.).

1. De naturâ et formis variis Animantium terræ, simulque de Vitâ universali. Marb. 1835, 8º.

2. Betrachtungen über Erfahr. u. Theorie in der Naturwissenschaft. 8º. Leipz. 1838.

BAZIN (A.).

1. Du Système nerveux, de la Vie animale et de la Vie végétale. 4º. Paris, 1841.

2. De la structure de la Membrane scléreuse, sousposée à la plèvre pulmonaire, et de l'Hypertrophie de cette membrane.—Ann. d'Anat. I. p. 28, fig.

3. Sur l'enveloppe propre du Poumon.—Ann. d'Anat. I. p. 317.

4. Analyse de l'ouvrage du Prof. *Panizza*, intitulé, *Sopra il Sistema linfatico dei Rettili*, etc.—Ann. d'Anat. II. p. 361.

5. Essai sur la différence du degré de certitude que présentent l'Idéologie et la Physique générale, et sur les procédés intellectuels qu'elles exigent. Paris et Londres, 8º.— *Val.* Repert. III. p. 11.

6. Sur une généralité des Marsupiaux, avec une description de l'Utérus du Kangurou dans l'état de Gestation.—Phil. Trans. 1834, II. fig.—Ann. d'Anat. I. p. 36.

7. Recherches sur la structure intime du Poumon de l'Homme et des Animaux vertébrés, suivies de considérations sur les fonctions et la pathologie de cet organe.—C. R. VIII. p. 879 (2e Mém.). IX. pp. 153, 234.

8. Lettre sur quelques petits Muscles qui sont restés inconnus jusqu'à ce jour.—Rev. Zool. 1839, p. 151.

9. Remarques sur le Nerf facial et ses rapports.—Rev. Zool. 1839, p. 65.—C. R. VIII. p. 337.

10. Mémoire sur les Muscles internes et sur l'appareil aquifère des branchies des Poissons.—Rev. Zool. 1839, p. 189.—C. R. VIII. p. 877.

11. Sur la structure des Bronches pulmonaires.—C. R. 1836, II. pp. 390, 515, 570.

12. Sur les connexions anatomiques, physiologiques et zoologiques du Système nerveux.—C. R. XI. pp. 289, 479, 569.

13. Sur le Ganglion céphalique, dit Glande pituitaire, et sur les connexions avec le Système nerveux de la vie organique.—C. R. IX. p. 507.

BAZIN (G. Aug.).

1. Observations sur l Accroissement du Corps humain. 1741, 8°
—Biogr. Un. III. p. 613.

2. Histoire naturelle des Abeilles. Strasb. 1744, 12°.—Par. 1747,
12°. 2 vols.—Bibl. Ent. I. p. 19.

3. Histoire Abrégée des Insectes. Par. 1747, 12°. 4 vols.—Par.
1750.—Bibl. Ent. I. p. 19.

4. Effet de l'Huile sur les Chenilles.—Acad. Sc. 1738, pp. 39, 54.
—Bibl. Ent. I. p. 19.

5. Observations sur les Plantes et leur analogie avec les Insectes;
ʜ avec un Discours sur la cause pour laquelle les Bêtes nagent
naturellement, et que l'Homme est obligé d'en étudier les moyens.
Strasb. 1741, 8°.—*Böhm.* Bibl. II. 1, p. 112.—Bibl. Ent. I. p. 19.

BEADLE (E. R.).

1. Barometrical Observations to ascertain the Level of the Dead Sea.
—Sill. Am. J. XLII. p. 214.

BEALE (Thom.).

1. A few Observations on the Natural History of the Sperm Whale.
Lond. 1835, 8°.— *Wiegm.* Arch. 1835, II. p. 290.

2. The Natural History of the Sperm Whale, etc.; to which is added
a Sketch of a South-Sea Whaling Voyage. 8°. Lond. 1838.—
Ann. of Nat. H. III. p. 118.

BEAMISH (N. L.).

1. On the apparent Fall or Diminution of Water in the Baltic, and
Elevation of the Scandinavian Coast.—Rep. Brit. Assoc. 1843,
Sect. p. 59.

BEAN (W.).

1. On *Fusus Turtoni* and *Limnæa lineata,* Bean.—Mag. Nat. Hist.
ser. 1, VII. p. 493.— *Wiegm.* Arch. 1835, I. pp. 324, 327.

2. On a Deposit of Fossil Shells at Bridlington Quay.— Mag. Nat.
Hist. ser. 1, VIII. p. 355.

3. On *Panopæa glycymeris* and *Anomia coronata.*—Mag. Nat. Hist.
ser. 1, VIII. p. 562.

4. On *Unio distortus* and *Cypris concentrica* from the Upper
Sandstone and Shale of Scarborough.—Mag. Nat. Hist. ser. 1,
IX. p. 376.

5. A Catalogue of the Fossils found in the Cornbrash Limestone of Scarborough, with figures and descriptions of some of the undescribed species.—Mag. Nat. Hist. ser. 2, III. p. 57.

BEATLEY ( ).

1. Entomological Tour in South Devon.—*Walk.* Ent. Mag. 1833, II. p. 180.—Bibl. Ent. I. p. 19.

BEAUFORT (Capt.).

1. Karamania, or a Description of the South Coast of Asia Minor. 8°. Lond. 1818.

BEAUFORT (L.).

1. Cosmopœia divina, s. Fabrica Mundi explicata. Lugd. Bat. 1656, 12°.—*Böhm.* Bibl. I. 1, p. 345.

BEAUGRAND (Nic.).

1. Le Maréchal expert, traitent du naturel des bons Chevaux, avec une Description de toutes les parties et Ossemens du Cheval. Lyon, 1633, 8°. fig.—Troyes, 1655, 8°.—*Böhm.* Bibl. II. 1, p. 415.

BEAUMOND (J.).

1. Two Letters written in Somersetshire concerning Rock-plants and their Growth.—Phil. Trans. II. p. 724; XIII. p. 276.—*Mod.* Bibl. Helm.

BEAUMONT (Alb. de).

1. Select Views of the Antiquities and Harbours in the South of France, with topographical and historical descriptions. Lond. 1794, fol. fig.

2. Travels through the Maritime Alps from Italy to Lyons, across the Col de Tende, &c., with topographical and historical descriptions; to which are added some Philosophical Observations on various Appearances in Mineralogy found in those countries. Lond. 1795, fol. fig.

3. Voyage pittoresque aux Alpes Pennines, précédé de quelques Observations sur les Hauteurs des Montagnes glacières, etc. Genève, 1787, 4°.—(Angl.) *A Picturesque Tour from Geneva to the Pennine Alps.* Lond. 1794, fol. fig.

4. Travels through the Rhætian Alps in the year 1786, from Italy to Germany through Tyrol. Lond. 1792, fol. fig.

5. Description des Alpes Grecques et Cottiennes, ou Tableau historique et statistique de la Savoye, etc.   Strasb. 1803, 2 vols. 4º. Atl. fol.

BEAUMONT (Elie de).

1. Note relative à une Observation de *M. Conybeare.*—Ann. Sc. n. 1831, XXII. p. 172.

2. Memoir on the Origin of Mount Etna.—Ed. N. Phil. J. Apr. 1836.—*Sill.* Am. J. XXXI. p. 168.

3. Note sur la Constitution géognostique des environs de Martigues, dépt. des Bouches du Rhône.—Mém. Soc. Linn. Norm. III. p. 138.—*Féruss.* Bull. 1828, XIII. p. 184.

4. Note sur la forme la plus ordinaire des objections relatives à l'origine attribuée à la Dolomie.—Ann. Sc. n. Nov. 1829.—*Féruss.* Bull. 1830, XXII. p. 2.

5. Notice sur les Mines de Fer et Forges de Framont et de Rothau (Vosges).—Bull. Soc. Philom. 1823, p. 76.

6. Sur un Gisement de Végétaux fossiles et de Graphite, situé au col du Chardonet (Hautes-Alpes).—Ann. Sc. n. 1828, XV. p. 353.

7. Sur un Gisement de Végétaux fossiles et de Bélemnites, situé à Petit-Cœur près Moûtiers, en Tarentaise.—Ann. Sc. n. 1828, XIV. p. 113.—Bull. Soc. Géol. VI. p. 96.

8. Thatsachen die Geschichte der Berge in Oisans erläuternd.—Ann. des Mines (3e S.), V. p. 3.—*L.* u. *Br.* N. Jahrb. 1836, p. 372.

9. Remarques sur une Lettre de *M. Gay* sur la Géologie du Chili. —C. R. VI. p. 918.

10. Remarques sur l'Évaluation de la Température de la surface du Globe pendant la période tertiaire, d'après la nature des débris organiques qui s'y rapportent.—Ann. Sc. n. 1836, VI. p. 313.—*L.* u. *Br.* N. Jahrb. 1837, p. 63.

11. Recherches sur quelques-unes des Révolutions de la surface du Globe. Mém. lu à l'Académie des Sciences. Par. 1830, 8º.—Ann. Sc. n. Sept. 1829 à Févr. 1830.—Ann. Phys. et Chim. Nov. 1839.—Revue Franç. XV. 1830.—*Féruss.* Bull. 1830, XXI. p. 338.

12. Recherches sur l'Age relatif des Montagnes. Extrait d'un Mém. lu à l'Académie des Sciences.—*Féruss.* Bull. 1829, XIX. p. 14.

13. Fragmens géologiques tirés de *Stenon,* de *Kazwini,* de *Strabon,* et du *Boun-Dehesch.*—Ann. Sc. n. 1832, XXV. p. 337.

14. Sur la Direction et l'Age relatif des Montagnes serpentineuses de la Ligurie. Réponse à une Note de *M. Laur. Paroto.*—Ann. Sc. n. 1830, XXI. p. 413.—Bull. Soc. Géol. I. p. 64.

15. Observations géologiques sur les différentes formations qui, dans le système des Vosges, séparent la formation houillère de celle du Lias. Par. 1828, 8°.—Ann. des Mines, 1827, I. p. 393 ; 1828, IV. p. 3.—*Féruss.* Bull. 1828, XII. p. 287 ; 1829, XIX. p. 1.

16. Sur les rapports qui existent entre le relief du sol de l'île de Ceylan et celui de certaines masses de montagnes qu'on aperçoit sur la surface de la Lune.—Ann. Sc. n. 1831, XXII. p. 88.

17. Note sur la construction géologique des îles Baléares.—Ann. Sc. n. Avr. 1827, p. 423.—*Féruss.* Bull. 1828, XIII. p. 35.

18. Lettre au sujet des observations faites par *M. Rozet* dans les Montagnes du Nord de l'Afrique.—Ann. Sc. n. 1831, XXII. p. 110.

19. Faits pour servir à l'histoire des Montagnes de l'Oisans. Par. 1834, 8°.—Ann. Sc. n. 1829, XVII. p. 66.

20. Note sur l'uniformité qui règne dans la constitution de la ceinture jurassique du grand Bassin géologique qui comprend Londres et Paris.—Ann. Sc. n. Juill. 1829.—*Féruss.* Bull. 1830, XX. p. 19.—*Leonh.* u. *Br.* Jahrb. 1833, p. 88 (*Gleichmässiges in der Zusammensetzung der Jura-Begränzung des grossen geol. Beckens, in welchem London u. Paris liegen*).

21. Recherches sur la structure et l'origine du Mont Etna.—C. R. I. p. 429.

22. Rapport sur les Recherches géologiques exécutées dans quelques parties de l'Asie Mineure par *M. Ch. Texier.*—C. R. II. p. 277.

23. Seconde Note sur l'âge géologique de Calcaire de Château-Landon.—C. R. V. p. 8.

24. Rapport sur un Mémoire de *M. Paillette*, concernant des Observations géologiques relatives à la partie occidentale de la Bretagne.—C. R. V. p. 83.

25. Instructions concernant la Géologie, pour l'Expédition scientifique du Nord de l'Europe.—C. R. VI. p. 549.

26. Rapport sur quatre Mémoires de *M. Rozet*, relatifs aux Montagnes qui séparent la Saône de la Loire.—C. R. XI. p. 255.

27: Leçons de Géologie pratique professées au Collège de France pendant l'année scolaire 1843–1844. Paris, 1845, 8°. 1ᵉ vol.

28. Voyages en Scandinavie, en Laponie, etc.

29. Note sur le Terrain qui contient le Tripoli de Bilin, en Bohème.—C. R. VII. p. 501.

30. Rapport sur un Mémoire de *M. Puillon-Boblaye*, relatif à la Géologie des provinces de Bone et de Constantine.—C. R. VII. p. 238.

31. Extrait d'une Série de Recherches sur quelques-unes des Révolutions de la surface du Globe. Paris, 1835, 8°.

32. Coup-d'œil sur les Mines. Paris, 1824, 8°. fig.

33. Rapport sur un Mémoire de *M. Al. d'Orbigny*, intitulé, Considérations générales sur la Géologie de l'Amérique méridionale. 4°.

34. Instructions géologiques pour l'exploration de l'Algérie.—C. R. VII. p. 142.

35. Rapport sur un Mémoire de *M. de Castelnau* relatif au Système Silurien de l'Amérique septentrionale. 1843, 4°.

36. Remarques sur deux points de la Théorie des Glaciers, et Note sur les pentes de la limite supérieure de la Zone erratique, etc. 1842, 8°.

37. Rapport sur un Mémoire de *M. A. Bravais* relatif aux lignes d'ancien Niveau de la Mer dans le Finmark. 1842, 4°.

38. Rapport sur un Mémoire de *M. Durocher*, intitulé, Observations sur le Phénomène diluvien du Nord de l'Europe. 1842, 4°.

39. *Macrobiotus Hufelandii*, animal e crustaceorum classe novum, reviviscendi post diuturnum Asphyxium et ariditatem potens.—Rev. Zool. 1838, p. 275.

40. Rapport sur la partie géologique et minéralogique de la campagne de la *Vénus*. C. R. XI. p. 336.

BEAUMONT (J. T. B.).

1. On the Origin of the Vegetation of our Coal-fields and Wealdens.
—Proc. Geol. Soc. III. 152.—Phil. Mag. ser. 3, XVII. p. 67.

BEAUMONT (W.).

1. Experiments and Observations on the Gastric Juice and the Physiology of Digestion. Plattsb. 1833. (Deutsch. übers. v. *B. Luden*, Leipz. 1834, 8°. fig.)—Amer. Journ. XXVI. p. 193.—*Müll.* Arch. 1835, p. 131 ; 1836, p. 66.

BEAUMONT (   ).

1. Auffindung filarienartiger Schmarotzer in der Leibeshöhle von *Blaps mortisaga*.—Instit. No. 139, p. 3.—*Wiegm.* Arch. 1837, IV. p. 254.

BEAUNEZ (L. de).

1. Propositions sur l'Anatomie de la Sangsue. 4°. Par. 1819.

BEAUPLAIN (De).

1. Description de l'Ukraine, etc. Rouen, 1660, 12º.—Par. 1661.—
(German.) *Beschreibung der Ukraine, der Krim und deren Ein-*
wohner, aus dem Franz. übers. u. mit einem Anhange, der die
Ukraine u. die Budziakische Tartarey betrifft, v. J. W. Möller.
Bresl. 1780, 8º.—*Böhm.* Bibl. I. 1, p. 630.

BEAURIEU (Gasp. Guillard de).

1. Abrégé de l'Histoire des Insectes. 1764, 2 vols. 12º.—Biogr. Un.
Ḥ III. p. 653.
2. Cours d'Histoire Naturelle. 7 vols. 12º. Liège, 1770.

BEAUVAIS (De).

1. On a new species of Siren.—Trans. Amer. Phil. Soc. IV. p. 277.
—Phil. Mag. IX. p. 118.
2. On Amphibia.—Trans. Amer. Phil. Soc. IV. p. 362.

BEAUX ( ).

1. Physiologie de la Glande lacrymale. Paris, 1821, 8º.

BECCARI (J. Barth.).

1. De Bononiensi Arenâ quâdam.—Act. Bonon. I. p. 62.—*Mod.*
Bibl. Helm.
2. De Luce Dactylorum fossilium.—Comment. Bonon. II. p. 248.

BECCHETTI (M.).

1. Teoria generale della Terra. Roma, 1782, 8º.

BECHER (Dav.).

1. Neue Abhandlung vom Carlsbade. Prag. 1766, 4º. (1 Thl.)
—*Ibid.* 1772, 3 Thle, 8º. fig.—*Böhm.* Bibl. V. p. 276.

BECHER (J. J.).

1. Physica Subterranea, profundam Subterraneorum Genesin, è
principiis hucusque ignotis ostendens. Francf. 1664 et 1666,
8º; 1681, 8º.—Lips. 1703; 1738, 8º. (accedit Specimen Beche-
rianum, sistens fundamenta, documenta, experimenta quibus
principia mixtionis subterraneæ demonstrantur, edit. a *G. E.*
Stahlio).—*Böhm.* Bibl. IV. 1, p. 33.
2. Thier-, Kräuter- und Bergbuch. Ulm, 1663, fol. fig.

BECHER (J. P.).

1. Mineralogische Beschreibung des Westerwaldes, insbesondere der beiden Holzkohlenbergwerke zu Stockhausen u. Horn. Berl. 1786, 8°. Karte.

2. Mineralogische Beschreibung der Oranien-Nassauischen Lande, nebst einer Geschichte des Siegenschen Hütten- u. Hammer-wesens. Marb. 1789, 8°. fig.

BECHSTEIN (J. M.).

1. Gemeinnützige Naturgeschichte Deutschlands. Leipz. 4 vols. 8°. 1789–95.—ed. 2, 1801–1809, 4 vols. 8°. fig. (Säugeth. u. Vögel.)—*Cuv.* R. An. III. p. 335.

2. Getreue Abbildungen naturhistorischer Gegenstände. Leipzig, 1793–1810, 8 vols.—*Dum.* et *Bib.* I. p. 350.—Isis, 1817, I. p. 113.

3. Manuel de l'amateur des Oiseaux de volière, etc. Trad. de l'Allemand, avec les addit. du trad. 2ᵉ éd. Paris, 1828, 8°.— *Féruss.* Bull. 1828, XV. p. 300.

4. Naturgeschichte der in- und ausländischen Insecten. Nürnb. 1793, 8°.—Bibl. Ent. I. p. 19.

5. *Lacépède's* Naturgeschichte der Amphibien, aus dem Franz. mit Anmerkungen u. Zusätzen. 5 Bde. Weimar, 1802.—N. Act. Nat. Cur. XIV. p. xvi.

6. Ueber den wahren Ursprung des fliegenden Sommers.—*Licht.* u. *Voigt's* Mag. VI. 1, p. 53.—Phil. Mag. IV. p. 119.

7. Bemerkungen über die Motacillen.—Naturforscher, st. 27, 1793, p. 38.

8. Von den Vuchuchen in Deutschland.—*Licht.* u. *Voigt's* Mag. VI. st. 1, 1789, p. 60.

9. Naturgeschichte der Stubenthiere, Säugethiere, Amphibien, Fische, Insekten und Würmer. Gotha, 1797, 8° ; 1807.

10. Naturgeschichte der Stubenvögel. Gotha, 1812, fig.—(Gall.) *Histoire naturelle des Oiseaux de chambre.* Genève, 1825, 8°. —(Angl.) *Nat. Hist. of Cage Birds,* 8°. Lond. 1845.

11. Forstinsectologie, etc., neu bearbeitet von *D. E. Müller.* Gotha, 1829–35, 2 vols. fig.

12. Ornithologisches Taschenbuch von u. für Deutschland. Leipz. 1803, 8°. 3 vols. fig.

13. Kurzgefasste Naturgeschichte des In- und Auslandes. Leipz. 1792–94, 2 vols. 8°. fig.

14. Kurze gründliche Musterung aller mit Recht oder Unrecht von dem Jäger als schädlich geachteten oder getödteten Thiere. Gotha, 1792 ; 1805, 8°. fig.

15. Naturgeschichte der schädlichen Waldinsecten.   Nürnb. 1798;
1800, fig. (1 einziges Heft.)

16. Zweckmässiges naturhistorisches Bilderbuch.   4°. Nürnberg.

BECHSTEIN (J. M.) u. SCHARFENBERG (   ).

1. Vollständige Naturgeschichte aller schädlichen Forstinsecten;
3 Thl. mit illum. Kupf. Leipz. 1803–1805, 4°.—Gotha, 1818,
4°.—Gotha, 1829–1835, 2 Th. 8°. mit Kupf. 4°. (besorgt v. *Dr.*
Desberger.)—*Wiegm.* Arch. 1835, II. p. 10.

BECICHEMUS (Mar.).

1. In Plinium prælectio.   Ferrar. 1504, fol.—Lutet. 1519, fol.—
Böhm. Bibl. I. 1. p. 206.

BECK (Abr.).

1. Beschreibung der Grasraupen, die in ungewöhnlicher Menge
erscheinen.—Schwed. Abh. 1742, p. 51.—*Fuessl.* N. Ent. Mag.
II. p. 347.—Bibl. Ent. I. p. 20.

BECK (Dav. van der).

1. Experimenta et Meditationes circa Rerum naturalium Principia.
Hamb. 1674; 1684; 1703, 8°.—*Böhm.* Bibl. II. 1, p. 128.

BECK (Dom.).

1. Briefe eines Reisenden von Salzburg durch München u. Nürnberg nach Sachsen, über verschiedene Gegenstände der Naturlehre u. Mathematik. Salzb. 1781, 8°.—*Böhm.* Bibl. I. 1, p. 585.

BECK (   ).

1. Versuch einer Naturgeschichte des Preussischen Bernsteins.
Königsb. 1797.

BECK (H.).

1. Index Molluscorum præsentis ævi Musei principis augustissimi
Christiani Frederici. Fasc. I. Hafn. 1838, 4°.— *Wiegm.* Arch.
1839, II. p. 218.

2. On the Geology of Denmark.—Proc. Geol. Soc. II. 217.—Phil.
Mag. ser. 3, VIII. p. 553.

3. *Rostellaria occidentalis,* de la Baie de St. Laurent.—Mag. Zool.
V. p. 72.

4. Geognostische Bemerkungen über einige Theile des Münsterlandes, mit besonderer Rücksicht auf das Steinsalz·Lager welches

die Westphälischen Soolen erzeugt.—*Karst.* Arch. VIII. p. 275.
—*L.* u. *Br.* N. Jahrb. 1836, p. 224.

BECK (H. J. le).

1. Beschreibung eines langarmigen, ungeschwänzten Affen aus dem Innern von Bengalen.—Naturforscher, st. 29, 1802, p. 1.

BECK (Lew. C.).

1. On the New Brunswick Tornado or Waterspout of 1835.—*Sill.* Am. Journ. XXXVI. p. 115.

2. On Mineralogical and Chemical Survey of State of New York. —*Sill.* Am. Journ. XXXVI. p. 1.—Nat. Hist. of New York, part 3.

3. Notices of the Native Copper, &c. found in the vicinity of New Brunswick, N. Jersey.—*Sill.* Am. Journ. XXXVI. p. 107.

BECK (L. v.).

1. Beiträge zur Bayer schen Insecten-Fauna.   Augsb. 1817, 8°. fig. —Bibl. Ent. I. p. 20.

BECK (T. S.).

1. On the Nerves of the Uterus.—Phil. Trans. CXXXVI. p. 213.

BECKER (Dav.).

1. Bericht von der Berliner Blutquelle, wie dieselbe ohne Aberglauben zu betrachten; item von andern blutrothen Wunderwassern.   Berl. 1677, 4°.—*Böhm.* Bibl. V. p. 45.

BECKER (F. Guill.).

1. De Glandulis thoracis lymphaticis atque Thymo, Specim. patholog.   Berol. 1826, 4°. fig.—Bibl. med.-ch. p. 31.

BECKER (G. H. M.).

1. De Hermaphroditismo.   Jenæ, 1842, 8°.

BECKER (G. W.).

1. Neue Untersuchungen über die Lebenskraft organischer Körper. 2 vols. 8°.   Liegnitz, 1802.

BECKER (H. F.).

1. Topographische Beschreibung des heiligen Dammes bei Dobberau u. Rehdewisch in Mecklenburg.   Schwerin, 1792, 8°.

BECKER (J. Herm.).

1. Der Magen, in seinem gesunden u. kranken Zustande betrachtet. 1ʳ Th. Stend. 1836, 8º.—Bibl. med.-ch. p. 32.

2. Beitrag zur Naturgeschichte der gemeinen Klapperschlange (*Crotalus horridus*).—Isis, 1826, XI. p. 1132.

BECKER (J. H.).

1. De Adulteriis Brutorum et Confusione Specierum Brutis ab Hominibus imperatâ. Rost. 1731, 4º.—*Böhm.* Bibl. II. 1, p. 139.

BECKER (J. W.).

1. Beschreibung des in der Grafschaft Saarbrück und im Oberamte Heerkivesen befindlichen Gesundbrunnen zu Neuweyer genannt. Saarbr. 176.—*Böhm.* Bibl. V. p. 332.

BECKER (Nic. Guil.).

1. De Ascaridibus Uteri.—Eph. Nat. Cur. Dec. I. Ann. 8. p. 121.

BECKER (Pet.).

1. Hypothesis nova de duplici Visionis et Organo et Modo, dioptrico altero, altero catoptrico, quorum hoc Insectis, illud verò Animalibus reliquis concessisse Natura videtur. Rost. 1730, 4º.—*Böhm.* Bibl. II. 1, p. 114.

BECKER (W. G.).

1. Der Plauische Grund bei Dresden, etc. 4º. Nuremb. 1799.

2. Ueber die Flötzgebirge im südlichen Polen. 8º. Freyberg, 1830.

BECKETT (H.).

1. On a Fossil Forest in the Parkfield Colliery near Wolverhampton. —Journ. Geol. Soc. I. 41.

BECKHER (Dan.).

1. Bedenken von dem Schwefelregen, wie auch der unnatürlichen und vielfältigen Mäuse auf dem Felde, Ursachen und Bedeutungen. Hamb. 1634, 4º.—*Böhm.* Bibl. II. 1, p. 348.

BECKMANN (Joh.).

1. Grundriss zu Vorlesungen über die Naturlehre. Gött. 1779, 8º —*Böhm.* Bibl. I. 1, p. 306.

2. Beytrag zu Naturgeschichte des *Monoculus Polyphemus*, L.— Naturforscher, st. 6, 1775, p. 35.

3. Anfangsgründe der Naturhistorie. Brem. u. Gött. 1767. (Fortsetz.) Tref. u. Leipz. 1785, 8°. (Verm. v *Scholz*) Bresl. 1814, 8°.—Bibl. Ent. I. p. 20.—*Böhm.* Bibl. I. 1, p. 306.

4. De Historiâ naturali Veterum libellus primus. Petropol. et Gotting. 1766, 8°.—*Böhm.* Bibl. I. 1, p. 169.

5. Sur le Fic des Poissons.—Mag. Hannov. 1769.—*Cuv.* et *Val.* I. p. 169.

6. Eine bequemere Einrichtung der Insecten-Sammlungen.—Besch. d. Berl. Naturf. II. p. 69.—Bibl. Ent. I. p. 20.

7. Kermes, Cochenille.—Beitr. z. Gesch. der Erfindungen, III. p. 1. —Bibl. Ent. I. p. 20.

8. Commentatio de historiâ Aluminis.—Comment. Gœtt. 1778, I. p. 111.

9. Commentatio de reductione rerum fossilium ad genera naturalia Protyporum.—N. Comm. Gœtt. 1771, II. p. 68; 1772, III. p. 95; IV.—*Böhm.* Bibl. IV. 2, p. 247.

10. Commentatio de laccis *Rubiæ tinctoriæ* et *Phytolaccæ decandræ.*—Comm. Gœtt. 1779, II. p. 65.

11. Commentatio de Spumâ maris, e quâ capitula ad fistulas nicotianas finguntur.—Comm. Gœtt. 1781, IV. p. 46.

12. Linneische Synonymie zu Klein's verbesserte Historie der Vögel.—Naturforscher, st. 1, 1774, p. 65.

13. Kleiner Beytrag zur Naturgeschichte des Murrachen (*Mergus serrator,* Lin.). Beschäft. der Berlin. Gesells. Naturforsch. Freunde, Bnd I. 1775, p. 170.

BECKMANN (J. Chr.).

1. Diss. de Prodigiis Sanguinis. Francf. s. Od. 1676, 4°.—*Böhm.* Bibl. V. p. 45.

BECKMANN (J. Chr.) u. BERNHARD (L.).

1. Beschreibung der Cur- und Mark Brandenburg. Berl. 1751, fol. —*Böhm.* Bibl. I. 1, p. 604.

BECKS (   ).

1. Ueber das Vorkommen fossiler Knochen in dem aufgeschwemmten Boden des Münsterlandes.—*Karst.* Arch. 1835, VIII. p. 390, fig. —*L.* u. *Br.* N. Jahrb. 1837, p. 237.

2. Bemerkungen über neue Höhle in Westphalen.—*L.* u. *Br.* N. Jahrb. 1841, p. 143, fig.

3. Ueber tertiäre Ablagerungen in den Niederländischen Provinzen Gelderland und Ober-Yssel.—*L.* u. *Br.* N. Jahrb. 1843, p. 257.

4. Kerne und Krystall-Drüsen in Kreide Echiniden.—*L.* u. *Br.* N. Jahrb. 1843, p. 168.

5. Ueber fossile Fährten, besonders jene am Ister-Berge.—*L.* u. *Br.* N. Jahrb. 1843, p. 188.

BECKWITH (John).

1. The History and Description of four new species of Phalæna.— Tr. Linn. Soc. Lond. II. p. 1, fig. col.

2. A Memoir on the Natural Walls, or Solid Dykes, in the State of North Carolina; about which there have been debates, whether they were basaltic, or of some other formation.—*Sill.* Am. Journ. V. p. 1.—*Féruss.* Bull. 1823, II. p. 51 (*Mémoire sur les Murs naturels ou filons de la Caroline du Nord*, etc.).

BÉCLARD (F. A.).

1. Uebersicht der neuern Entdeckungen in der Anatomie und Physiologie. Aus d. Franz. v. *Cerutti.* Leipz. 1823, 8°.

2. Élémens d'Anatomie générale. Paris, 1823, 8°.—2e éd. 1827. —Biogr. Un. Suppl. LVII. p. 445.

3. Nouveau Manuel d'Anatomie descriptive, d'après ses cours. 2e éd. Paris, 1836, 8°.—*Val.* Repert. II. p. 19.

BECQUEREL (εδη).

1. De l'Argile plastique d'Auteuil, et des substances qui l'accompagnent.—Ann. Chim. et Phys. 1823, Avr. p. 348.—*Féruss.* Bull. 1823, III. p. 40.

2. Observations sur la présence de Sables aurifères dans le gisement de la Galène de St. Santin-Cantalés, et sur le gisement des Sables aurifères en général.—C. R. XI. p. 129.

3. Sur une couche de Lignite trouvée à Auteuil. 4°. 1819.

4. Électricité animale.—Ann. Sc. n. 1831, XXII. p. 15.

BECQUEREL (   ) et BRESCHET (G.).

1. Première Mémoire sur la Chaleur animale.—Ann. Sc. n. 1835, III. p. 257.—*Müll.* Arch. 1836, V. p. cxix (*Ueber die thierische Wärme*).

2. Seconde Mémoire sur la Chaleur animale.—Ann. Sc. n. 1835, IV. p. 243.—C. R. I. p. 28.

3. Recherches expérimentales physico-physiologiques sur la Température des Tissus et des Liquides animaux.—Ann. Sc. n. 1837, VII. p. 94.—C. R. III. p. 771 (3e Mémoire).

4. Nouvelles observations sur la mesure de la Température des Tis-

sus organiques du corps de l'Homme et des Animaux, au moyen des effets thermo-électriques.—Ann. Sc. n. 1838, IX. p. 271.

5. Expériences sur la Torpille.—Ann. Sc. n. 1836, VI. p. 123.

BEDDE (Sam. Siegfr.).

1. Diss. de Verme Tæniâ dicto. Viennæ, 1767, 8°.—*Böhm.* Bibl. II. 2, p. 404.

BEDDEVOLE (Dom.).

1. Dissertatio de Hominis Generatione in Ovo. 4°.—Biogr. Un. IV. p. 37.

2. Essais d'Anatomie, où l'on explique clairement la construction des Organes. Leyde, 1684, 12°.—Biogr. Un. IV. p. 37.

BEDDOES (T.).

1. On the Affinity between Basaltes and Granite.—Phil. Trans. LXXXI. p. 48.

BEDEMAR (Vargas).

1. Ueber die Kalk- und Kreide-Formation von Texoë, Stevens- und Moëns-Klint.—*Leonh.* Tasch. 1820, I. p. 40, fig.

2. Ueber vulkanische Erzeugnisse aus Island.—*Leonh.* Tasch. 1819, I. p. 105.

3. Die Insel Bornholm in geognostischer Hinsicht. Frankf. 1819, 8°.—Cat. Bibl. Turic.—*Leonh.* Tasch. 1820, I. p. 3.

BEDFORD (W.).

1. On the Strata of Lincoln.—Mag. Nat. Hist. ser. 2, III. p. 553.

BEDINELLI (Fr. P.).

1. Observatio perfectæ androgyneæ Structuræ. Pisauri, 1755, 8°.—*Böhm.* Bibl. II. 1, p. 140.

BEDOYA Y PAREDES (P. G. de).

1. Hist. universal de las Fuentes minerales de Espanna. San Jago, 1764, 4°. (Germ. *Chr. Pluer*) *Büsch.* Mag. III.—*Böhm.* Bibl. I. 1, p. 500.

BEECHEY (F. W.).

1. Voyage to the Pacific and Behring's Strait for the purpose of Discovery, and of cooperating with the Expeditions under Capt. Parry and Franklin, performed in H.M.S. Blossom, in the years 1825, 26, 27 and 28. London, 1831, 4°.

Beechey ( ).

1. Results of Deep Dredging off the Mull of Galloway.– Ann. and Mag. N. H. X. p. 21.

Been (J.).

1. Dissert. de ultimo Incendio Heclæ. Hafn. 1694, 4º.—(Germ.) Hamb. Mag. VI. p. 97.—*Böhm.* Bibl. IV. 2, p. 376.

Been (J. Nic.).

1. De Pisce qui Jonam deglutivit. Hafn. 1698, 4º.—*Böhm.* Bibl. II. 1, p. 77.

2. Dissert. Quaternarium Quæstionum. (III. Num Cygnus morti vicinus suaviter canat.) Hafn. 1706, 4º.—*Böhm.* Bibl. II. 1, p. 564.

Beer (F. W.).

1. Abhandl. Akad. Paris; *vide sup.* Pars I. p. 28.

Beer (K. E. v.).

1. Beiträge zur Kenntniss der niedern Thiere.—Nov. Act. Ac. Leop. XIII. 2, 1827.

Beermann (Sig.).

1. Historische Nachrichten u. Anmerkungen von der Grafschaft Pyrmont u. ihren berühmten Sauerbrunnen. Frankf. u. Leipz. 1706, 8º.—*Böhm.* Bibl. V. p. 352.

Begbie (P. J.).

1. The Malayan Peninsula. 8º. Madras, 1834.

Begg (H. v.).

1. Der Borkenkäfer in Gallizien.—*Liebisch* Aufmerks. Forstm. Prag. 1827, II. 2, p. 107.—*Eis.* p. 196.

Beggiato (Fr.).

1. Delle Terme Euganee. Padova, 1833, 8º.

Bégin (Em. Aug.).

1. Connaissance physique et morale de l'Homme, ou Manuel d'A-natomie physiologique, avec des règles d'Hygiène ,etc. Nanci, 1837, 8º. fig.

BEHLEN (St.).

1. Lehrbuch der Forst- und Jagdthiergeschichte. Leipz. 1826, 8°.
—Isis, 1827, X. p. 886.

2. Lehrbuch der Gebirgs- und Bodenkunde, in Beziehung auf das
Forstwesen. Gotha u. Erf. 1826, 8°. fig.—Isis, 1827, X. p. 882.

BEHN (F. W. G.).

1. Ueber den Einfluss des Pulses auf die Bewegung unserer Körpertheile.—*Müll.* Arch. 1835, p. 516.

2. Entdeckung eines von den Bewegungen des Rückengefässes unabhängigen, und mit einem besondern Bewegungsorgane verse-
henen Kreislaufes in den Beinen halbflügelichter Insekten.—
Müll. Arch. 1835, p. 554, fig.—*Wiegm.* Arch. 1836, p. 298;
1837, p. 330.—Ann. Soc. Entom. Fr. 3ᵉ trim. p. LX.—Ann. Sc.
n. 1835, IV. p. 5 (*Découverte d'une circulation de fluide nutritif
dans les pattes de plusieurs Insectes hémiptères, etc.*).

3. *George Cuviers* Briefe an *C. H. Pfaff* aus den Jahren 1788 bis
1792 (vid. *Cuv.*).

BEHR (F.).

1. Bildung von Gyps-Krystallen in Toscana.—*L.* u. *Br.* N. Jahrb.
1843, p. 483.

BEHRENS (Fr. Chr.).

1. Dissert. de Monocerote. Lips. 1667, 4°.—*Böhm.* Bibl. II. 1,
p. 482.

BEHRENS (G. H.).

1. *Hercynia curiosa*, oder Curieuser Harzwald; d. i. Verzeichniss
und Beschreibung derer curieusen Höhlen, Seen, Brunnen, Bergen u. vieler andern an und auf dem Harze vorhandenen Sachen;
mit physikalischen Anmerkungen. Nordh. 1703, 4°; 1712, 4°;
1720, 4°.—*Böhm.* Bibl. I. 1, p. 593.—(Tr. Angl. *J. Andrée.*) 8°.
Lond. 1730.

BEHRENS (H.).

1. De Lumbricis effractoribus. Halæ, 1740, 4°.

BEICKY (J. St.).

1. De Vermibus nasalibus. Rud. 1782, 8°.

BEIER (Adr.).

1. Abbildung der Jenaischen Gegend, Grund u. Bodens. Jena, 1672, 8°.—*Böhm.* Bibl. I. 1, p. 598.

BEIREIS (G. Chr.).

1. Beschreibung eines bisher unbekannt gewesenen amerikanischen Frosches.—Schr. d. Berlin. Ges. Naturf. Freunde, IV. p. 178.— Dict. Sc. nat. XV. p. 256.
2. Programma de utilitate et necessitate Historiæ naturalis. Helmst. 1759, 4°.—*Böhm.* Bibl. I. 1, p. 177.

BEITHNER (J. Th. Ant.).

1. Beschreibung der Böhmischen Flüsse nach ihrem Ursprunge und Laufe bis zum Austritte in fremde Länder; mit mineralischen Anmerkungen. Prag. 1771, 8°.—*Böhm.* Bibl. V. p. 77.

BEKE (Ch. T.).

1. On the Historical Evidence of the Advance of the Land upon the Sea at the Head of the Persian Gulf.—Phil. Mag. and Journ. ser. 3, VII. p. 40.—Cat. R. Soc. Lond.
2. On the Alluvia of Babylonia and Chaldæa.—Phil. Mag. ser. 3, XIV. p. 426.

BEKKER (F.).

1. Populäre Darstellung der Naturgeschichte. 8°. Wien, 1845.

BÉLANGER (Ch.).

1. Voyage aux Indes Orientales par le Nord de l'Europe, etc. Paris, 1834–40, 8°. Zoologie.—*Wiegm.* Arch. II. 260.—Isis, 1834, XII. p. 1219.—Mag. Zool. and Bot. I. p. 269.—*Féruss.* Bull. XXV. p. 291.

BELCHER (E.).

1. Narrative of a Voyage round the World. Lond. 1843, 2 vols.— Ann. and Mag. N. Hist. XIII. p. 126.
2. On some Geological Specimens from the West Coast of Africa. —Proc. Geol. Soc. II. 188.

BELCHER (Capt.), BOWERS (Lieut.), and CUMING (H.).

1. On the effects produced at Valparaiso by the Earthquake of Nov. 1822.—Proc. Geol. Soc. II. 213.—Phil. Mag. ser. 3, VIII. p. 159.

BELERIO (G.).

1. Περὶ τῶν ἐντόμων. Upsal, 1675.

BELÈZE (G.).

1. L'Hist. nat. mise à la portée des Enfans. 18°. Paris, 1844.

BELHOMME (   ).

1. Troisième Mémoire sur la localisation des Fonctions cérébrales et de la Folie, considéré chez les Animaux et chez l'Homme. Par. 1839, 8°.

BELIUS (Matth.).

1. Hungariæ antiquæ et novæ Prodromus. Norib. 1723, fol. fig.— *Böhm.* Bibl. I. 1, p. 642.
2. Notitia Hungariæ historico-geographica. Viennæ, 1735–42, fol. fig.—*Böhm.* Bibl. I. 1, p. 642.

BELKMEER (Corn.).

1. Natuurkundige Verhandeling of waarneminge betreffende de Houtnytras pende en doorboorende Zeeworm. Amst. 1733, 8°; 1753, 8°. fig.—*Böhm.* Bibl. II. 2, p. 489.

BELKNAP (J.).

1. On the White Mountains in New Hampshire.—Trans. Am. Phil. Soc. II. p. 42.

BELL (Benj.).

1. Strictures on the Hypothesis of *M. Jos. du Commun* on Volcanos and Earthquakes.—*Sill.* Am. J. XVI. 1, p. 51.

BELL (Ch.).

1. On the Nervous System.—Phil. Trans. CXXX. p. 245.
2. Exposition du Système naturel des Nerfs du Corps humain, trad. de l'Angl. par *J. Genest.* Paris, 1825, 8°. fig.
3. Darstellung der Arterien ; zum Unterricht für Aerzte u. Wund-ärzte. Nach den Engl. bearbeitet etc. v. *H. Robbi.* Leipz. 1819, 8°. fig.—Descriptio Arteriarum, icon. illustr. Latinè etc. ab *H. Robbi.* Lips. 1819, 8°.—Bibl. med.-ch. p. 36.
4. Darstellung der Nerven, zum Unterricht für Aerzte u. Wund-ärzte. Nach der 3ten orig. Ausg. bearb. etc. v. *H. Robbi.* Leipz. 1820, 8°. fig.—Bibl. med.-ch. p. 36.

5. The Nervous System of the Human Body. Lond. 1830, 4º.— *Müll.* Arch. 1837, II. p. 263.

6. The Hand, its Mechanism and Vital Endowments as evincing Design. 8º. Lond. 1833.—Bridgewater Treatises, IV. (*Die menschliche Hand. Aus dem Engl. übersetzt von H. Hauff.*) Stuttg. 1836, 8º. fig.—*Val.* Repert. II. p. 22.

7. Beitrag zur Anatomie der Hirnfaserungen.—Phil. Trans. 1834, II.—*Müll.* Arch. 1835, p. 12.

8. Untersuchungen über die Structur des Gehirns. (Fortsetz.)— Phil. Trans. 1835, p. 255.—*Müll.* Arch. 1837, VII. p. xxiii.

9. Zergliederung einiger Theile des menschlichen Körpers. Aus dem Engl. v. *Rosenmüller.* Leipz. 1817, 8º. 2 B. fig.—Bibl. med.-ch. p. 36.

10. Physiologische u. pathologische Untersuchungen des Nervensystems. Aus d. Engl. v. *M. H. Romberg.* Berl. 1836, 8º. fig.— Bibl. med.-ch. p. 36.

11. On the Motions of the Eye, in illustration of the Uses of the Muscles of the Orbit.—Phil. Trans. 1823.—Isis, 1825, VII. L. A. p. 53.—*Féruss.* Bull. III. p. 84.

12. Essais anatomiques et physiologiques sur la Physionomie.— Ann. Sc. n. 1826, VIII. p. 245.

13. On the Comparison of the Nerves of the Spine with those of the Encephalon.—Ann. of Nat. H. II. p. 68.

BELL (C. M.).

1. Geological Notes on part of Mazunderān.—Trans. Geol. Soc. ser. 2, V. p. 577.—Proc. Geol. Soc. II. 591.—Phil. Mag. ser. 3, XII. p. 571.

BELL (J.).

1. Travels from St. Petersburg in Russia to diverse parts of Asia. Glasg. 1763, 4º. 2 vols.—(Gall.) *Voyages depuis St. Pétersbourg dans diverses contrées de l'Asie, etc., on y a joint une Description de la Sibérie.* Par. 1766, 12º. 3 vols.—*Böhm.* Bibl. I. 1, p. 674.

BELL (J. II.).

1. The Anatomy of the Human Body. Lond. 8º. 3 vols. 1793-1802. —(Deutsch. umgearb. v. *Rosenmüller* u. *Heinroth,* Leipz. 1806, 2 vols. 8º. fig. *Zergliederung des menschlichen Körpers.*)—Biogr. Un. Suppl. LVII. p. 489.—Bibl. med.-ch. p. 35.

2. Engravings of the Arteries illustrating the 2nd vol. of the Anatomy of the Human Body. Lond. 1801, 8°.—Biogr. Un. Suppl. LVII. p. 489.

3. Engravings explaining the Anatomy of the Bones, Muscles and Joints. Lond. 1794, 4°.—Biogr. Un. Suppl. LVII. p. 489.

BELL (Th.).

1. Remark on the Animal nature of Sponges.—Zool. Journ. I. p. 202. —*Féruss.* Bull. 1826, VIII. p. 165.

2. Observations on the Neck of the Three-toed Sloth (*Bradypus tridactylus*, Linn.).—Proc. Zool. Soc. Lond. III. p. 99.—Trans. Zool. Soc. I. p. 113.—Ann. Sc. n. 1824, II. p. 382.

3. Observations on the Structure of the Throat in the genus *Anolis.* —Zool. Journ. IV. 1826, p. 11.—*Féruss.* Bull. VII. p. 117.—Isis, 1830, VIII. p. 829.— Ann. Sc. n. VII. p. 191, fig.

4. On the structure and use of the Submaxillary Odoriferous Glands in the genus *Crocodilus.*—Phil. Trans. CXVII. p. 132.—*Féruss.* Bull. 1828, XV. p. 164.—*Heus.* Zeitschr. III. p. 140.

5. History of British Quadrupeds. Lond. 1836, 8°. fig.—Mag. Zool. and Bot. 1837, I. p. 280.

6. Account of a pair of living Acouchies (*Dasyprocta Acouchy*).— Proc. Zool. Soc. Lond. I. p. 6.

7. Description of a new species of *Phalangista* (*Ph. gliriformis*).— Linn. Tr. XVI. 1, p. 121, fig.—Isis, 1830, IX. p. 914.—*Féruss.* Bull. 1829, XVII. p. 433.

8. On the genus *Galictis*, with Description of a new species.—Trans. Zool. Soc. II. p. 201.—Genre nouveau dans la famille des Mustélides.—Phil. Mag. 1838, p. 390.—Rev. Zool. 1838, p. 292.

9. Note on the supposed identity of the genus *Isodon* of *Say* with *Capromys.*—Zool. Journ. I. 1825, p. 230.

10. A Monograph of the Tortoises having a moveable Sternum, &c. —Zool. Journ. 1826, II. p. 299.—Isis, 1830, X. p. 1030.—*Féruss.* Bull. IX. p. 96.

11. Characters of the Order, Families and Genera of *Testudinata.* —Zool. Journ. III. p. 513.—*Féruss.* Bull. 1828, IV. p. 306.— Isis, 1830, XII. p. 1241.

12. On *Hydraspis*, a new genus of Freshwater Tortoises of the family *Emydidæ.*—Zool. Journ. III. 1828, p. 511.—*Féruss.* Bull. 1828, XV. p. 305.—Isis, 1830, XII. p. 1240.

13. Description of a new species of *Terrapene*, with further obser-

vations on *T. carolina* and *T. maculata.*—Zool. Journ. II. p. 484. —*Féruss.* Bull. 1827, XII. p. 271.

14. Characters of a new genus of Freshwater Tortoise (*Cyclemys*) —Proc. Zool. Soc. Lond. IV. p. 17.—*Wiegm.* Arch. 1835, II. p. 294.

15. On two new genera of Land Tortoises (*Pixys* and *Kinixys*).— Linn. Tr. XV. 2, p. 392, fig.—Isis, 1829, XII. p. 1275.

16. Descriptions of three new species of Land Tortoises.—Zool. Journ. III. p. 419.—*Féruss.* Bull. 1828, XV. p. 400.

17. Monograph of the Testudinata. fol. Lond. 1833, fig.

18. Description of a new species of Lizard (*Uromastyx acanthi-nurus*).—Zool. J. I. p. 457, fig.—Isis, 1830, VIII. p. 821.

19. On a new genus of Iguanidæ (*Amblyrhynchus*).—Zool. J. II. 1826, p. 204.—Isis, 1828, IX. p. 940.

20. Description of a new species of *Agama* brought from the Columbia River.—Linn. Trans. XVI. p. 105, fig.—*Féruss.* Bull. 1829, XVII. p. 437.—Isis, 1830, IX. p. 910.

21. Characters of two new genera of Reptiles (*Anops, Lerista*).— Proc. Zool. Soc. Lond. III. p. 98.

22. Description of a new species of *Anolius* and a new species of *Amphisbæna*, collected by *MacLeay* in the Island of Cuba.— Zool. Journ. III. 1828, p. 235.

23. On a new Fossil species of *Chelydra* from Œningen.—Trans. Geol. Soc. ser. 2, IV. p. 379.—Proc. Geol. Soc. I. 342.—Phil. Mag. ser. 2, XI. p. 282.—*L.* u. *Br.* N. Jahrb. 1833, p. 614.

24. On *Leptophina*, a group of Serpents comprising the genus *Dryinus*, Merr., and a newly formed genus proposed to be named *Leptophis.*—Zool. Journ. 1826, II. p. 321.—*Féruss.* Bull. 1826, IX. p. 104.—Isis, 1830, X. p. 1035.

25. Reptiles, in *Darwin*, Zool. Voy. Beagle.

26. A History of British Reptiles. Lond. 1829, 8°. fig.—Ann. of Nat. Hist. I. p. 222.

27. On *Aranea domestica.*—Zool. Journ. I. 2, p. 282.—Isis, 1830, IV. p. 418.

28. Abstract of a Memoir on the Physiology of *Helix pomatia* by *B. Gaspard*, with Notes.—Zool. Journ. I. pp. 93, 174.

29. Description of a new species of *Emarginula.*—Zool. Journ. I. p. 52.

30. Observations on the genus *Cancer* of *Dr. Leach* (*Platycarcinus,*

Latr.), with descriptions of three new species.—Trans. Zool. Soc. I. 1835, p. 335.

31. Account of Crustacea of South America, with Descriptions of new genera and species.—Beschreibungen und Abbildungen mehrerer Krabben aus der Tribus der *Oxyrhynchi*.—Trans. Zool. Soc. II. 1, p. 40, fig.— *Wiegm.* Arch. 1837, VI. p. 244.

32. On the *Thalassina Emerii*, a fossil Crustacean from New Holland.—Journ. Geol. Soc. I. 93.

33. History of British Crustacea. 8º. Lond. 1847.

34. On Crustacea found by *Prof. E. Forbes* and *Mr. MacAndrew* on their cruises round the coast.—Rep. Brit. Assoc. 1846, Sect. p. 80.

BELL (T.), CHILDREN (J. G.), SOWERBY (J. De C.), and SOWERBY (G. B.).

1. Zoological Journal; *vide sup.* Pars I. p. 78

BELL (W.).

1. On the Double-horned Rhinoceros of Sumatra. Phil. Trans. LXXXIII. p. 3.

2. On a species of Chætodon, called by the Malays *Ecan bonna.*— Phil. Trans. LXXXIII. p. 7.—*Cuv.* et *Val.* I. p. 138.

BELLAMY (J. C.).

1. The Natural History of South Devon. Plym. 1839, 8º.—Ann. and Mag. N. Hist. VII. p. 209.

2. Description of two Green-streaked Wrasses (*Labrus lineatus*, Flem.).—Ann. and Mag. N. Hist. XIII. p. 77.

BELLAMY (P. F.).

1. A brief Account of two Peruvian Mummies in the Museum of the Devon and Cornwall Nat. Hist. Society.—Ann. and Mag. N. Hist. X. p. 95, fig.

BELLANI (Ang.).

1. Sul mezzo di prevenire il danno che arrecano gli Insetti al Frumento.—Ann. d'Agric. R. d'Ital. XIX.

2. Della indefinibile durabilita della vita nelle Bestie. 8º. Milan, 1836.

BELLARDI (Lud.).

1. Estratto della Mem. in cui proponesi un mezzo facile ed econo-

mico per nutrire i Bachi da seta in mancanza della foglia recente di Mori.—Opusc. scelti X. p. 179.—Bibl. Ent. I. p. 21.

2. Genus *Borsonia*, etc.—Bull. Soc. Géol. Fr. X. p. 30.

3. Description des Cancellaires fossiles des Terrains tertiaires du Piémont. Turin, 1841, 4°. fig.—Mém. Acad. Tur. III. ser. 2.— Rev. Zool. 1841, p. 352.—*L. u. Br.* N. Jahrb. 1840, p. 343.

BELLARDI (L.) e MICHELOTTI (G.).

1. Saggio orittografico sulla classe dei Gasteropodi fossili dei Terreni terziarii del Piemonte. Torino, 1840, 4°. fig.

BELLER ( ).

1. Description géographique de la Guyane. Par. 1763, 4°. fig. et cartes.—*Böhm.* Bibl. I. 1, p. 751.

BELLERMANN (J. J.).

1. Von dem Werthe des Studiums der Naturwissenschaften auf Gymnasien. Erfurt, 1797, 4°.—N. Act. Nat. Cur. XVII. p. xx.

2. Ueber das bisher bezweifelte Daseyn des Rattenkönigs. Berlin, 1820, 8°.—N. Act. Nat. Cur. XVII. p. xx.

3. Versuch einer gleichförmigen systematischen Aufstellung der Conchilien, nach Classen, Ordnungen u. Gattungen; mit beigefügten deutschen Namen.—Berl. Mag. VII. p. 83.—Isis, 1818, IX. p. 1478.

BELLERS (F.).

1. On the several Strata of Earth, Stone, Coal, &c. found in a Coalpit at the west end of Dudley in Staffordshire.—Phil. Trans. XXVII. p. 541.

BELLERY ( ).

1. Dissert. sur la Tourbe de Picardie, qui a remporté le prix de l'Académie d'Amiens. Am. 1755, 8°. fig.—*Böhm.* Bibl. IV. 1, p. 199.

BELLEVAL (P. Richer de).

1. Opuscules, publiés par *Broussonnet.* Paris, 1785, 8°.—Trans. Linn. Soc. Lond. V. p. 278.

BELLINGERI (C. F.).

1. Experimenta physiologica in Medullam spinalem.—Mem. Tur. XXX. p. 293.—Isis, 1834, I. p. 83.

2. Note sur l'Électricité du Sang dans les maladies.—Bull. Soc. Philom. 1823, p. 189.

3. Ragionamenti, sperienze ed osservazioni pathologiche comprovanti l'Antagonismo nervoso. Torin. 1833, 8°.—N. Act. Nat. Cur. XVII. p. xx.—Isis, 1834, I. p. 81 (*Versuche über den Antagonismus der Nerven*).

4. Paragone fra la Vista e l' Udito.—Giorn. Sc. Med. Tor. IV.

5. Dell' influenza del Cibo e della Bevanda sulla Fecondità e sulla Proporzione dei Sessi nelle Nascite del Genere umano.—Giorn. Sc. Med. Tor. VIII.

6. Della Fecondità e della Proporzione dei Sessi nelle Nascite degli Animali vertebrati. Tor. 1840, 4°.

7. Mastologia, con Considerazioni anatomico-fisiologiche sul numero e posizione delle Mammelle. Tor. 1840, 4°.

8. De Medullâ spinali Nervisque ex eâ prodeuntibus, Annotationes anatomico-physiologicæ. 4°. Aug. Taur. 1823.—Cat. R. Soc. L.

9. Sulla struttura e posizione degli Organi dell' Udito e della Vista nei principali generi dei Mammiferi.—Mem. Accad. Tor. I. ser. 2.

10. Sugl' Emispheri cerebrali dei Mammiferi. Torino, 1838, 8°.

11. De Nervis Faciei, Anatomes et Physiologia. Diss. inaug. Taur. 1818, 8°.

12. Table de la Fécondité des Mammifères, précédé d'une analyse détailleé.—Rev. Zool. 1839, p. 220.—C. R. VII. p. 948; IX. p. 338.

BELLINGHAM (O'B.).

1. Catalogue of the Entozoa indigenous to Ireland.—Mag. Nat. Hist. ser. 2, IV. p. 343.—Ann. and Mag. Nat. Hist. XIII. pp. 101, 167, 254, 335, 422; XIV. pp. 162, 251, 317, 396, 471.

2. On a Specimen of the *Orthagoriscus mola* (Sun Fish), caught off the Irish coast in June 1839.—Mag. Nat. Hist. ser. 2, IV. p. 235.

3. *Trichocephalus dispar.*—Brit. Assoc.—*Sill.* Am. Journ. XXXIV. 1, p. 10.

4. Description d'une espèce d'*Ascaris* qu'il a découverte et qu'il propose de nommer *A. alata.*—Rev. Zool. 1839, p. 126.

BELLINI (Laur.).

1. Discorsi di Anatomia, colla Prefazione di *Ant. Cocchi*. Fir. 1741-1744, 3 vols. 8º.

2. De structurâ et usu Renum, ut et de Gustûs organo; ed. *Blasius*. Lugd. 1714, 4º.

BELLMISSERIUS (P.).

1. Carmen de Historiâ Animalium, s. Elegiæ Latinæ XXXVI. in priores Libros *Aristotelis* de Animalibus. Romæ, 1534.

BELON (P.).

1. Les Observations de plusieurs singularitez et choses mémorables, trouvées en Grèce, Asie, Judée, Égypte, Arabie et autres pays estranges. Par. 1553, 4º. fig.—(Revu et augm.) Par. 1554, 4º. fig.; 1555, 4º; 1558, 4º; 1585, 4º.—Anv. 1555, 8º; 1588, 8º.——(Lat.) *Observationes Historiæ naturalis et memorabilium rerum in Græciâ, Asiâ*, etc. Antw. 1589. 12º. fig.—*Böhm.* Bibl. I. 1, p. 652.

2. De la Nature et de la Diversité des Poissons, avec leurs portraicts en bois. Paris, 1555, 8º.

3. Portraits d'Oiseaux, Animaux, Serpens, Herbes et Arbres, Hommes et Femmes d'Arabie et d'Egypte. Paris, 1557, 4º; 1618, 4º.—*Dum. et Bib.* I. p. 305.—*Böhm.* Bibl. I. 1, p. 652.

4. Histoire naturelle des Oiseaux, avec leurs descriptions et naïfs pourtraicts, retirez du naturel, écrite en sept livres. Paris, 1555, fol.—Biogr. Un. IV. p. 133.

5. Histoire naturelle des étranges Poissons marins, traitant de leur nature et propriété, avec les portraicts d'iceux. Par. 1551, 1553, 1555, 4º. fig.—*Cuv.* R. an. III. p. 335.—*Klein* Disp. Echin. p. VIII.—*Böhm.* Bibl. II. 2, p. 51.

6. De Aquatilibus, Lib. II. Paris, 1553, 8º. fig.—Trans. Linn. Soc. Lond. VII. p. 216.

BELOW (Jac. Fr.).

1. De Generatione Animalium equivocâ. Lund. Goth. 1706, 4º.—Biogr. Un. IV. p. 136.

2. Diss. de Vermibus Intestinorum. Ultraj. 1691, 4º.—*Böhm.* Bibl. II. 2, p. 376.

BELPAIRE (B.).

1. Mémoire sur les changemens que la Côte d'Anvers à Boulogne a subis, tant à l'intérieur qu'à l'extérieur, depuis la conquête de César jusqu'à nos jours.—Ann. Sc. nat. XVI. 1829.

BELPAIRE (B.) et QUETELET (L. A. J.).

1. Rapport sur les Observations des Marées faites en 1835, en différens points des côtes de Belgique.—N. Mém. Brux. XI.

BELTRAMI ( ).

1. Lettre sur un Lézard bicephale.—Ann. Sc. Nat. 1831, XXIII. pp. 41, 54, 59.
2. L'Italie et l'Europe.—Bull. Soc. Géol. Fr. VI. p. 14.

BEMBI (Petr.).

1. De Ætnâ, ad *A. Chabrielem* Liber. Venet. 1495, 4°.—*Böhm.* Bibl. IV. 2, p. 383.

BENDISCIOLI (Gius.).

1. Monographie des Serpens de la province de Mantoue.—Giorn. Fisic. Chim. etc. 1826, p. 413.—*Féruss.* Bull. 1828, XIII. p. 133.

BENDZ (H.).

1. Bidrag til den Anatomi af Nervus glossopharyngeus, etc., hos Reptilierne.—Vid. Selsk. naturvidensk. Afhand. 1843.
2. Ueber die Orbitalhaut bei den Haussäugethieren.—*Müll.* Arch. 1841, p. 196.
3. De connexu inter Nervum vagum et accessorium. Hafn. 1833, 4°.—*Müll.* Arch. 1837, V. p. xxiii.
4. De anastomosi Jacobsonii et ganglio Arnoldi. Hafn. 1833.— *Müll.* Arch. 1837, p. 288; 1834, p. 13.

BENEDEN (Van). Vid. VAN BENEDEN.

BENEDICTUS (Lib.).

1. De Principiis Naturæ et Artis Liber aureus, oder güldenes Büchlein, so das beschreibt, wie die Metalle in den Klüften der Erde durch die Natur in ihren Minern geboren, etc. Francof. 1623, 1630, 8°.

BENEDIX ( ).

1. De Myelomalaciâ. Berol. 1841, 8°.

BENEKE (Fr. Ed.).

1. Lehrbuch der Psychologie. Berl. 1833, 8°.—Bibl. med.-ch. p. 37.
2. Psychologische Skizzen. I. Bd. Gött. 1825, 8°.—Bibl. med.-ch. p. 37.

BENICKEN ( ).

1. Beiträge zur nordischen Ornithologie.—Isis, 1824, VIII. p. 877.
2. Beiträge zur Naturgeschichte einiger Wasservögel.—Ann. der Wetteravischen Gesellschaft, III. 1814, p. 137.

BENIGNI (Call.).

1. Catalogo dei Fossili di Monte Mario presso Roma. Roma, 1781.

BENIGNI (Fost.).

1. Sugli Insetti distruggitori delle Viti. Milano, 8°.

BENING (B. F.).

1. Diss. inaug. de Hirudinibus. 4°. Hardervici, 1776.

BENNATI ( ).

1. Notice Physiologique sur *Paganini*.—Ann. Sc. nat. XXIII. 1831
2. Du Mécanisme de la Voix humaine pendant le Chant.—Ann. Sc. nat. XXIII. p. 32.—(Trad. all.) Ilmenau, 1833, 8°. fig.—Bibl. med. ch. p. 38.
3. Etudes Physiologiques et Pathologiques sur la Voix humaine. Par. 1833, 8°. fig.
4. Recherches sur le Mécanisme de la Voix humaine, précédé du Rapport de *MM. Cuvier*, de *Prony*, et *Savart*. Paris, 1832, 8°. fig.

BENNET (Ethelr.).

1. A Catalogue of the Organic Remains of the County of Wilts. Warminst. 1831, 4°. fig.

BENNET (H. G.).

1. On a Whin Dyke traversing Limestone in the county of Northumberland.—Trans. Geol. Soc. ser. 1, IV. p. 102.
2. Some Account of the Island of Teneriffe.—Trans. Geol. Soc. ser. 1, II. p. 286.
3. On the Geology of Madeira.—Trans. Geol. Soc. ser. 1, I. p. 391. —Phil. Mag. XXXVIII. p. 284.

BENNETT (E. T.).

1. Characters of new Species of Mammalia from California.—Proc. Zool. Soc. Lond. I. p. 39.
2. Characters of a new Species of Antilope (*Antilope Mhorr*).— Proc. Zool. Soc. Lond. I. p. 1.—Trans. Zool. Soc. 1835, p. 1.

3. On a Specimen of an Antilope, probably the young of *Antilope Cervicapr* , Pall.—Proc. Zool. Soc. Lond. I. p. 12; IV. p. 34.—J. A. S. B. III. p. 304.—Madras Journ. Lit. V. p. 406.

4. Characters of two new species of the genus *Mus*, Linn., collected by Colonel *Sykes* in Dukhun.—Proc. Comm. Zool. Soc. Lond. II. p. 121.

5. Characters of a new genus of *Lemuridæ*, presented by Mr. *Telfair*. —Proc. Comm. Zool. Soc. Lond. II. p. 20.

6. On a specimen of the *Manis Temminckii*, Smuts, from South Africa.—Proc. Zool. Soc. Lond. II. p. 81.—*Wiegm.* Arch. 1835, II. p. 332.

7. *Paradoxurus Grayi*.—Proc. Zool. Soc. III. p. 118.—*Wiegm.* Arch. 1836, II. p. 282.

8. Characters of two new species of Hedgehog (*Erinaceus*, Linn.) from the Himalaya Mountains.—Proc. Comm. Zool. Soc. Lond. II. p. 123.

9. Characters of a new species of Hedgehog (*Erinaceus*, Linn.) from South Africa.—Proc. Comm. Zool. Soc. Lond. II. p. 193.

10. Characters of a new species of Spider-Monkey (*Ateles*, Geoffr.). —Proc. Comm. Zool. Soc. Lond. I. p. 38.

11. Characters of two species of Mammalia (one constituting a new genus) from Sierra Leone (*Perodicticus* et *Aulacodus*).—Proc. Comm. Zool. Soc. Lond. I. p. 109.—Isis, 1834, VIII. p. 838.

12. On the *Mus barbarus*, Linn.—Zool. J. 1829, IV. p. 472, fig.—Isis, 1831, XII. p. 1366.—*Féruss.* Bull. 1831, XXIV. p. 75.

13. On Mammalia from Algoa Bay.—Proc. Comm. Zool. Soc. II. p. 122.

14. Characters of a new genus of Rodent Mammalia from Chili (*Octodon*), presented by Mr. *Cuming*.—Proc. Comm. Zool. Soc. Lond. II. p. 46.—Trans. Zool. Soc. II. p. 75, fig.—Isis, 1835, IV. pp. 365, 374.

15. Observations on a species of *Paradoxurus*, probably *P. prehensilis*, Gray.—Proc. Zool. Soc. II. p. 33.—*Wiegm.* Arch. 1835, II. p. 343.

16. On the *Chinchillidæ*, a family of herbivorous Rodentia, and on a new genus referable to it.—Tr. Zool. Soc. Lond. I. p. 35, fig. —Proc. Zool. Soc. I. p. 5 —Ann. Sc. n. 1834, I. p. 375.—Isis, 1835, IV. p. 530.

17. Some account of *Macropus Parryi*.—Proc. Zool. Soc. 1834, p. 152.—Trans. Zool. Soc. I. p. 295, fig.—*Wiegm.* Arch. 1835, II. p. 335.

18. *Pteropus Whitii*.—Trans. Zool. Soc. II. 1.—*Wiegm.* Arch. 1837, IV. p. 182.

19. On *Macropus penicillatus*, Gray.—Proc. Zool. Soc. III. p. 1.

20. On the *Vultur auricularis*, Daud.—Proc. Comm. Zool. Soc. Lond. I. p. 66.

21. Characters of a new species of *Polyborus*? (since ascertained to be the young of *Vultur Angolensis*, Penn.).—Proc. Comm. Zool. Soc. Lond. I. p. 13.

22. On the History and Synonymy of the *Cereopsis Novæ Hollandiæ*, Lath.—Proc. Comm. Zool. Soc. Lond. I. p. 26.

23. Characters of new genera and species of Fishes from the Atlantic Coast of Northern Africa, presented by Capt. *Belcher*.—Proc. Comm. Zool. Soc. Lond. I. p. 146.—Isis, 1835, IV. p. 354.

24. Characters of several new species of Fishes from Ceylon, presented by Dr. *Sibbald*.—Proc. Comm. Zool. Soc. Lond. II. p. 182. —Isis, 1835, V. p. 449.

25. Characters of new species of Fishes, collected by Lieut. *Allen* in Western Africa.—Proc. Zool. Soc. Lond. II. p. 45.—*Wiegm.* Arch. 1835, II. p. 261.

26. Characters of some new species of Fishes collected by Mr. *Cuming*.—Proc. Comm. Zool. Soc. Lond. II. p. 4.

27. Observations on a collection of Fishes from the Mauritius, presented by Mr. *Telfair*, with Characters of new genera and species. —Proc. Comm. Zool. Soc. Lond. I. pp. 59, 61, 126, 165; II. p. 184. —Proc. Zool. Soc. I. p. 32; III. p. 206.

28. New Fish from Trebizond, collected by *Keith Abbott*.—Proc. Zool. Soc. III. p. 91.—*Wiegm.* Arch. 1836, II. p. 238.

29. Observations on the Fishes contained in the Collection of the Zoological Society.—Zool. Journ. III. p. 371; IV. p. 31.—*Féruss.* Bull. 1829, XVI. p. 131; 1828, XVII. p. 439.—Isis, 1830, XI. p. 1170.

30. Observations on a collection of Fishes formed during the voyage of H.M.S. *Chanticleer*, with characters of two new species.— Proc. Comm. Zool. Soc. Lond. I. p. 112.

31. Characters of a new species of *Pterois* (*Pt. Russelii*).—Proc. Comm. Zool. Soc. Lond. I. p. 128.

32. Characters of a new species of *Aphrophora* (*A. Goudoti*) from Madagascar.—Proc. Zool. Soc. Lond. I. p. 12.

33. Description of an hitherto unpublished species of *Buccinum*, recently discovered at Cork.—Zool. Journ. I. p. 398.—*Féruss.* Bull. 1826, VII. p. 259.

34. Notice on a particular Property of a species of *Echinus*.—Tr. Linn. Soc. Lond. XV. p. 74.—Isis, 1828, X. p. 1064.

35. General Observations on the Anatomy of the Thorax in Insects,

and on its functions during Flight.—Zool. J. 1825, I. p. 391.—
Isis, 1830, IV. p. 422.—*Féruss.* Bull. 1826, VIII. p. 292.

36. On the Lachrymal Sinus of the Antelopes.—Proc. Zool. Soc. IV.
p. 41.—Lond. Edinb. Phil. Mag.— *Wiegm.* Arch. 1837, V. p. 181,
Not. 1, p. 52.

37. On several Animals recently added to the Society's Menagerie.
—Proc. Zool. Soc. Lond. I. p. 118 ; II. pp. 41, 110.

38. Characters of a new species of Otter (*Lutra,* Erxl.), and a new
species of Mouse (*Mus,* L.), collected in Chili by Mr. *Cuming.*—
Proc. Comm. Zool. Soc. Lond. II. p. 1.

39. On a new species of Kangaroo (*Macropus major*).—Proc. Zool.
Soc. Lond. II. p. 151.—Ann. Sc. n. 1835, III. p. 379.

40. Notice of a new genus of Viverridous Mammalia from Madagascar, constituting a new form among the Viverridous Carnivora.
—Proc. Zool. Soc. Lond. I. p. 46; IV. p. 13.—Tr. Zool. Soc.
I. 2, p. 137, fig.—Ann. Sc. n. III. 1835, p. 50.— *Wiegm.* Arch.
1835, II. p. 343.

41. Characters of a new species of Monkey (the Malbrouck of
Buffon), hitherto confounded with the *Simia Faunus,* Auct.—
Proc. Zool. Soc. Lond. I. p. 109.

42. Description of a young Nylghau (*Antilope picta,* Pall).—Proc.
Comm. Zool. Soc. Lond. I. p. 37.

43. Characters of a new species of Cat (*Felis viverrinus*) from the
continent of India, presented by *J. M. Heath,* Esq.—Proc. Zool.
Soc. Lond. I. p. 68.—J. A. S. B. III. p. 306.

44. Characters of a new species of Deer (*Cervus humilis*).—Proc.
Comm. Zool. Soc. Lond. I. p. 27.

45. Characters of two new species of Monkeys in the Society's Collection (*Semnopithecus Nestor, Cercopithecus Pogonias*).—Proc.
Zool. Soc. Lond. I. p. 67.

46. Characters of a new species of Lemur (*Lemur rufifrons*).—Proc.
Zool. Soc. Lond. I. p. 106.

47. The Tower Menagerie. 8°. London.

48. The Gardens and Menagerie of the Zoological Society. 2 vols.
8°. Lond.--Ann. Sc. nat. XXII.

49. Evidences in proof of certain statements in the " Gardens and
Menagerie of the Zoological Society."—Mag. Nat. Hist. ser. 1,
IV. p. 199.

50. Observations on the Hyacinthine Maccaw.—Mag. Nat. Hist.
ser. 1, IV. p. 211.

51. On the Habits of a Mexican Bee, with a description of the Insect and its Hive.—Beechey's Voy. to the Pacific, p. 613.

52. Description of a new species of *Julis.*—Zool. Journ. III. 1828, p. 577.

53. Additional Remarks on the genus *Lagotis,* with some account of a second species (*L. pallipes*) referable to it.—Trans. Zool. Soc. I. 1835, p. 331, cum tab. 1.—Proc. Zool. Soc. III. p. 67.

54. On the genus *Cryptoprocta.*—Proc. Zool. Soc. II. p. 13.

55. On Mammalia from Travancore.—Proc. Zool. Soc. III. p. 66.

56. On Mammalia from Trebizond.—Proc. Zool. Soc. III. p. 89.

57. Characters of a species of *Acanthurus.*—Proc. Zool. Soc. III. p. 119.

58. On a new species of Crocodile.—Proc. Zool. Soc. III. p. 128.— *Wiegm.* Arch. 1836, II. p. 256.

59. On a remarkable Pteropine Bat from the Gambia.—Proc. Zool. Soc. III. p. 149.

60. On a new *Ctenomys* and other Rodents from the Straits of Magellan.—Proc. Zool. Soc. III. p. 189.

BENNETT (F. D.).

1. On the Light emitted by a species of *Pyrosoma.*—Proc. Zool. Soc. Lond. I. p. 79 ; V. p. 51.

2. On the Larynx of the Albatros (*Diomedea exulans,* Linn.).— Proc. Zool. Soc. Lond. I. p. 78.

3. On the Anatomy of the Spermaceti Whale.—Proc. Zool. Soc. IV. p. 127 ; V. p. 39.— *Wiegm.* Arch. 1837, V. p. 192.

4. Physalia.—*Fror.* Notiz. XLII. p. 183.— *Wiegm.* Arch. 1835, I. p. 31.

5. Narrative of a Whaling Voyage round the World. 2 vols. 8°. London, 1840.

BENNETT (G.).

1. Wanderings in New South Wales, Batavia, Pedir Coast, Singapore and China, during the years 1832–1834. 2 vols. 8°.— *Wiegm.* Arch. 1835, II. p. 256.

2. On the Nasal Gland of the wandering Albatros (*Diomedea exulans,* Linn.).—Proc. Zool. Soc. Lond. II. p. 151.—*Müll.* Arch. 1836, III. IV. p. LXIII.

3. On the Phosphorescence of the Ocean.—Proc. Zool. Soc. V. p. 1.

4. Ueber den Dingo (wilden Hund Neuhollands).—Wand. I. p. 232. —*Fror.* Notiz. XLII. p. 168.— *Wiegm.* Arch. 1835, II. p. 341.

5. Ueber einen in Gefangenschaft gehaltenen Wombat (*Phascolomys*).—Wand. I. p. 330.— *Wiegm.* Arch. 1835, II. p. 335.

6. *Trachyglossus (Echidna)*.—Wand. I. p. 299.— *Wiegm.* Arch. 1835, II. p. 334.

7. Einige Notizen über einen zahmen Orang-Utan (Wand. I. p. 366), und über einen Ungka-Affen (ibid. II. p. 343).—*Fror.* Notiz. XLIII. No. 12 u. 13.— *Wiegm.* Arch. 1835, II. p. 347.

8. Ueber das Känguruh (*Halmaturus giganteus*).—Wand. I. p. 283. — *Wiegm.* Arch. 1835, II. p. 335.

9. Lebensweise von *Dacelo gigantea.*—Wand. I. p. 122.—*Fror.* Notiz. XLII.— *Wiegm.* Arch. 1835, II. p. 304.

10. On the Habits of a species of Horned Pheasant (*Tragop. Temminckii*, Gray).—Proc. Zool. Soc. Lond. II. p. 33.—Wand. II. p. 60.— *Wiegm.* Arch. II. p. 256.

11. Benehmen des Paradies-Vogels, den er in der Gefangenschaft sah.—Wand. II. p. 37.—*Fror.* Notiz. XLIII. 8.— *Wiegm.* Arch. 1835, II. p. 307.

12. *Pelecanus Onocrotalus*, L.—Pr. Zool. Soc. Lond. II. p. 49; 1834, p. 19.— *Wiegm.* Arch. 1835, II. p. 316.

13. On the Habits of the King Penguin (*Aptenodytes patagonica*, Gmel.).—Proc. Zool. Soc. Lond. II. p. 34.—Instit. 81.—*Fror.* Notiz. XLI. p. 248.— *Wiegm.* Arch. 1835, II. p. 317.

14. Observations on *Glaucus hexapterygius.*—Proc. Zool. Soc. IV. p. 113.— *Wiegm.* Arch. 1837, II. p. 269.

15. Notes on the Natural History and Habits of the *Ornithorhynchus paradoxus.*—Trans. Zool. Soc. vol. 1, 1835, p. 229.—Proc. Zool. Soc. I. p. 82 ; II. p. 141.— *Wiegm.* Arch. 1835, II. p. 333.—Isis, 1835, XII. p. 1056.

16. On the *Simia syndactyla*, or Ungka Ape of Sumatra.—Mag. Nat. Hist. ser. 1, V. p. 131.

17. Fregattvögel (*Tachypetes aquila*).—Wand. II. p. 254.— *Wiegm.* Arch. 1835, II. p. 315.

18. On *Physalia pelagica.*—Proc. Zool. Soc. V. p. 43.

BENNETT (J. A.).

1. Liste des Vers qui existent dans les Pays-Bas.—Natuurk. Verh. Holl. M. d. Wet. XV. 2, 1826.—*Féruss.* Bull. 1827, XI. p. 128.

BENNETT (J. A.) et OLIVIER.

1. Traité sur les Insectes des Pays-Bas.—Natuurk. Verhand. Harlem. 1825, XVI.—*Eis.* Insect. p. 159.

BENNETT (J. W.).

1. A Selection of the most remarkable and interesting of the Fishes

found on the Coasts of Ceylon.—Lond. 1828–1830, 4º.—Lond. Lit. Gaz. 1830, p. 223.—*Féruss.* Bull. 1829, XVI. p. 131 ; 1830, XXI. p. 325.

BENNETT (T. H.).

1. On the parasitic Vegetable Structures found growing in Living Animals.—Tr. Roy. Soc. Edinb. XV.—Ann. and Mag. Nat. Hist. XI. p. 126.

BENNINGSEN-FÖRDER (Rud. v.).

1. Erläuterungen zur geognostischen Karte der Umgegend von Berlin. Berl. 1843, 4º.
2. Geognostische Beobachtungen im Luxemburgischen. 8º.
3. Das Zahlengesetz in den Gestein-Formationen, in Bezug auf Vertheilung von Thälern, Quellen, fliessenden u. stehenden Ge- wässern, Erhöhungen u. Ortschaften, vornehmlich in N. Frank- reich, etc. Berl. 1843, 4º.

BENNOUX (M. T.).

1. Ueber die fossilen Menschenknochen von Darfourt.—Le Temps, 1830, 10 Juill.—*L.* u. *Br.* Jahrb. III. p. 350.

BENOISET (S.).

1. Discours véritable d'une Fontaine, ornée de merveilleuse propri- étés et vertus, trouvée près de Dié, en Dauphiné. Dié, 1610, 4º. —*Böhm.* Bibl. V. p. 180.

BENOISTON DE CHATEAUNEUF ( ).

1. Notice sur l'intensité de la fécondité en Europe au commence- ment du dix-neuvième siècle.—Ann. Sc. nat. 1826, IX. p. 431.

BENOIT (L.).

1. Ornitologia Siciliana, osia Catalogo ragionato degli Uccelli che si trovano in Sicilia.—Messina, 1840, 8º.

BENROTH (Otto).

1. Vom Ermslebischen Steine, der im Halberstädtischen gefunden wird. 1625, 4º.—*Böhm.* Bibl. IV. 1, p. 224.

BENSELER (G. E.).

1. Geschichte Freibergs und seines Bergbaues. 1. Lief. Freib. 1843, 8º.

BENSELLAER (J. van.).

1. On the Skeleton of a Mastodon from the Delaware and Hudson Canal.—*Sill.* Amer. J. XIV. p. 31.—*Fisch.* Mamm. p. 405.

BENSON (Ch.).

1. Die charakteristischen Parallelköpfe des *J. B. della Porta,* worin die Aehnlichkeit des Menschen mit gewissen Thieren dargestellt wird. 8°. Leipzig, 1812.

BENSON (W. H.).

1. On the genus *Scaphula.*—Proc. Zool. Soc. 1834, p. 91.— *Wiegm.* Arch. 1836, II. p. 210.

2. Note on the Importation of a living *Cerithium Telescopium,* Brug.—Proc. Zool. Soc. Lond. II. p. 91.

3. Conchological Notices, chiefly relating to the Land and Freshwater Shells of the Gangetic Provinces of Hindostan.—Proc. Zool. Soc. II. p. 89.—Zool. Journ. V. p. 458.—J. A. S. B. I. p. 76.

4. On a new genus of Land Snails, allied to *Cyclostoma* of Lamarck. —J. A. S. B. I. p. 11.

5. On *Oxygyrus,* a new genus of Pelagian Shells, allied to *Atlanta* of Lesueur.—J. A. S. B. IV. p. 173.

6. On two new species of *Carinaria,* discovered in the Indian Ocean. —J. A. S. B. IV. p. 215.

7. On Land and Freshwater Shells from the Gangetic Provinces of India.—Proc. Zool. Soc. 1834, p. 89.—J. A. S. B. IV. p. 528.

8. Corrected character of the genus *Cuvieria* of Rang.—J. A. S. B. IV. p. 698.

9. Descriptive Catalogue of Terrestrial and Fluviatile Testacea, chiefly from the North-East Frontier of Bengal.—J. A. S. B. V. p. 350.

10. Descriptive Catalogue of a Collection of Land and Freshwater Shells, chiefly contained in the Museum of the Asiatic Society.-- J. A. S. B. V. p. 741.

11. Description of the Shell and Animal of *Nematura,* a new genus of Mollusca inhabiting situations subject to alternations of fresh and brackish water.—J. A. S. B. V. p. 781.

12. On *Balantium,* a genus of the Pteropodous Mollusca; with the characters of a new species inhabiting the Southern Indian Ocean. —J. A. S. B. VI. p. 150.

13. On the Genera *Oxygyrus* and *Bellerophon.*—J. A. S. B. VI. p. 316.

14. On the affinities of *Galathea* of Lamarck (*Potamophila* of Sowerby), a genus of Fluviatile Testacea.—J. A. S. B. VII. p. 420.

BENVENISTI (M.).

1. Saggio di Notomia fisiologica e patologica delle Vene.—Bull. Sc. med. di Bol. (ser. 3ᵉ) I. p. 126.
2. Gangliorum Anatomia. Patav. 1840.

BENVENUTI (Gius.).

1. Instituzioni di Mineralogia con la maggior chiarezza disposte. Parma, 1790, 8°.

BENZ (von).

1. Verzeichniss der im Königreich Würtemberg gefundenen Schalthiere.—Corresp. Bl. Würt. Landw. Ver. XVII.

BENZA (P. M.).

1. On the Geology of the country between Madras and the Neilgherry Hills.—Madras Journ. Lit. IV. p. 1.—Ind. Rev. I. pp. 256, 305; II. p. 1.
2. On the Geology of the Neelgherry and Koondah Mountains.—Madras Journ. Lit. IV. p. 241.—Ind. Rev. I. p. 621.
3. Geological Notes of a Journey through the Northern Circars.—Madras Journ. Lit. V. p. 43.—Ind. Rev. II. pp. 227, 301, 377.
4. Geological Sketch of the Neilgherries.—J. A. S. B. IV. p. 413.

BENZEL (Laur.).

1. Diss. de Re Metallicâ Sueco-Gothorum. Ups. 1703, 8°. fig.—*Böhm.* Bibl. IV. 1, p. 151.

BENZENBERG (F.).

1. Mémoire sur les Sources chaudes d'Aix-la-Chapelle.—Jahrb. f. Min. 1831.—Bull. Soc. Géol. Fr. I. p. 170.
2. Die Sternschnuppen sind Steine. Bonn, 1834, 8°. fig.

BENZON (Laur. Jac.).

1. De Pecudibus Jacobi artificiosis particula physica. Hafn. 1763, 4°.—*Böhm.* Bibl. II. 1, p. 142.

Bérard (A.).

1. Texture et Développement des Poumons. Paris, 1836, 8º.— *Val.* Repert. II. p. 23.

Béraud de l'Oratoire (J. J.).

1. Mémoire sur la description d'une Machine propre à pêcher le Corail, etc.—Journ. de Phys. XVII.—*Lamx.* Pol. corall. p. 521.

Beraud ( ).

1. Mémoire sur l'Education des Abeilles. 1787, 12º. fig.

Beraz ( ).

1. Lehrbuch der Anatomie des Menschen mit physiologischen Zusätzen. Landsh. 1839, 8º.

Berch (A.).

1. Dissertatio : Jamtelands Djur-fänge. Resp. *Æ. Nordhlom.* Upsal, 1749, 8º.—Tr. Linn. Soc. Lond. V. p. 278.

Berch (Chr.).

1. Afhandling i Swenska Bergs-lagfarenheten om Forfattningar, etc. Ups. 1778, 4º.—*Böhm.* Bibl. IV. 1, p. 153.

Berchelmann (D.).

1. Fragmente zur Arzney- und Naturkunde ; *vide sup.* Pars I. p. 19.

Berchem (B. van).

1. On the Wild Goat of the Alps.—Mém. de la Soc. des Sciences Phys. de Lausanne, II.—Phil. Mag. XII. p. 153.

Berchem (Van) et Struve (H.).

1. Principes de Minéralogie, ou Exposition succincte des Caractères extérieurs des Fossiles, etc. Paris, 1794, 8º.

Berck (F. H. van).

1. Verhandeling ten bewijze, dat de olifants-ofsnuittorretjes de bedervers zijn der vruchtboomen, benevens middel daartegen. 8º. Haarlem, 1807.—Bibl. Ent. I. p. 25.

Berdin ( ).

1. Notice sur des Os fossiles de grands Mammifères trouvés à la

Croix-Rousse près de Lyon, en Avril 1824.—Arch. hist. et stat. du Rhone, déc. 1824, p. 97 ; 1825, pp. 206, 291, 386, 426, 443 ; 1827, pp. 157, 337.—*Féruss.* Bull. 1827, III. p. 125 ; X. p. 389.

BERDOT (D. C. E.).

1. Observ. de Lumbricis e Cubito erumpentibus.—Act. Helvet. VII. p. 177.

BERELIO (G.).

1. Dissertatio de Insectis.   Upsal, 1675, 8°.—Bibl. Ent. I. p. 21.

BERENDT (C.).

1. De Atmosphærâ Nervorum sensitivâ Comment. 4°. Danz. 1816. —Isis, 1817, I. p. 116.

BERENDT (G. C.).

1. Die im Bernstein befindlichen organischen Reste der Vorwelt gesammelt.   fol. Berlin, 1845.

2. Die Insecten in Bernstein, ein Beitrag zur Thiergeschichte der Vorwelt. 1tes H. Danzig, 1830, 4°.—Isis, 1831, XI. p. 1231 ; XII. p. 1368.—*Féruss.* Bull. 1831, XXIV. p. 107 (*Les Insectes qui se trouvent dans l'Ambre*).

3. Mémoire pour servir à l'Hist. des Blattes antédiluviennes, trad. de l'All. par *Heller.*—Ann. Soc. Ent. Fr. V. p. 539, fig.—*Erichs.* in *Wiegm.* Arch. 1837, VI. p. 333. (*Ueber in Bernstein einge-schlossene Blatten.*)

BÉRENGER (Jac.).

1. Commentaria cum amplissimis additionibus super Anatomiâ *Mundini.* Bol. 1521–22, 4°.—Lond. 1664, 12°. (Angl.)—Biogr. Un. IV. p. 237.

2. Isagogæ breves in Anatomiam corporis humani, cum aliquot fig. anatom.   Bol. 1522–25, 4°.—Venice, 1523–25, 4°.—Cologne, 1529, 8°.—Strasb. 1530, 8°.—Biogr. Un. IV. p. 237.

BERG (E. von).

1. Die Biologie der Zwiebelgewächse.   Neustrelitz, 8°.—*Val.* Rep. III. p. 12.

BERGA (A. J. Th.).

1. Der Naturforscher, o. Unterhalt aus dem Thier- Pflanzen- u. Mineralreich. 2 vols. 8°. Berlin, 1818.

BERGE (F₁).

1. Käferbuch, Allgemeine und Specielle Naturgeschichte der Käfer, mit vorzüglicher Rücksicht auf die europäischen Gattungen, etc. Stuttg. 1844, 8°.

2. Die Fortpflanzung europaischer u. aussereurop. Vögel. Stuttg. 1840–43, 16°.

3. Die Vertebraten Würtembergs zusammengestellt. 8°. Stuttg. 1840.

4. Schmetterlingsbuch, oder allgemeine und besondere Naturgeschichte der Schmetterlinge, etc. Stuttg. 1842, 8°. fig.

BERGEN (Ch. A. de).

1. Anatomes experimentalis, pars 1ª et 2ª.   Francf. 1755, 1758, 8°. —Biogr. Un. IV. p. 248.

2. Pentas observationum anatomico-physiologicarum.   1743, 4°.— Biogr. Un. IV. p. 248.

3. Programma de Nervis quibusdam Cranii ad novem paria hactenùs non relatis.   Francf. 1738.—Biogr. Un. IV. p. 248.

4. De Nervo intercostali.   1731.—Biogr. Un. IV. p. 248.

5. Programma de Piâ Matre.   Nürnb. 1736, 4°.—Biogr. Un. IV. p. 248.

6. De Membranâ cellulosâ.   1732.—Biogr. Un. IV. p. 248.

7. Icon nova Ventriculorum Cerebri.   Francf. 1734.—Biogr. Un. IV. p. 248.

8. Methodus Cranii Ossa dissuendi, et Machinæ hunc in finem constructæ per figuras ligno incisas delineatio. 1741, 4°.—Biogr. Un. IV. p. 248.

9. Elementa Physiologiæ juxta selectiora experimenta.   Geneva, 1749, 8°.—Biogr. Un. IV. p. 248.

10. Observationes de Ranarum Anatome.—Comm. litter. Norimb. 1738, p. 131.—Dict. Sc. n. XV. p. 258.

11. Progr. de Alchimillâ, gramineo et baccis quæ circa radices ejus reperiuntur.   Francf. 1739, 4°.—Bibl. Ent. I. p. 22.

12. Epistola de Alchimillâ pupinâ ejusque Coccis.   Francf. 1748, 4°.—Bibl. Ent. I. p. 28.

13. Classes Conchyliorum.   Nurimb. 1760, 4°.—N. Act. Nat. Cur. II. App. p. 1.—Tr. Linn. Soc. Lond. VII. p. 241.—*Böhm.* Bibl. II. 2, p. 447.

14. Diss. de Animalibus hieme sopitis.   Francof. 1752, 4°.—*Böhm.* Bibl. II. 1, p. 145.

15. Diss. de Dentibus Hippopotami. Francof. O. 1747, 4°.—*Böhm.* Bibl. II. 1, p. 299.

16. Oratio de Rhinocerote. Francof. O. 1746, 4°.—*Böhm.* Bibl. II. 1, p. 479.

17. De Microcosmo, belluâ marinâ omnium vastissimâ.—Act. Nat. Cur. II. p. 143.—*Mod.* Bibl. Helm.

BERGER (Chr. Ph.).

1. Versuch einer gründlichen Erläuterung merkwürdiger Begebenheiten in der Natur. Lemgo, 1737–39, 8°. fig.—*Böhm.* Bibl. I. 1, p. 67.

BERGER (D. M.).

1. Faits relatifs à la construction d'une Echelle de degrés de la Chaleur animale.—Mém. Soc. Phys. et Hist. Nat. VI. p. 257.

BERGER (Ferd.).

1. Handbuch zum Gebrauch für das anatomische Studium des menschlichen Körpers, besonders für bildende Künstler, etc. Berl. 1842, fig.

BERGER (H. A. C.).

1. Die Versteinerungen der Fische u. Pflanzen im Sandstein der Coburg-Gegend. Cob. 1832, 4°. fig.—Isis, 1832, X. p. 1114.—
Leonh. u. *Br.* Jahrb. 1833, p. 225.—Bull. Soc. Géol. Fr. VI. p. 62.

BERGER (J. Eric).

1. Allgemeine Grundsätze der Wissenschaft der Natur u. des Menschen. Altona, 1817–27.—Biogr. Un. Suppl. LVIII. p. 31.

BERGER (J. F.).

1. On the Physical Structure of Devonshire and Cornwall.—Trans. Geol. Soc. ser. 1, I. p. 93.

2. On the Geology of some parts of Hampshire and Dorsetshire.—
Trans. Geol. Soc. ser. 1, I. p. 249.—Schr. d. Gesellsch. f. Min. zu Dresd. II. p. 1.—*Leonh.* Tasch. 1822, III. p. 756.

3. Mineralogical Account of the Isle of Man.—Trans. Geol. Soc. ser. 1, II. p. 29.—*Leonh.* Tasch. 1820, I. p. 65.

4. On the Geological Features of the North-Eastern Counties of

Ireland, with an Introduction and Remarks by the *Rev. W. Conybeare.*—Trans. Geol. Soc. ser. 1, III. p. 121.

5. On the Dykes of the North of Ireland.—Trans. Geol. Soc. ser. 1, III. p. 233.

6. Expériences et remarques sur quelques animaux qui s'engourdissent pendant la saison froide.—Mém. Mus. XVI. p. 201.—*Féruss.* Bull. XXV. p. 110.

7. Theoretische Erklärung der Krümmung der Kalksteinlager, welch das Jura-Gebirge bilden.—Geol. Soc.—Isis, 1818, IV. p. 594.

BERGER (J. G.).

1. De arteriæ Aortæ divisione in ramos carotides et subclavios.—Act. Erudit. XVII. p. 295, fig.

2. Physiologia medica, sive de Naturâ humanâ Liber bipartitus. Vitemb. 1701. 4°. fig.—Act. Erudit. XX. p. 492.

3. Prodromus Commentationis de Carolinis Bohemiæ Fontibus. Witteb. 1708, 4°.—*Böhm.* Bibl. V. p. 271.

BERGER (P.).

1. Mém. sur les Egagropiles des Bêtes bovines et ovines.   8°. Versailles, 1836.

BERGERON (   ).

1. Voyages faits en Asie dans les XII–XV siècles par plusieurs célèbres Voyageurs.   Leyde, 1729, 4°. 2 vols.—La Haye, 1735, 4°. 2 vols.—*Bohm.* Bibl. I. 1, p. 669.

BERGHAUS (H.).

1. Beobachtungen über die Ostsee-Küste von der Weichsel-Mündung bis zur Grenze von Pommern.—Ann. d. Erdkunde, 1838, XVIII. p. 48.—*L. u. Br.* N. Jahrb. 1839, p. 108.

2. Erdbeben vom 23 Jan. 1838 in Ost-Europa.—Ann. d. Erdk. 1838, XVIII. p. 56.—*L. u. Br.* N. Jahrb. 1839, p. 473.

3. Mesures barométriques faites pendant un voyage de Dresde à Tœplitz, Karlsbad et Franzensbad, avec une carte du Défilé de Nollendorf.—Hertha, IX. 6, p. 475.—*Féruss.* Bull. 1828, XIV. p. 190.

4. Nivellement barométrique du Fichtelgebirge, d'Eger à Bayreuth. —Hertha, VIII. 3, p. 123.—*Féruss.* Bull. 1829, XVII. p. 28.

5. Résumé des Mesures de Hauteur faites en Espagne.—Hertha, XII. 3, p. 418.—*Féruss.* Bull. 1829, XVII. p. 176.

6. Deutschlands Höhen, 2te Aufl. Berl. 1833, 8°.

BERGHOLZ (S. Th.).

1. De Monstro duplici per Implantationem ac de Duplicitate. Berol. 1840, 8°.

BERGIUS (P. J.).

1. Sur la Couleur et le changement de couleur des Animaux. Hendl. 1761.—Biogr. Un. IV. p. 257.

2. Lettre sur l'histoire naturelle et la translation des Poissons.— Berl. Beschäft. II.—Biogr. Un. IV. p. 257.

3. Anmärkningar öfver herbarier, etc. (*Remarques sur les Herbiers et les Collect. d'Insectes*).—Vetensk. Ac. Handl. 1786, p. 302.— Bibl. Ent. I. p. 22.

4. Tal om kalla Bad i gemen, och Loka Badningar i synnerhet. Stockh. 1763, 8°.—*Böhm.* Bibl. V. p. 387.

5. Beskrifning på et Sjökräk, som är et slags *Teredo*, etc.—Act. Holm. XXVI. p. 225.—*Mod.* Bibl. Helm.

BERGLUND (E. H.).

1. Anthracides Sueciæ. 4°. Lund. 1814.

BERGMANN (C. G. H. B.).

1. Dissert. inaug. anatomica et physiologica de Glandulis suprarenalibus. Gœtt. 1839, 8°. fig.

BERGMANN (C.).

1. De Placentæ fatalis resorptione. Gœtt. 1838, 8°.

2. Zur Vergleichung des Unterschenkels mit dem Vorderarm.— *Müll.* Arch. 1841, p. 201.

3. Zur Verständigung über die Dotterzellenbildung.—*Müll.* Arch. 1842, p. 92.

4. Ueber die Bewegungen von Radius u. Ulna am Vogelflügel.— *Müll.* Arch. 1839, p. 296.

5. Die Zerklüftung und Zellenbildung im Froschdotter.—*Müll.* Arch. 1841, p. 89.

6. Untersuchungen über die Struktur der Mark- u. Rindensubstanz des grossen u. kleinen Gehirns.—*Müll.* Arch. 1841, p. 126, fig.

7. Beobachtungen über die Skelettsystem der Wirbelthiere. 8°
Göttingen, 1846.

8. Neue Untersuchungen über die innere Organisation des Gehirns,
etc. Hannov. 1831, 8°. fig.—Isis, 1832, IV. p. 443.

BERGMANN et NOGGERATH.

1. Sur les Dents de Mammouth de Liedberg.—Jahrb. f. Phys. u.
Chem. 1828, 2, pp. 145, 157.—*Féruss.* Bull. 1829, XVIII. p. 336.

BERGMANN (Gust.).

1. Geschichte von Liefland, nach Bossuetischer Art entworfen.
Leipz. 1776, 8°.—*Böhm.* Bibl. I. 1, p. 639.

BERGMANN (J.).

1. Was die Thiere gewiss nicht, u. was sie am wahrscheinlichsten
seyen. 8°. Mainz, 1784.

2. Anfangsgründe der Naturgeschichte. Mainz, 1774–78; 1782–
83, 3 vols. 8°.—*Böhm.* Bibl. I. 1, p. 323.—Cat. Bibl. Turic.

3. Tabellarischer Entwurf der Naturgeschichte. Mainz, 1778, fol.
—*Böhm.* Bibl. I. 1, p. 323.

4. Conspectus brevis universæ Physicæ generalis et particularis.
Mogunt. 8°.—*Böhm.* Bibl. I. 1, p. 323.

BERGMANN (Torb.).

1. Classes Larvarum Insectorum.—N. Act So R. Upsal. I. p. 58.
—Opusc. V. p. 131.

2. Beschreibung einiger neuen Insecten.—*Jacqu.* Misc. Austr. II.
—Bibl. Ent. I. p. 22.

3. Anmärkningar, etc. (*Observations sur les fausses Chenilles et
Mouches à Scie*).—Vet. Acad. 1763, p. 154.—(Allem.) p. 165.—
Fuessl. N. Mag. d. Ent. III. p. 53.—Bibl. Ent. I. p. 22.

4. Bref engäende Anmärkn. som utkommit öfver det swar på frägan
om skadaliga, etc. (*Lettre contenant quelques remarques sur la
réponse à faire aux questions proposées par l'Acad. Roy.*: *Quel
dommage les Chenilles peuvent faire aux arbres*, etc.). Stockh.
1764, 8°.—Swar på, etc. (*Réponse sur le même sujet*). Stockh.
1769, 8°. (avec celles de *Lund* et *Modeer*).—Swar på, etc. (*Rép.
sur le même sujet*). Stockh. 1783, 8°. (avec celles de *Leche, Schröder, Nelin, Linné* et *Cüdbeck*).—Bibl. Ent. I. p. 23.—*Böhm.* Bibl.
II. 2, p. 184.

5. Huru kunna Mascar, etc. (*Sur les dommages que les Chenilles
peuvent causer aux fruits*, etc.). Stockholm, 1763.—Bibl. Ent.
I. p. 22.—*Böhm.* Bibl. II. 2, p. 183.

6. Uplysning, etc. (*Eclaircissemens sur une Chenille nuisible du pin* : B. pythiocampa ?)—Vet. Acad. 1769, p. 272.—Opusc. V. p. 171. —*Fuessl.* N. Mag. d. Ent. III. p. 69.—Bibl. Ent. I. p. 23.

7. Anmärkningar, etc. (*Rem. sur les Abeilles, et de la manière d'évaluer par le poids les différences qui se trouvent dans leur quantité de miel.*)—Vet. Acad. 1779, p. 300.—(Allem.) p. 266.—Opusc. V. p. 176.—Bibl. Ent. I. p. 23.

8. Supplementum Historiæ Reaumur. Tenthredinum.—N. Act. Ups. III. 1767, p. 166, fig.—Opusc. V. p. 146.—Bibl. Ent. I. p. 23.

9. Ett sällsamt Galle-äpte (*sur une nouvelle Noix de Galle*, Cynips).—Vet. Acad. 1762, p. 139. (Allem.) p. 140.—(Lat.) Opusc. V. p. 141.—Bibl. Ent. I. p. 22.

10. De Tenthredinibus earumque larvis.—Schwed. Ak. Abh. 1763, p. 165.—*Eis.* Ins. p. 219.

11. Von den Insecten, welche die Fichten zerstören.—Schwed. Ak. Abh. XXXI. p. 270.—*Eis.* Ins. p. 214.

12. Outlines of Mineralogy, translated by *W. Withering.* 8°. Birmingh. 1783. —Cat. R. Soc. Lond.

13. Anmerkungen über die Westgothischen Berge.— Schwed. Ak. Abh. XXX. p. 329.—*Böhm.* Bibl. I. 1, p. 620.

14. Physisk Beskrifning öfver Jord Klotet på Cosmographiska Sällskapets Wagnar författed *J. D. B. Brandt.* Ups. 1766, 1774, 8°. fig. 2 vols.—(Dan.) Kop. 1771, 8°. fig.—(German.) *Physikalische Beschreibung der Erdkugel* ; *aus dem Schwed. v. L. H. Röhl.* Greifsw. 1769, 8°. fig. ; 1780, 8°. 2 vols.; 1792, 8°.— *Böhm.* Bibl. I. 1, p. 304.

15. Auszug aus seiner Physikalischen Erdbeschreibung, nebst einem kurzen Abrisse der Naturgeschichte. Leipz. 1781, 8°. fig.—*Böhm.* Bibl. I. 1, p. 304.

16. Sciagraphia Regni mineralis, secundum principia proxima digesti. Lips. 1782, 8°.—(Gall.) *Manuel du Minéralogiste*, etc. trad. et augm. par *Mongez.* Par. 1784, 8°.—*Böhm.* Bibl. IV. 1, p. 83.—(Germ. *X. C. Lippert*): *Grundriss des Mineralreichs, in einer Anordnung nach den nächsten Bestandtheiler der Körper.* Wien, 1787, 8°.—Propädeut. Miner. p. 240.

17. Rön om *Coccus aquaticus.*—Act. Holm. XVII.—*Mod.* B. Helm.

18. Afhandling om Iglar.—Act. Holm. XVIII.—*Mod.* B. Helm.

BERGNER (A.).

1. Ueber die Bildung der Oberfläche auf beiden Seiten des Firingebirgs in Thüringen.—*Ballenst.* Arch. IV. p. 77.

BERGSON (J.).

1. De Prosopodysmorphiâ sive novâ Atrophiæ facialis specie. Berol. 1837, 8º.—*Val.* Repert. III. p. 30.

BERGSTRÆSSER (H. W.).

1. Sphingum Europ. Larvæ, quotquot adhuc innotuerunt, ad Lin-
næanum, Fabricianum et Viennensium imprimis catalogos recen-
sitæ, etc. Han. 1782, 4º. fig.—Bibl. Ent. I. p. 24.—*Böhm.* Bibl.
II. 2, p. 250.

BERGSTRÆSSER (J. A. Ben.).

1. Entomologia Erxlebiana scholarum in usum concinnata.—Hanau,
1776, 8º.—Hannov. 1784.—Bibl. Ent. I. p. 24.—*Böhm.* Bibl. II.
2, p. 146.

2. Nomenclatur u. Beschreib. der Insekten in der Grafschaft Hanau-
Münzenberg, wie auch der Wetternau u. der angränzenden Nach-
barschaft. Hanau, 1778–1780, 4º. fig.—Bibl. Ent. I. p. 24.—
Böhm. Bibl. II. 2, p. 166.

3. Ergänzungen des *Rösel*'schen Insectenwerks. 1783?—Bibl. Ent.
I. p. 24.

4. Naturgeschichte der europäischen Tagefalter. 3 H. fig.—Bibl.
Ent. I. p. 24.

5. Etwas von der Naturgeschichte der *Phalæna fimbria,* Linn.—
Schr. Berl. Naturf. I. p. 297, fig.—Bibl. Ent. I. p. 23.

6. Ueber den Weissdornspanner, nach *Sepp.*—Besch. Berl. Naturf.
IV. p. 29.—Bibl. Ent. I. p. 24.

7. Ueber die Insecten mit Flügeldecken ; ausz. aus *Degeer.*—Han.
Mag. I. p. 105.—Bibl. Ent. I. p. 24.

8. Icones Papilionum diurnorum, quotquot adhuc in Europâ occur-
runt, descriptæ, etc. Han. 1779–81, 4º.—*Böhm.* Bibl. II. 2,
p. 239.—Bibl. Ent. I. p. 24.

BERINGER (J. B. Ad.).

1. Lithographiæ Wirceburgensis, ducentis Lapidum figuratorum à
potiori Insectiformium prodigiosis imaginibus exornatæ Specim.
I. Wirceb. 1726, fol. fig. Francof. et Lips. 1767, id.—*Böhm.*
Bibl. IV. 1, p. 121.—Gött. Gel. Anz. 1767, p. 654.—*Leonh.* etc.
Propæd. p. 277, No. 176.

BERINGER (J. L. Chr.).

1. Diss. de Lumbricis in Duplicaturâ Omenti repertis. Heidelb.
1744, 4º. fig.—*Böhm.* Bibl. II. 2, p. 389.

BERKELEY (M. J.).

1. On the Existence of a Second Membrane in the Asci of Fungi. —Mag. Zool. and Bot. 1838, II. p. 222, fig.

2. On *Hæmatococcus sanguineus*, Ag.—Ann. and Mag. N. Hist. XV. p. 372.

3. On some British Serpulæ.—Mag. Nat. Hist. ser. 1, VII. p. 420.

4. Description of the Animal of *Voluta denticulata*, Mont., and *Assiminia Grayana*, Leach.—Zool. J. V. p. 427.

5. On the Occurrence of *Dreissena polymorpha* in Northamptonshire.—Mag. Nat. Hist. ser. 1, IX. p. 572.

6. *Serpula tubularia, Mülleri, vermicularis, triquetra.*—Zool. J. Septemb. p. 420.—*Wiegm.* Arch. 1835, I. p. 343.

7. A Description of the Anatomical Structure of *Cyclostoma elegans.* —Zool. J. IV. p. 278.—Isis, 1830, XII. p. 1263.

8. On *Guilding's* description of *Ancylus.*—Zool. Journ. V. p. 268. —Isis, 1832, VI. p. 668.—*Féruss.* Bull. XXV. p. 360.

9. A Short Account of a new species of *Modiola*, and of the animal inhabitants of two British *Serpulæ.*—Zool. J. 1828, III. p. 229. —*Féruss.* Bull. 1828, XV. p. 316.—Isis, 1830, XI. p. 1163.

10. On the Internal Structure of *Helicolimax* (*Vitrina*) *Lamarckii.* —Zool. J. V. p. 305.

11. Description of the Anatomical Structure of *Cerithium telescopium*, Brug.—Zool. J. V. p. 431.

12. On *Gloionema paradoxum.*—Ann. and Mag. N. H. VII. p. 449, fig.

13. On the Preservation of Objects of Natural History for the Microscope.—Ann. and Mag. N. Hist. XV. p. 104.

14. On the *Dentalium subulatum* of *Deshayes.*—Zool. J. V. p. 424.

BERKENHOUT (J.).

1. Outlines of the Natural History of Great Britain and Ireland. Lond. 1769–72, 8°. 3 vols.—Bibl. Ent. I. p. 25.—*Böhm.* Bibl. I. 1, p. 519.

2. Synopsis of the Natural History of Great Britain and Ireland. Lond. 1789, 8°. 2 vols.—Bibl. Ent. I. p. 25.

BERKHEY (J. Lefr. van).

1. Over en hard geschaald Ey van een Zee-Hoorn.—Act. Vliss. III. p. 576, fig.—*Mod.* Bibl. Helm.

2. Lettre sur la Génération des Testacés.—Mém. Flessing. III.— Biogr. Un. IV. p. 268.

3. Natuurlyke Historie van Holland. Amst. 1769–74, 3 vols. 8°.
fig.—(Germ.) *Naturgeschichte von Holland.* Leipz. 1779-82,
2 vols. 8°.—*Böhm.* Bibl. I. 1, p. 534.—(Gall. *Jansen.*) Par. 1782,
4 vols. 12°. fig.

BERLACK (J.).

1. Symbola ad Anatomiam Vesicæ Natatoriæ Piscium. Regiom. 8°.
T. 1.—*Müll.* Arch. 1836, p. 240.

BERLÈZE ( ).

1. Déstruction de la Larve du Hanneton. 8°. Paris, 1828.

BERLINGHIERI, SILVESTRE, ROBILLIARD et BRONGNIART.

1. Premier rapport des expériences faites, d'après M. l'abbé *Spallanzani*, sur la génération des Grénouilles.—Ann. de Chim. XII.
p. 77.—Médec. éclairée, etc. III. p. 137.—Dict. Sc. nat. XV.
p. 259.

BERNAERT (M. B. F.).

1. Notice sur la Baleine échouée près d'Ostende. 8°. Paris, 1829.

BERNARD (C. A.).

1. Die Functionen des elektrischen Fluidums, vorzüglich in Hinsicht
des menschlichen Körpers im gesunden u. kranken Zustande.
Wien. 1838, 8°.

BERNARD (J. Et.).

1. Introduction anatomique (Anonyme). Nomenclature des parties
du corps (*Hypatus*). 1744.—Biogr. Un. IV. p. 295.
2. Histoire et figure du Cochenille de l'Olivier.—Mém. II. p. 275,
fig.—*Féruss.* Bull. III. p. 251.
3. Mémoires pour servir à l'Histoire naturelle de la Provence. Par.
1787, 3 vols. 12°. fig.

BERNARD DESCHAMPS ( ).

1. Recherches microscopiques sur l'organisation des ailes des Lépidoptères. 8°. Paris, 1835.

BERNARDI ( ).

1. Rapport sur des Ossemens fossiles trouvés récemment dans le
voisinage de Palerme.—Giorn. offic. Palerm. 1830.—*Féruss.*
Bull. 1830, XXII. p. 31.

BERNATOWITZ ( ).

1. Mémoire sur la Chenille de l'Alizier, qui fait des ouates.—Bibl. Univ. Févr. 1825.—Bibl. Ent. I. p. 25.

BERNEAUD (Thiéb. de).

1. Voyage à l'île d'Elbe. Paris, 1808, 8°. fig.—Ed. Angl. Lond. 1814, 8°.—Bibl. Ent. I. p. 25.
2. Travaux de la Société Linnéenne de Paris, depuis sa réorganisation jusqu'à 1821. Par. 1822, 8°.—Verz. Bibl. d. Schw. Naturf. Ges.
3. Rapport sur le Fossile trouvé au long Rocher dans la Forêt de Fontainebleau. 8°.

BERNER (G. E.).

1. Exercitatio de applicatione mechanismi, cum observ. de punctura Araneæ. 8°. Amst. 1720.

BERNHARD (J.).

1. Der kleine Buffon. 8°. Karlsruhe, 1843-44.

BERNHARD (W.).

1. Repetitorium der Naturwissenschaften, oder Abriss der Physik, Chemie, Botanik, Zoologie u. Mineralogie. 8°. Berlin, 1839.

BERNHARDI (B.).

1. Thier Fährten auf Flächen des bunten Sandsteins bei Hildburghausen.—L. u. Br. N. Jahrb. 1834, p. 642.

BERNHARDI (J. J.).

1. Annalen des Nationalmuseums zu Paris ; vide sup. Pars I. p. 24.
2. Ueber die Grenzen der Mineralogie. Erfurt. 1809, 8°.
3. In wiefern gibt es Individuen im Mineralreiche.—Leonh. Tasch. III. p. 60.

BERNHARDI (R.).

1. Darstellung des gegenwärtigen Zustandes der Geologie. Eine gekrönte Preisschrift. Harlem, 1832.—L. u. Br. N. Jahrb. 1835, p. 220.
2. Alter des Hildburghäuser Sandsteins mit Fährten ; erratische Blöcke durch Polar-Eis u. Gletscher bewegt.—L. u. Br. N. Jahrb. 1841, p. 455.

BERNHARDT ( ).

1. Symbolæ ad Ovi Mammalium historiam ante prægnationem. Vratisl. 4°.—*Müll.* Arch. 1835, p. 228.

BERNIARD ( ).

1. Mém. contenant l'analyse chymique de l'os trouvé à Paris dans une cave rue Dauphiné, etc.—Journal de Physique, Oct. 1781.

BERNIER (Fr.).

1. Voyages, contenant la description des Etats du Grand-Mogol. Amst. 1699, 12°.—*Fisch.* Cat.

BERNINI (Clem.).

1. Ornitologia dell' Europa meridionale.   Parma, 1772, fol. fig.— *Böhm.* Bibl. II. 1, p. 509.

BERNITZ (M. B. v.).

1. Gammarus alatus, seu Papilio elegans et rarus signaturam Gammari habens.—Misc. Nat. Cur. 1671, Dec. p. 171.—Bibl. Ent. I. p. 25.
2. De usu et utilitate Cocci Polonici.—Ephem. Nat. Cur. ann. III. Dec. p. 143.—Bibl. Ent. I. p. 25.

BERNON (Léon.).

1. Recueil des Pièces curieuses apportées des Indes, d'Egypte et d'Ethiopie, qui se trouvent dans son Cabinet. 8°. Par. 1670.— *Böhm.* Bibl. I. 1, p. 377.

BERNOULLI (Chr.).

1. Ueber das Leuchten des Meeres.   Gött. 1803, 8°.—Verz. Bibl. d. Schw. Naturf. Ges.
2. Geognostische Uebersicht der Schweiz, nebst einem systematischen Verzeichnisse aller in diesem Lande vorkommenden Mineralkörper u. deren Fundörter.   Basel, 1811, 8°.
3. Sur les Mines des Grisons.—Schweiz. Arch. f. Statist. 1827, I. p. 40.—*Féruss.* Bull. 1828, XIV. p. 412.
4. Grundriss der Mineralogie.   Bas. 1821, 8°.—Verz. Bibl. d. Schw. Naturf. Ges.

BERNOULLI (D. J. F.).

1. Dissertatio de actione Fluidorum in corpora solida et motu Soli-

dorum in fluidis.—Comm. Ac. Petr. 1727, II. p. 304; 1728, III. p. 214.

2. Tentamen novæ de Motu Musculorum Theoriæ.—Comm. Ac. Petr. 1726, I. p. 297.

BERNOULLI (Eph.).

1. Versuch einer physischen Anthropologie; oder Darstellung, etc. Halle, 1804, 8°. 2 vols.—Bibl. med.-ch. p. 41.

BERNOULLI (J.).

1. Lettres sur différens sujets, ecrites pendant le cours d'un Voyage par l'Allemagne, la Suisse, la France méridionale et l'Italie en 1774–75, avec des Additions et des Notes concernant l'Histoire naturelle, l'Astronomie, etc. Berl. 1777–79, 3 vols. 8°. fig.— *Böhm.* Bibl. I. 1, p. 488.

2. Reisen durch Brandenburg, Pommern, Preussen, Curland, Rusland u. Polen, in den Jahren 1777–78. Leipz. 1779–80, 6 vols. 8°.—(Gall.) *Voyages de Brandenbourg*, etc. Vassov. 1782, 8°.— *Böhm.* Bibl. I. 1, p. 488.

3. Sammlung kurzer Reisebeschreibungen u. anderer zur Erweiterung der Länder- u. Menschenkenntniss dienenden Nachrichten. Berl. 1781–83, 9 vols. 8°. fig.—*Böhm.* Bibl. I. 1, p. 488.

4. Zusätze zu den neuesten Reisebeschreibungen, nach *Volkmann's* Ordnung. Leipz. 1777, 3 vols. 8°.—*Böhm.* Bibl. I. 1, p. 556.

5. Archiv zur neuern Geschichte, etc; *vide sup.* Pars I. p. 27.

6. Observatio de quorundam Lepidopterum facultate Ova sine prægresso coitu fœcunda excludendi.—N. Mém. Ac. Berl. 1772, p. 24. —J. de Phys. XIII. p. 104.—(*Observations sur les Œufs des Papillons.*)—Opusc. scelti, II. p. 217.—N. Hamb. Mag. 96 St. p. 504.—Bibl. Ent. I. p. 26.

7. De Motu Musculorum; de Effervescentiâ et Fermentatione. Accedunt *P. A. Michelotti* Animadversiones X. ad ea quæ *J. Keil* protulit, etc. 4°. Venet. 1721.—Cat. R. Soc. L.

BERNSTEIN (J. G.).

1. Antitypographus, oder Widerlegung der Meinung, dass der Borkenkäfer an der Wurmtrockniss der Waldungen Schuld sei. 8°. Leipz. 1793.—*Eis.* Insect. p. 195.

BEROALDI (Phil.).

1. De Terræ Motu et Pestilentia; cum Annotamentis *Galeni.* 4°. Argent. 1510.—Cat. R. Soc. Lond.

BEROLDINGEN (Fr. v.).

1. Beobachtungen, Zweifel u. Fragen, die Mineralogie überhaupt u. insbesondere ein natürliches Mineralsystem betreffend. Hannov. 1778, 2 vols. 8°.—Hann. u. Osnab. 1792–94, id.

2. Physikalisch-chemische Beschreibung des in dem Bisthum Paderborn gelegenen Gesundbrunnens zu Driburg. Hildesh. 1783, 8°.—Cat. Bibl. Turic.

3. Die Vulkane älterer u. neuerer Zeiten, physikalisch u. mineralogisch betrachtet. Mannh. 1791, 8°.—Biogr. Un. IV. p. 335.

4. Bemerkungen auf einer Reise durch die Pfalzischen u. Zweibrückischen Quecksilber-Bergwerke. Berl. 1788, 8°.—Trans. Linn. Soc. Lond. IX. p. 326.

BERQUEN (Rob.).

1. Les Merveilles des Indes Orientales et Occidentales, ou Traité des Pierres et Perles, contenant leur vraie nature. Paris, 1669, 4°.—*Mod.* Bibl. Helm.

BERRES (Chr.).

1. Anthropotomie, oder Lehre von dem Baue des menschl. Körpers. Lemb. 1821, 8°. 4 Bde.—2te Aufl. Wien, 1834, 8°. 1ter Bd. fig.—Bibl. med.-ch. p. 42.

BERRES (J.).

1. Anthropotomie, oder Lehre von dem Baue des menschlichen Körpers. (2te Aufl.) Wien, 1841, 8°.

2. Anatomie der mikroskopischen Gebilde des menschlichen Körpers. I.-XII. Wien, 1836–42, fol. fig.

3. Microscopische Beobachtungen über die innere Bauart der Nerven und Centraltheile des Nervensystems.—Med. Jahrb. d. Oesterr. St. p. 274.—*Müll.* Archiv. 1836, II. p. xiv. 155.

4. Untersuchungen über die verschiedenen Formen der Capillargefässe.—Med. Jahrb. d. Oesterr. St. XV.—*Müll.* Arch. p. 8.

BERRUTI (S.).

1. Sulla Generazione spontanea e sulla natura dei Zoospermi. Tor. 1842, 8°.—Giorn. Sc. med. Tor. XVI. p. 129.

2. De Luce; de Oculi Globo; de Visu. Aug. Taur. 1823, 8°.

3. Theses physiologicæ ad usum Prælectionum academicarum. Taur. 1836, 1838, 1842, 8°.

4. Sulla Fosforescenza in genere, e più particolarmente da quella dei Corpi organici.—Giorn. Sc. med. Tor. V. 1839.

5. Note alle Lettere fisiologiche del Prof. *Medici.*—Giorn. Sc. med. Tor. VII. 1840.

6. Cenni sulla Ovologia e sulla Embriologia.—Giorn. Sc. med. Tor. XI. 1841.

7. Note alla Fisiologia del Sistema nervoso del Prof. *Müller.*—Giorn. Sc. med. Tor. XI. e XII. 1842.

8. Considerazioni ed Esperienze sulla origine e sulla funzioni del Nervo intercostale.—Giorn. Sc. med. Tor. XIII. 1842.

9. Esperienze sulla esistenza delli correnti elettrofisiologiche negli Animali a sangue caldo.—Giorn. Sc. med. Tor. 1840.

BERRYAT (J.).

1. Recueil de Mémoires ; *vide sup.* Pars I. p. 43.

BERSANDIER (A.).

1. Disc. sur le dégât que les Sauterelles firent en Provence en 1613–14, 12°. Paris.

BERTALDI (J. L.).

1. Confectio de Hyacintho et Conf. Alchermes.   Tur. 1613, 1619.—Bibl. Ent. I. p. 26.

BERTELE (G. A.).

1. Handbuch der Minerographie einfacher Fossilien, für seine Vorlesungen bestimmt.   Landsh. 1804, 8°.

BERTEREAU (Mart. de).

1. Véritable déclaration faite au Roi des riches et inestimables Trésors nouvellement découverts dans le Royaume de France. 1632, 8°.—*Böhm.* Bibl. IV. 1, p. 97.

BERTERO (   ).

1. Notice sur l'Histoire naturelle de l'île Juan-Fernandez.—Ann. Sc. nat. 1830, XXI. p. 344.

BERTEZEN (S.).

1. Thoughts on the different kinds of Food given to young Silkworms, and the possibility of their being brought to perfection in the Climate of England ; founded on Experiments made near the Metropolis.   Lond. 1789, 8°.

BERTHELOT (Jos.).

1. Dissert. de venenatis Galliæ Animalibus. Monsp. 1763, 4º.—
Böhm. Bibl. II. 1, p. 63.

BERTHELOT (Sabine).

1. Considérations sur l'acclimatement et la domestication. 8º Paris,
1844.
2. *Cryptella Canariensis.*—Mag. Zool. V. p. 63.
3. Remarques sur la Cochenille du Nopal.—Rev. Zool. 1839, p. 216.
4. Allgemeine Bemerkungen über die Canarischen Inseln, besonders
naturgesch. Inhalts.—Isis, 1826, X. p. 960.

BERTHEY (　).

1. Naturgeschichte von Holland. Aus d. Holländ. Leipz. 1779,
2 vols. fig.

BERTHIER (Jos. Et.).

1. Histoire des premiers temps du Monde, prouvée par l'accord de
la Physique avec la Génèse. Paris, 1778, 12º.

BERTHIER (P.).

1. Notice géologique sur les environs de Nemours, Puiseaux et
Château-Landon. Avec une carte et une planche de coupes
géognostiques.—Ann. d. Min. I. (2ᵉ Sér.) p. 287.—*Féruss.*
Bull. 1828, XIII. p. 9.

BERTHOLD (A. A.).

1. Beiträge zur Anatomie, Zoologie u. Physiologie. Gött. 1831, 8º.
fig.—Isis, 1832, VIII. p. 906.
2. Ueber den Fabricischen Beutel der Vögel.—N. Act. Nat. Cur.
XIV. p. 903.—*Féruss.* Bull. 1830, XXI. p. 132 (*Sur la bourse
de Fabricius chez les Oiseaux*).
3. Ueber die Kopfknochen der Nagethiere.—Isis, 1825, VIII. p. 907;
IX. p. 983.—*Féruss.* Bull. 1826, VII. p. 240 (*Mémoire sur les
Os de la tête des Rongeurs*).
4. Einige Bemerkungen über die Schädelknochen und deren Näthe.
—Isis, 1830, II. p. 196.
5. Lehrbuch der Physiologie des Menschen und der Thiere. 2 Th.
Götting. 1829, 8º.—N. Act. Nat. Cur. XIV. p. XIII.—Isis, 1830,
IV. p. 429.

6. Das Aufrechterscheinen der Gesichtsobjecte trotz des umgekehrt stehenden Bildes derselben auf der Netzhaut des Auges. Gött. 1830, 8°.—Isis, 1832, VIII. p. 908.

7. Bedeutung u. Nutzen der Luftröhrenringe.—Isis, 1827, IX. p. 761.

8. Ueber die Bedeutung der Bauchmuskeln.—Isis, 1826, IV. p. 416.

9. De Gravitate Halitus. Götting. 1833, 4°.—N. Act. Nat. Cur. XVII. p. xx.

10. Nachtrag zur Lehre vom Hirschgeweih.—Isis, 1834, V. p. 532, fig.

11. Etwas zur Naturgeschichte des gemeinen Igels, u. über dessen Urachus.—Isis, 1827, II. p. 168.—*Féruss.* Bull. 1828, XV. p. 269 (*Rem. sur le Hérisson commun et son ouraque*).

12. Neue Versuche über die Temperatur der kaltblütigen Thiere. Gött. 1835.—*Wiegm.* Arch. 1837, IV. p. 134.—*Müll.* Arch. 1836, p. 119.—Ann. d'Anat. II. p. 174 (*Nouvelles expériences sur la température des animaux à sang froid*).

13. *Latreille*'s Natürl. Familien des Thierreichs; mit Anmerk. und Zusätzen übersetzt. Weim. 1827, 8°.—Isis, 1828, III. p. 386.— *Féruss.* Bull. 1828, XIV. p. 218 (*Familles naturelles du R. anim. de Latreille, trad. Allem. par Berthold*).

14. Einige Notizen aus der Anatomie und Physiologie des Spechtes im Allgemeinen, des Grasspechtes (*Picus viridis*) aber insbesondere.—Isis, 1824, V. p. 555.

15. Darstellung sämmtlicher Saügethierarten. 4°. Gött. 1832.

16. Ueber den Hasen.—Isis, 1825, IV. p. 446; V. p 601.

17. Einleitung in die Zergliederung des Hasen und des Caninchens. —Isis, 1825, II. p. 220.

18. Beiträge zur Anatomie des krebsartigen Kiefenfusses (*Apus cancriformis*).—Isis, 1830, VI. p. 655.—*Féruss.* Bull. 1831, XXIV. p. 97 (*Sur l'anatomie de l'Apus cancriforme*).

19. Ueber die Formveränderung des Schädels der gem. Fischotter nach der Geburt.—Isis, 1830, VI. p. 570.—*Féruss.* Bull. 1830, XXIII. p. 264 (*Sur les changemens de forme que subit le crâne de la Loutre commune*).

20. Austritt des Eyes aus dem Eyerstocke des Hundes.—Isis, 1830, VI. p. 574.

21. Bildung u. Regeneration der Eyerschalenhaut.—Isis, 1830, VI. p. 573.

22. Ueber die Bildung u. den Nutzen der Hagel (*Chalazæ*) im Vogeley.—Naturf. in Berl. 1828.—Isis, 1829, IV. p. 404.

23. Ueber den Nervenhalsband einiger Mollusken.—*Müll.* Arch. 1835, p. 378, fig.

24. Ueber verschiedene neue oder seltene Amphibienarten. Gött.
1842, 4°. fig.

25. Ueber verschiedene neue Reptilien aus Neu-Grenada. 4°. Gött.
1846.

26. Ueber den Bau des Wasserkalbes (*Gordius aquaticus*). Gött.
1842, 4°. fig.

27. Ueber einen Schädel aus den Gräbern der alte Palaste von
Mitla.—Nov. Act. Ac. Leop. XIX. 2, 1842.

28. Lehrbuch der Zoologie. 8°. Gött. 1845.

29. Mémoire sur diverses espèces d'Amphibies nouvelles ou peu
connues.—Soc. Sc. Goett.—Rev. Zool. 1840, p. 286.

30. Der gespaltene Unterkiefer, eine Hemmungsbildung, beobachtet
an einem Kalbe.—Nov. Act. Ac. Leop. XIX. 1.

31. Ueber ein linsenförmiges Knöchelchen im Musculus Stapedius
mehrerer Säugthiere.—*Müll.* Arch. 1838, p. 46, fig.

32. Einige Versuche über die Aufsaugungsthätigkeit (Inhalation)
der Haut.—*Müll.* Arch. 1838, p. 177.

BERTHOLLD (Dan. Gotth.).

1. Dissert. de Rebus petrificatis, earumque Divisione, observationes
varias continens. Witteb. 1767, 4°.—*Mod.* Bibl. Helm.

BERTHOLDI (   ).

1. Ueber einen in Tauris Fossil gefundenen Hai-Zahn. —Bull. Natur.
Mosc. 1833, VI.

BERTHOUT v. BERGHEM (J. P.).

1. Betrachtungen über den wilden Ursprung der Hausziege.—
Höpfn. Mag. II. p. 23.

2. Beschreibung und Naturgeschichte des Steinbocks der Savo-
ischen Alpen.—*Höpfn.* Mag. IV. p. 333.

3. Excursion dans les Mines du Haut-Faucigny, et Description de
deux nouvelles routes sur le Buet et le Breven, etc. Laus. 1787,
8°.

4. Itinéraire de la Vallée de Chamouny et d'une partie du Bas-
Vallais et des Montagnes avoisinantes à Lausanne. Laus. 1790,
12°.

BERTIN (   ).

1. Sur la Structure de l'Estomac du Cheval.—Acad. Sc. Par.
1746.

BERTIN (Ex. Jos.).

1. Traité d'Ostéologie. 1754, 12°. 4 vols. (Trad. en All. par *Pflug*). *Osteologie, oder Knochenlehre.* Copenh. 1777.—Biogr. Un. IV. p. 365.—Bibl. med.-ch. p. 43.

2. Lettres sur le nouveau Système de la Voix et sur les Artères lymphatiques. 1748 (1ère éd. du Syst. de la Voix ; La Haye, 1745, 8°).—Biogr. Un. IV. p. 365.

BERTINATTI ( ).

1. Anatomia fisiologica applicata alle Belle Arti figurative. Tor.1837.

BERTINI (Bern.).

1. Idrologia Minerale del Piemonte. 8°. Tor. 1822.

BERTOLONI (Ant.).

1. Amœnitates Italicæ, sistentes Opuscula ad Rem Herbariam et Zoologiam Italiæ spectantia. Bonon. 1819, 4°.

2. Rariorum Italiæ Plantarum decas tertia ; accedit specimen Zoophytorum portus Lunæ. Pisæ, 1810, 8°.—Dict. Sc. n. LX. p. 536.

BERTOLONI (Gius.).

1. Lettera al conte *Fil. Re* su varii Insetti nocivi all' Agricoltura (Rapp. a S.E. il Ministro dell' Interno, etc.). Milano, 1812, 8°.

2. Della Galleruca che distrugge gli Olmi.—Ann. d'Agric. R. d'Ital. XXI.

3. Memoria sopra due rare Farfalle trovate nel Territorio Lunese. Bol. 1829, 8°. fig.

4. Descriptio novæ speciei e Coleopterorum ordine (*Nebria fulviventris*). Bonon. 1837, 4°. fig.—N. Comm. Bonon. III. p. 83.

5. Dissert. de Insectis quæ hieme et vere ann. 1832–33 sata Tritici vastarunt in arvis Italiæ. Bonon. 1837, 4°. fig.—N. Comm. Bonon. III. p. 195.

6. Modo facile di distruggere l'Insetto divoratore delle foglie dell' Olmo (*Galleruca calmariensis*, Linn.).—N. Ann. Sc. nat. Bol. Fasc. IX.

7. Storia di due Insetti nocivi all' Agricoltura.—N. Ann. Sc. nat. Bol. 1841.

8. Fauna Insettologica del Territorio Bolognese.

9. Commentarius de Bupreste Fabricii. Bol.1841, 4°.fig.—N. Comm. Bonon. V. p. 89.

Bertolotti (D.).

1. Raccoglitore; *vide sup.* Pars I. p. 67.

Bertolotti (Phil. Mar.).

1. De Gustu et Saporibus, Olfactu et Odoribus; De Cerebri Fabricâ Usibusque, etc. (Theses pro aggreg.) Taurini, 1744, 8°.

Bertolotto ( ).

1. The History of the Flea, with Notes and Observations (2d ed.). Lond. 1834, 8°. fig.—(Gall.) 12°. Par. 1834.

Berton ( ).

1. Recherches sur l'Hydrocéphale aigu, sur une variété de Pneumonie et sur la dégénérescence tuberculeuse. Paris, 1834.— *Müll.* Arch. 1835, p. 242.

Bertossi (Gius.).

1. Delle Terme Padovane, Trattato. Venez. 1759, 4°.

Bertram ( ).

1. Observations sur les Vèrs à Soie qui naissent dans l'Amerique Septentrionale.—*Rozier,* Journ. de Phys. II. 1773, p. 51.

Bertrand (Al.).

1. Lettres sur les Révolutions du Globe. 3e éd. Paris, 1824, 12°.— *Féruss.* Bull. 1824, I. p. 305.—4e éd. augm. etc. par *MM. Arago, E. de Beaumont, Al. Brongniart,* etc. Par. 1839, 8°. fig.— (Germ. *P. v. Maack*): *Die Revolutionen des Erdballs, nach der* 5ten *Ausg. des Franz. Originals,* etc. Kiel, 1844, 8°. fig.

2. Phænomena of the Earth, the Revolutions of the Globe familiarly described; with an Appendix giving a succinct Account of every Theory from that of *Ray* in 1692 to the present time. 12°. Lond. 1835.—Edinb. 1837, 8°.

Bertrand (Bern. Nic.).

1. Elémens de Physiologie. 1756, 12°.—Biogr. Un. IV. p. 376.

2. Elémens d'Oryctologie. Neuch. 1770, 8°.—Biogr. Un. IV. p. 376.

Bertrand (Elie).

1. Elémens d'Oryctologie, ou Distribution méthodique des Fossiles. Bâle, 1779, 8°.

2. Essais sur les usages des Montagnes, avec une Lettre sur le Nil. Zür. 1754, 4°.—Biogr. Un. IV. p. 377.

3. Mémoires sur la structure intérieure de la Terre. Zür. 1752, 8°.; 1760, 8°.—Biogr. Un. IV. p. 377.—*Böhm.* Bibl. IV. 1, p. 58.

4. Recueil de divers traités sur l'Histoire naturelle de la Terre et des Fossiles. 1766, 4°.—Biogr. Un. IV. p. 377.

5. Dictionnaire universel des Fossiles propres et des Fossiles accidentels. La Haye, 1763, 8°. 2 vols.—Avign. 1764, 8°.—*Klein.* Disp. Echinod. p. viii.—Biogr. Un. IV. p. 377.

6. Mémoires pour servir à l'histoire des Tremblemens de terre de la Suisse, principalement pour l'année 1755. 8°. Berne, 1756 ; La Haye, 1757, 8°.—Biogr. Un. IV. p. 377.

BERTRAND (J. Bapt.).

1. Lettre sur le mouvement des Muscles et sur les esprits animaux. —Biogr. Un. IV. p. 375.

BERTRAND (L.).

1. Renouvellemens périodiques des Continens terrestres. Paris, 1799, 8°. fig.—Genève, 1803, 8°.

BERTRAND (Phil.).

1. Nouveaux principes de Géologie. Paris, 1798, 8°.—2e éd. corrigée, 1804, 8°.—Biogr. Un. Suppl. LVIII. p. 162.

2. Nouveau système sur les Granits, les Schistes, les Mollasses et autres pierres vitreuses, etc. Paris, 1794, 8°.—Biogr. Un. Suppl. LVIII. p. 162.

3. Lettre à *M. de Buffon,* ou critique et nouvel essai sur la théorie générale de la Terre. Besanç. et Paris, 1780, 12°.— 2e éd. augm. d'un Supplément, etc. Ibid. 1782, 8°.—Biogr. Un. Suppl. LVIII. p. 161.

BERTRAND-GESLIN (Ch.).

1. Aperçu géognostique sur le Bassin gypseux d'Aix, Dép. des Bouches-du-Rhône.—Mém. Soc. d'Hist. Nat. Paris, I. p. 273.— *Féruss.* Bull. 1823, II. p. 43 ; 1824, II. p. 112.

2. Considérations géognostiques générales sur le Terrain de transport du Val-d'Arno supérieur.—Ann. Sc. n. 1828, XIV. p. 363.— *Féruss.* Bull. 1829, XVII. p. 30.

3. Nòte sur la Caverne à ossemens de Banwell (Somersetshire).— N. Bull. Soc. Philom. Août, 1826, p. 118.—*Féruss.* Bull. 1827,

XII. p. 26.—*Leonh.* Zeitschr. 1827, I. p. 554 (*Ueber die Höhle mit Thier-Gebeinen zu Banwell*).

4. Note sur la Caverne à ossemens d'Adelsberg en Carniole. Extr. d'une lettre à *M. Brongniart.*—*Féruss.* Bull. 1826, VIII. p. 12. —Isis, 1834. VIII. p. 858 (*Ueber die Knochenhöhle bei Adelsberg in Krain*).

5. Notice sur le Gisement du Zircon-Hyacinthe.—Bull. Soc. Phil. 1821, p. 10.

6. Lettre sur les Porphyres pyroxéniques et à Coquilles fossiles.— Bull. Soc. Géol. Fr. VI. p. 8.

7. Beschreibung des Knochen-Schuttlandes im oberen Arno-Thale. —Mém. Soc. Géol. Fr. I. p. 161, fig.—*L. u. Br.* N. Jahrb. 1833, p. 689.

8. Notice géognostique sur l'île de Noirmoutiers, departement de la Vendée. In-4, fig.

BERTRAND-GESLIN (Ch.), TRETTENERO et MARASCHINI.

1. Observations sur les Roches Pyrogènes de la Vallée de Fiemme. (*V. Maraschini*).

BERTRAND-ROUX DE DOUE (J. M.).

1. Mémoire sur les Ossemens fossiles de Saint-Privat-d'Allier, et sur le Terrain basaltique où ils ont été découverts.—Ann. Soc. d'Agric., etc. du Puy, 1829, fig.—*Féruss.* Bull. 1830, p. 206.— *Brewst.* Journ. Science, Ser. 2, II. p. 276.

2. Description géognostique des environs du Puy en Velay, et particulièrement du Bassin au milieu duquel cette ville est située. Paris, 1828, 8°. fig.—*Féruss.* Bull. 1824, III. p. 1.—*Leonh.* Zeitschr. 1825, I. pp. 214, 283, 406.—Ed. 2, Par. 1833.

BERTRANDI (Ambr.).

1. Opere anatomiche. Tor. 1786–99, 10 vols. 8°.

2. Dissert. anatomicæ de Hepate et de Oculo. Aug. Taur. 1748, 8°.

BERTRANDI (J. A. Mar.).

1. Dissertatio de Hepate. 1747.—Biogr. Un. IV. p. 378.

2. De glanduloso Ovarii corpore, de Placentâ, et de Utero gravido. Ac. R. d. Sc. I.—Biogr. Un. IV. p. 379.

BERTUCH (F. J.).

1. Bilderbuch für Kinder. Weimar, 1792-1812, Atl. 7 vols. 4°;

Text v. *Funke,* 10 vols. 8°.—Isis, 1822, III. p. 336 ; XII. p. 1337 ; 1823, V. p. 486.

2. Die prächtige Maenura (*Mænura superba*).—*Voigt's* Magaz. IV. 1802, p. 689, fig.

3. Ueber die beste Art grosse Quadrupeden für naturhistorische Sammlungen aufzusetzen.—*Voigt's* Mag. IX. 1805, p. 269.

4. Tafeln der allgemeinen Naturgeschichte. 4°. Weimar, 1807.

5. Der Wombat, *Didelphys wombat* oder *ursina*.—*Voigt's* Mag. IV. 1802, p. 681, fig.

6. Ueber die Mittel ·Naturgeschichte gemeinnützig zu machen. Weim. 1799, 4°. fig.

BERZELIUS (J. J.).

1. An Attempt to establish a pure scientific System of Mineralogy ; translated from the Swedish by *John Black*. Lond. 1814, 8°.— Tr. Linn. Soc. Lond. XI. 2, p. 422.

2. Jahresbericht über die Fortschritte der physischen Wissenchaften. Eingereicht an die Schwed. Ak. d. W. den 31 März, 1838. Deutsch herausgeg. v. *F. Wöhler*. Tüb. 1839, 8°.

3. On Meteoric Stones.—Vetensk. Acad. Handl. 1835, p. 115.— *Sill*. Am. J. XXXVII. p. 93.

4. Ueber Metamorphosen der Gebirgsarten.—*L.* u. *Br*. N. Jahrb. 1843, p. 219.

5. Ueber *Fuchs*' Neptunische Theorie der Urgebirge.—*L.* u. *Br*. N. Jahrb. 1843, p. 817.

6. Mémoire sur la composition des Fluides animaux. 8°. Par. 1814. —Cat. R. Soc. Lond.

7. Neue Mineralien in Schweden.—*L.* u. *Br*. N. Jahrb. 1841, p. 682.

BESCHERER (J.).

1. Methodik des naturwissenschaftlichen Unterrichts für Schulen überhaupt. Dresd. 1838, 8°.

BESCHERER (W.).

1. Kleine unterhalt. Erzählungen aus der Thierwelt. 12°. Nürnb. 1818.

BESEKE (J. M. G.).

1. Beyträge zur Naturgeschichte der Vögel Kurlands. Mittau et Leipz. 1792, 8°.—Schr. Berl. Naturf. Fr. VII. p. 446.

2. Microscopische Beobachtungen über Thiere des süssen Wassers.
—Leipz. Mag. III. p. 316.—*Mod.* Bibl. Helm. (Add.)

3. Versuch einer Geschichte der Hypothesen über die Erzeugung
der Thiere. Mittau, 1797, 8°.

4. Versuch einer Geschichte der Naturgeschichte. 8°. Mittau,
1802.

5. Ein Zuruf an die Naturforscher. 8°. Leipz. 1786.

BESEL (Dr.).

1. Monographie einiger neuen mit *Holothuria maculata, Cham.* et
Eysenh. verwandten Arten.—*Wiegm.* Arch. 1835, I. p. 42.

BESLER (Basil).

1. Fasciculus rariorum et adspectu digniorum varii generis Historiæ
naturalis, cum figuris æneis. Norimb. 1606, fol.; 1616, 4°.—Tr.
Linn. Soc. Lond. VII. p. 243.—*Mod.* Bibl. Helm.

2. Continuatio rariorum et aspectu digniorum varii generis, etc.
Norib. 1622, fol.—*Klein.* Disp. Echinod. p. VIII.

BESLER (M. R.).

1. Gazophylacium rerum naturalium, ex Regno vegetabili, animali
et minerali depromptarum. Norimb. 1642, fol. fig.—Leipzig,
1733.—Tr. Linn. Soc. Lond. VII. p. 230.—*Böhm.* Bibl. I. I,
p. 374.—*Mod.* Bibl. Helm.

2. Rariora Musei Besleriani quæ olim *Bas.* et *M. R. Besler* col-
legerunt, etc. Commentario illustrata à *J. H. Lochner.* Nürnb.
1716, fol.—Biogr. Un. IV. p. 389.—*Böhm.* Bibl. I. 1, p. 374.—
Mod. Bibl. Helm.

3. Admirandæ fabricæ humanæ mulieris partium .... et fœtûs,
fidelis, 5 tabulis ad magnitud. natur. typis æneis, hactenùs nunquam visa delineatio. Nurnb. 1640, fol.—Biogr. Un. IV. p. 389.

4. Observatio anatomico-medica Mulieris cujusdam Calend. Jan.
1644 tres filios viventes enixæ. 4°. Norimb.—Cat. R. Soc. L.

BESNARD (Ant.).

1. Ueber den Unterschied zwischen Genus, Species u. Varietas.
8°. München, 1835.

BESSER (J. W.).

1. Additamenta et Observatiunculæ in Tentyrias et Opatra, collectionis *Steven.*—Mém. Natur. Mosc. VIII.—N. Mém. II. p. 1, fig.
—Bibl. Ent. I. p. 26.

BESSERER (A.).

1. Observatio de Unguium anatomiâ atque pathologiâ. Diss. inaug. Bonn. 1834, fig.—*Müll.* Arch. 1835, p. 36.

BESSIERES (G. L.).

1. Introduction à l'étude philosophique de la Phrénologie, et nou- velle Classification des Facultés cérébrales. Paris, 1835, 8°.

BESSON ( ).

1. Sur les moyens de rendre utiles les voyages des naturalistes. 8°. Paris, 1792.

2. Manuel pour les Savans et les Curieux qui voyagent en Suisse ; avec des Notes par *M. Wyttenbach.* Laus. 1786, 2 vols. 8°.

BEST (C. C.).

1. Briefe über Ost-Indien, das Vorgebirge der Guten Hoffnung, etc. 4°. Leipz. 1807.

BETTI (Zacc.).

1. Memorie intorno alla Ruca de meli. Ver. 1760, 8°. fig.—Bibl. Ent. I. p. 26.

2. Il Baco da seta (poëma), con annotaz. Ver. 1765, 4°.—Bibl. Ent. I. p. 26.—*Böhm.* Bibl. II. 2, p. 262.

3. Descrizione d'un maraviglioso Ponte naturale nei Monti Veronesi. Ver. 1766, 4°.—Giorn. d'Ital. II. p. 401.—*Böhm.* Bibl. I. 1, p. 566.

4. Pensieri tratti dalla Storia naturale a difesa dell' Uomo contro i dubbi della falsa Filosofia, con le necessarie Notizie intorno ad alcuni Animali citati nell' opera. Ver. 1772, 8°.—*Böhm.* Bibl. I. 1, p. 319.

BEUDANT (F. S.).

1. Mémoire sur la Structure des parties solides des Mollusques, des Radiaires et des Zoophytes.—Ann. Mus. XVI. p. 66.

2. Mémoire sur la possibilité de conserver en vie dans l'eau salée les Mollusques d'eau douce, et vice versâ.—Acad. Sc. Par. 1816. —J. de Phys. LXXXIII. p. 268.—Isis, 1817, VI. p. 705.

3. Versuche über die Versetzung der Süsswasser-Schnecken in Salzwasser.—Ann. de Chim. II.—Phil. Mag. XLVII. p. 223.— Isis, 1834, V. p. 449.

4. Note sur trois espèces nouvelles de Mollusques gastéropodes aquatiles.—Ann. Mus. XV. p. 199.—Ann. Sc. n. XV. p. 199.

5. Notice sur le Dépôt salifère de Villiczka en Gallicie.—Bull. Soc. Philom. 1819, p. 65.

6. Notice sur le Gisement des Anthracites de Schœnfeld, en Saxe. —Bull. Soc. Philom. 1819, p. 42.

7. Observations sur les Bélemnites.—Ann. Mus. XVI. p. 76, fig.

8. On the Calcareous Tufas of Hungary.—Edinb. Phil. Mag. VIII. p. 29.

9. Geognostische Skizze von Ungarn (Voy. minér. en Hongrie).— *Leonh.* Tasch. 1823, IV. p. 838.

10. Traité élémentaire de Minéralogie. Par. 1824, 8°. fig. ; 1830, 2 vols. 8°. fig.—Cat. R. Soc. Lond.—*Quér.* Littér. Franç.

11. Voyage minéralogique et géologique en Hongrie pendant l'année 1818. Par. 1822, 3 vols. 4°. Atl.—Cat. Bibl. Turic.

BEUDANT (F. S.), DE JUSSIEU et MILNE EDWARDS.

1. Cours élémentaire d'Histoire naturelle. Par. 1841, 12°.—(Germ.) *Populäre Naturgeschichte der drei Reiche.* Stuttg. 1844, 16°.

BEULAC (J. P.).

1. Manuel de Physiologie, ou Description complète des fonctions que remplissent les diverses parties qui constituent le corps humain. 18°. Par. 1826.—Cat. R. Soc. L.

BEUMER (P. J.).

1. Der Kleine *Raff*, oder, Vater Gotthold's Unterhaltungen mit seinen Kindern über die Reiche der Natur. Wesel, 1841, 12°. fig.

BEUNIE (J. B. de).

1. Mémoire sur une Maladie produite par des Moules vénimeuses (avec un Abrégé d'Hist. nat. des Astéries et des Moules).—Mém. Ac. Brux. III. p. 187.

2. Réflexions sur quelques pièces de bois pétrifiés, trouvées dans les environs de Bruges.—Mém. Acad. Brux. V. p. XVII. (Extrait).

BEURARD (J.).

1. Dictionnaire allemand-français, contenant les termes propres à l'exploitation des Mines, à la Mineralogie et à la Metallurgie. Par. 1809, 8°.

BEURER (J. Ambr.).

1. De Lapide Osteocolla Inquisitio.—Phil. Trans. XLIII. p. 373.
2. De rarioribus quibusdam Fossilibus (Montis Mauritiani).—Act. Nat. Cur. X. p. 372.
3. De natura Succini.—Phil. Trans. XLII. p. 322.

BEUST (Fr. C. v.).

1. Kritische Beleuchtung der *Werner*schen Gang-theorie aus dem gegenwärtigen Standpunkte der Geognosie.   Freiberg, 1840, 8°.
2. Geognostische Skizze der wichtigsten Porphyr-gebilde zwischen Freiberg, Frauenstein Tharandt u. Nossen. Freiburg, 1825 ; 1835, 8°.—*L.* u. *Br.* N. Jahrb. 1838, p. 480.

BEUTEL (Tob.).

1. Cedretum electorale Saxonicum, s. brevis delineatio Elect. Saxon. operum regalium, nimirùm illius ornatissimi Theatri rerum artific. et aliorum operum quæ in electorali Dresdæ sunt. (Lat. et Germ.). Dresd. 1671 ; 1683, 4°.; 1703, 8°.—*Böhm.* Bibl. I. 1, p. 377.

BEUTH (Fr.).

1. Juliæ et Montium Subterranea, sive Fossilium variorum per utrumque Ducatum hinc inde repertorum Syntagma, in quo singula breviter recensentur ac describuntur, quæ quidem collecta huc usque servantur in Museo Auctoris.   Dusseld. 1776, 8°. fig. —*Böhm.* Bibl. IV. 1, p. 129.

BEVAN (E.).

1. The Honey-Bee community.   Length of life allotted to its different members.—Mag. Zool. and Bot. I. p. 57.
2. The Honey-Bee ; its Natural History, Physiology and Management.—Ann. of Nat. H. II. p. 293.— *Walk.* Ent. Mag. 1834, VIII. p. 270.—Bibl. Ent. I. p. 26.—(Germ.) *Die Honigbiene,* &c. 8°. Stuttgart, 1828.

BEVILACQUA-LAZISE (   ).

1. Illustrazioni Storico-mineralogiche e statistiche alla Carta del Dip^{to} dell' Adige.   Verona, 1812, 8°.
2. Dei Combustibili fossili del Veronese.   Verona, 1816, 8°.

BEVILLE (   ).

1. Traité de l'Éducation des Abeilles et de leur conservation. Paris, 1804, 8°. fig.

BEWICK (Th.).

1. General History of Quadrupeds. 8°. Newcastle, 1790; ed. 5, 1807; ed. 8, 1824.—Biogr. Un. Suppl. LVIII. p. 219.

2. History of British Birds. The figures engraved on wood. 2 vols. 8°. Newcastle, 1797-1804; 1805; 1809; 1816; 1821; 1832; 1847.—Suppl. 1821.—Isis, 1833, IX. p. 906.

BEYER (Ad.).

1. Nachricht von alten Bergwerken in Chursächsischen Landen. Leipz. 1734, 4°.—*Böhm.* Bibl. IV. 1, p. 142.

2. Anmerkungen über *Woltersdorf's* Mineralsystem (1748) und *Pott's* dagegen gerichtete Fortsetzung seiner Lithogeognosie (1751).—*Grundig* Nat. u. Kunstgesch. XX. p. 737.—*Böhm.* Bibl. IV. 1, p. 54.

3. *Otia metallica*, oder Bergmännische Nebenstunden, darinnen verschiedene Abhandlungen von Bergsachen, aus den Geschichten etc. enthalten sind. Schneeb. 1748-58, 3 vols. 8°. fig.—*Böhm.* Bibl. IV. 1, p. 55.

4. Geseegnetes Marggrafthum Meissen an unterirrdischen Schätzen u. Reichthum an allen Metallen u. Mineralien, etc. Dresd. 1732, fol.—*Böhm.* Bibl. IV. 1, p. 142.

BEYRICH (E.).

1. Mémoire sur les Goniatites qui se trouvent dans les Terrains de transition du Rhin. Berl. 1837, 4°. fig. (allem.).—Trad. en franç. par *H. Lecoeq.* Ann. Sc. n. 2ᵉ S. X. p. 65.—*On the Goniatites found in the Transition Formations of the Rhine.*—Ann. of Nat. H. III. pp. 9, 155, fig.

2. Beiträge zur Kenntniss der Versteinerungen des Rheinischen Uebergangs-Gebirges. Berl. 1837, 4°. fig.—*L.* u. *Br.* N. Jahrb. 1837, p. 497.

3. Ueber einige böhmische Trilobiten. 4°. Berlin, 1845.

BEYSCHLAG (J. F.).

1. De Ebore fossili suevico-halensi. 4°. Halæ-Magdeb. 1734.

BIANCANI ( ).

1. Ossa fossili di Monte Maggiore.—Comm. Bonon. IV. p. 133.

BIANCHI (J. B.).

1. Historia hepatica, seu de Hepatis structurâ, usibus et morbis.

Tur. 1710, 4°. ; 1716.—Geneve, 1725, 4°. 2 vols. fig.—Biogr. Un. IV. p. 440.

2. De Lacteorum Vasorum positionibus et fabricâ. Tur. 1743, 4°. —Biogr. Un. IV. p. 440.

3. Ductus lacrymales novi, eorum anatome, usus, morbi, etc. Turin, 1715, 4°. Leyde, 1823.—Biogr. Un. IV. p. 440.

4. Lettera sull' Insensibilità. Tur. 1755, 8°.—Biogr. Un. IV. p. 440.

5. De naturali in humano corpore vitiosâ morbosâque generatione, Historia. 8°. Aug. Taur. 1741, 8°. fig.—Cat. R. Soc. L.—*Mod.* Bibl. Helm.

6. Modo di distruggere la Scarabeo Mangiaviti.—Opusc. Scelt. V. p. 280.

BIANCHI (Giov.).

1. De Bagni di Pisa posti a pie del Monte di San Giuliano. Firenze, 1757, 8°. fig.—*Böhm.* Bibl. V. p. 253.

2. Su i Corni d'Ammone. Lettera al Sig. *Breyn.*—Mem. di Lucca, 1742, I. p. 204. (V. *Plancus,* Janus.)

BIANCONI (G.).

1. Rapporto su di un progetto di una Carta geognostica ed orittognostica della Toscana e di un Viaggio da lui fatto per l'Appennino Modonese.—Bull. Soc. med. Bol. II.

2. Sopra alcuni Zoofiti descritti sotto i nomi di *Cliona celata* Grant, *Vioa* Nardo, e *Spongia terebrans* Duvern.—N. Ann. Sc. nat. Bol. 1841.

3. Sulla Determinazione delle Foglie fossili.—Bull. Sc. med. di Bol. VII.

4. Dei Fenomeni geologici prodotti dal Gaz idrogene, e dell' Origine di esso Gaz.—N. Ann. Sc. Nat. 1840.

5. Congetture sopra l'origine del calore delle Acque termali.—N. Ann. Sc. Nat. II.

6. Storia naturale dei Terreni ardenti, dei Vulcani fangosi, delle Sorgenti inflammabili, dei Pozzi idropirici e di altri Fenomeni geologici operati del Gaz hydrogene. 8°. fig.

BIASOLETTO (B.).

1. Di alcune Alghe microscopiche, Saggio. Triest. 1832, 8°.—N. Act. Nat. Cur. XVI. p. xv.

BIBERG (J. J.).

1. Œconomia Naturæ, præside *Linnæo.* Upsaliæ, 1749, 4º.—
Amœnit. Academ. II. p. 1.

BIBIENA (F.).

1. De Hirudine Sermones quinque. fig.—Comm. Bonon. VII.
2. Spicilegium de Bombyce.—Comment. Bonon. V. p. 9, fig.—Bibl.
Ent. I. p. 26.

BIBRA (E. v.).

1. Ueber Elmsfeuer und Erderschütterungen in Franken.—*Pogg.*
Ann. XLVI. p. 655.—*L.* u. *Br.* Jahrb. 1839, p. 719.
2. Chemische Untersuchungen über die Knochen u. Zähne des
Menschen u. der Wirbelthiere. 8º. Schweinfurt, 1844.

BIBRON ( ).

1. On some British specimens of the genus *Triton.*—Proc. Zool.
Soc. VI. p. 23.—Ann. of Nat. Hist. II. p. 229.

BICHAT (Xav.).

1. Recherches physiologiques sur la Vie et la Mort. Par. 1800, 8º.
1818, id. 1844, 18º. (*Cerise*). (Trad. allem. par *Veizhans*, Tübing.
1802.)—Biogr. Un. IV. p. 467.—(5e éd. *Magendie*) 1829, 8º.
2. Physiologische Untersuchungen über Leben und Tod ; in Auszug
mit Anmerk. von *Herhold.* Aus dem Dänisch. von *Pfaff.* Copenh. 1802.—Bibl. med.-ch. p. 47.
3. Anatomie générale appliquée à la Physiologie et à la Médecine.
Paris, 1801, 8º. 4 vols. ; 1812, id. ; 1831, id. (Addit. de *Béclard*
et *Blandin*). (Trad. en all. par H. *Pfaff, Allgemeine Anatomie
angewandt auf die Physiologie und Arzneiwissenschaft.* 3 vols.
8º. Leipz. 1802.)—Biogr. Un. IV. p. 468.—Bibl. med.-ch. p. 47.
4. Anatomie pathologique ; avec une Notice sur la Vie de *Bichat*
par *F. G. Boisseau.* 8º. Par. 1825.—Aus dem Franz. v. *A. W.
Pestel.* Leipz. 1827, 8º.—Bibl. med.-ch. p. 47.
5. Traité des Membranes en général. 1800, 8º. (Trad. en all. par
Dörner, Abhandl. Ueber die Häute. Leipz. 1802, 8º. Tub. 1802).
—Biogr. Un. IV. p. 467.—Bibl. med.-ch. p. 47.—(Notes par *Magendie.*) Par. 1827, 8º.
6. Traité d'Anatomie descriptive. Paris, 1831, 5 vols. 8º.

BICHENO (J. E.).

1. On Systems and Methods in Natural History.—Tr. Linn. Soc.

Lond. XV. 2, p. 479.—Phil. Mag. ser. 2, III. pp. 213, 265.—
Féruss. Bull. 1829, XVIII. p. 205.

2. An Address delivered at the Anniversary Meeting of the Zool.
Club of the Linn. Society. 1826, 8°.—Bibl. Ent. I. p. 27.

BIDART (A.).

1. Observations sur la Carbonization du Bois résultant de son séjour
prolongé dans un Terrain de troisième Formation.—Bull. Soc.
Géol. Fr. VI. p. 11.

BIDAUT (E.).

1. De la Houille et son exploitation en Belgique, spécialement dans
la province de Namur ; avec une carte géolog. Brux. 1837.

BIDDER (F. H.).

1. Zur Anatomie der Retina, insbesondere zur Würdigung der stabförmigen Körper in derselben.—*Müll.* Arch. 1839, p. 371 ; 1841,
p. 248.

2. Neurologische Beobachtungen. Dorpat, 1836, 8°.— *Val.* Repert.
II. p. 21.—*Müll.* Arch. 1827, III. p. xxvi.

3. Einige Bemerkungen über Entstehung, Bau, u. Leben der
menschlichen Haare.—*Müll.* Arch. 1840, p. 538.

4. Neue Beobachtungen über die Bewegungen des weichen Gau-
mens u. über den Geruchsinn. Dorp. 1838, 4°.

5. Versuche über die Möglichkeit des zusammenheilens funktionnell
verschiedener Nervenfasern.—*Müll.* Arch. 1842, p. 102.

6. Zur Histogenese der Knochen.—*Müll.* Arch. 1843, p. 372, fig.

7. Ueber das Vorkommen zweier *Ovula* in einem Graafschen Folli-
kel.—*Müll.* Arch. 1842, p. 86.

BIDDER (F. H.) u. VOLKMANN (A. W.).

1. Die Selbstständigkeit des Sympathischen Nervensystems, durch
anatomische Untersuchungen nachgewiesen. Leipz. 1842, 4°.—
Val. Repert. VIII. pp. 7, 34.—*Müll.* Arch. 1844, p. 359.

BIDLOO (God.).

1. Opuscula omnia anatomico-chirurgica. Leyd. 1715, 1725, 4°.
fig.—Biogr. Un. IV. p. 472.

2. Anatomia corporis humani, 105 tabulis per artificiosissimum *G.
de Lairesse* ad vivum delineatis, demonstrata, etc. Amst. 1685,
fol. Leyd. 1739, fol. Utrecht, 1750, fol. c. suppl.—Act. Erudit.
IV. p. 295.

3. Vindiciæ quarundam Delineationum anatomicarum contra Animadversiones *Fr. Ruyschii.* 1697, 4º. fig.—Act. Erudit. XVII. p. 32.

4. Brief over de Dieren die man in Lever der Schaapen vind. Delft, 1698, 4º. fig.—(Lat.) *Observationes de Animalculis in ovillo Hepate et aliorum animalium detectis.* Leyd. 1698, 4º.—Biogr. Un. IV. p. 472.—*Böhm.* Bibl. II. 1, p. 396 ; II. 2, p. 400.

5. Van den Garnaet, Krabbe en Kreeft. Delft, 1704, 4º. fig.—*Böhm.* Bibl. II. 2, p. 360.

6. De oculis et visu variorum Animalium. 4º. Lugd. Bat. 1715.

BIE (Al. de).

1. Diss. de Lepore. Amst. 1660, 4º.—*Böhm.* Bibl. II. 1, p. 339.

BIELING (C.).

1. Geschichte der Entdeckung, auch Darstellung des geognostischen Vorkommens der bei dem Dorfe Thiede gefundenen merkwürdigen Gruppe fossiler Zähne u. Knochen urweltlicher Thiere. Wolfenb. 1818, 4º. fig.

BIELZ (Mich.).

1. Karpathen-Sandstein, etc.—*L.* u. *Br.* N. Jahrb. 1834, p. 403.

BIENAYMÉ (P. Fr.).

1. Mémoire sur les Abeilles ; nouvelle méthode de construire les Ruches, etc. Metz et Paris, 1804, 8º. (1e éd. 1780.)—Biogr. Un. Suppl. LVIII. p. 242.

BIERAVIUS (J. Ern.).

1. Oratio de Fodinarum Hasso-Darmstadinarum origine, progressu et usu. Giess. 1705.—*Böhm.* Bibl. IV. 1, p. 127.

BIERING (J. Alb.).

1. Historische Beschreibung des sehr alten Mannsfeldischen Bergwerkes, von den Schmelzhütten, Seygerhütten, Fossilibus, u. dgl. m. Leipz. 1734, fol.—*Böhm.* Bibl. IV. 1, p. 140.

BIERKANDER (Cl.).

1. Om Hvitax-Masken (*Ver blanc*).—Vet. Ac. XL. p. 289.—Bibl. Ent. I. p. 27.

2. Slö Hafre Masken (*Larve qui détruit l'Avoine*).—Bibl. Ent. I. p. 27.

3. Om Rot-Masken (*Larve qui attaque les racines du Chou-rave*). —Vet. Ac. 1779, p. 161.—Journ. f. d. Gartenk. VII. p. 437.— Bibl. Ent. I. p. 28.

4. Beskrifning, etc. (*Larve très-nuisible aux racines*): Elater se-getis.—Vet. Ac. 1779, p. 284.—Bibl. Ent. I. p. 28.—Schw. Acad. 1779, p. 254.

5. Insectenkalender für die Jahre 1781, 1784 u. 1790.—N. Abh. Schw. Acad. III. p. 115 ; V. p. 319.—*Eis.* Insect. p. 126.

6. Biens Flora.—Vet. Acad. XXXVI. p. 20.—*Füessl.* N. Ent. Mag. III. p. 80.—Bibl. Ent. I. p. 27.

7. Om Maskar, etc. (*Sur une Larve et une Mouche qui en provient, nuisibles aux Abeilles*).—Vet. Ac. XXXVII. p. 226.—Bibl. Ent. I. p. 27.

8. Om Rot-Masken (*Sur une Larve qui attaque les racines*).—Vet. Ac. XXXIX. p. 29.—*Licht.* Mag. II. p. 101.—Bibl. Ent. I. p. 27.

9. Beschr. der Rockenzwergmade (*Larve qui attaque le seigle* : Musca Pumilionis).—Schw. Akad. XL. p. 231.—Bibl. Ent. I. p. 27.

10. *Phalæna secalis.*—Vet. Acad. XL. p. 240 et 277.—Bibl. Ent. I. p. 27.

11. Beskrifning, etc. (*Métamorphoses d'une Larve qui se trouve à la racine des choux*).—Vet. Ac. 1780, p. 194.—Bibl. Ent. I. p. 28.

12. Beskrifning, etc. (*Nouv. Chenille du framboisier* : Tinea).— Vet. Ac. 1781, p. 20.—N. Schwed. Akad. 1781, p. 19.—Bibl. Ent. I. p. 28.

13. Hafre Masken (*Larve qui attaque l'avoine*).—Vet. Ac. 1781, p. 171.—Bibl. Ent. I. p. 28.

14. Berättelse, etc. (*Larve trouvée dans la crème* : Musca vomitoria). Vet. Ac. 1781, p. 171.—N. Schw. Akad. 1781, p. 174.—Bibl. Ent. I. p. 28.

15. Beskrifning, etc. (*Larve qui attaque le seigle en automne*).— Vet. Ac. 1783, p. 152.—Bibl. Ent. I. p. 28.

16. Beskrifning, etc. (*Larve qui vit dans la framboise*).—Vet. Ac. 1783, p. 246.—N. Schwed. Akad. 1783, p. 239—Bibl. Ent. I. p. 28.

17. Anmarkning om Socker på Gran (*Obs. sur le sucre des Pins ; divers Pucerons.*)—Vet. Ac. 1784, p. 238.—N. Schwed. Akad. 1784, p. 241.—*Crell* Chem. Annal. 1786, I. p. 351.—Bibl. Ent. I. p. 28.

18. Beskrifning, etc. (*Descr. de deux Larves qui font du tort aux fleurs et aux fruits*).—Vet. Ac. 1785, p. 156.—Bibl. Ent. I. p. 29.

19. Berättelse, etc. (*Rem. sur la couleur brune de certaines feuilles des arbres*).—Vet. Ac. 1787, p. 237.—Bibl. Ent. I. p. 29.

20. Om Maskar, etc. (*Sur une Larve qui fait du tort au blé*).—Vet. Ac. 1789, p. 232.—Bibl. Ent. I. p. 29.

21. Om en Thrips, etc. (*Sur le Thrips et les dégâts qu'il cause au blé*).—Vet. Ac. 1790, p. 226.—Bibl. Ent. I. p. 29.

22. Beskrifning, etc. (*Descr. de deux nouv.Phalènes et d'un Ichneumon*).—Vet. Ac. 1790, p. 132.—N. Schw. Akad. 1790, p. 124.—Bibl. Ent. I. p. 29.

23. *Musca subcutanea*, eller en ny och obeskrefven fluga uti Kornbladen.—Vet. Ac. 1793, p. 57.—Bibl. Ent. I. p. 29.

24. Sätt at döda Natt-Fjärilar, etc. (*Moyen de faire périr les Papillons de nuit lorsqu'ils sont en état de larve*).—Vet. Ac. 1793, p. 298.—Bibl. Ent. I. p. 29.

25. Sätt at döda Waggloss (*Moyen de détruire les nids de feuilles* : Tineites).—Vet. Ac. 1794, p. 233.—Bibl. Ent. I. p. 30.

26. *Phalæna Ekebladella*, en ny Natt-fjäril beskrifven.—Vet. Ac. 1795, p. 58.

27. Vom Apfelschäler (*Sur l'éplucheur des pommes* : Tineite).—*Wissenb.* Wochenbl. XII. p. 81.—Bibl. Ent. I. p. 30.

28. Anmärkningar huru tidigt trän, buskar och örter blommade, foglar och insecter fram kommo uti Mart, April och Mai innevarande år (1794).—Vet. Ac. 1794, p. 197.

BIERKOWSKI (L. J.).

1. Diss. sistens Moschi historiam naturalem et medicam. 8º. Leipz. 1830.

BIERLING (C. T.).

1. Serpens Vaccam emulgens.—Miscel. Acad. Nat. Cur. Dec. 1, A. 2, 1671, p. 244.

BIERLING (C. Th.).

1. Thesaurus theoretico-practicus, etc. cum Præfatione *Jac. Wolff.* —Act. Erudit. XIII. p. 1.

BIERMANN (J. C. Ad.).

1. Abhandlungen naturhistorischen, gerichtlichen u. medicinischen Inhalts. Leipz. 1828, 8º.—Bibl. med.-ch. p. 48.

BIERMAYER (Laur.).

1. Musæum anatomico-pathologicum Nosocomii Univers. Vindo-bonensis, quod ordine systematico descripsit. Vindob. 1816. 12°. —Bibl. med.-ch. p. 48.

BIET (Ant.).

1. Voyage de la France équinoxiale en l'Isle de Cayenne. Par. 1664, 4°.—*Böhm.* Bibl. I. 1, p. 753.

BIGELOW (J.).

1. Account of the White Mountains of New Hampshire. 1816, 8°. —Tr. L. Soc. Lond. XII. 2, p. 589.—Isis, 1818, XII. p. 2005 (*Ueber die Weissen Berge v. New Hampshire*).

BIGG (H.).

1. On a species of Bee from the Brazils, found living on splitting a log of peach-wood containing its Comb.—Proc. Zool. Soc. Lond. II. p. 118.

BIGGE (E.).

1. Observations on the Natural History of two species of Wasps. 8°. Oxford, 1835.—Ind. Rev. I. p. 127.

BIGGE (H.).

1. On a new site of Coal in Upper Assam.—J. A. S. B. VI. p. 243.

BIGONI (Aug.).

1. Vero Rapporto del Fisico e del Morale dell' Uomo. 1820, 6 vols. 8°.

BIGOT DE MOROGUES (P. M. S.).

1. Mémoire historique et physique sur les chûtes des Pierres tom-bées sur la surface de la Terre à diverses époques. Orléans, 1812, 8°.
2. Notice sur le Kaolin de Dignac (Charente).—Bull. Soc. Phil. 1823, p. 81.—Ann. Min. 1822, p. 589.—*Féruss.* Bull. 1823, III. p. 44.
3. Notice minéralogique et géologique sur le Quartz fétide des en-virons de Nantes.—Ann. Mus. IX. p. 392.

4. Observations minéralogiques et géologiques sur les principales Substances des Dép. du Morbihan, du Finistèrre et des Côtes-du-Nord.—Bull. Soc. Géol. Fr. II. p. 110.

BIGSBY (J. J.).

1. On the Geography and Geology of Lake Huron.—Trans. Geol. Soc. ser. 2, I. p. 175.—Amer. Journ. III. 2, p. 254.—Isis, 1835, X. Litt. A. p. 107.

2. On the Topography and Geology of Lake Ontario.—Phil. Mag ser. 2, V. pp. 1, 81, 263, 339, 424.—Féruss. Bull. 1830, XXIII. p. 163.

3. On the Geography and Geology of Lake Superior.—Journ. Roy. Inst. XVIII. pp. 1, 228.—Féruss. Bull. 1826, VII. p. 8.—Leonh. Zeitschr. 1826, II. p. 189.

4. Description générale du Lac Erie.—Quart. Journ. of Sc. 1828, p. 358.—Ann. Sc. n. 1829, XVI.—Rev. Bibl. p. 35.—Féruss. Bull. XXV. p. 227.

5. Topographie de la rivière du Niagara.—Quart. Journ. of Sc. Mars 1829, p. 39.—Féruss. Bull. 1830, XXI. p. 52.

6. On the fixed Rocks of the Valley of the St. Lawrence, in North America.—Proc. Geol. Soc. I. p. 23.—Phil. Mag. ser. 2, II. p. 217. —Féruss. Bull. 1829, XVI. p. 47.

7. On the Geology of Quebec and its vicinity.—Proc. Geol. Soc. I. p. 37.—Phil. Mag. ser. 2, III. p. 132.—Féruss. Bull. 1829, XVI. p. 388.

8. Outline of the Mineralogy, Geology, &c. of Malbay in Lower Canada.—Sill. Am. Journ. V. 2, p. 205.—Féruss. Bull. 1823, II. p. 52. (Esquisse de la Minéralogie, de la Géologie, &c. de Malbay dans le Bas-Canada.)

9. Remarks on the environs of Carthage Bridge, near the mouth of the Genesee River.—Sill. Amer. Journ. II. 2, p. 250.

10. A List of Minerals and Organic Remains occurring in the Canadas.—Sill. Amer. Journ. VIII. 1, p. 60.

11. Notice sur une Grotte à ossemens dans de Lanark, Canada supér.—Sill. Am. Journ. Juin 1825, p. 354.—Féruss. Bull. 1826, VII. p. 302.

12. Beschreibung der Bergkalk-Formazion im nördlichen Amerika. —Sill. Am. Journ. VIII. p. 77.—Leonh. Tasch. 1824, IV. p. 922.

BIGUET (F.).

1. Considérations sur les Bélemnites. 8°. Lyons, 1819.

BILBERG (J.).

1. Disputatio de naturâ Montium.   Holm. 1681, 8°.

BILDER (F. H.) u. VOLKMANN (A. W.).

1. Die Selbstständigkeit des Sympathischen Nervensystems durch
anatomische Untersuchungen nachgewiesen.   Leipz. 1842, 4°.

BILDERDYK (Guill.).

1. Traité de Géologie (en holland.).—Biogr. Un. Suppl. LVIII.
p. 261.

BILDERO (G. J.).

1. Synopsis Faunæ Scandinaviæ.   12°. Holmiæ, 1827.

BILLARD (Ch. M.).

1. Traité de la Membrane muqueuse gastro-intestinale dans l'état
sain et dans l'état inflammatoire, ou Recherches d'Anatomie
pathologique sur les divers aspects sains et morbides que peuvent
présenter l'estomac et les intestins.   Paris, 1825, 8°.—Biogr. Un.
Suppl. LVIII. p. 270.

BILLARDIÈRE (La).

1. Relation du Voyage à la recherche de *La Peyrouse* pendant les
années 1791–1794.   Paris an VIII. 2 vols. 4°.—Bibl. Ent. I. p. 30.

2. Extrait d'un Mémoire ayant pour titre : Mélanges d'Histoire
naturelle, ou Observations faites dans un voyage au Levant.—
Ann. Mus. XVIII. p. 453.

3. Note sur les mœurs des Bourdons.   Lue à la classe des Sc. phys.
de l'Institut le 6 Déc. 1813.—Mém. Mus. I. p. 55.

BILLATE (N.).

1. Dissertation historique sur les Eaux minérales de Provins.   Prov.
1738, 12°.—*Böhm.* Bibl. V. p. 204.

BILLAUDEL (   ).

1. Essai sur le gisement, la nature, l'origine et l'emploi des Cailloux
roulés qui servent à la construction des routes dans le Dép. de la
Gironde.—Ann. Sc. n. 1831, XXII.; Rev. bibl. p. 17.

2. Essai sur la détermination de quelques Ossemens fossiles trouvés
dans le Dép. de la Gironde, et sur les conséquences de cette
découverte. fig.—Bull. Soc. Linn. Bord. I. pp. 1, 60, 95, 113, 319.
—*Féruss.* Bull. X. p. 391 ; XIII. p. 427.

3. Note sur quelques Ossemens fossiles de Palæotherium, recüeillis dans le Dép. de la Gironde. Bord. 1829, 8°. fig.

BILLBERG (J. Gust.).

1. Monographia Mylabridum; c. tab. æn. col. Holmiæ, 1812, 8°. fig.—*Eis.* Insect. p. 186.

2. Novæ Insectorum species.—Mém. Acad. St. Pét. VII. 1820, fig. —*Eis.* Insect. p. 184.

3. Insecta ex ordine Coleopterorum, Decas 1ª.—Mém. Mosc. 1806, I. p. 282.—N. Act. Ups. VII. p. 271.—*Eis.* Insect. p. 184.

4. Enumeratio Insectorum in Museo suo. Holm. 1820, 4°.—Bibl. Ent. I. p. 30.

5. Synopsis Faunæ Scandinaviæ. 12°. Stockholm, 1827.

BILLBERG

1. Dissertatio: Locustæ. Upsal, 1690, 8°.—Bibl. Ent. I. p. 30.

2. Dissertatio: Historiola de Formicis. Ups. 1690, 8°.—Bibl. Ent. I. p. 30.

BILLERBECK (H. L. J.).

1. De Strigibus ab Aristotele, Plinio, cæteraque veterum grege commemoratis. 4°. Hildesheim, 1809.

BILLIET (Al.).

1. Aperçu géologique sur les environs de Chambéry.—Mém. Soc. Acad. Sav. I. p. 135.—*Féruss.* Bull. 1828, XIV. p. 35.

2. Notice sur les Tremblemens de terre que l'on a éprouvés en Maurienne depuis le 19 Déc. 1838, jusqu'au 18 Mars 1840. 4°.

BILLISTEIN (Ch. L. A. de).

1. État physique et agricole de la Lorraine (Essai sur les Duchés de Lorraine et de Bar; Amst. 1762, 8°. tom. VI.).—*Böhm.* Bibl. I. 1, p. 514.

BILLY (E. de).

1. Beobachtungen über das Versteinerungenführende Uebergangs-Gebirge der Bretagne.—Mém. Soc. H. Nat. Strasb. I. 11.—*L.* u. *Br.* N. Jahrb. 1835, p. 94.

BILS (L. de).

1. Epistolica Dissertatio quâ verus Hepatis circa chylum et pariter ductûs chyliferi hactenùs dicti usus docetur. Rotterd. 1659, 4°. —Biogr. Un. IV. p. 497.

2. Exemplar fusioris codicilli in quo agitur de verâ corporis humani Anatomiâ. Rotterd. 1659, 4°.—Biogr. Un. IV. p. 497.

3. Epistola ad omnes veræ Anatomiæ Studiosos. Rotterd. 1660, 4°. —Biogr. Un. IV. p. 497.

4. Auditûs organi Anatomia. Rotterd. 1661, 4°.—Biogr. Un. IV. p. 497.

5. Responsio ad admonitiones *Jo. ab Horne*, ut et ad animadversiones *P. Barbette* in Anatomiam Bilsianam. Rotterd. 1661, 4°.— Biogr. Un. IV. p. 497.

6. Specimina anatomica cum clariss. et doctiss. virorum epistolis aliquot et testimoniis. Rotterd. 1661, 1665, 4°.—Biogr. Un. IV. p. 497.

7. Responsio ad epistolam *Tob. Andreæ*, quâ ostenditur diversus usus Vasorum hactenùs pro lymphaticis habitorum. Marp. 1654, 4°. Rotterd. 1669, 4°; 1678, 4°.—Biogr. Un. IV. p. 497.

8. Inventa anatomica antiquo-nova, cum clariss. virorum epistolis, etc. interprete *Ged. Buenio*. Amst. 1692, 4°.—Biogr. Un. IV. p. 497.

BINA (Andr.).

1. Ragionamento sopra la cagione de Terremoti, ed in particolare di quello della Terra di Gualdo di Nocera nell' Umbria, seguito l'an. 1751. 4°. Perug. 1751.—Cat. R. Soc. Lond.

BINDSEIL (Th. F. C.).

1. De Ankylosi. Halæ, 1836, 8°.— *Val.* Repert. II. p. 25.

BINET ( ).

1. De Tæniâ.—*Roux* Journ. Méd. XXXIV. p. 217.—*Mod.* B. H.

BINGE (N. A.).

1. Description de deux dépôts remarquables de Tourbe.--N. Schr. d. Miner. Soc. in Jena, 1825, II. p. 161.—*Féruss.* Bull. 1827, X. p. 213.

2. Description des couches de Tourbe des environs du Klostersee (Holstein).—Marburg. Ges. d. ges. Naturw. 1823, I. p. 167.— N. Schr. d. Jenaer Miner. Soc. II. 1825, p. 131.—*Féruss.* Bull. 1824, III. p. 282; 1828, XIII. p. 25.

3. Couche marneuse près Lensahn.—N. Schr. d. Miner. Ges. in Jena, II. p. 278.—*Féruss.* Bull. 1828, XIII. p. 25.

4. Courte Description de la caverne de Hackershöhle au Harz.— N. Schr. d. Miner. Ges. in Jena, 1825, II. p. 175.—*Féruss.* Bull. 1827, X. p. 36.

5. Fragment sur la Patomité du Granite.—N. Schr. d. Miner. Ges. in Jena, II. p. 284.—*Féruss.* Bull. 1826, IX. p. 213.

6. Versuch einiger Beyträge zur Naturkunde u. Oekonomie. 8°. Altona, 1817.

BINGLEY (W.).

1. Animal Biography, or popular Zoology. Lond. 1803 and 1813, 3 vols. 8°. (4th ed.)—4 vols. 12°. Lond. 1829.—Bibl. Ent. I. p. 30. (Germ. v. *J. A. Bergk.*) 8°. Leipz. 1804–10.

2. Memoirs of British Quadrupeds. 2 vols. 8°. London, 1809.— Biogr. Un. Suppl. LVIII. p. 301.

3. Useful Knowledge, or a familiar account of the various Productions of Nature. 3 vols. 8°. London, 1818.

BINNEY (Amos).

1. Report on *Storer's* Reports on the Fishes, Reptiles and Birds of Massachusetts.—Proc. Bost. Soc. Nov. 1839.—*Sill.* Am. Journ. XXXVIII. p. 393.

2. Descriptions of American Limacidæ. Boston, 1842.—Rev. Zool. Soc. Cuv. VII. 1842, p. 221.

3. On the Influence of Physical Causes on the Geographical Distribution of the Genera and Species of Terrestrial Mollusks of the United States.—Proc. Bost. Soc. N. H. 1843, p. 142.

4. On some of the species of naked Pneumonobranchous Mollusca of the United States.—Proc. Bost. Soc. N. H. 1841, p. 51.

5. Results of his observations made during two successive summers at Nahant, on the Habits of the *Orthagoriscus mola,* or short Sunfish.—Proc. Bost. Soc. N. H. 1842, p. 93.

6. Report upon two works of Sig. *Michelotti* of Turin.—Proc. Bost. Soc. N. H. 1842, p. 85.

7. Critical Notice of the species found in the United States, which at present are described as constituting the genus *Pupa.*—Proc. Bost. Soc. N. H. 1843, p. 104.

BINNEY (E. W.).

1. On the relation of the New Red Sandstone to the Carboniferous Strata in Lancashire and Cheshire.—Journ. Geol. Soc. II. 12.

2. On the Dukinfield Sigillaria.—Journ. Geol. Soc. II. 390.

3. Aufrechte Stigmarien im Steinkohlen-Gebirge bei St. Helens.— Geol. Soc. Manch.—Instit. XII. 1844, p. 182.—*L. u. Br.* N. Jahrb. 1844, p. 871.

4. On the Great Lancashire Coal-Field.—Rep. Brit. Assoc. 1842, Sect. p. 49.

5. Report on the Excavation made at the junction of the Lower New Red Sandstone with the Coal-measures at Collyhurst, near Manchester.—Report Brit. Assoc. 1843, p. 241.

BINNINGER (Gasp.).

1. La véritable cause du Flux et Reflux de la Mer découverte par, &c. Halle, 1749, 8°.—(Germ.) *Entdeckung der wahren Ursache von Ebbe und Fluth auf dem Meere, dabey zugleich von der innern Beschaffenheit der Erde, den unterirdischen Wassern, Feuer, Luft u. dem Ursprunge der Quellen behandelt wird.* Bresl. 1761, 8°. —*Böhm.* Bibl. V. p. 84.

BINNINGER (L. Reinh.).

1. Oryctographiæ agri Buxovillani et viciniæ Specimen. Argentor. 1762, 4°.—*Böhm.* Bibl. IV. 1, p. 126.

BIÖRKLUND (Chr.).

1. Geographiska och physiska Anmarkningar, giorda på en resa ifrän St. Petersburg til Poltava.—Vet. Acad. 1773, p. 193.— Schwed. Acad. Abhand. 1773, p. 181.

BJÖRN (S.).

1. De indole et origine Aërolithorum. 8°.—Isis, 1817, I. p. 115.

2. Vom Subbaltischen Ufer, oder von der Enstehung des Bernsteins. Danzig, 1808, 8°.

BJÖRNSTÄHL (J. J.).

1. Resa til Frankryke, Italien, Schweitz, Tyskland, Holland, Aengland, Turkiet och Grekeland. 8° Stockh. 1777; 8°. 1780.— (Germ.) *Briefe auf seinen ausländischen Reisen, mit Anmerkungen J. Ern. u. Chr. H. Groskurk.* Strals. u. Leipz. 1778. 8° 5 vols. (vol. VI.) Leipz. u. Rost. 1783.—*Böhm.* Bibl. I. 1, p. 487.

BIOT (Ed.).

1. Sur la cause probable des anciens Déluges rapportés dans les Annales historiques des Chinois.—Sur les Tremblemens de terre, Affaissemens et Soulèvemens de montagnes observés en Chine depuis les temps anciens jusqu'à nos jours.—C. R. VIII. p. 705.

2. Mémoire sur quelques Phénomènes géologiques observés en Chine.—C. R. X. p. 787.

3. On Earthquakes.—L'Institut 21 Mai 1840.—Edinb. New Phil. Journ. XXIX.

BIOT (J. B.).

1. Relation d'un Voyage fait dans le Département de l'Orne, pour

constater la realité d'un Météore observé à Aigle le 26 Flor. an 11. Paris, 1804, 4°.—Tr. Linn. Soc. Lond. VIII. p. 360.

2. Sur les Insectes qui vivent en société.—Ac. Sc. Par. 1817.—Isis, 1818, III. p. 414.

3. Sur les Insectes tenus dans la vide pendant plusieurs jours.—N. Bull. Soc. Philom. 1817, p. 44.—Bibl. Ent. I. p. 31.

4. Tables barométriques portatives, donnant les différences du Niveau par une simple soustraction, avec une Instruction contenant l'histoire de la Formule et sa démonstration complète par les simples élémens de l'Algèbre, etc. Paris, 1811, 8°.

BIOW (H.).

1. Gemälde der Naturgeschichte der Säugthiere. Nach Cuvier's Classification. Hamburg, 4°. 1841.

BIRBEK (Chr.).

1. Relatio de fœminâ, maximam fœtûs partem per umbilicum excludente.—Act. philos. Angl. 1702, N° 275.—Act. Erudit. XXII. p. 143.

BIRCH (S.).

1. On the Monkeys known to the Chinese, from the native authorities.—Mag. Nat. Hist. ser. 2, III. p. 587; IV. p. 35.

BIRCH (Thomas).

1. History of Roy. Soc. of London; *vide sup.* Pars I. p. 77.

BIRCHEROD (J. J.).

1. Dissertationes XVI. (de Metallis, Lapidibus, Origine Fontium, æstu Maris, etc.). Hafn. 1650–51, 4°.—*Böhm.* Bibl. I. 1, p. 48.

2. Anti-Burnetius, s. Tractatus in quo Opus Creationis diei tertii explicatur, et Telluris forma, Aquarum altitudo et copia illustratur. Hafn. 1668, 8°.—*Böhm.* Bibl. IV. 2, p. 229.

3. Tractatus curiosus de Terrâ et Aquâ, in quo Terræ origo, forma et mutationes ante et post Diluvium, nec non Marium, Fluviorum, Aquarumque Fluxus, copia et altitudo traditur et contra Novatores vindicatur. Francf. 1694, 8°.—*Böhm.* Bibl. IV. 2, p. 229.

BIRCHEROD (Th. Br.).

1. Historia naturalis quatuor Costarum bubularum, quibus quæ superinducta caro fuerat in Os est conversa; cum aliis Observationibus hûc pertinentibus ed. à filio *J. Bircherod.* Hafn. 1723, 4°. fig.

BIRD ( ).

1. History and present state of Virginia, in four parts. Lond. 1705, 8°. fig.—(Gall.) *Histoire de la Virginie.* Amst. 1707, 12°. fig. —*Böhm.* Bibl. I. 1, p. 745.

BIRD (C. S.).

1. On the want of Analogy between the Sensations of Insects and their Actions.— *Walk.* Ent. Mag. 1833, N° 2, p. 105.—Bibl. Ent. I. p. 31.

2. Capture of Insects at Burghfield.— *Walk.* Ent. Mag. 1834, N° 6, p. 39.—Bibl. Ent. I. p. 31.

BIRD (Fr.).

1. Notizen aus dem Gebiete der psychischen Heilkunde. Berl. 1835, 8°.—N. Act. Nat. Cur. XVII. p. XXI.—Isis, 1835, VII. p. 672.

2. Ueber Einrichtung u. Zweck der Krankenhäuser für Geistes-kranke. Berlin, 1835, 8°.—N. Act. Nat. Cur. XVII. p. xx.

3. Sur les Proportions relatives des diverses Parties du Corps de l'Homme.—Zeitschr. f. Anthrop. I. p. 330.—*Féruss.* Bull. 1823, IV. p. 231.

BIRD (Is.).

1. Notice of various facts relating to Palestine.—*Sill.* Am. J. 1827, XII. p. 145.—*Feruss.* Bull. 1828, XIII. p. 308. (*Notice sur plus. faits relatifs à la Palestine.*)

2. Notice of Minerals, etc., from Palestine, Egypt, etc.—Amer. Journ. X. p. 21.

BIRD (J.).

1. Geological Memoir of the country from Punah to Kittor.—Madras Journ. Lit. VI. p. 375.—Journ. Roy. As. Soc. II. p. 65.

BIRKE (Th.).

1. Adlerspredigt, darinnen die Eigenschaften des Adlers nach Matth. XXV. vorgestellt werden. Tüb. 1590, 4°.—*Böhm.* Bibl. II. 1, p. 552.

BIRKHOLTZ (J. Chr.).

1. Oekonomische Beschreibung aller Fische, welche in den Gewäs-sern der Churmark gefunden werden. Berl. u. Strals. 1770, 8°. —*Böhm.* Bibl. II. 2, p. 65.—*Cuv.* et *Val.* I. p. 140.

BIRKMEYER (J. M.).

1. De Filariâ Medinensi Commentatio propriis observationibus illustrata. Onoldi, 1838, 8°. fig.—*Val.* Repert. IV. p. 185.—*Wiegm.*
Arch. 1839, II. p. 154.

BIRNBAUM (F. H. G.).

1. Ueber die Veränderungen des Scheidentheiles und des unteren
Abschnittes der Gebärmutter in der zweiten Hälfte der Schwan-
gerschaft. Bonn, 1841, 8°.

BIRON (C.).

1. Curiosités de la Nature et de l'Art, apportées de deux Voyages
des Indes, en Occident 1698–99, en Orient 1701–2, avec une
Relation abrégée des deux Voyages. Par. 1703, 12°. fig.—*Böhm.*
Bibl. I. 1, p. 453.

BISCHOF (C. A. L.).

1. Betrachtungen der vornehmsten Gegenstände der Natur. Nürnb.
1805, 2 Th. fig.

BISCHOF (J. Chr.).

1. Betrachtung des Weltgebäudes und einiger Merkwürdigkeiten
der Natur. Danz. 1764, 8°. f.g.—*Böhm.* Bibl. I. 1, p. 299.

BISCHOFF (Gust.).

1. Recherches chimiques sur es eaux minérales de Geilnau, Fachin-
gen et Selters, suivies d'observations générales sur les sources
minérales volcaniques, et en particulier sur leur origine, leur composition, et leur rapport avec les formations. Bonn, 1826, 8°.—
Féruss. Bull. 1828, XIII. p. 385.

2. Sur l'origine des Sources minérales, par le procédé simple de la
dissolution.—*Schweig.* Journ. f. Chem. u. Phys. LV. 2, p. 221.—
Féruss. Bull. 1826, IX. p. 4.

3. Ueber fossile, halb-fossile und nicht fossile Knochen.—*L. u. Br.*
N. Jahrb. 1842, p. 145.

4. Ueber die Quellen-Verhältnisse des östlichen Abhanges des Teutoburger Waldes.—Journ. f. prakt. Chem. I. p. 321.—*L. u. Br.*
N. Jahrb. 1837, p. 54.

5. On the Nat. History of Volcanos and Earthquakes.—Ed. N.
Phil. J. XXVI. pp. 25, 347.—*Sill.* Am. Journ. XXXVI. p. 230;
XXXVII. p. 41.

6. Die Wärmelehre des Innern unseres Erdkörpers, ein Innbegriff
aller mit der Wärme in Beziehung stehender Erscheinungen in

und auf der Erde, nach physikalischen, chemischen und physiolo gischen Untersuchungen. Leipz. 1837, 8°.

7. Populäre Vorlesungen über naturwissenschaftliche Gegenstände aus dem Gebiete der Geologie, Physik und Chemie. Bonn, 1843, 8°.

8. On the Terrestrial Arrangements connected with the Appearance of Man on the Earth.—Edinb. New Phil. Journ. XXXVII. p. 44.

9. Researches on the Internal Heat of the Globe. 8°. Lond. 1841.

10. Ueber die Entstehung der Quarz- und Erz-Gänge.—*L.* u. *Br.* N. Jahrb. 1844, pp. 257, 341.—Edinb. New Phil. Journ. XXXVIII. p. 344; XXXIX. p. 125; XL. p. 220.

11. Einige Bemerkungen über die Bildung der Gang-Massen.— *Pogg.* Ann. LX. p. 285.—*L.* u. *Br.* N. Jahrb. 1844, p. 100.

12. Das Felsen-Labyrinth zu Adersbach in Böhmen.—Köln. Zeit. 1844, N° 98, 99.—*L.* u. *Br.* N. Jahrb. 1844, p. 482.

13. Die vulkanischen Mineralquellen Deutschlands und Frankreichs, deren Ursprung, Mischung und Verhältniss zu den Gebirgsbildungen, etc. Bonn, 1826, 8°.

14. On the cause of the Temperature of Hot and Thermal Springs ; and on the bearings of this subject, as connected with the general question regarding the Internal Temperature of the Earth.— Edinb. New Phil. Journ. XX. p. 329; XXIII. p. 330; XXIV. pp. 132, 252.

15. Further Reasons against the Chemical Theory of Volcanos.— Edinb. New Phil. Journ. XXX. p. 14.

16. Die Bedeutung der Mineralquellen u. der Gaz-Exhalationen bei der Bildung u. Veränderung der Erdoberfläche, etc.—*Schw. Seid.* N. Jahrb. Phys. IV. p. 376.—*L.* u. *Br.* N. Jahrb. 1833, pp. 355, 558.

17. Ueber die Quellen-Verhältnisse des westlichen Abhanges vom Teutoburger Walde.—*Schweigg.* N. Jahrb. VIII. p. 249.---*L.* u. *Br.* N. Jahrb. 1834, p. 55.

18. Die Gletscher in ihren Beziehungen zur Hebung der Alpen, zur Kontraktion krystallinischer Formationen u. zu den erratischen Geschieben.—*L.* u. *Br.* N. Jahrb. 1843, p. 505.

19. On the Foundation of a New Geology.—L'Institut.—Edinb. N. Phil. Journ. XLIII. p. 304.

BISCHOFF (G.) et NÖGGERATH (J.).

1. Sur les Sels efflorescens des roches volcaniques, en particulier sur ceux du Trass des environs du lac de Laach et des laves de Bertrich.—*Die Gebirge im Rheinlande Westphalen*, IV. p. 238.— *Féruss.* Bull. 1827, X. p. 38.

BISCHOFF (G. W.), BLUM (R.), BRONN (H. G.), etc.

1. Naturgeschichte der drei Reiche. Stuttg. 1832 *et seq.* 8°.

BISCHOFF (J. R.).

1. Grundzüge der Naturlehre des Menschen, von seinem Werden bis zu dem Tode. Wien, 1838-39, 8°.—*Müll.* Arch. 1839, p. CXXIII.

BISCHOFF (Th. L. W.).

1. Beiträge zur Lehre v. den Eyhüllen des menschlichen Fœtus. Bonn, 1834, 8°. fig.—Bibl. med.-ch. p. 50.

2. Anatomisch-physiologische Untersuchungen über die Eihüllen des Menschen. Bonn, 8°.—*Müll.* Arch. 1835, p. 33.

3. Commentatio de novis quibusdam experimentis chemico physiologicis ad illustrandam doctrinam de respiratione institutis. Heidelb. 1837, 4°.— *Val.* Repert. III. p. 24.—*Müll.* Arch. 1838, p. CLIX.

4. Nervi accessorii Willisii Anatomia et Physiologia. Darmst. 1832, 4°. fig.—Isis, 1834, III. p. 324; 1835, I. p. 95.

5. Ueber den Bau des Crocodil-Herzens, besonders von *Crocodilus lucius.*—*Müll.* Arch. 1836, p. 1, fig.—*Wiegm.* Arch. 1837, VI. p. 230.

6. De verâ vasorum plantarum spiralium structurâ et functione Commentatio. Bonn, 1829, 8°. fig.—Isis, 1834, IX. p. 880.

7. Bericht über die Fortschritte der Physiologie im Jahre 1838; *Müll.* Arch. 1839, p. CXXIII.; im Jahre 1839, l. c. 1840, p. XCV.; im Jahre 1840, l. c. 1841, p. I.; im Jahre 1841, l. c. 1842, p. LXI.; im Jahre 1842, l. c. 1843, p. LXXXVIII.; im Jahre 1843, l. c. 1844, p. 68.

8. Beweis der von der Begattung unabhängigen periodischen Reifung u. Loslösung der Eier der Saügethiere u. des Menschen als der ersten Bedingung ihrer Fortpflanzung. 4°. Giessen, 1844.

9. Einige physiologisch-anatomische Beobachtungen an einem Enthaupteten.—*Müll.* Arch. 1838, p. 486.

10. Ueber den Bau der Magenschleimhaut.—*Müll.* Arch. 1838, p. 503, fig.

11. Sur le mouvement rotatoire qu'exécute le Vitellus de l'Œuf des Mammifères dans son passage à travers l'Oviducte.—Arch. f. Physiol. 1841, No. 1.—Ann. Sc. n. (2e S.) XVI. p. 298.

12. Ein Fall von *Trichina spiralis.* Heidelb. 1840, 8°.

13. Entwickelungsgeschichte der Saügthiere u. des Menschen. Leipz. 1842, 8°. (7ter Bd. der n. Aufl. von *Sömmerring*).—*Val.* Repert. VIII. pp. 16, 41.

14. *Lepidosiren paradoxa,* anatomisch untersucht und beschrieben. Leipz. 1840, 4°. fig.—Ann. Sc. n. (2ᵉ S.) XIV. p. 116, fig. (*Description anatomique du Lepidosiren paradoxa*).—*Müll.* Arch. 1840, p. CLXXIX.

15. Anatomisch-physiologische Bemerkungen.—*Müll.* Arch. 1838, p. 351.

16. Commentatio de novis quibusdam Experimentis chemico-physiologicis ad illustrandam doctrinam de Respiratione institutis. Heidelb. 1837.

17. Ueber das Drehen des Dotters im Säugethiereie während dessen Durchgang durch den Eileiter.—*Müll.* Arch. 1841, p. 14, fig.

18. Ueber electrische Ströme in den Nerven.—*Müll.* Arch. 1841, p. 20.

19. Observations sur le Détachement et la Fécondation de l'Œuf humain et des Œufs des Mammifères (Extr. d'une Lettre à *Mr. Bresch*).—Ann. Sc. n. (2ᵉ Sér.) XX. p. 93.

20. Entwickelungsgeschichte des Kaninchen-Eies (Gekr. Preisschr.). Braunschw. 1842, 4°. fig.—*Val.* Repert. VIII. pp. 17, 41.

21. Entwickelungsgeschichte des Hunde-Eies. 4°. Braunschw. 1845, fig.

22. Mémoire sur la maturation et la chûte périodique de l'Œuf de l'Homme et des Mammifères, indépendamment de la fécondation (Trad. de l'Allem.).—Ann. Sc. n. (3ᵉ S.) II. p. 104, fig.

23. Traité du Développement de l'Homme et des Mammifères, suivi d'une Hist. du Développement de l'Œuf du Lapin. Trad. de l'All. par *A. J. L. Jourdan.* Paris, 1843, 8°. Atl. 4°. (Encycl. Anatomique.)

24. Ueber die erste Bildung des Centralnervensystems bei Säugthieren, mit Berücksichtigung der Kritischen Beleuchtung meiner Beobachtungen durch *Hn. Dr. Reichert.*—*Müll.* Arch. 1843, p. 252.

BISHOP (John).

1. Experimental Researches into the Physiology of the Human Voice. 8°. Lond. 1836.—Cat. R. Soc. L.

BISOZZI (Giac.).

1. Die menschliche Stimme und ihr Gebrauch, für Sänger und Sängerinnen. Leipz. 1838, 8°.

BISSATI ( ).

1. Osservazione sulla educazione de Bachi da Seta.—Opusc. scelti XII. p. 179.—Bibl. Ent. I. p. 31.

BITTERMANN ( ).

1. Dissertatio de Vermibus. Vindob. 1763.

BITTNER (J.).

1. Aquæ naturales ad Aigen prope Salisburgum descriptæ. Salisb. 1601, 4°.—*Böhm.* Bibl. V. p. 302.

BIUMI (P. Jér.).

1. Esamina di alcuni Canaletti chiliferi che dal fondo del ventricolo per le tonache del omento sembrano penetrare nel fegato, etc. Milano, 1717, 8°.—Biogr. Un. IV. p. 534.

BIURSₗEDT (J.), HEDENBORG (J.), RUNGREN (J. E.), et WER-TERLING (C. U.).

1. Diss. entomologicæ de Hemipteris rostratis Capensibus. 4°. Upsal, 1822.

BIVONA (A. Bern.).

1. Nuovi generi e nuove specie di Molluschi.—Eph. scient. e litt. per la Sicil. Palermo, 1832, 8°.

BIVORT (J. B.).

1. Compte rendu de la Soc. du Hainaut ; *vide sup.* Pars I. p. 38.

BIWALD (Leop.).

1. Physica generalis et particularis, etc. præmissâ Dissert. de Studii physici naturâ ejusque perficiendi mediis. Græcii, 1767, 4°. fig. —*Böhm.* Bibl. I. 1, p. 309.

BIZET ( ).

1. Mémoire sur la Tourbe. Amiens, 1758, 12°.—*Böhm.* Bibl. IV. 1, p. 199.

BIZIO (Bart.).

1. Scoperta del Principio purpureo nei due *Murex brandaris* e *trunculus,* Linn.—Ann. Sc. R. Lomb.-Ven. 1833, p. 346.

2. Investigazioni chimiche sul *Murex brandaris.*—Ann. Sc. R. Lomb.-Ven. 1835.

BIZOT ( ).

1. Recherches sur le Cœur et le Système artériel chez l'Homme.— Mém. Soc. méd. d'Obs. 1836.

BLACK (J.).

1. On some appearances inferred to have been connected with the Antediluvian Congelation of the Interstitial Water of Rocks.—Edinb. New Phil. Journ. XXXI. p. 38.

2. On a Fossil Stem of a Tree recently discovered near Bolton-le-Moor.—Proc. Geol. Soc. II. 670.—Phil. Mag. ser. 3, XIII. p. 229.

3. On a Slab of New Red Sandstone from the Quarries at Weston, near Runcorn, Cheshire, containing the Impressions of Footsteps and other markings.—Journ. Geol. Soc. II. p. 65.

BLACKADDER (A.).

1. On the Superficial Strata of the Forth District.—Mem. Wern. Soc. V. p. 424.—*Féruss.* Bull. 1826, IX. p. 13.

BLACKWALL (J.).

1. On two Birds, hitherto uncharacterized, belonging to the genera *Crex* and *Rallus.*—Brewst. Journ. of Science, ser. 2, VI. p. 77.

2. Researches in Zoology. Lond. 1840, fig.—Rev. Zool. 1841, p. 202.

3. Remarks on the Pulvilli of Insects.—Trans. Linn. Soc. XVI. p. 487.—Phil. Trans. 1816, p. 323.

4. On the Diving of Aquatic Birds.—Phil. Mag. ser. 3, I. p. 23.

5. On the Manners of the Grenadier Grosbeak (*Loxia Oryx,* Linn.) when in captivity.—Phil. Mag. ser. 2, XI. p. 97.

6. Characters of some undescribed genera and species of *Araneidæ.*—Phil. Mag. ser. 3, III. pp. 104, 187, 344, 436; V. p. 50; VIII. p. 481; X. p. 100.—Isis, 1835, VI. p. 574.

7. On the Structure and Functions of Spiders.—Rep. Brit. Assoc. 1833, p. 444.

8. On the Manner in which the Geometric Spiders construct their Nets.—Zool. Journ. V. p. 181.—Isis, 1832, VI. p. 659.

9. On the House-Spider.—Phil. Mag. ser. 3, I. p. 95.

10. On a species of *Arachnida* belonging to the family *Araneidæ* (*Dysdera Latreillii*).—Phil. Mag. ser. 3, I. p. 190.

11. Characters of a new genus (*Deletrix*) and some undescribed species of *Araneidæ.*—Phil. Mag. ser. 3, X. p. 100.

12. On the Palpi of Spiders.—Rep. Brit. Assoc. 1842, Sect. p. 66.

13. Account of a species of Ichneumon whose Larva is parasitic on Spiders.—Rep. Brit. Assoc. 1842, Sect. p. 68.—Ann. and Mag. N. Hist. XI. p. 1.

14. Report on some recent Researches into the Structure, Functions and Œconomy of the *Araneidea* made in Great Britain.—Rep. Brit. Assoc. 1844, p. 62.—Ann. and Mag. N. Hist. XV. p. 221.

15. An Examination of *Virey's* observations on Aëronautic Spiders, published in the Bulletin des Scienc. Nat.—Phil. Mag. X. 1831, p. 180.

16. Observations and Experiments made with a view to ascertain the Means by which the Spiders that produce Gossamer effect their aërial excursions.—Tr. Linn. Soc. Lond. XV. 2, p. 449.—Isis, 1829, XII. p. 1277.—*Féruss.* Bull. 1829, XVIII. p. 131.

17. On the Means by which various Animals walk on the vertical Surfaces of highly-polished Bodies.—Ann. and Mag. N. Hist. XV. p. 115.

18. Descriptions of some newly discovered species of *Araneidea.*—Ann. and Mag. N. Hist. XIII. p. 179.

19. Notice of several cases of defective and redundant Organization observed among the *Araneidea.*—Ann. and Mag. N. Hist. XI. p. 165.

20. Notice of several recent Discoveries in the Structure and Œconomy of Spiders.—Trans. Linn. Soc. XVI. 1833, p. 471.

21. On the Mammulæ employed by Spiders in spinning.—Ann. of Nat. Hist. I. p. 478.

22. Descriptions of newly discovered Spiders.—Linn. Soc. June 1839.—Ann. of Nat. Hist. IV. p. 66.

23. Three tribes of *Araneidea* : 1. *Octonoculata* ; 2. *Senoculina* ; 3. *Binoculina.*—Linn. Soc. Apr. 1840.—Ann. and Mag. N. Hist. VI. p. 229.

24. A Catalogue of Spiders, either not previously recorded or little known as indigenous to Great Britain ; with Remarks on their Habits and Œconomy.—Linn. Soc. Apr. 1842.—Ann. and Mag. N. Hist. X. p. 407.

25. On a remarkable formation of the Bill observed in several species of Birds.—Mag. Nat. Hist. ser. 1, III. p. 402.

26. On a new species of *Lamprotornis.*—Brewst. Journ. of Science, ser. 2, V. p. 332.

27. Manners and Œconomy of the Pied Flycatcher.—Mag. Nat. Hist. ser. 1, p. 331.

28. On the capture of *Chrysomela cerealis* in N. Wales.—Mag. Nat. Hist. ser. 1, IV. p. 23.

29. Facts relating to the Natural History of the Cuckoo.—Zool.

Journ. IV. p. 294.—Isis, 1831, I. p. 106.—*Féruss.* Bull. XXII. p. 126.

30. Observations tendantes à compléter l'histoire du Coucou.—Tr. Phil. Soc. of Manch. (2ᵉ S.) IV.—*Féruss.* Bull. 1826, VIII. p. 106.

31. Observations on a newly described species of Swan (*Cygnus Bewickii*).—Zool. Journ. V. p. 189.—Isis, 1832, VI. p. 661.

32. On an undescribed Bird of the family *Falconidæ.*—Phil. Mag. ser. 2, X. p. 264.

33. On the Notes of Birds.—Mem. Lit. and Phil. Soc. Manchester, IV.—Phil. Mag. LXVI. p. 14.—Isis, 1834, V. p. 469.

34. Notes on the Salmon.—Ann. and Mag. of N. Hist. XI. p. 409.

35. On the Instincts of Birds.—Mem. Lit. and Scient. Soc. Manchester, ser. 2, V.—Edinb. New Phil. Journ. XIV. p. 241.— *Féruss.* Bull. 1826, VII. p. 97.

36. Extracts from a Zoological Journal kept at Crumpsall, near Manchester.—Zool. Journ. V. p. 10.

BLADH (A. J.).

1. Fundamenta Entomologiæ dissert. sub præsid. *Linnæi.* Upsaliæ, 1767, 4º.—Amœnit. Acad. VII. p. 129.

BLADON (J. B.).

1. On the Theory of Spontaneous Generation.—Mag. Nat. Hist. ser. 2, IV. pp. 280, 339.

BLAES ( ).

1. Mémoires pour servir à l'histoire naturelle des Animaux. Paris, 1676, fol.—Dict. Sc. nat. XXII. p. 505.

BLAINVILLE (H. D. de).

1. Uebersicht der vorzüglichsten Arbeiten in den Naturwissenschaften, während des Jahres 1817.—J. de Phys. 1818.—Isis, 1819, I. p. 8.

2. Bericht über Zoologie, Anatomie und Physiologie, im Jahr 1818. —Isis, 1820, Litt. Anz. p. 409.

3. Bericht über die naturwissenschaftlichen Arbeiten im Jahr 1819. Isis, 1821, Litt. A. p. 1.

4. Essai sur une nouvelle Classification des Animaux.—Bull. Soc. Philom. 1816.—Ann. Sc. n. II.

5. Prodrome d'une nouvelle Distribution systématique du Règne animal.—J. de Phys. LXXXIII. p. 244.—N. Bull. Soc. Phil. 1816, p. 105.—Bibl. Ent. I. p. 31.

6. Rapport sur le Mémoire de *M. Jacobson* ayant pour titre : Observations sur le développement prétendu des œufs des Mulettes et des Anodontes dans leurs branchies.—Ann. Sc. n. Mai 1828, p. 22.—*Féruss.* Bull. 1828, XIV. p. 369.—Isis, 1830, II. p. 217.

7. Mémoire sur la nature du produit femelle de la Génération dans l'Ornithorynque.—N. Ann. Mus. II. p. 369, fig.

8. Rapport sur un Mémoire de *M. de Quatrefages* intitulé : Sur la vie intrabranchiale des petites Anodontes.—Ann. Sc. n. 1835, IV. p. 283.

9. Anatomie des Coquilles polythalames siphonées récentes, pour éclaircir la structure des espèces fossiles.—N. Ann. Mus. III. p. 1, fig.

10. Sur l'Anatomie des Bivalve de *Bojanus.*—Journ. de Phys. Août 1819.

11. Quelques observations sur l'animal de la Spirule, et sur l'usage du Siphon des Coquilles polythalames.—Ann. d'Anat. I. p. 369.

12. Rapport sur un Mémoire de *M. Deshayes*, relatif à l'anatomie du genre Ambrette.—Ann. Sc. n. 1830, XXI.—Rev. Bibl. p. 109.

13. Lettre à MM. les Rédacteurs des Annales d'Anatomie et de Physiologie sur la Poulpe de l'Argonaute.—Ann. d'Anat. I. p. 188, fig.

14. Sur le Cachalot.—Ann. d'Anat. II. p. 335, fig.

15. Mémoire sur le Squale Pélerin.—Ann. Mus. XVIII. p. 88, fig.

16. De l'Organisation des Animaux, ou Principes d'Anatomie comparée. Paris, 1822, 8°.—Isis, 1823, IX. p. 971.—*Féruss.* Bull. 1823, I. p. 371.

17. Sur un nouveau caractère ostéologique servant à distinguer les Animaux quadrupèdes en deux Sections.—Journ. de Phys. LXXXIX. p. 157.—Bull. Soc. Phil. 1819, p. 41.—Isis, 1820, V. Litt. A. p. 509.

18. Considérations générales sur le Système nerveux.—Bull. Soc. Phil. 1821, p. 39.

19. Sur la dégradation du Cœur et des gros Vaisseaux dans les Ostéozoaires, ou Animaux vertébrés.—Bull. Soc. Phil. 1819, p. 146.

20. Ostéographie, ou Description iconographique comparée du Squelette et du Système dentaire des cinq classes d'Animaux vertébrés récens et fossiles, pour servir de base à la Zoologie et à la

Géologie. Par. 1839 à suiv. 4°. et fol.—*Val.* Repert. V. p. 19 ; VI. p. 27 ; VII. p. 35 ; VIII. p. 35.—Rev. Zool. 1839, pp. 63, 335 ; 1840, p. 360.

21. Sur l'emploi de l'Opercule dans l'établissement des genres de Coquilles univalves.—Bull. Soc. Philom. 1825.

22. Cours de Physiologie générale et comparée, publ. par *Hollard.* Paris, 1833, 3 vols. 8°.

23. Sur quelques Crânes d'Hommes trouvés en Allemagne, et description d'une Tête de Momie.—Journ. de Phys. XCIV. 1822, p. 396.

24. Sur quelques Crânes de Phoques et les espèces de ce genre.— Journ. de Phys. XCI. 1820, pp. 286, 419.

25. Rapport sur les Collections zoologiques recueillies par *M. Ad. Delessert* dans les Indes-Orientales.—C. R. XI. p. 385.

26. Note sur les Vertèbres cervicales de l'Aï (*Bradypus tridactylus,* L.).—C. R. IX. p. 762.

27. Communication des Veines avec les Vaisseaux lymphatiques.— Ann. Sc. n. 1829, XVIII.—Rev. Bibl. p. 120.

28. Note sur la forme des extrémités articulaires du corps des Vertèbres dans les Ostéozoaires ou Vertébrés.—Ann. d'Anat. I. p. 138, fig.

29. Sur quelques anomalies du Système dentaire dans les Mammifères. Paris, 1838, 8°. fig.—Ann. d'Anat. I. p. 285, fig.

30. Sur le Système dentaire du *Sorex aquaticus,* ou du genre *Scalops.*—Bull. Soc. Phil. 1820, p. 130.—Rev. Zool. 1838, p. 81.

31. Mémoire sur les Lernées.—Journ. de Phys. XCV. 1822, pp. 372, 437.—*Féruss.* Bull. 1823, II. p. 462 ; III. pp. 265, 418.

32. Note sur la structure et l'analogie de la Plaque dorso-céphalique des Rémoras ou Echéneis.—Bull. Soc. Phil. 1822, p. 119.

33. Beschreibung der Geschlechtstheile des *Ornithorhynchus.*—N. Ann. Mus. 1833.—*Müll.* Arch. 1835, p. 36.

34. Sur les Mammelles de l'Ornithorinque femelle, et sur l'Ergot du mâle.—N. Bull. Soc. Philom. 1826, p. 138.—*Féruss.* Bull. 1827, XI. p. 107.

35. Ueber den Kiemendeckel der Fische.—Bull. d. Sc. Juill. 1817. —Isis, 1818, VIII. p. 1412.

36. Note sur la différence des sexes dans une espèce de Gélasime. —*Féruss.* Bull. 1828, XV. p. 180.

37. Note sur la forme des extrémités du corps des Vertèbres.— Ann. Sc. n. 1837, VIII. p. 58.

38. Sur l'existence de véritables Ongles à l'aile de quelques espèces

d'Oiseaux.—Journ. de Phys. LXXXIX. p. 156.—Bull. Soc. Phil. 1819, p. 41.—Isis, 1820, V. Litt. A. p. 508.

39. Sur l'organe appellé Galète (*Galea*) dans les Orthoptères.—N. Bull. Soc. Phil. 1820, Juin, p. 85.—Bibl. Ent. I. p. 31.

40. Sur la cause organique de la ponte du Coucou dans un nid étranger.—Ann. d'Anat. I. p. 249.

41. Influence de la Circulation sur la Chaleur animale.—Ann. Sc. N. 1831, XXIII.—Rev. Bibl. p. 54.

42. Recherches sur l'ancienneté des Mammifères insectivores à la surface de la terre; précédées de l'histoire de la science à ce sujet, des principes de leur classification et de leur distribution géographique actuelle.—Ann. Sc. n. 1838, X. p. 118.—C. R. VI. p. 738.

43. Note sur les Carnassiers insectivores.—Ann. d'Anat. I. p. 315.

44. Sur la distribution géographique des Mammifères primates (Quadrumanes).—Ann. d'Anat. II. p. 358.

45. Recherches sur l'ancienneté des Cheiroptères ou des animaux de la famille des Chauve-souris à la surface de la terre, précédées de l'histoire de la science à leur sujet, des principes de leur classification et de leur distribution géographique actuelle.—Ann. Sc. nat. 1838, IX. p. 357.—*L. u. Br.* N. Jahrb. 1843, p. 854.—Ed. New Phil. Journ. XXV. p. 21.

46. Ueber mehrere Arten Säugethiere aus der Sippschaft der Wiederkäuer.—Bull. d. Sc. 1816, p. 73.—Isis, 1819, VII. p. 1090.

47. Rapport sur un Mémoire de *M. Jourdan*, concernant deux nouvelles espèces de Mammifères de l'Inde.—Ann. Sc. n. 1837, VIII. p. 270.

48. Ueber die verschiedenen Arten von Nashörnern.—Bull. d. Sc. 1818.—Isis, 1819, II. p. 264.

49. Orang-Outang.—Ann. Sc. n. V. p. 60.—Isis, 1819, I. p. 133.—*Wiegm.* Arch. 1837, II. p. 146.—C. R. 1836, II. p. 73.

50. Grauer Bär von America.—Bull. d. Sc. 1811, Nov.—Isis, 1818, VII. p. 1197.

51. *Rupicapra americana.*—Bull. d. Sc. 1817, p. 157.—Isis, 1819, VII. p. 1102.

52. Ueber den *Wapiti*, eine Hirschart aus dem nördlichen America. —Bull. d. Sc. 1817, p. 37.—Isis, 1819, VII. p. 1098.

53. Note sur un Cétacé échoué au Hâvre, et sur un Ver trouvé dans sa graisse.—Bull. Soc. Philom. Sept. 1825.—*Féruss.* Bull. 1826, VII. p. 370.

54. *Bradypus ursinus,* Shaw.—Bull. d. Sc. XVII.—Isis, 1818, II. p. 319.

55. Sur une nouvelle espèce de Rongeur fouisseur du Brésil.—Bull. Soc. Philom. Avril 1826, p. 62.—Ann. Sc. n. IX. p. 97.—*Féruss.* Bull. 1826, VIII. p. 266.

56. Description de l'Écureuil à bandes (*Sciurus vittatus,* Desm.).— Bull. Soc. Philom. 1820, p. 116.

57. Sur l'espèce de Rongeur à laquelle *Shaw* a donné le nom de *Mus bursarius.*—Bull. Soc. Philom. 1821, p. 138.

58. Plan d'un cours de Physiologie générale et comparée, fait à la Faculté des Sciences de Paris pendant les années 1829–1832.— Rev. Zool. 1839, p. 235.

59. Dissertation sur la place que la famille des Ornithorhinques et des Echidnés doit occuper dans la Série naturelle. Paris, 1812, 4°.

60. Observations sur l'organe appellé Ergot dans l'Ornithorinque.— Journ. de Phys. LXXXIV. 1817, p. 318.

61. Sur l'emploi et la forme du Sternum et de ses annexes pour l'établissement ou la confirmation des familles naturelles parmi les Oiseaux.—Journ. de Phys. XCII. 1821, p. 185.

62. Mémoire sur les caractères distinctifs des espèces de Cerfs.— Journ. de Phys. XCIV. 1822, p. 254.—*Féruss.* Bull. 1823, II. p. 60.

63. Sur la structure de la Plaque dorso-céphalique du Rémora ou Echéneis.—Journ. de Phys. XCV. 1822, p. 132.

64. Mémoire sur les traces qu'ont laisseés à la surface de la terre, les Édentés terrestres.—Rev. Zool. 1839, p. 33.—C. R. VIII. p. 139.

65. Mémoire sur l'ancienneté des Mammiferès du sous-ordre des Édentés terrestres à la surface du globe.—Rev. Zool. 1839, p. 2. —C. R. VIII. p. 65.

66. Rapport sur les résultats zoologiques du Voyage autour du Monde de la Bonite.—Rev. Zool. 1838, p. 51.—C. R. VI. p. 445.

67. Rapport sur un nouvel envoi d'Ossemens fossiles des environs d'Auch, fait par *M. Lartet.*—C. R. VI. p. 889.

68. Rapport sur des Ossemens fossiles de Mammifères présentés par *M. d'Hombres-Firmas.*—C. R. XI. p. 13.

69. Rapport sur deux Mémoires de *M. Puel,* concernant des Ossemens fossiles de Rennes et de divers autres animaux, trouvés dans le Dépt. du Lot.—C. R. XI. p. 390.

70. *Ctenomys brasiliensis.*—Ann. Sc. n. IX. p. 97.—Isis, 1834, IX. p. 897.

71. Quelques observations sur la distinction des espèces en Ornithologie.—Bull. Soc. Philom. 1826, Oct. p. 156.—*Féruss.* Bull. 1828, XIII. p. 432.

72. Mémoire sur le Ganga, ou Gelinotte des Pyrénées (*Tetrao alchata*, L.).—Anal. des trav. de l'Acad. pendant l'année 1829, p. 100.—*Féruss.* Bull. 1830, XXII. p. 122.

73. Sur le Fou de Bassan, *Sula alba*, Meyen.—Bull. Soc. Philom. 1826, Janv.—*Féruss.* Bull. 1827, X. p. 155.

74. Sur la place du Touraco dans la classe des Oiseaux.—Bull. Soc. Philom. 1826, Mars.—*Féruss.* Bull. 1826, IX. p. 93.

75. Mémoire sur le genre *Chionis*, ou Bec-en-fourreau, et sur la place qu'il doit occuper dans le Système ornithologique.—C. R. III. p. 155.—Ann. Sc. n. 1836, VI. pp. 97, 123.

76. Mémoire sur le Dronte ou Dodo (*Didus ineptus*, L.).—Ann. Sc. n. XXI. p. 109.—N. Ann. Mus. IV. p. 1, fig.—*Wiegm.* Arch. 1836, II. p. 270.

77. Sur la patrie du Choquard, ou Choucas des Alpes (*Corvus pyrrhocorax*, L.).—Bull. Soc. Philom. 1821, p. 140.

78. Remarque qu'un certain nombre d'Ophidiens offre de véritables ongles.—Pr. d'Anat. comp. I. p. 14.—*Féruss.* Bull. 1826, VII. p. 445.

79. Observations sur plusieurs Serpens du genre Pithon, vivant à Paris dans le mois de Janvier 1823.—Journ. de Phys. XCVI. p. 271.—Bull. Soc. Philom. 1823, p. 49.

80. Description de quelques espèces de Reptiles de la Californie, précédée de l'analyse d'un Système général d'Erpétologie et d'Amphibiologie.—N. Ann. Mus. IV. p. 233, fig.—*Wiegm.* Arch. 1836, II. p. 256.

81. Note sur un Crocodile du Nil, vu vivant à Paris en Janvier 1823.—J. de Phys. 1823, Mai, p. 282.—Bull. Soc. Philom. Févr. 1823, p. 24.—*Féruss.* Bull. 1824, II. p. 83.

82. Rapport sur le *Sarcoptes exulcerans.*—Instit. No. 74, p. 330.—*Fror.* Notiz. XLII. No. 11.—*Wiegm.* Arch. 1835, I. p. 354.

83. Rapport sur le Ciron de la Gale (*Acarus scabiei*).—N. Ann. Mus. IV. p. 213.

84. Sur la concordance des Anneaux des Entomozoaires hexapodes adultes.—N. Bull. Soc. Phil. 1820, Mars, p. 33.—Bibl. Ent. I. p. 31.

85. Essai d'une Monographie de la famille des Hirudinées. Paris, 1827, 8°.—*Cuv.* R. An. III. p. 336.

86. Note sur la génération de l'Hydre verte.—Bull. Soc. Philom. 1826, Mai, p. 77.—*Féruss.* Bull. 1826, IX. p. 375.

87. Ueber die Classe der Setipoden, eine Abtheilung der rothblütigen Würmer.—Bull. d. Sc. 1818.—Isis, 1818, XII. p. 2061.

88. Sur quelques petits Animaux qui, après avoir perdu le mouve-

ment par la dessication, le reprennent comme auparavant quand on vient à les mettre dans l'eau.—Ann. Sc. n. IX. p. 104.—N. Bull. Soc. Philom. Juin 1826.—*Féruss.* Bull. 1827, X. p. 425.— Isis, 1834, IX. p. 897.

89. Manuel de Malacologie et de Conchyliologie, etc.   Paris, 1825, 8ᶜ **ad.** et fig.—*Féruss.* Bull. 1826, VIII. p. 127.—Isis, 1829, X. p. 1109.

90. Faune française, ou Histoire générale et particulière des Animaux qui se trouvent en France, etc. 18ᵉ Livr. Mollusques. Paris, 1828.—*Féruss.* Bull. 1829, XVI. p. 117.

91. Methodische Classification der Mollusken.—Bull. d. Sc. 1814. —Isis, 1818, X. p. 1676.

92. Disposition méthodique des espèces récentes et fossiles des genres Pourpre, Ricinule, Licorne et Concholépas de *M. de Lamarck*, et Description des espèces nouvelles ou peu connues, faisant partie de la collection du Muséum d'Hist. nat. de Paris.—N. Ann. Mus. I. p. 189, fig.—*L.* u. *Br.* N. Jahrb. 1834, p. 375.

93. Sur la *Patella distorta* de *Montagu.*—Bull. Soc. Phil. 1819, p. 72.

94. Sur l'animal de la *Patella ombracula* de *Chemnitz.*—Bull. Soc. Phil. 1819, p. 178.

95. Mollusken aus der Ordnung der *Pterodibranchia.*—Bull. d. Sc. 1816.—Isis, 1818, X. p. 1682.

96. Mollusken aus der Ordnung der Polybranchen.—Bull. d. Sc. 1816.—Isis, 1818, X. p. 1685.

97. Mollusken aus der Ordnung der Cyclobranchen.—Bull. d. Sc. 1816.—Isis, 1818, X. p. 1687.

98. Structure des Coquilles à plusieurs chambres.—N. Ann. Mus. III. p. 1, fig.—*Wiegm.* Arch. 1835, I. p. 329.

99. Mémoire sur les espèces du genre Calmar (*Loligo*, Lam.).— Journ. de Phys. 1823, Mars, p. 116.—*Féruss.* Bull. 1824, III. p. 90.

100. Mémoire sur le genre *Hyale.*—Journ. de Phys. XCIII. 1822, p. 81.

101. Observations sur les différences de la coquille d'individus de sexes différens dans les Mollusques céphalés.—Journ. de Phys. XCIV. p. 92.

102. Remarques à l'occasion de la Lettre de *M. Baillon* sur le Cygne de Bewick.—C. R. VII. p. 1022.

103. Note sur l'analogue du Peigne des Oiseaux dans les Reptiles et les Poissons.—Journ. de Phys. XCV. 1822, p. 72.

104. Rapport sur l'importance des résultats obtenus par *M. Lartet* dans les fouilles qu'il a entreprises pour rechercher des Ossemens fossiles.—Rev. Zool. 1838, p. 132.—C. R. VII. p. 100.

105. Rapport sur les Myodaires de *Robineau-Desvoidy.* Paris, 1826, 8°.

106. Note sur les Animaux articulés.—Journ. de Phys. LXXXIX. 1819, p. 467.

107. Sur l'animal de la Spirule.—Ann. d'Anat. et de Phys. III. p. 82.—Rev. Zool. 1839, p. 244.

108. Note sur le *Stylophorus chordatus* de *Shaw.*—Journ. de Phys. LXXXVII. 1818, p. 68.

109. Recherches sur l'ancienneté des Édentés terrestres à la surface de la terre.—Ann. Sc. n. sér. 2, XI. p. 113.

110. Sur le Poulpe habitant de l'Argonaute.—Journ. de Phys. LXXXV. p. 72 ; LXXXVI. pp. 366, 434 ; LXXXVII. p. 47.

111. Note sur le Chameau fossile et sur le *Sivatherium* des Sous-Himalayas méridionaux.—C. R. IV. pp. 71, 166, fig.

112. Note sur la tête de *Dinotherium giganteum* apportée à Paris par *MM. Kaup* et *Klipstein.*—C. R. IV. pp. 421, 427.

113. Rapport sur la découverte de plusieurs Ossemens fossiles de Quadrumanes dans le Dépôt tertiaire de Sansan, près d'Auch, par *M. Lartet.*—C. R. IV. p. 981.—Rev. Zool. 1838, p. 101.

114. Rapport sur un Mémoire de *MM. de Laizer* et *de Parieu,* concernant un nouveau Carnassier fossile (*Hyænodon lep o rhychus*).—C. R. VII. p. 1004.

115. Rapport en réponse à une Lettre du Ministre de l'Instr. publ. concernant de nouvelles fouilles à faire dans la Caverne à ossemens de Fouvent-le-Bas.—C. R. VII. p. 1014.

116. Note sur l'organisation de l'animal de l'Ampullaire.—Journ. de Phys. XCV. 1822, p. 459.—*Féruss.* Bull. III. p. 288.

117. Notice sur l'animal du genre Scarabæus de Denys de Montfort (*Helix scarabæus,* Linn.).—Journ. de Phys. XCIII. 1822, p. 304.

118. Mémoire sur l'organisation d'une espèce de Mollusque nu, de la famille des Limacinés.—Journ. de Phys. XCVI. 1823, p. 175. —*Féruss.* Bull. IV. p. 45.

119. Principes d'Anatomie comparée. 2 vols. 8°. Paris, 1822-23.

120. Rapport sur la partie zoologique du Voyage de *l'Astrolabe* et *la Zélée.*—C. R. 4 Oct. 1841.—Rev. Zool. 1841, p. 318.

121. Rapport sur plusieurs mémoires de Paléontologie.—C. R. X. p. 925, 1840.—Rev. Zool. 1840, p. 187.

122. Description et détermination d'une mâchoire appartenant à un

Mammifère jusqu'à présent inconnu, *Hyænodon leptorhynchus.*—Rev. Zool. 1838, p. 306.

123. Doutes sur le prétendu Didelphe fossile de Stonefield, ou à quelle classe, à quelle famille, à quel genre doit on rapporter l'animal auquel ont appartenu les ossemens fossiles, à Stonefield, désignés sous les noms de *Didelphis Prevostii* et *Didelphis Bucklandii* par les paléontologistes.—Rev. Zool. 1838, pp. 161, 244.—C. R. VII. pp. 402, 727, 749.—Mag. Nat. Hist. ser. 2, II. p. 639; III. p. 49.

124. Rapport sur un Mémoire de *M. le Dr. Bazin*, sur la structure intime des Poumons chez les Animaux vertébrés.—Ann. Sc. n. (2ᵉ S.) XII. p. 148.—Rev. Zool. 1839, p. 250.—C. R. IX. p. 234.

125. Notice historique sur la place assignée aux Cécilies dans la Série zoologique.—Ann. Sc. n. (2ᵉ S.) XII. p. 360.—C. R. IX. p. 663.

126. Rapport sur un Mémoire de *M. Dufo*, intitulé : Observations sur les Mollusques marins, terrestres et fluviatiles des îles Séchelles et des Amirantes.—Ann. Sc. n. (2ᵉ S.) XIII. p. 198.—C. R. X. p. 392.

127. Rapport sur un Mémoire de *Jacobson* intitulé : Observations sur le Développement prétendu des Œufs des Moulettes, etc. Paris, 1827, 4°.

128. Notice sur le cinquième fascicule de l'Ostéographie des Vertébrés.—C. R. X. 1840, p. 890.—Rev. Zool. 1840, p. 181.—(Sur le 6ᵉ fascicule) C. R. XI. p. 12.—(7ᵉ fascic.) C. R. XI. p. 802.

129. Note sur le Plan de son Ostéographie des Primates.—C. R. IX. p. 782.

130. Rapport sur un Mémoire de *M. Foville*, concernant la structure du Cerveau et ses rapports avec le Crâne.—C. R. X. p. 734.

131. Rapport sur les résultats concernant l'Histoire naturelle obtenus dans l'Expédition de la *Vénus.*—C. R. XI. p. 339.

132. Note sur de prétendues Empreintes de pieds d'un Quadrupède, dans le Grès-bigarré de Hildburghausen en Saxe.—C. R. 1836, II. p. 454.

133. Rapport sur un Mémoire de *M. Jules de Christol* intitulé : Recherches sur divers Ossemens fossiles attribués par Cuvier à deux Phoques, au Lamantin et à deux espèces d'Hippopotames, et rapportés au Metaxytherium, nouveau genre de Cétacés de la famille des Dugongs.—Rev. Zool. 1841, p. 86.

134. Description du Polype *Cristatella Mucedo.*—Dict. Sc. n. XI. p. 611.

135. Malacozoaires et Poissons de la Faune française. Paris, 8°. 1820-30.

136. Prodrome d'une Monographie des Ammonites. 8°.

137. Sur les Ichthyolites ou les Poissons fossiles. Paris, 1818, 8°. —(Germ. *J. F. Kruger*): *Die versteinerten Fische, geologisch geordnet*. Quedl. 1822.—*Féruss*. Bull. 1824, II. p. 94.

138. Rapport sur une Note de *M. Rang* concernant la Poulpe de l'Argonaute.—Ann. Sc. n. 1837, VII. p. 172.—C. R. IV. p. 602.

139. Manuel d'Actinologie et de Zoophytologie. Paris, 1834, 8°. Atl.—Ann. Sc. n. 1835, III. p. 255.—*Wiegm.* Arch. 1835, I. p. 302.

140. Animal de la Bélemnite.—Ann. Sc. n. 1830, XXI.; Rev. Bibl. p. 140.

141. Bericht über die Geologie im Jahr 1818.—Isis, 1820, Litt. Anz. p. 404.

142. Mémoire sur l'Animal fossile d'Eichstœdt.—Bull. Soc. Philom. 1822, p. 101.

143. Note sur une tête de Chameau fossile trouvée dans le Grès des Sous-Himalaya.—Ann. Sc. n. 1836, VI. p. 317.—C. R. III. p. 528.

144. Mémoire sur les Ossemens fossiles attribués au prétendu géant *Theutobochus,* roi des Cimbres.— C. R. IV. p. 633.—N. Ann. Mus. IV. p. 38, fig.—*L.* u. *Br.* N. Jahrb. 1835, p. 498.

145. Sur les Ichthyolites. Paris, 1818, 8°.—Tr. Linn. Soc. Lond. XII. 2, p. 589.

146. Rapport sur un Mémoire de *M. Deshayes* intitulé : Observations générales sur le genre Bélemnite.—Ann. Sc. n. 1836, VI. p. 364.—C. R. III. p. 690.

147. Ueber die Belemniten.—Ann. Sc. n. VII. p. 428.—Isis, 1834, VIII. p. 855.—*Leonh.* Zeitschr. 1826, II. p. 286.

148. Mémoire sur les Bélemnites, considérées zoologiquement et géologiquement. Paris, 1827, 4°. fig.—*Féruss.* Bull. 1826, IX. p. 367 ; 1829, XVIII. p. 136.

149. Rapport sur des ossemens d'Éléphans provenant d'un terrain attenant à l'hospice Necker.—Rev. Zool. 1838, p. 308.—C. R. VII. p. 1051.

150. Tableau analytique des subdivisions du Règne animal.—Journ. de Phys. 1816, p. 151.

BLAINVILLE (De) et EDWARDS.

1. Mémoire sur les Animaux vivans trouvés dans les corps solides, et observations à ce sujet, communiquées à la Société Philomatique.—*Féruss.* Bull. 1827, p. 129.

BLAINVILLE (De).

1. Travels through Holland, Germany, Switzerland, specially Italy; translated from the Author's own Manuscript by *D. Turnebull* and others. Lond. 1742–45, 3 vols. 4°. fig.; 1767, 4°.—(Germ.) übers. v. *J. Tob. Köhler.* Lemgo, 1764–66, 4°. 3 vols.—(Fortsetz.) *Zu des Hrn von Blainville Reisebeschreibung, besonders durch Italien, erster Zusatz, v. Ed. Wright; aus dem Engl. übers. u. mit vielen Anmerk. vermehrt v. Köhler.* IV. Bd. Lemgo, 1768, 4°.—*Zweiter u. letzter Zusatz* (V. Bd.). Lemgo, 1767, 4°.—*Böhm.* Bibl. I. 1, p. 462.

BLAIR (P.).

1. On the Organ of Hearing in the Elephant.—Phil. Trans. XXX. p. 859.—Biogr. Un. IV. p. 552.—*Böhm.* Bibl. II. 1, p. 298.

2. Osteographia Elephantina: or, A full and exact description of all the Bones of an Elephant.—Phil. Trans. XXVII. p. 52.—Biogr. Un. IV. p. 552.—*Böhm.* Bibl. II. 1, p. 298.

BLAIR (T.).

1. On the Habits of *Testacellus scutulum.*—Mag. Nat. Hist. ser. 1, VI. p. 43.

BLAKE (J. H.).

1. Meteorological Observations made at the Mines of San Fernando, in the Partido de la Manicaraqua, Isl. of Cuba.—*Sill.* Am. J. XLII. p. 292.

2. On Coal Mines in Cuba.—*Sill.* Am. Journ. XLII. p. 388.

BLAKE (Rob.).

1. Disputatio medica inaug. de Dentium formatione et structurâ in Homine et in variis Animalibus. 8°. Edinb. 1798.—Cat. R. Soc. Lond.

BLAKIE (J.).

1. On the Slate Quarries of Aberdeenshire.—Prize Essays Highland Soc. X. p. 98.

BLANC ( ).

1. Sur les Travaux au Glacier de Giétroz. Laus. 1825, 8°.—Verz. Bibl. d. Schw. Naturf. Ges.

BLANC (L. G.).

1. Handbuch des Wissenswürdigsten aus der Natur. 8°. Halle, 1841.

BLANCAERT (St.).

1. Anatomia nova reformata, seu Prosectio corporis humani, veris-
simis ac curiosissimis ævi hujus observatis superstructa, multisque
figuris æneis exornata. Nec non Tractatus de Balsamatione corporum nunquam anteà ità evulgata. Amst. 1686, 8°.—Leyd.
1665, 8°.—Leipz. 1691, 4°. (Allem.)—Amst. 1688 (Franç.)—
Lond. 1690 (Angl.).—Act. Erudit. XIV. p. 389; XVI. p. 162.—
Biogr. Un. IV. p. 560.

2. De Circulatione Sanguinis per fibras, et de Valvulis in iis repertis.
Amst. 1676, 12°.—Biogr. Un. IV. p. 560.

3. Schouburg der Rupsen, Wormen, Maden, en vliegende Dierkens
daaruit voortkomende. Amst. 1688, 8°. fig.—(Germ.) Schau-
platz der Raupen, etc., aus dem Niederländ. übers. durch J. Chr.
Rodoch. Leipz. 1690, 8°. fig.—Böhm. Bibl. II. 2, p. 128.—
Theatrum Erucarum, Vermium, Termitum, et quæ ex his oriuntur
Bestiolarum volatilium, propriâ experientiâ collectum. Amst.
1688, 8°. fig.—Act. Erudit. IX. p. 55.—Trans. Linn. Soc. Lond.
V. p. 278.

4. Anatomia practica rationalis, sivè variorum Cadaverum morbis
denatorum anatomica inspectio. Amst. 1688, 12°.—(Deutsch.)
Hannover, 1692, 8°.—Biogr. Un. IV. p. 561.

BLANCANUS (Jac.).

1. Iter per montana quædam Agri Bononiensis loca.—Act. Bonon.
V. 1, p. 18, 2, p. 151, fig.—Mod. Bibl. Helm.

2. De quibusdam animalium exuviis lapidefactis (in agro Bononiensi).—Comm. Bonon. IV. C. p. 41; O. p. 133.

BLANCHARD (Em.).

1. Monographie du genre Ommexecha, de la famille des Acridiens.
—Ann. Soc. Ent. Fr. V. p. 603, fig.—Erichs. in Wiegm. Arch.
1837, VI. p. 309.

2. Monographie du genre Phoraspis, de la famille des Blattiens;
précédée de quelques Observations sur les Blattes des anciens.—
Ann. Soc. Ent. Fr. VI. p. 271, fig.

3. Beitrag zur Naturgeschichte der Gattung Cantharis.—Guér.
Mag. de Zool. IX. p. 165, fig.—Erichs. in Wiegm. Arch. 1837,
VI. p. 294.

4. Abbildung von der Larve und Nymphe von Staphylinus olens.
Guér. Mag. de Zool.—Erichs. in Wiegm. Arch. 1837, VI. p. 293.

5. Mantis chlorophæa.—Guér. Mag. de Zool. IX. p. 135.—Burm.
in Wiegm. Arch. 1836, II. p. 323.

6. Note sur l'Ascalaphe italique.—Bull. Soc. Linn. Bord. I. p. 40.—
Féruss. Bull. 1826, IX. p. 123.

7. Sur la coquille de l'Argonaute.—Bull. Soc. Linn. Bord III. 1829, p. 195.—*Féruss.* Bull. 1829, XIX. p. 120.

8. Histoire naturelle des Insectes (*S. à B. Dumeril*). III. Orth. Neur. Hémipt. Hyménopt. Lépid. et Dipt. Par. 1840.

9. Hist. des Insectes, traitant de leurs mœurs, etc. 8°. Paris, 1845.

10. Notice sur les Métamorphoses des Coléoptères du genre Téléphore.

11. Extrait d'un nouvel Ouvrage sur les *Hémiptères hétéroptères*, par *Max. Spinola.*—Ann. Sc. n. (2e Sér.) 1839.

12. Le Buffon de la jeunesse. Paris, 1804, 5 vols. fig.

13. De la distribution géographique des Animaux articulés.—C. R. —Rev. Zool. 1841, p. 205.

BLANCHET (Rod.).

1. Le Mécanisme des Sensations. Laus. 1843, 8°.

2. Aperçu de l'Histoire géologique des Terrains tertiaires du Canton de Vaud. Vevey, 1843, 8°.

3. Notice sur l'Histoire naturelle des environs de Vevey. 8°.

4. Essai sur l'Histoire naturelle des environs de Vevey. Vev. 1843, 8°.

BLANCHET ( ) u. SELL ( ).

1. Zusammensetzung einiger organischen Substanzen. Heidelb. 1833, 8°.—Verz. Bibl. d. Schw. Naturf. Ges.

BLAND (R.).

1. Observations on Human and Comparative Parturition. 8°. Lond. 1794.—Cat. R. Soc. Lond.

BLAND ( ).

1. Traité élémentaire de Physiologie physique. 3 vols. 8°.

BLAND (W.).

1. On Delhi Point, Pulo-Tinghie. etc., and on some Pelagic Fossil Remains found in the Rocks of Pulo-Lédah.—J. A. S. B. V. p. 575.

2. On the genus *Pterocyclos* of *Mr. Benson* and *Spiraculum* of *Mr. Pearson.*—J. A. S. B. V. p. 783.

3. Note on the Malay Woodpecker.—J. A. S. B. VI. p. 952.

4. On the Influence of Season over the Depth of Water in Wells.—Proc. Geol. Soc. I. 339.—Phil. Mag. ser. 2, XI. p. 58.

BLANDIN (Ph.).

1. Traité d'Anatomie topographique, ou Anatomie des Régions du Corps humain, etc. Paris, 1826; 1834, 8º. atl. fol.

2. Mémoire sur la structure et les mouvemens de la Langue dans l'Homme.—Arch. gén. Méd. 1823, Avr.—*Féruss.* Bull. 1823, III. p. 81.

3. Nouveaux Élémens d'Anatomie descriptive. Paris, 8º. 1837-1838.

4. Anatomie du Système dentaire, considéré dans l'Homme et les Animaux. Paris, 1836, 8º. fig.— *Val.* Repert. II. p. 23.

5. Note sur la distinction des Nerfs rachidiens en Nerfs sensitifs et Nerfs moteurs.—Ann. Sc. n. (2e S.) XI. p. 311.

BLANGY (Ducarne de).

1. Traité de l'éducation économique des Abeilles. 12º. Paris, 1771. —Suppl. 1776.

BLANK (J. B.).

1. Handbuch der Mineralogie. 8º. Würzb. 1810.

2. Handbuch der Zoologie. 8º. Würzb. 1811.

3. Naturalien-cabinet im Minoriten-Kloster zu Würzburg. 8º. Würzb. 1795; 1803; 1810.

BLANKEN (Gér.).

1. Catalogue de ce qu'on voit de plus remarquable dans la Chambre de l'Anatomie de Leyde. Leyde, 1713, 4º; 1718, 4º.—*Böhm.* Bibl. I. 1, p. 381.

BLANQUET (Sam.).

1. Epistola de Aquâ quæ in Saxa obrigescit. Mimati, 1731, 4º.— *Böhm.* Bibl. IV. 2, p. 365.

BLASIUS (E.).

1. Diss. inaug. de tractus intestinorum formatione in Mammalium embryonibus. 8º. Berlin, 1823.

BLASIUS (Gerh.).

1. Zootomiæ, seu Anatomes variorum Animalium pars 1ma. Amst. 1676, 12º. fig.—Biogr. Un. IV. p. 575.—*Böhm.* Bibl. II. 1, p. 81.

2. Anatome Animalium terrestrium variorum, Aquatilium, Serpentum, Insectorum Ovorumque structuram naturalem, ex veterum, recentiorum, propriisque observationibus proponens. Amst. 1681, 4°. fig.—Act. Erudit. I. p. 11.—*Böhm.* Bibl. II. 1, p. 81.

3. Observata anatomica in Homine, Simiâ, Equo, Vitulo, Testudine, Echino, Glire, Serpente, Ardeâ, variisque Animalibus aliis. Accedunt extraordinaria in Homine reperta, praxim medicam æquè ac anatomen illustrantia. Leyd. et Amst. 1674, 8°.—Biogr. Un. IV. p. 575.—*Böhm.* Bibl. II. 1, p. 81.

4. Anatome contracta in gratiam Discipulorum conscripta et edita. Amst. 1666, 12°.—(Belg.) 1675, 8°.—Biogr. Un. IV. p. 574.

5. Anatome compilatitia Animalium terrestrium variorum, volatilium, aquatilium, etc. Amst. 1681, 4°. fig.— *Cuv.* et *Val.* I. p. 72. —Biogr. Un. IV. p. 575.

6. Observationes anatomicæ selectiores, editæ è Collegio Medicorum privatorum Amstelodamensi. Amst. 1667.—Biogr. Un. IV. p. 575.

7. Commentarius in Syntagma anatomicum *J. Weslingii*, atque Appendix ex veterum, recentiorum, propriisque observationibus. Amst. 1659, 1666, 4°.—Utr. 1669, 4°. fig.—Biogr. Un. IV. p. 574.

8. Anatome Medullæ spinalis et Nervorum indè provenientium. Amst. 1666, 12°.—Biogr. Un. IV. p. 574.

9. Zoologia, s. Anatome Hominis Brutorumque variorum. Amst. 1673, 8°. fig.—*Böhm.* Bibl. II. 1, p. 81.

BLASIUS (J. H.).

1. Reise im südlichen Russland.—*Wiegm.* Arch. 1845, II. p. 6.

BLASSIÈRE (J. J.).

1. Histoire naturelle de la Reine des Abeilles. 1771.—*Eis.* Insect. p. 226.—Gemeinnütz. Arb. d. Oberlaus. Bienenges. I. pp. 21, 47. —Bibl. Ent. I. p. 32 (*Einleitung zu der entdeckten neuen natürl. Geschichte der Bienenkönigin*, etc.).

BLATIN (H.).

1. Des Enveloppes du Fœtus et des Eaux de l'Amnios. Par. 1840, 8°.

2. De la formation et du mode d'accroissement des Dents, de l'influence de leur arrangement sur les Arcs alvéolaires. Par. 1840, 8°.

BLAUD (P.).

1. Traité élémentaire de Physiologie philosophique, ou Élémens de

la science de l'Homme ramenée à ses véritables principes. Paris,
1830, 3 vols. 8°.—Cat. R. Soc. L.

BLAVIER (Ed.).

1. Essai d'une Statistique minéralogique et géologique du dép. de
la Mayenne. 8°. Par. 1837.—C. R. 1836, II. p. 415.

2. Notice statistique et géologique sur les Mines et le Terrain à
Anthracite du Maine. Par. 1834, 8°.—Ann. des Mines (3e S.)
VI.—Bull. Soc. Géol. VI. p. 62.

BLAZE (E.).

1. Hist. du Chien chez tous les peuples du monde. 8°. Paris, 1842.

BLÉGNY (Nic. de).

1. De quelques Papillons qui paraissent une fois tous les ans sur les
bords de la Meuse.—Nouv. Découv. dans la Médecine, IIe ann.
p. 188.—Bibl. Ent. I. p. 32.

2. Nouvel Abrégé d'Ostéologie. 12°. Paris, 1681.—Cat. R. Soc. L.

3. Journal des nouvelles Découvertes concernant les Sciences et les
Arts qui font partie de la Médecine. 3 vols. 12°. Par. 1681.—
Cat. R. Soc. L.

4. Histoire anatomique d'un Enfant qui a demeuré 26 ans dans le
ventre de sa mère. Paris, 1679, 12°.—Biogr. Un. IV. p. 578.

5. Zodiacus medico-gallicus, s. Miscellaneorum medico-physicorum
Gallicorum Annus II. et III. Genev. 1679–1682, 4°.—Act. Erudit. I. p. 220.

BLESSON (L.).

1. Observations on the *Ignis fatuus.*—*Walk.* Ent. Mag. 1833, IV.
p. 353.—Bibl. Ent. I. p. 32.

2. Bemerkungen über Sand u. Dünen (*Observations sur le Sable et
les Dunes*).—Hertha, XI. 2, p. 177; III. p. 279; IV. p. 416.—
Féruss. Bull. 1828, XV. p. 347; 1829, XVII. p. 164.—*L.* u. *Br.*
N. Jahrb. 1834, p. 109.

BLEULAND (J.).

1. Descriptio Musei anatomici Academiæ Rheno-Trajectinæ. Traj.
ad Rh. 4°. 1826.—Bibl. med.-ch. p. 52.

2. Icones anatomico-pathologicæ partium Corporis humani quæ in
Descriptione Musei Academiæ Rheno-Traject. inveniuntur. Fasc.
I. II. Traj. ad Rh. 1826, 1827, 4°.—Bibl. med.-ch. p. 52.

3. Icones Anatomiæ comparatæ. 4°. Traj. ad Rh. 1827–33.

4. Observationes anatomico-medicæ de sanâ et morbosâ Œsophagi structurâ. Lugd. Bat. 1785, 8°.—Bibl. med.-ch. p. 51.

5. Vasculorum in Intestinorum tenuium tunicis, subtilior. Anatom. oper. detegend. Düsseld. 1797, 4°.—Bibl. med.-ch. p. 52.

6. Icones tunicæ villosæ Intestini Duodeni. Traj. ad Rh. 1790, 4°. —Bibl. med.-ch. p. 52.

7. Experimentum anatomicum de Arteriolarum lymphaticarum existentiâ. Lugd. Bat. 1784, 4°. fig.—Bibl. med.-ch. p. 51.

8. Otium Academicum, descriptio partium Corporis humahi et Animalium. 4°. Utrecht, 1828.

9. Icones Hepatis Fœtûs octimestris. Traj. ad Rh. 1789, 4°. fig.

BLEY (   ).

1. Chemische Untersuchung menschlicher Gallensteine.—*Erdm.* Journ. f. prakt. Chemie, 1834, II.—*Müll.* Arch. 1836, VI. p. ccxxiv.

BLICHFELD (H. Fr.).

1. Relatio de Lithanthrace Bornholmiæ. Hafn. 1770, 8°.—Act. Œcon. Soc. Reg. Dan. I. 1776.—*Böhm.* Bibl. IV. 1, p. 487.

2. Kort Efterretning om Bergverket i Sundhornlehn udi Bergens Stift i Norge. Kiöb. 1771, 8°.

BLISSON (J. F. J.).

1. Essai sur une méthode à faciliter la recherche des Larves et des Lépidoptères. 8°. Mans. 1840.

BLOCH (M. El.).

1. Beschreibung der Schleicheidechse (*Lacerta serpens*).—Besch. Berl. Naturf. Fr. II. p. 28, fig.

2. Nachricht von der Dosenschildkröte.—Schr. Berl. Naturf. Fr. I. p. 131 ; II. p. 16, fig.

3. Ichthyologie, ou Histoire générale et particulière des Poissons. Berl. 1785–1796, 12 parts, fol. fig.—*Cuv.* R. An. III. p. 336.— Biogr. Un. IV. p. 585.

4. Naturgeschichte der ausländischen Fische. Berl. 1784–1795, 9 Th. 4°. fig. 4°. & fol.—Tr. Linn. Soc. Lond. V. p. 278.—*Bohm.* Bibl. II. 2, p. 64.—Cat. Bibl. Turic.

5. Oekonomische Naturgeschichte der Fische Deutschlands. Berlin, 1782–1784, 4°. 3 Thl. fig.—Tr. Linn. Soc. Lond. V. p. 278.— *Böhm.* Bibl. II. 2, p. 64.

6. Von den vermeinten doppelten Zeugungsgliedern der Rochen und Haye.—Schrift. der Gesells. Naturf. Fr. VI. 1785, p. 377, fig.

7. Abhandlung von den vermeinten Mänlichen Gliedern des Dornhayes. Schrift. der Gesells. Naturforsch. Freunde, VIII. 1788, p. 9, fig.

8. Systema Ichthyologiæ, ed. *J. G. Schneider.* 8°. Berl. 1801.—*Cuv.* R. An. III. p. 336.

9. Beschreibung zweier neuen Fische, der Pennantsche Barsch u. der Borstenlachs.—Schr. Berl. Naturf. X. p. 424, fig.

10. Naturgeschichte der Marene.—Besch. Berl. Naturf. IV. 1779, p. 61, fig.

11. Pleuronectarum duplex species.—N. Act. Ac. Petr. 1785, III. Suppl. p. 139.

12. Beiträg zur Naturgeschichte des Kopals.—Besch. Berl. Naturf. II. p. 91, fig.—Bibl. Ent. I. p. 32.

13. Beiträge zur Naturgeschichte der Würmer, welche in andern Thieren leben.—Besch. Berl. Naturf. IV. p. 534, fig.—Bibl. Ent. I. p. 32.

14. Abhandlung von der Erzeugung der Eingeweidewürmer, u. den Mitteln wider dieselben. Berl. 1782, 4°. fig.—*Böhm.* Bibl. II. 2, p. 383.—(*Traité sur la Génération des Vers intestinaux.*) Strasb. 1798, 8°. fig.

15. Naturgeschichte der Fische in den Preussischen Staaten.—Schrift. Berl. Naturf. Fr. I. p. 231.—*Böhm.* Bibl. II. 2, p. 64.

16. Nachricht von einem ästigen Punktcorall mit pfriemenförmigen Oefnungen.—Berl. Beschäft. III. p. 415, fig.

17. Beytrag zur Naturgeschichte der Blasenwürmer.—Schr. Berl. Naturf. Fr. I. p. 335, fig.

18. Von den in Kopal eingeschlossenen Insekten.—Beschäft. der Berlin Gesellsch. Naturf. Fr. II. 1776, p. 156.

19. Ornithologische Rhapsodien.—Besch. Berl. Naturf. Fr. IV. pp. 5, 79.—Schriften der Berlin Gesells. Naturf. Fr. III. 1782, p. 372.

BLOCK (P. N.).

1. Lehrbuch der Naturkenntniss. 8°. Gött. 1823.

2. Observationes Entomologiæ. 4°. Lund. 1807.

BLOCQUEL (S.).

1. Beautés de l'Hist. nat. des Reptiles, etc. 16°. Lille, 1826.

2. Ichthyologie de la Jeunesse. 16°. Lille, 1826.

3. Ornithologie du jeune âge. 2 vols. 16°. Lille, 1826.

Blöde (G. v.).

1. Beiträge zur Geologie des südlichen Russlands.—*L.* u. *Br.* 1841, p. 505.

2. Geognostische Beschreibung des Gouvernements Charkow. Petersb. 1840, 8°.—Bull. Natur. Mosc. 1841, p. 34.—*L.* u. *Br.* N. Jahrb. 1842, p. 246.

3. Geologische Schilderung des grössten Theiles vom Gubernium Poltawa.—*L.* u. *Br.* N. Jahrb. 1842, p. 198.

4. Ergebnisse einer Reise von Charkow nach dem Donetz.—*L.* u. *Br.* N. Jahrb. 1842, p. 253, fig.

5. Nachträge zu meiner Schrift über die Uebergangs-Formation im Königreich Polen, etc. mit Berücksichtigung der Abhandlungen der Hrn. *Schneider* u. *Beker.*—*L.* u. *Br.* N. Jahrb. 1833, p. 129.

6. Ueber die Uebergangs-Gebirgsformation im Königreich Polen, nebst einer vorangehenden Uebersicht der sämmtlichen Gebirgsformationen von Polen, u. einer nachfolgenden Aufstellung der hierin vorkommenden Mineralien. 8°. Breslau, 1830.—Bull. Soc. Géol. Fr. I. p. 56 (*Sur la Formation intermédiaire de la Pologne, suivi d'un Aperçu de tous les Terrains de ce pays*).

7. Ueber plutonischen (?) Kalkstein bei Zagdainsko in Polen.—*L.* u. *Br.* N. Jahrb. 1834, p. 34.

8. Versuch einer Theorie über die Bildung des Geyer'schen Stockwerkes.—*Leonh.* Tasch. 1816, I. p. 3.

Blöde (K. A.).

1. *Galls* Lehre über die Verrichtungen des Gehirns. (2te Aufl.) Dresd. 1806, 8°.—Verz. Bibl. Schw. Naturf. Ges.

Blöm (C. Magn.).

1. Descriptiones quorundam Insectorum nondum cognitorum ad Aquisgranum anno 1761 detectorum.—Act. Helvet. V. p. 154.—Biogr. Un. Suppl. LVIII. p. 366.

2. Beskrifning, etc. (*Description d'une petite Phalène nuisible aux Abeilles*).—Vet. Acad. 1764, p. 12.—*Fuessl.* N. Mag. III. p. 62.—Bibl. Ent. I. p. 33.

3. Ytterligare rön och anmärkningar om masken *Ascaris lumbricoides.*—Vet. Acad. 1776, p. 313. (Allem.) p. 314.

4. Beskrifning pä en helt lins-grä, eller nästan huit, orrhöna, *Tetrao tetrix* fœm. *Linn* —Vet. Acad. 1785, p. 230. (Allem.) p. 225.

Blom (P. J.).

1. Abhandlungen über die Auscultation oder den Gebrauch des

Laënnecschen Stethoskopes, angewandt auf die Geburtshilfe. Aus dem Holländ. übers. durch *F. W. Schröder*. Lond. 8°.— *Val.* Repert. III. p. 27.

BLOME (Rich.).

1. The Present State of His Majesty's Isles and Territories in America. Lond. 1687, 8°. fig.—(Gall.) *L'Amérique Anglaise, ou Description*, etc. Amst. 1688, 12°. fig.—(Germ.) *Englisches Amerika*, etc. Leipz. 1697, 12°.—*Böhm.* Bibl. I. 1, p. 726.

BLOND (J. B. le).

1. Reise nach den Antillen u. nach Sudamerika. 8°. Hamb. 1815.

BLONDEAU ( ).

1. Observations sur les Mouches communes.—J. de Phys. IV. p. 155. —Bibl. Ent. I. p. 33.

2. Manuel de Minéralogie. Par. 1827, 12°.

BLONDEL (Franç.).

1. Observation sur les serpens qui ne sont pas vénineux dans quelques Isles et qui le deviennent quand on les porte à la Martinique tandis que ceux de cette Isle perdent leur vénine, si on les transporte ailleurs.—Mém. Acad. Sc. Paris, I. p. 362.

2. Observation sur les Montagnes des Pyrénées séparées par les tremblemens de terre.—Mém. Acad. Sc. Paris, I. p. 362.

3. Observations sur des pierres qu'on trouve près de Toulon qui étant cassées présentent des huitres bonnes à manger.—Mém. Acad. Sc. Paris, I. p. 363.

BLONDEL (G. Ferd.).

1. De Navigatione Salomonis in Ophir, de Purpurâ et Cocco. Hamb. 1660, 8°.—*Mod.* Bibl. Helm.

BLONDEL (Hip.).

1. Mémoire sur une nouvelle espèce de Brachélytre du genre Prognathe.—Ann. Sc. n. X. p. 412, fig.—*Féruss.* Bull. 1828, XIII. p. 257.

2. Note sur la découverte d'Ossemens fossiles dans un terrain dépendant de hopital Necker.—Rev. Zool. 1838, p. 308.—C. R. VII. p. 1027

BLONDEL (J. Aug.).

1. Dissertation physique sur la force de l'imagination des Femmes enceintes sur le Fœtus. Leyd. 1737, 8°.—Biogr. Un. IV. p. 595.

BLONDLOT (N.).

1. Traité analytique de la Digestion considérée particulièrement dans l'Homme et dans les Animaux vertébrés. Par. et Nancy, 1843, 8°.

BLOT (F.).

1. Sur les propriétés des Insectes des environs de Caen.—Mém. Soc. Linn. Calv. I. p. 85.

2. Mémoire sur un nouveau genre et une nouvelle espèce de Diptère (*Myopites Blotii*, Breb.).—Mém. Soc. Linn. Calv. III. p. 101, fig.—*Féruss.* Bull. 1828, XV. p. 208.

3. Mémoire sur le Puceron lanifère. Caen, 8°.—Bibl. Ent. I. p. 33.

BLOXAM (A.).

1. Land Birds met with at Sea on a voyage from England to South America.—Mag. Nat. Hist. ser. 1, IV. p. 145.

2. Land and Freshwater Shells met with in Norfolk and Derbyshire.—Mag. Nat. Hist. ser. 1, VI. p. 324.

3. Zoological Appendix to *Byron's* Voy. to Sandwich Islands.

BLUETT ( ).

1. Méthode facile pour obtenir le Squelette des petits Poissons.—Phil. Mag. 1830, Febr. p. 151.—*Féruss.* Bull. XXII. p. 157.

BLUFF (M. Jos.).

1. Ueber die Krankheiten als Krankheiturssachen. Aach. u. Leipz. 1829, 8°.—N. Act. Nat. Cur. XIV. p. XIII.

2. Merkwürdiger Fall von Elephantiasis.—N. Act. Nat. Cur. XVII. p. 409, fig.

3. Ueber die Heilkräfte der Küchengewächse. 1828, 8°.—N. Act. Nat. Cur. XIV. p. XIII.

4. Entwickelungs-Combinationen organischer Wesen. Köln, 1827, 8°.—Bibl. med.-chir. p. 52.—*Féruss.* Bull. 1828, XV. p. 62 (*Combinaisons de développement des Etres organiques*).

BLUM (R.).

1. Mineralogische Notizen.—*L.* u. *Br.* N. Jahrb. 1837, p. 34, fig.

2. Ueber einige geologische Erscheinungen in der Nagelflue.—*L.* u. *Br.* N. Jahrb. 1840, p. 525, fig.

3. Lehrbuch der Oryktognosie. Stuttg. 1834, 8°.

4. Pinite trouvée dans le Granite des environs d'Heidelberg.—Ann. Sc. n. 1829, XVI.—Rev. Bibl. p. 21.

BLUMENBACH (C. W.).

1. Neueste Landkunde des Erzherzogthums Oesterreichs unter der Ems. Wien, 1816.—*Féruss.* Bull. 1824, II. p. 13 (*Description récente de l'Archiduché d'Autriche sous l'Ems*).

BLUMENBACH (J. Fr.).

1. Beiträge zur Naturgeschichte. Gött. (1790), 1806–11, 2 vols. 8°. fig.

2. Beiträge zur Naturgeschichte der Vorwelt.—*Voigt's* Mag. VI.

3. Einige naturhistorische Bemerkungen bei Gelegenheit einer Schweizer Reise.—*Voigt's* Mag. V.

4. Verbessertes System der Saügthiere.—*Voigt's* Mag. für den Neuesten Zustand der Naturk. III. 1802.—Phil. Mag. XVI. p. 68.

5. Beobachtungen an einem lebendigen Beutelthier (*Didelphys Marsupialis*).—*Voigt's* Mag. III. 1802, p. 683.—Phil. Mag. XVI. p. 68.

6. Einige Anatomische Bemerkungen über den *Ornithorhynchus paradoxus* aus New Sudwallis.—*Voigt's* Mag. II. 1800, p. 184. —Mém. de la Soc. Méd. d'Emulation, 4ᵐᵉ année.—Phil. Mag. XI. p. 366.

7. Zergliederung eines Casuars.—*Voigt's* Mag. I. 1800, p. 299.

8. Ueber den Bildungstrieb und das Zeugungsgeschäfte. Gött. 1781, 8°.—*Böhm.* Bibl. II. 1, p. 135.—(*Essay on Generation*, tr. by *A. Crichton*.) Lond. 1792, 12°.—Tr. Linn. Soc. Lond. V. p. 278.

9. Handbuch der Naturgeschichte. Götting. 1779, 8°. fig.; 1782, 8°. fig.; 1788 u. 1791; 1807; 1814.—Götting. 1825. (11ᵉ éd.). —Götting. 1830 (12ᵉ éd.).—Trad. Franç. par *Artaud* (*Manuel d'Histoire naturelle*). Paris, 1803, 2 vols. 8°. fig.—(Dan. *O. J. Mynster*) Kop. 1793, 8°.—*Féruss.* Bull. 1826, VII. p. 189; 1831, XXIV. p. 350.—Isis, 1831, XII. p. 1305.—(Angl *Gore*) 8°. Lond. 1825.—(Belg. *J. A. Bennet*) 8°. Leyd. 1802.—(Ital.) 2 vols. 8°. Lugano, 1825; 6 vols. 12°. Milan, 1826.

10. Specimen Historiæ naturalis antiquæ, artis operibus illustratæ, eaque vicissim illustrantis. 8°. fig.—Comm. Gœtt. XVI. p. 169. —Isis, 1817, I. p. 115.

11. Abbildungen naturhistorischer Gegenstände. Götting. 1796–1810, 10 H.—Tr. Linn. Soc. Lond. V. p. 278; Suppl. VI. p. 391. —*Cuv.* R. An. III. p. 337.

12. Beiträge zur Naturgeschichte der Schlangen.—*Voigt's* Mag. V. 1, p. 1.—Dict. Sc. n. XV. p. 253.

13. Anzeige verschiedener, vorzüglicher Abbildungen v. Thieren in älteren Kupferstichen u. Holzschnitten.—Gött. Mag. 2ter Jahrg. IV. p. 136.

14. De quorundam Animantium coloniis, sive sponte migratis, sive casu aut studio ab hominibus aliorsum translatis.—Comm. Gœtt. 1823, 4°.—Bibl. med.-ch. p. 53.

15. Handbuch der vergleichenden Anatomie. Gött. 1805, fig. (3te Aufl.); Gött. 1824, 8°. fig.—Bibl. med.-ch. p. 53.—(Angl. *W. Lawrence*): *A Manual of Comparative Anatomy.* 8°. Lond. 1827.

16. De Animalium quorundam, per hiemem dormientium, Vasis cephalicis et Aure internâ, epistola.—N. Act. Nat. Cur. XIII. p. 23.

17. Prolusio anatomica de Sinubus frontalibus. Gött. 1779, 4°.——Bibl. med.-ch. p. 53.

18. Geschichte u. Beschreibung der Knochen des menschlichen Körpers. Gött. 1807, fig. (2te Aufl.).—Bibl. med.-ch. p. 53.

19. De anomalis et vitiosis quibusdam Nisus formativi aberratiouibus commentatio.—Comm. recent. Gœtt. II. p. 3.

20. Ueber die Zauberkraft der Klapper-Schlange.—*Voigt's* Mag. der Naturkunde, II. 1798.—Phil. Mag. II. p. 251.—Tr. Linn. Soc. Lond. VI. p. 391.

21. De Oculis Leucæthiopum et Iridis motu. fig.—Comm. Gœtt. VII. p. 29.

22. De Generis humani Varietate nativâ. Gött. 1776; 1781, 8°. fig. —(3e éd.) Gött. 1795, 8°. fig.—Deutsch übers. v. *Gruber* (*Ueber die natürl. Verschiedenheiten im Menschengeschlechte*). Leipz. 1798. —Trad. Fr. par *Chardel* (*De l'Unité du Genre humain et de ses Variétés*). Paris, 1808, 8°.—Bibl. med.-ch. p. 53.

23 Collectio Craniorum diversarum gentium illustr. c. fig. Dec. I.- VII. Gött. 1790-1826, 4°.—*Féruss.* Bull. 1827, X. p. 393; 1829, XVI. p. 453.—Isis, 1829, V. p. 554.—Bibl. med.-ch. p. 53.

24. Specimen Archæologiæ telluris, terrarumque inprimis Hannoveraniarum. Gött. 1803.—Comm. Gœtt. XV. p. 132.—Isis, 1817, I. p. 115.

25. Institutiones physiologicæ. Gött. 1787; 1789, 8°. fig.; 1821, 8°. fig.—Bibl. med.-ch. p. 53.

26. Physiologie des menschlichen Körpers. Aus dem Latein, v. *J. Eyerel.* Wien, 1795, 8°.—Bibl. med.-ch. p. 53.

27. Specimen physiol. compar. inter Animantia calidi et frigidi sanguinis. Gött. 1787, 4°. fig.—Comm. Gœtt. VIII. p. 69.—Bibl. med.-ch. p. 53.

28. Specimen physiol. compar. inter Animantia calidi sanguinis, vivipara et ovipara. Gött. 1789, 4°.—Comm. Gœtt. IX. p. 108. —Tr. Linn. Soc. Lond. V. p. 278.—Bibl. med.-ch. p. 53.

29. Commentatio de vi vitali Sanguinis. Gött. 1788, 4°.—Comm. Gœtt. IX. p. 3.—Bibl. med.-ch. p. 53.

30. De vi vitali Sanguinis negandâ, etc. Gött. 1795, 4°.—Bibl. med.-ch. p. 53.

31. Nuperæ Observationes de Nisu formativo et Generationis negotio. Gött. 1787, 4°. fig.—Comm. Gœtt. VIII. p. 41.—Bibl. med.-ch. p. 53.

32. Von den Federbuschpolypen in den Göttingischen Gewässern. —Gött. Mag. I. p. 117.

33. Versuche mit einigen Gattungen von Käfern die statt der sogenannten Spanischen Fliegen zum Blasen ziehen taugen (*Lytta*). —*Voigt's* Mag. II. 1801, p. 641.

34. Miscellaneous Notices in Natural History.—Ed. Phil. Journ. 1823, p. 259.—*Féruss.* Bull. IV. p. 69 (*Mélanges sur l'Histoire naturelle*).

35. Beschreibung einer merkwürdigen Gebirgsart (mitgetheilt von *v. Hoff*).—*Leonh.* Tasch. VI. p. 353.

36. Kleine Schriften zur vergleich. Physiol. Anat. u. Naturgesch. übers. v. *J. G. Gruber.* 8°. Leipz. 1804.

37. Gebirgsart von der *Côte du Mole* an der *Grande Terre* von Guadeloupe, in welcher neuerlich die fossilen Menschen-Gerippe entdeckt wurden.—Allg. Anz. d. Deutsch. 1815, No. 312.—*Leonh.* Tasch. 1816, I. p. 237.

38. Ueber die fossilen Gebeine von Elephanten und Mammutsthieren, u. über andere präadamitische Thier u. Pflanzenreste, besonders aus den Hannöverschen Landen.—*Leonh.* Tasch. 1814, II. p. 540.

39. Ueber künstliche oder zufällige Verstümmelungen am thierische Körper die mit der Zeit zum erblichen Schlag ausgeartet.— *Licht.* u. *Voigt's* Mag. VI. 1, 1789, p. 13.

40. Von den Negern.—*Licht.* u. *Voigt's* Mag. IV. 3, 1787, p. 1.

41. Ueber Menschen- Racen und Schweine-Racen.—*Licht.* u. *Voigt's* Mag. VI. 1, 1789, p. 1.

BLUMENBACH (J. Fr.) u. BORN (Ig. v.).

1. Ueber die Nutritionskraft. Zwei gekr. Abhandl. nebst Erläuter. v. *C. F. Wolff.* Petersb. 1789, 4°.—Bibl. med.-ch. p. 53.

BLUMENGARTEN (S.).

1. De Tetano. Berol. 8°.— *Val.* Repert. III. p. 19.

BLUMENTHAL (G. A.).

1. De externis oculi integumentis quorundam Animalium. 4°. Berl. 1812.

BLUMHARDT ( ), DUVERNOY (G. L.) et SEEGER ( ).

1. Medicin. Correspondenzblatt; *vide sup.* Pars I. p. 34.

BLUMHOF (T. G. L.).

1. Nachricht von einen un gewöhnlichen Zuge von Insekten in Schweden.— *Voigt's* Mag. III. 1801, p. 276.

2. Uebersicht der Beobachtungen älterer und neuerer Naturforscher über das Leuchten des Meerwassers.— *Voigt's* Mag. I. 4, 1799, p. 1.

BLUMRICH (G.).

1. Die Anatomie in einer Nuss. Nürnb. 1835, 32º. (2te Aufl.)— Bibl. med.-ch. p. 54.

BLYTH (E.).

1. Considerations pertaining to Classification.—Mag. Nat. Hist. ser. 1, VI. p. 485.

2. On the Songs of the Bramble Finch, Mountain Linnet, and Tree Sparrow.—Mag. Nat. Hist. ser. 1, VII. p. 487.

3. Attempt to classify the "Varieties" of Animals.—Mag. Nat. Hist. ser. 1, VIII. p. 40.

4. On Hybrids.—Mag. Nat. Hist. ser. 1, VIII. pp. 198, 228.

5. On the Cuckoo.—Mag. Nat. Hist. ser. 1, VIII. p. 325.—*Fror.* Notiz. XLV. pp. 129, 145, 161.— *Wiegm.* Arch. 1836, II. p. 269.

6. On a remarkable individual of *Lanius collurio.*—Mag. Nat. Hist. ser. 1, VIII. p. 364.

7. On the arrival of British summer Birds of passage.—Mag. Nat. Hist. ser. 1, VII. p. 338; IX. p. 551.

8. On the seasonal and other changes which regularly take place in Birds.—Mag. Nat. Hist. ser. 1, IX. pp. 392, 610.

9. On the affinities of the Feathered Race, and the nature of specific distinctions.—Mag. Nat. Hist. ser. 1, IX. p. 504.

10. Birds observed near Tooting.—Mag. Nat. Hist. ser. 1, IX. p. 622.— *Wiegm.* Arch. 1837, p. 195.

11. On the Psychological Distinctions between Man and all other Animals, and the consequent Diversity of Human Influence over the inferior ranks of Creation, from any mutual and reciprocal Influence exercised among the latter.—Mag. Nat. Hist. ser. 2, I. pp. 1, 77, 131.

12. On the Habits and Peculiarities of the common Bottletit or Mufflin (*Mecistura vagans,* Leach ; *Parus caudatus* of Linnæus). —Mag. Nat. Hist. ser. 2, I. p. 199.

13. On the Reconciliation of certain apparent Discrepancies observable in the mode in which the seasonal and progressive Changes of Colour are effected in the Fur of Mammalians and Feathers of Birds ; with various Observations on Moulting.—Mag. Nat. Hist. ser. 2, I. pp. 259, 300.

14. On Woodcocks, Fieldfares and Redwings building within the British Islands.—Mag. Nat. Hist. ser. 2, I. p. 439.

15. On the Plumage of Birds.—Mag. Nat. Hist. ser. 2, I. p. 477.

16. On the Pern or Honey Buzzard.—Mag. Nat. Hist. ser. 2, I. p.536.

17. On the Counterfeiting of Death, as a means to escape from danger, in the Fox and other animals.—Mag. Nat. Hist. ser. 2, I. p. 566.

18. Outlines of a new arrangement of Insessorial Birds.—Mag. Nat. Hist. ser. 2, II. pp. 256, 314.

19. Analytical Descriptions of the Groups of Birds composing the Order *Insessores Heterogenes.*—Mag. Nat. Hist. ser. 2, II. pp. 351, 420.

20. Analytic Descriptions of the Groups of Birds composing the Order *Strepitores.*—Mag. Nat. Hist. ser. 2, II. p. 589; III. p. 76.

21. On some additional species of the genus *Equus* to those currently admitted by Zoologists.—Mag. Nat. Hist. ser. 2, IV. pp. 81, 368.

22. On a peculiarity in the Structure of the Feet in the *Trogonidæ.* —Proc. Zool. Soc. VI. p. 20.—Ann. of Nat. H. II. p. 227.

23. A summary Monograph of the species of the genus *Ovis.*—Proc. Zool. Soc. VIII. pp. 12, 62.—Ann. and Mag. N. H. VI. p. 302.

24. An amended List of the species of the genus *Ovis.*—Ann. and Mag. N. H. VII. pp. 195, 248.

25. On some Animals of North Africa.—Proc. Zool. Soc. IX. p. 64. —Ann. and Mag. N. H. IX. p. 62.

26. List of Birds which are found both in India and Europe, with observations upon them.—Proc. Zool. Soc. X. p. 93.—Ann. and Mag. N. H. XI. p. 477.

27. On the Affinities of *Glareola torquata.*—Ann. and Mag. N. H. XII. p. 74.

28. List of Birds obtained in the vicinity of Calcutta, &c.—Ann. and Mag. N. H. XII. pp. 90, 165, 229.

29. Further Notice of the species of Birds occurring in the vicinity of Calcutta.—Ann. and Mag. N. H. XIII. p. 113.

30. Descriptions of some new species of Birds found in the neighbourhood of Calcutta.—Ann. and Mag. N. H. XIII. p. 175.

31. Further Observations on the Ornithology of the neighbourhood

of Calcutta; with Notes by *H. E. Strickland.*—Ann. and Mag. N. H. XIV. pp. 34, 114.

32. Remarks on *Mr. Gray's* Catalogue of Mammalia and Birds presented by *B. H. Hodgson,* Esq. to the British Museum.— Ann. and Mag. N. Hist. XX. p. 313.

33. Remarks on *Prof. Sundevall's* paper on Birds from Calcutta.— Ann. and Mag. N. Hist. XX.

34. On the Osteology of the Great Auk.—Proc. Zool. Soc. V. p. 122.

35. On the Plumage and progressive changes of the Cross-bill and Linnet.—Proc. Zool. Soc. VI. p. 115.

36. On various Quadrupeds chiefly collected in Little Thibet.— Proc. Zool. Soc. VIII. p. 79.

37. A general review of the species of True Stag, or elaphoid form of *Cervus,* comprising those more immediately related to the Red Deer of Europe.—Journ. As. Soc. Beng. X. p. 736.

38. On another new species of Pika (*Lagomys*) from the Himalaya. —Journ. As. Soc. Beng. X. p. 816.

39. A Monograph of the species of Wild Sheep.—Journ. As. Soc. Beng. X. p. 858.

40. On three Indian species of Bat, of the genus *Taphozous.*—Journ. As. Soc. Beng. X. p. 974.

41. On various Indian and Malayan Birds, with descriptions of presumed new species.—Journ. As. Soc. Beng. XI. p. 160.

42. On the predatory and sanguinivorous habits of the Bats of the genus *Megaderma,* with some remarks on the Blood-sucking propensities of other *Vespertilionidæ.*—Journ. As. Soc. Beng. XI. p. 255.

43. A Monograph of the species of Lynx.—Journ. As. Soc. Beng. XI. p. 740.

44. On the Bat described as *Taphozous longimanus* by *Gen. Hardwicke.*—Journ. As. Soc. Beng. XI. p. 784.

45. A Monograph of the Indian and Malayan species of *Cuculidæ.* —Journ. As. Soc. Beng. XI. pp. 897, 1095 ; XII. p. 240.

46. On various Mammalia, with descriptions of many new species. —Journ. As. Soc. Beng. XIII. p. 463.—Ann. and Mag. N. Hist. XV. p. 449.

47. "On the Leiotrichane Birds of the Subhemalayas," by *B. H. Hodgson, Esq.*; with some additions and annotations,—a Synopsis of the Indian *Pari,* and of the Indian *Fringillidæ.*—Journ. As. Soc. Beng. XIII. p. 993.

48. On various new or little-known species of Birds.—Journ. As. Soc. Beng. XIV. pp. 173, 546; XV. p. 1.

49. On *Caprolagus,* a new genus of Leporine Mammalia.—Journ. As. Soc. Beng. XIV. p. 247.

50. Drafts for a Fauna Indica (comprising the Animals of the Himalaya Mountains, those of the Valley of the Indus, of the Provinces of Assam, Sylhet, Tipperah, Arracan, and of Ceylon, with occasional notices of species from the neighbouring countries).—Journ. As. Soc. Beng. XIV. p. 845.—Ann. and Mag. N. H. 1846.

51. On the Fauna of the Nicobar Islands.—Journ. As. Soc. Beng. XV. p. 367.

52. Reports of the Curator of the Museum of the Asiatic Society of Bengal.—Journ. As. Soc. Beng. X. pp. 836, 917, 936; XI. pp. 95, 199, 444, 585, 788, 865, 880, 969, 1202 ; XII. pp. 166, 925 ; XIII. p. 361.

Boag (W.).

1. On the Poison of Serpents.—Asiat. Res. VI. p. 103.— *Voigt's* Mag. III. p. 269.

Boase (H. S.).

1. On the Alluvial Formations of the Western part of Cornwall.— Trans. Geol. Soc. Cornw. III. p. 17.—*Féruss.* Bull. 1829, XVI. p. 368.

2. On the Sand-Banks of the Northern shores of Mount's Bay.— Trans. Geol. Soc. Cornw. III. p. 166.—*Féruss.* Bull. 1829, XVI. p. 366.

3. On the Nature of the Structure of Rocks.—Phil. Mag. ser. 3, VII. p. 376.

4. Contributions towards a Knowledge of the Geology of Cornwall. —Trans. Geol. Soc. Cornw. III. p. 166.

5. A Treatise on Primary Geology. Lond. 1834, 8°.

6. On two species of Fishes taken in Mount's Bay, Cornwall.—Pr. Zool. Soc. I. p. 114.

7. Remarks on *Mr. Hopkins's* " Researches in Physical Geology." —Phil. Mag. ser. 3, IX. p. 4 ; X. p. 14.

Boate (Ger.).

1. Ireland's Natural History, being a Description of its situation, greatness, shape, and nature of its Hills, Woods, etc. Lond. 1652, 8°.—Dubl. 1726, 4° ; 1753, 1755, 1799.—(Gall.) Trad. de *P. Briot,* Par. 1666, 12°. 2 vols.—(Lat.) 12°.—*Böhm.* Bibl. I. 1, p. 531.—Tr. Linn. Soc. Lond. V. p. 278.

Bobba ( ).

1. Un mot sur les Idées du *Dr. Gall.* Milan, an. I. Rép.

Bobe-Moreau ( ).

Sur les Termites observées à Rochefort. 8°. Saintes. 1843.

Böber (J. de).

1. Observations sur la famille de Papillons connue sous le nom de Damies ou Fritillaires.—Natur. de Mosc. III. p. 1.—Bibl. Ent. I. p. 34.
2. Nouveaux Papillons en Sibérie.—Natur. de Mosc. II. 1806, p. 20 ; III. p. 396.—*Eis.* Insect. p. 205.

Böbert (K. F.).

1. Analogie der Glanzkobalt-Lagen bei Skuterud in Norwegen u. bei Vena in Schweden.—*Karst.* Arch. IV. p. 280 (277).—*L.* u. *Br.* 1833, pp. 213, 558.
2. Ueber das Hervortreten des grauen Uebergangs-Kalksteins zwischen den Alaun-Schiefer-Schichten.—*Karst.* Arch. IV. p.274. *L.* u. *Br.* N. Jahrb. 1833, p. 563.

Boblaye (Puillon).

1. Modifications de certains Terrains de Sédiment par le voisinage des Roches ignées.—C. R. VI. p. 168.
2. Notice sur la Géologie des environs d'Alger.—C. R. XI. p. 348.
3. Carte topographique et géognostique de l'Ile d'Égine.—Bull. Soc. Géol. Fr. I. p. 82.
4. Notice sur l'altération des Roches calcaires du Littoral de la Grèce.—Bull. Soc. Géol. Fr. I. p. 160.
5. Lettre à *M. de Férussac* (sur la Géologie de la Morée), de Modon, 18 Sept. 1829.—*Féruss.* Bull. XIX. p. 34.
6. Lettre à *M. Arago,* sur la Température de la Grèce.—C. R. IV. p. 237.
7. Observations sur la constitution géologique de la Morée.—Ann. Sc. n. 1831, XXII. p. 113.—*L.* u. *Br.* N. Jahrb. 1834, p. 97 (*Beobachtungen über die geognost. Beschaff. v. Morea*).
8. Essai sur la configuration et la constitution géologique de la Brétagne.—Mém. Mus. XV. p. 49.—*Féruss.* Bull. XIII. p. 179.
9. Mémoire sur la Formation jurassique dans le Nord de la France

—Ann. Sc. n. 1829, XVII. p. 35.—Bull. Soc. Géol. VI. p. 69.—
L. u. *Br.* N. Jahrb. 1833, p. 97 (*Jura-Gebilde im nördlichen Frankreich*).

BOBLAYE (Puillon) et VIRLET (Th.).

1. Expédition scientifique de Morée. fol. Par. 1833 (Livr. I.-VI.).
Bull. Soc. Géol. Fr. VI. p. 69 (Descript. géognostique).
2. Ueber die Emporhebungen der Bergketten in Griechenland.—
Bull. Soc. Géol. Fr. V. p. 207.—*L.* u. *Br.* N. Jahrb. 1836, p. 381.

BOCCA ( ).

1. Traité complet des Abeilles, avec une méthode nouvelle de les
gouverner. Paris, 1790, 8°. 3 vols.—Bibl. Ent. I. p. 34.

BOCCONE (P. Silv.).

1. Curieuse Anmerkungen über ein und andere natürliche Dinge,
aus seinem noch nie in Druck gegebenen *Museo experimentali
physico* zusammengezogen, und im Durchreisen durch Deutschland zum Andenken seiner in deutscher Sprache zum Druck
hinterlassen. Frankf. u. Leipz. 1697, 12°. fig.—*Böhm.* Bibl. I. 1,
p. 249.

2. Recherches et Observations naturelles sur la production de plusieurs Pierres, sur la pétrification de quelques parties d'Animaux,
sur les principes des Glossopètres, sur la Pierre étoilée et sur
l'embrasement du Mont Etna. Par. 1671, 12°.—Cat. R. Soc.
Lond.

3. Museo di Fisica e di esperienze, variato e decorato di osservazioni naturali, etc. Ven. 1697, 4°. fig.—*Böhm.* Bibl. I. 1, p. 249.
—*Mod.* Bibl. Helm.

4. Osservazioni naturali, ove si contengono materie medico-fisiche,
etc. Bon. 1684, 8°.—Biogr. Un. IV. p. 626.—*Böhm.* Bibl. I. 1,
p. 249.—*Mod.* Bibl. Helm.

5. Recherches et Observations d'Histoire naturelle touchant le Corail, la Pierre étoilée, etc. Paris, 1670, 12°. fig.—Amst. 1674,
8°. fig.—*Böhm.* Bibl. I. 1, p. 248.—(Belg.) Amst. 1744, 8°. fig.
(*Natuurkundige Naspeuringen, etc.*)—*Mod.* Bibl. Helm.

6. Observatio circa nonnullas Plantas imperfectas, uti Fucos, Co-
rallinas, Zoophyta, Fungos et similes, earumque originem.—Misc.
Nat. Cur. Dec. III. Ann. 4, 1696, p. 142, App.—*Mod.* Bibl. Helm.

7. De materia simili lithomargæ agricolæ, aut agarico minerali Fer-
rantis Imperati, quæ in cavitate quorundam saxorum aut silicum
in districtu civitatis Rothomagensis et portus Gratiæ in Nor-
mannia invenitur.—Ephem. Ac. Nat. Cur. Cent. 1 et 2, p. 5.

BOCHART (Sam.).

1. Hierozoicon, seu de Animalibus sacræ Scripturæ Compendium, etc. Franequ. 1619, 4°.—Lond. 1633 et 1663, 2 vols. fol.—Francf. 1675, fol.—Lugd. Bat. 1722, fol.—(Recens. *E. F. C. Rosenmüller*), 3 vols. 4°. Leipz. 1793–99.—*Wiegm.* Arch. 1837, I. p. 3.—*Böhm.* Bibl. II. 1, p. 73.

BOCHAUTE (Van).

1. Mémoire sur le Cuivre de Hongrie.—Mém. Ac. Brux. IV. p. 315.

2. Essai sur la Reproduction des Etres organisés et la continuation de leurs espèces.—Mém. Ac. Brux. IV. p. 49.

BOCK (A. C.).

1. Handbuch der praktischen Anatomie des menschl. Körpers, oder vollständige Beschreibung desselben, nach der natürlichen Lage seiner Theile, etc. Meiss. 1820–22; 1831, 2 vols. 8°.—Bibl. med.-ch. p. 54.

2. Katechismus der praktischen Anatomie für Aerzte u. Wundärzte, etc. Leipz. 1826, 1828, 8°.—Bibl. med.-ch. p. 54.

3. Der Prosector, oder Unterricht zur praktischen u. technischen Zergliederungskunst, für solche, welche sich vorzüglich der praktischen Zergliederung widmen wollen, etc. Leipz. 1829, 8°. fig. —Bibl. med.-ch. p. 55.

4. Tabellarische Uebersicht der gesammten Anatomie, nach der Lage der Theile. Leipz. 1817, fol.—Bibl. med.-ch. p. 54.

5. Darstellung der Venen des menschlichen Körpers, nach ihrer Struktur, Vertheilung und Verlauf, etc. Leipz. 1823, 8°. fig. col. —Bibl. med.-ch. p. 54.

6. Darstellung der Saugadern des menschlichen Körpers, nach ihrer Struktur, etc. Leipz. 1828, 8°. fig.—Bibl. med.-ch. p. 54.

7. Darstellung des Gehirns, des Rückenmarkes u. der Sinneswerkzeuge, sowie auch des menschl. Körpers überhaupt nach seinem äussern Umfange. Leipz. 1824, 8°. fig.—Bibl. med.-ch. p. 54.

8. Accurata Nervorum Spinalium Descriptio. (Trad. Lat. par *Fr. Hænel.*) Leipz. 1828, 8°. fig.—Bibl. med.-ch. p. 55.

9. Die Rückenmarksnerven nach ihrem ganzen Verlaufe, etc. Leipz. 1827, fol. fig.—Bibl. med.-ch. p. 54.

10. Beschreibung des fünften Nervenpaares u. seiner Verbindungen mit andern Nerven, vorzügl. mit dem Ganglionsystem. Meissen, 1817, fol. fig.—Nachtrag zu diesem Werke, Meiss. 1821, fol. fig. —Bibl. med.-ch. p. 54.

11. Darstellung der weiblichen Geburtsorgane sowohl im unbe-

schwängerten als im beschwängerten Zustande, etc. Leipz. 1825, 8°. fig.—Bibl. med.-ch. p. 54.

12. Darstellung der Organe der Respiration, des Kreislaufes, der Verdauung, des Harnes u. der Fortpflanzung, etc. Leipz. 1825, 4°. fig.—Bibl. med.-ch. p. 54.

13. Der menschliche Körper nach seinem äussern Umfange, etc. Leipz. 1824, fol. fig.—Bibl. med.-ch. p. 45.

14. Chirurgisch-anatomische Tafeln, oder Abbildung der Theile des menschlichen Körpers in Bezug auf chirurgische Krankheiten u. Operationen. 1833, fol. 13 pl. (*Tabulæ chirurgico-anatomicæ, seu icones partium corporis humani, ratione perpetua habita morborum et operationum chirurgicarum.*)

BOCK (C. E.).

1. Handbuch der Anatomie des Menschen. Leipz. 1838, 2 vols. 8°. (ed. 2ᵃ). Leipz. 1840, 8°. (ed. 3ᵃ). Ibid. 1842, 8°.—(Belg. *P. H. Pool*): *Handboek der Ontleedkunde van den Mensch*, etc. Amst. 1842, 3 vols. 8°.

2. Anatomisches Taschenbuch, enthaltend die Anatomie des Menschen, systematisch, im ausführlichen und übersichtlichen Auszuge zur schnellern u. leichtern Repetition bearbeitet. Leipz. 1839, 16°.—(2ᵗᵉ Aufl.) Leipz. 1841, 8°.

3. Hand-Atlas der Anatomie des Menschen, nebst einem tabellarischen Handbuche der Anatomie. Leipz. 1841, 8°.

BOCK (Fr. Sam.).

1. Nachricht von einem Preuss. Naturalienkabinet so sich in dem Saturguschen Garten zu Königsberg befindet. 8°. Königsb. 1764.

2. Beschreibung einer noch unbekannten vielkammerigen Seetulpe (*Echinis*).—Naturf. XII. 1778, p. 168, fig.—*Mod.* Bibl. Helm.

3. Versuch einer wirthschaftlichen Naturgeschichte von dem Königreiche Ost- und Westpreussen (*Essai d une Histoire naturelle de la Prusse orientale et occidentale*). Dessau, 1782–1784, 5 vols. 8°. fig. (Allem.)—Biogr. Un. IV. p. 630.

4. Preussische Ornithologie.—Naturf. VIII. p. 39 ; IX. p. 39 ; XII. p. 131 ; XIII. p. 201 ; XVII. p. 66.—Biogr. Un. IV. p. 630.

5. Versuch einer vollständigen Natur- u. Handlungsgeschichte der Heringe (*Essai sur l'Histoire naturelle et le Commerce des Harengs*). Königsb. 1769, 8°. (Allem.)—Biogr. Un. IV. p. 630.— *Böhm.* Bibl. II. 2, p. 111.

6. Betrachtungen über das Nutzbare und Angenehme der Naturgeschichte. Königsb. 1767, 8°.—*Böhm.* Bibl. I. 1, p. 176.

7. Versuch einer kurzen Naturgeschichte des Preussischen Bern-

steins und einer neuen wahrscheinlichen Erklärung seines Ur-
sprungs. Königsb. 1767, 8°.—*Böhm.* Bibl. IV. 1, p. 472.

Bock (Sam.).

1. Dissertatio de Membranâ deciduâ *Hunteri.* Bonn, 1831, 4°. fig.
—Bibl. med.-ch. p. 55.

2. Ueber Analogien der Knochen und Muskeln. Nürnb. 1832, 8°.
fig.—Bibl. med.-ch. p. 55.

Boctis (Gaët. de).

1. Raggionamento istorico del Incendio del Vesuvio accaduto nel
mese d' Ottobre. Nap. 1768, 4°.—*Böhm.* Bibl. IV. 2, p. 380.

2. Raggionamento istorico intorno a nuovi Vulcani comparsi nella
fine dell' anno scorso 1766, nello territorio della Torre del Greco.
Nap. 1761, 4°.—*Böhm.* Bibl. IV. 2, p. 374.

Boddaert (P.).

1. Dierkundig Mengelwerk. Utrecht, 1776, 4°.—Bibl. Ent. I. p. 35.

2. Notice des principaux Ouvrages zoologiques culuminés, avec la
table des planches de *Daubenton.* Utr. 1783, fol.—Bibl. Ent. I.
p. 35.

3. Elenchus Animalium. Vol. I. sistens Quadrupedia. Rotterd.
1785, 8°.—*Cuv.* R. An. III. p. 337.

4. Abhandlung von Amphibien.—Schr. Berl. Ges. Naturf. Fr. II.
p. 369.

5. Relation d'une Tortue à cuir trouvée dans la mer de Toscane.—
Gaz. de Santé, 1761, No. 6.—*Dum.* et *Bib.* I. p. 422.

6. Epist. de Ranâ bicolore. Amst. 1770.—Biogr. Un. Suppl. LVIII.
p. 412.

7. Epistola de Testudine cartilagineâ. Amsterd. 1770, 4°. fig.—
Dum. et *Bib.* I. p. 424.

8. Specimen novæ Methodi distinguendi Serpentia.—N. Act. Nat.
Cur. VII. p. 12.—Dict. Sc. n. XV. p. 252.

9. Epist. de Chætodonte diacantho. Amst. 1772.—Biogr.Un. Suppl.
LVIII. p. 412.

10. Epist. de Chætodonte Argo ex Museo *Schlosseri.* Amst. 1770,
fig.—Biogr. Un. Suppl. LVIII. p. 412.

11. Natuurkundige beschouwing der Dieren. 8°. Utrecht, 1778.

12. Verhandeling over de Insecten algem. geneeskund, IV. p. 157.
(Ouvr. période.)— Bibl. Ent. I. p. 35.

13. Brief aan den Schryver der Bedenkingen over den dierlyken

Oorsprong der Koral-Gewassen. Utr. 1771, 8°.—Dict. Sc. n. LX. p. 536.

14. List der Plantdieren beschryven door der *Herr Pallas*, mit Anmerkingen. Utrecht, 1768, 8°.—Dict. Sc. n. LX. p. 536.

15. Abhandlung ueber den Affen mit dem Schweinkopfe (*Simia porcaria*, B.).—Naturf. XXII. 1787, p. 1.

16. Beschreibung Zweyer merkwürdiger Fische (*Sparus palpebratus* et *Muræna colubrina*).—Neue Nordische Beytr. II. 1781, p. 55, cum fig.

17. Natuurbeschouwer, of Verzameling van de nieuwste verhandelingen over de drie ryken der natuur, welke thans in Duitschland uitgegeven worden. 2 vols. 8°. 's Gravenhage, 1779–81.

18. Kort begrip van het zamenstel der natuur van *C. Linnæus*, met zeer veele zoorten vermeerded. 8°. Amst. 1783.

19. Beschreibung einiger seltenen Amboinischen Fische und des *Sargus palpebratus.*—Schr. Berl. Ges. Naturf. Fr. III. p. 458.

BODDINGTON (J.).

1. On some Bones found in the Rock of Gibraltar.—Phil. Trans. LX. p. 414.

BODEI ( ) e BRIGNOLE ( ).

1. Alcuni Cenni sulle Produzioni naturali del Dipartimento del Metauro. Urbino, 1813, 8°.

BODEN (Alb.).

1. Die Anatomie des Menschen nach den besten Hilfsmitteln tabellarisch zusammengestellt, nebst Anleitung zur Präparation der einzelnen Körpertheile. Leipz. 1840, fol.

BODICHON ( ).

1. Notice sur un Renard à longues oreilles, apporté d'Alger en 1836. —C. R. IV. p. 649.

2. Observations sur une espèce du genre *Canis*, habitant le désert de Sahara et certaines vallées de l'Atlas.—C. R. III. p. 251.— Ann. Sc. n. 1836, VI. p. 156.—*Wiegm.* Arch. 1837, II. p. 160 (*Ueber einen Fuchs in der Wüste Sahara*).

BODINUS (J.).

1. Universæ Naturæ Theatrum, in quo rerum omnium effectrices causæ et fines contemplantur et discutiuntur. Lugd. 1596, 8°. —Francof. 1597, 8°.—Hanov. 1605, 8°.—(Gall.) *Le Théâtre de la Nature.* Lyon, 1597, 8°.—*Böhm.* Bibl. I. 1, p. 237.

BOECK (Chr.).

1. Observations sur le *Vibrio Lunula,* et sur les limites entre les règnes animal et végétal.—Mag. för Naturv. 1826, I. p. 40, fig. —*Féruss.* Bull. 1827, XII. p. 365.

2. Notice sur les Trilobites.—Mag. för Naturv. 1827, I. p. 11, fig. —*Féruss.* Bull. 1828, XIV. p. 146.

3. Uebersicht der bisher in Norwegen gefundenen Formen der Trilobiten-Familie.—*Keilh.* Gæa Norw. 1838, I. p. 138.—*L.* u. *Br.* N. Jahrb. 1841, p. 724.

BOECK (C. J. A.).

1. De Spinis Hystricum. 4º. Berlin, 1834.

BOECK (J. A.).

1. Beschreibung der Bienen. Neustadt, 1709, 8º.—Bibl. Ent. I. p. 35.

BOECLER (J.).

1. Dissertatio de Pluviâ. Argent. 1710, 4º.—*Böhm.* Bibl. V. p. 52.

BOER (H. X.).

1. Versuch einer Darstellung des kindlichen Organismus in physischer, pathologischer u. therapeutischer Hinsicht. Wien, 1813; 1818, 8º.—Bibl. med.-ch. p. 56.

BOERHAAVE (Kaau).

1. Observationes anatomicæ.—N. Comm. Ac. Petr. 1748, I. p. 353.

2. Dissertatio de cohæsione solidorum in corpore animali.—N. Comm. Ac. Petr. 1753, IV. p. 343.

3. Observatio anatomica Musculi in pectore præternaturalis et varii in diversis corporibus inventi.—N. Comm. Ac. Petr. II. p. 257, fig.

4. Historia anatomica Ovis pro hermaphrodito habiti.—N. Comm. Ac. Petr. 1748, I. p. 315, fig.

BOERHAAVE (Herm.) et RUYSCH (F.).

1. Opusculum anatomicum de fabricâ Glandularum in Corpore humano. 4º. Lugd. Bat. 1722.—Cat. R. Soc. L.

BOERHAM (H.).

1. Essays upon the Silkworm. Lond. 4º.—*Böhm.* Bibl. II. 2, p. 264.

Boethius (Jac.).

1. Alvearia et arbores à Formicis tuenda.—Schwed. Abh. 1763,
p. 34.—Bibl. Ent. I. p. 36.

Boetius (Ans.).

1. Gemmarum et Lapidum Historia, quam posteà *Adr. Follius* recensuit. Lugd. Bat. 1636; 1647, 8°. fig.—(Han. 1609, 4°. fig.)
Burgæ, 1609, 8°.—*Klein*, Disp. Echinod. p. viii.—*Mod.* B. Helm.

Bogdanus (Mart.).

1. Apologia pro Vasis lymphaticis *Bartholini* adversus insidias se-
cundò structas ab *Ol. Rudbeck.* Cop. 1654, 4°.—Biogr. Un. IV.
p. 673.
2. *Rudbeckii* Insidiæ structæ Vasis lymphaticis *Th. Bartholini.*
Francf. et Cop. 1654, 4°.—Biogr. Un. IV. p. 673.

Bogg (E.).

1. On the Geology of the Lincolnshire Wolds.—Trans. Geol. Soc.
ser. 1, III. p. 392.

Bögner (J.).

1. Die Entstehung der Quellen und die Bildung der Mineral-Quel-
len, nebst einem Berichte über die kürzlich bei Assmannshausen
gefundene warme u. die bei Weilbach gefundene Kalte Mineral-
quelle. Frankf. 1843, 8°.

Bogros (J. A.).

1. Mémoire sur la structure des Nerfs.—Ann. Sc. n. XIII. 1828,
p. 5.—Isis, 1834, X. p. 1006 (*Ueber die Struktur der Nerven*).
2. Note sur des Canaux découverts dans les Nerfs.—Ann. Sc. n.
1825, V. p. 325.—Isis, 1831, VII. p. 802 (782) (*Ueber Canäle in
den Nerven*).

Bohadsch (J. B.).

1. Nachricht von einer besondern Versteinerung.—*Licht.* Mag. I.
2. Diss. de veris Sepiarum Ovis. Prag. 1752, 4°. fig.—Biogr. Un.
IV. p. 679.—*Böhm.* Bibl. II. 2, p. 425.—*Mod.* B. H.
3. De quibusdam Animalibus marinis eorumque proprietatibus vel
nondum vel minus notis, etc. Dresdæ, 1761, 4°. fig.—(Germ.)
Ludwig, mit Anmerk. v. *Leske*; Dresd. 1776, 4°. fig. (*Abhandlung v. verschiedenen bisher wenig oder nicht bekannter Seethierer*).
—*Mod.* B. H.—*Böhm.* Bibl. II. 1, p. 43.

4. Positiones zoologicæ.—*Klink.* Coll. Dissert. I. Nᵒ 4.—*Böhm.* Bibl. II. 1, p. 43.

5. Bericht über seine Reise nach dem Oberösterreichischen Salzkammerbezirk.—Prag. Ath. V. p. 91.—*Mod.* B. H.

BÖHAIMB (C. W.).

1. Allgemeine Naturgeschichte. 16°. Augsb. 1829.

BOHEMAN (Ch. H.).

1. Observations sur la Métamorphose de quelques Insectes.—Vetensk. Acad. Handl. 1828, p. 164.—*Féruss.* Bull. 1831, X. p. 110.

2. Novæ Coleopterorum species.—N. Mém. Natur. Mosc. I. p. 101. —Bibl. Ent. I. p. 36.

3. En ny art af Insect-slägtet *Pimpla.*—Vetensk. Acad. Handl. 1821, p. 335.

4. *Calodromus*, genus è familia Curculionidum.—Mém. Ac. Stockh. 1837, fig.—Rev. Zool. 1840, p. 311.

5. Skandinaviska Pteromaliner.—Vetensk. Acad. Handl. 1833, p. 329 ; 1835, p. 222.

6. Observationes in *Derbe* genus, una cum specierum quinque novarum descriptionibus.—Mém. Acad. Stockh. 1837, 8°.—Rev. Zool. 1840, p. 311.

7. Arsberättelse om framstegen i Insekternas, Myriapodernas och Arachnidernas historia under åren 1840–1842, Stockholm, 1844 ; 1843 och 1844. 8°. Stockholm, 1845.

8. Beskrifning af de i Sverige funne arterue af Insekt-slägtet *Ceraphron.*—Vetensk. Acad. Handl. 1831, p. 322.

BOHL (J. Chr.).

1. Dissertatio sistens Historiam naturalem Viæ lacteæ Corporis humani, per extispicia animalium olim detectæ, nunc insolito ductu chilifero genuino auctæ, cum notis, etc. Königsb. 1741, 4°. fig. —Biogr. Un. Suppl. LVIII. p. 438.

2. Dissertatio epistolaris de usu novarum Cavæ propaginum in Systemate chylopæo. Amst. 1727, 4°.—Biogr. Un. Suppl. LVIII. p. 437.

BÖHLER ( ).

1. Die Hirsche in verschiedenen Stellungen nach dem Leben gezeichnet. fol. Leipz. 1802.

BÖHM (A.).

1. Abhandl. Seeländ. Gesellsch.; *vide sup.* Pars I. p. 20.

z 2

Böhm (L.).

1. De Glandularum intestinalium structurâ penitiori. Berl. 1835, 4°. fig.—*Val.* Repert. I. p. 8.—Bibl. med.-chir. p. 55.—*Müll.* Arch. 1836, p. xxxv (*Untersuch. über die in der Schleimhaut des Darmcanals vorkommenden Drüsen*).

Böhm (M. J.).

1. Nachtrag zu v. *Hoffmansegg* alphabetische Verzeichnisse von *Hübners* Papilionen.—*Illig.* Mag. V. 1806, p. 176.

2. Vorschlag zu einer neuen Tödtungsmethode hartschaliger Insecten.—*Illig.* Mag. III. p. 222.—Bibl. Ent. I. p. 35.

Böhmer (G. R.).

1. Systematisch-litterärisches Handbuch der Naturgeschichte, Oeconomie u. anderer damit verwandten Wissenschaften und Künste. Leipz. 1785–1789, 8°. 9 vols. (*Bibliotheca scriptorum Historiæ naturalis, Œconomiæ*, etc.)—Bibl. Ent. I. p. 36.—Biogr. Un. IV. p. 653.

2. De Experimentis *Reaumurii* ad Digestionis modum in variis animalibus declarandum institutis. Witteb. 1757, 4°.—Biogr. Un. IV. p. 654.—*Böhm.* Bibl. II. 1, p. 530.

3. Programma de Rebus naturalibus, Vermibus intestinalibus, etc. Witteb. 1796–1797, 4°.

Böhmer (J. K. H.).

1. Entdeckung des Blasenwurms im Gehirn der Schafe.—N. Oekon. Nachr. Schles. Ges. I. p. 241.

Böhmer (Ph. Ad.).

1. Institutiones osteologicæ. Halæ, 1751, 8°. fig.—Bibl. med.-ch. p. 57.

2. Observationes rariores anatomicæ. Observationum anatomicarum rariorum Fasciculus (I. et II.), notabilia circa Uterum humanum continens, cum figuris ad vivum expressis. 2 vols. fol. Hal. 1752–1756.—Bibl. med.-ch. p. 57.

Bohn (J.).

1. Circulus anatomico-physiologicus, seu Œconomia corporis animalis, etc. Lips. 1680; 1686; 1697; 1710, 4°.—Act. Erudit. V. p. 225.—Biogr. Un. IV. p. 685.

2. Observationes quædam anatomicæ, structuram Vasorum biliariorum et motum bilis spectantes. Lips. 1682–1683, 4°.—Act. Erudit. I. p. 20.—Biogr. Un. IV. p. 685.

3. Observatio circa Bilis motum ab Hepate ad Vesicam biliariam et intestinum.—Act. Erudit. II. p. 126.

4. Observatio circa proportionem partis purpureæ ac substantiæ seroso-gelatinosæ Sanguinis, intra vasa animalium fluctuantis.— Act. Erudit. I. p. 126.

5. De renunciatione Vulnerum, seu Vulnerum lethalium Examen, exponens horum formalitatem et causam, tam in genere, quam in specie, ac per singulas corporis partes. Lips. 1689, 8°.—Act. Erudit. VIII. p. 101.

6. Observatio singularis circa Venæ pulmonalis propaginem, tussi sanguinolentâ rejectam.—Act. Erudit. II. p. 218.

7. Observatio atque Experimenta circa usum Spiritus Vini externum, in Hæmorrhagiis sistendis.—Act. Erudit. II. p. 153.

8. Dissert. de Lapide Ceraunio. Lips. 1661, 4°.—*Böhm.* Bibl. IV. 2, p. 318.

BOHNER (Leonh.).

1. Diss. de Varietate in formis Animalium externis, tanquam indice existentiæ divinæ. 1725, 4°.—*Böhm.* Bibl. II. 1, p. 80.

BOHTLINK (W.).

1. On the Principal Traces left by the last great Revolution which took place in the mountainous countries of Scandinavia.—Edinb. New Phil. Journ. XXXI. p. 253.

BOJANUS (L.).

1. Ein Wort über das Verhältniss der *Membrana decidua* und *decidua reflexa* zum Ei des menschlichen Embryo.—Isis, 1821, III. p. 268.

2. De Fœtûs canini Velamentis, imprimis de Allantoide.—Mém. Ac. Pétersb. V. 1815.—*Cuv.* in J. d. Sav. Janv. 1817.—Isis, 1817, VII. p. 876 (*Abhandl. über die Hüllen des Hundsfœtus*, etc.).

3. Observatio anatomica de Fœtu canino 24 dierum ejusque Velamentis.—N. Act. Nat. Cur. X. pp. 139, 723, fig.

4. Ueber die Darmblase des Schaafsfœtus, zum Beweise dass die *Vesicula umbilicalis* mit dem Darm unmittelbar zusammenhängt. —*Meck.* Arch. I. 1818.—Isis, 1818, X. p. 1623, fig.

5. Ueber die Darmblase des Pferdefœtus, zum Beweise, dass die *Vesicula umbilicalis* mit dem Darm unmittelbar zusammenhängt. —*Meck.* Arch. IV. p. 34, fig.—Isis, 1818, X. p. 1633, fig.

6. Dottergang im Fœtus des *Coluber berus.*—Isis, 1818, XII. p. 2093.—Journ. de Phys. LXXXIX. p. 65 (*Notice sur le Canal vitello-intestinal du Fœtus de la Vipère*).

7. Anatome Testudinis Europææ. Vilna et Leipz. 1819–1821, fol. fig.—Isis, 1819, XI. p. 1766; 1822, VIII. p. 886; 1823, VII. p. 750; 1824, IV. p. 465.

8. Craniorum Argalidis, Ovis et Capræ domesticæ Comparatio.— N. Act. Nat. Cur. XII. 1, p. 293, fig.—Isis, 1825, II. p. 202.— *Féruss.* Bull. 1826, VIII. p. 393.

9. Adversaria ad Dentitionem equini generis et Ovis domesticæ spectantia.—N. Act. Nat. Cur. XII. 2, p. 697, fig.—Isis, 1825, XII. p. 1318.—*Féruss.* Bull. 1826, IX. p. 340.

10. Nachtrag zu *Distoma hepaticum.*—Isis, 1821, III. p. 305.

11. Anatomie des Blutegels.—Isis, 1817, VII. p. 873.

12. Was wissen wir denn nun eigentlich vom Bau des Blutegels?— Isis, 1818, XII. p. 2089.

13. De Uro nostrate ejusque sceleto Commentatio, Bovis primigenii sceleto aucta.—N. Act. Nat. Cur. XIII. p. 411, fig.—*Féruss.* Bull. 1829, XVI. p. 121.—Isis, 1829, II. p. 131.

14. Bemerkungen aus dem Gebiete der vergleichenden Anatomie. —Russ. Samml. f. Naturw. u. Heilk. II. 4.—Isis, 1818, VIII. p. 1425.

15. Sendschreiben an *Cuvier,* über die Athem- u. Kreislaufwerkzeuge der zweischaaligen Muscheln, insbesondere des *Anodon cygneum.* 1818.—J. de Phys. LXXXIX. p. 108 (*Mémoire sur les Organes respiratoires et circulatoires des Coquillages bivalves en général, et de l'Anodonte en particulier ; et Observations à ce sujet par M. de Blainville*).—Isis, 1819, I. p. 41.

16. Antwort auf *Blainville*'s gemachte Einwendungen, in Betreff der Athemwerkzeuge der zweischaaligen Muscheln.—Isis, 1820, VII. p. 404.

17. Ueber die Nasenhöhle u. ihren Sack-Anhang in den Pricken.— Isis, 1821, XII. p. 1167.

18. Vasa chylifera Testudinis Europææ.—Isis, 1821, III. p. 270.

19. Versuch einer Deutung der Knochen im Kopfe der Fische.— Isis, 1818, III. p. 498.

20. Bemerkungen in Bezug auf die Deutung der Kopfknochen im Fische.—Isis, 1818, XII. p. 2095.

21. Weiterer Beitrag zur Deutung der Schädel-Knochen.—Isis, 1819, VIII. p. 1360.

22. Abermals ein Wort zur Deutung der Kopfknochen.—Isis, 1821, XII. p. 1145.

23. Kurze Nachricht über die Zerkarien u. ihren Fundort.—Isis, 1818, IV. p. 729.—*Sieb.* in *Wiegm.* Arch. 1835, I. p. 76.

24. Enthelmintica.—Isis, 1821, II. p. 162.—*Wiegm.* Arch. 1835, I. p. 69.

25. De Merycotherii sibirici, seu gigantei Animalis ruminantis dentibus, antediluviano quodam incerto Sibiriæ loco erutis, declarato vestigio, Commentatio.—N. Act. Nat. Cur. XII. 1, p. 263.—Isis, 1825, II. p. 201.—*Féruss.* Bull. 1824, III. p. 226 (*Dissert. sur des Dents trouvées en Sibérie, considérées comme ayant appartenu à un grand Ruminant antédiluvien, nommé Merycotherium sibiricum*).

26. Introductio in Anatomiam comparatam. 8°. Wilna, 1815.

27. Observations nouvelles sur l'organisation de la Sangsue (*Hirudo medicinalis*).—Journ. de Phys. LXXXVIII. 1819, p. 468.

Boid ( ).

1. A Description of the Azores or Western Islands from personal observation, comprising Remarks on their peculiarities, topographical, geological, statistical, etc.   Lond. 1835, fig.

Boie (Jan.).

1. Journal holden paa Reysen til China. Copenh. 1745, 8°.—*Böhm.* Bibl. I. 1, p. 693.

Boie (Fr.).

1. Bemerkungen über zu den Temminkschen Ordnungen Cursores, Grallatores, Pinnatipedes, und Palmipedes gehörige Vögel; mit besonderer Rucksicht auf die Herzogthümer Schleswig und Holstein.— *Wiedemann,* Zool. Mag. I. 3, 1819, p. 92.

2. Notices diverses relatives à l'Histoire naturelle.—Isis, 1827, VIII. et IX. p. 726.—*Féruss.* Bull. 1829, XVIII. p. 206.

3. Ueber die Abtheilungen im natürlichen Thiersystem.—Isis, 1828, III. p. 351.—*Féruss.* Bull. 1829, XVI. p. 271 (*Rem. sur les Coupes du Système naturel des Animaux*).

4. Tagebuch, gehalten auf einer Reise durch Norwegen im Jahr 1817. Schlesw. 1822, 8°.—Isis, 1823, I. p. 95.

5. Beiträge zur Naturgeschichte der Säugethiere. 1te Lief.—Isis, 1823, IX. p. 964.

6. Beiträge zur Naturgeschichte europäischer vierfüssiger Thiere. 2te Lief.—Isis, 1825, XI. p. 1199.—*Féruss.* Bull. 1828, XIII. p. 235 (*Rem. concernant l'hist. nat. des Mammifères de l'Europe*).

7. Neue Fledermausgattung (*Leuconoë*).—Isis, 1830, III. p. 256.

8. Beschreibung drei neu entdeckter Eulen.—Isis, 1835, IV. p. 323.

9. Generalübersicht der ornithologischen Ordnungen, Familien u. Gattungen.—Isis, 1826, X. p. 975.

10. Ueber Classification, insonderheit der europäischen Vögel.—
Isis, 1822, V. p. 545.

11. Fernere Bemerkungen über Classification der Vögel.—Isis,
1833, IX. p. 876.

12. Ornithologische Beiträge. I.-Vte Lief.—Isis, 1822, VII. p. 768;
VIII. p. 871; 1823, VI. p. 664; 1827, XI. p. 959; 1828, III.
p. 300; 1835, III. p. 251.—*Brehm*, Ornis. II. p. 97; III. p. 54,
151.—*Féruss.* Bull. 1829, XVI. p. 126 (*Notices ornithologiques*,
4e livr.).

13. Verschiedenheit der Nahrungsmittel bei Vögeln.—Isis, 1831,
X. p. 1097.

14. Bemerkungen über Species und einige ornithologischen Familien und Sippen.—Isis, 1831, V. p. 538.

15. Bemerkungen zu *Gloger*'s Abändern der Vögel, etc.—Isis, 1834,
VI. p. 385.—*Wiegm.* Arch. 1835, II. p. 300.

16. Veränderung des Farbenkleides der Gattung *Colymbus* durch
die Herbstmäuser.—Isis, 1830, III. p. 257.

17. *Cygnus Bewickii*, Yarr. bei Dünkirchen erlegt.—Isis, III. p. 262.
—*Wiegm.* Arch. 1836, II. p. 274.

18. Bemerkungen über mehrere neue Vögelgattungen.—Isis, 1828,
III. p. 312.—*Féruss.* Bull. 1829, XVII. p. 286 (*Remarques sur
plusieurs nouveaux genres d'Oiseaux*).

19. *Hydrobates, Lestris* u. *Puffinus.*—Isis, III. p. 253.—*Wiegm.*
Arch. 1836, II. p. 274.

20. *Anthus Richardi* et *Emberiza lapponica.*—Isis, 1834, IV. p. 385.

21. *Sterna leucopareia*, Natt., gesellig in der Gegend von Paris.—
Isis, III. p. 259.

22. Generalübersicht der Familien u. Gattungen der Ophidier.—
Isis, 1826, X. p. 981.

23. Zu *Kaup*'s Aufsatz über Lurche.—Isis, 1825, X. p. 1089.—
Bull. Un. Sc. VII. p. 345 (*Corrections au Mémoire de Kaup*, etc.).
—*Dum.* et *Bib.* I. p. 306.

24. Bemerkungen über *Merrem*'s Versuch eines Systems der Amphibien. 1 Lief. Ophidier. Marb. 1820.—Isis, 1827, VI. p. 508.
—*Féruss.* Bull. 1828, XIII. p. 357.

25. Ueber eine noch nicht beschriebene Art von *Cordylus*, Gron.
(*Cordylus cataphractus*, Boie). Mitgetheilt von *Max. Pr. zu
Wied.*—N. Act. Nat. Cur. XIV. p. 139.—*Féruss.* Bull. 1829,
XVII. p. 437.

26. Ueber das Leuchten einiger Batrachier.—Isis, 1827, IX. p. 726.

27. Beiträge zur Geschichte der Insecten. 1te Lief.—Isis, 1833,
VII. p. 663.

28. Notice sur l'Erpétologie de Java.—*Féruss.* Bull. 1826, IX. p. 233.

29. Caractères de quelques espèces de Reptiles du Japon.—Isis, 1826, II. p. 203.—*Féruss.* Bull. 1827, X. p. 160.

30. Lettre à *M. Wagler*, sur quelques Oiseaux et Reptiles de l'Ile de Java.—Isis, 1827, p. 724.—*Féruss.* Bull. XVI. p. 127.

31. Ueber Verwüstung der Wiesen durch Insecten-Larven.—Isis, 1835, IV. p. 331.

32. Entomologische Beiträge. 6te Lief. (Beobachtungen über mehrere Eulen).—Isis, 1835, IV. p. 319.—*Burm.* in *Wiegm.* Arch. 1836, II. p. 319.

33. Ueber *Mygale avicularia.*—Isis, 1827, IX. p. 729.

34. Ueber *Actora æstuum.*—Isis, 1827, IX. p. 728.

35. Fortegnelse over danske, slesvig holsteenske og lauenborgske Sommerfugle.—*Kröy.* Tidskr. ser. 1, I. pp. 505, 521 ; II. p. 127. —Isis, 1841, pp. 115, 199.

36. Zur Verwandlungsgeschichte inländischer Zweiflügler.—*Kröy.* Tidskr. ser. 1, II. p. 234.

37. Zur Geschichte inländischer Amphibien.—*Kröy.* Tidskr. ser. 1, II. p. 207.

38. Entomologische Beiträge.—*Kroy.* Tidskr. ser. 1, III. p. 315.

39. Ueber eine Race langhaariger Katzen.—*Kröy.* Tidskr. ser. 1, III. p. 325.

Boie (H.).

1. Kenteekenen van eenige Japansche Amphibien. Bijdragen tot de Natuurk. Wetensch.—Aul. II. 1827, p. 243.

2. Briefe aus Java an *Schlegel.*—Isis, 1828, pp. 724, 1025.—*Féruss.* Bull. 1829, XVI. p. 127 (*Lettres sur quelques Oiseaux et Reptiles de l'Ile de Java).—Dum.* et *Bib.* I. p. 307.

3. Merkmale eininger Japanischen Lurche.—Isis, 1826, II. p. 203. —*Féruss.* Bull. 1827, X. p. 160 (*Caractères de quelques espèces de Reptiles du Japon*).

4. Bemerkungen über die von Hrn. *v. Spix* abgebildeten brasilianischen Saurier.—Isis, 1826, I. p. 117.

Bojer (W.).

1. Letter on the Habits, etc. of *Cryptoprocta ferox,* Benn.—Proc. Zool. Soc. II. p. 13.

Böing (H.).

1. De Circulationis Mechanismo. Berol. 1839, 8°.

Bois (F. du). *Vide* Dubois de Montpereux (F.).

Boisduval (J. A.).

1. Genera et Index methodicus Europæorum Lepidopterorum. Par. 1840, 8°.—Ann. Sc. n. (2ᵉ S.) XVI. p. 379.—Rev. Zool. 1841, p. 16.

2. Entdeckung einer Art Tasche bei den Männchen mehrerer Colias-Arten am Vorderrande der Hinterflügel.—Ann. Soc. Ent. Fr. V. p. x.—*Erichs.* in *Wiegm.* Arch. 1837, VI. p. 327.

3. Partie entomologique de la Relation du Voyage autour du monde en 1826–1829 par *M. Dumont-d'Urville.* 5 livr. fol. Paris.—Bibl. Ent. I. p. 37.

4. Faune entomologique de l'Océanie, contenant la description de toutes les espèces de Coléoptères découvertes jusqu'à ce jour dans cette partie du monde et les espèces des autres ordres rapportées par l'expédition de l'Astrolabe. Paris, 1835, 2 vols. 8°. atl—Ann. Soc. Ent. Fr. IV. p. xxxv.—*Wiegm.* Arch. 1836, II. p. 297.

5. Faune entomologique de Madagascar, Bourbon et Maurice; partie des Lépidoptères, avec des notes sur leurs mœurs par *Sganzin.* Paris, 1833, 8°. fig. (par livr.)—N. Ann. Mus. II. p. 149 (*Description des Lépidoptères de Madagascar*).

6. Icones historiques des Lépidoptères d'Europe nouveaux ou peu connus. Paris (par livr.).—Ann. Soc. Ent. Fr. I. p. 116.

7. Spécies général des Lépidoptères. Paris, 1836, 8°.—Ann. Sc. n. 1836, VI. p. 128.—Mag. Zool. and Bot. I. p. 568.

8. Europæorum Lepidopterorum Index methodicus. Paris, 1829, 8°.—*Cuv.* R. An. III. p. 338.

9. Histoire naturelle des Insectes Lépidoptères (Suites à *Buffon*). Paris, 8°. fig.

10. Description de deux Lépidoptères nouveaux d'Espagne (*Argus Marchandii, Anthophila Sancti-Florentis*).—*Silberm.* Rev. Ent. II. p. 120, fig.—Bibl. Ent. I. p. 38.

11. Observations sur un Mémoire de *M. Zinken-Sommer*, sur des Lépidoptères de Java.—Ann. Soc. Ent. Fr. I. p. 416.

12. Monographie des Zygénides, suivie d'un Tableau méthodique de classification des Coléoptères. Par. 1828, 8°.—*Féruss.* Bull. 1828, XIII. p. 261.

13. Notice sur cinq espèces nouvelles de Lépidoptères d'Europe. Paris, 1827, 8°. fig.—Ann. Soc. Linn. Par. 1827, VI. 1, fig.—

Mém. Soc. Linn. Par. VI. 1828, p. 109, fig.—*Féruss.* Bull. 1827, XII. p. 179.

14. Description de quatre nouvelles espèces de Noctuélides.—Ann. Soc. Ent. Fr. II. p. 373, fig.

15. Notice sur un nouveau genre de Noctuélides (*Dianthœcia*).— *Silberm.* Rev. Ent. II. p. 245.—*Wiegm.* Arch. 1835, II. p. 61.

16. Anomalie du genre *Urania.*—Ann. Soc. Ent. Fr. II. p. 248.

BOISDUVAL (J. A.) et LACORDAIRE (   ).

1. Faune Entomologique des environs de Paris, ou Spécies général des Insectes de tous les ordres connus dans un rayon de 15 à 20 lieues de la capitale. Paris, 1835, 12°. fig.—Ann. Soc. Ent. Fr. IV. p. XLIX.—*Burm.* in *Wiegm.* Arch. 1836, II. p. 296.

BOISDUVAL (J. A.) et LECOMTE (J.).

1. Histoire générale et Iconographie des Lépidoptères et des Chenilles de l'Amérique septentrionale. Paris (par livr.).—Ann. Sc. n. 1829, XVII.—Rev. Bibl. p. 89.

BOISDUVAL (J. A.) et DEJEAN (   ).

1. Iconographie et Histoire naturelle des Coléoptères d'Europe. Paris, 1829, 8°.—Ann. Sc. n. 1829, XVIII.—Rev. Bibl. p. 106.

BOISDUVAL (J. A.), RAMBUR (P.) et GRASLIN (A.).

1. Collection iconographique et historique des Chenilles d'Europe. Paris, 8°. (par livr.)—Ann. Soc. Ent. Fr. I. p. 116.

2. Essai sur une Monographie des Zygénides. 8°. Paris, 1829.

BOISJUGAN (De).

1. Nouveau Traité des Abeilles, et nouvelles Ruches de paille. Caen, 1771, 8°.

BOISRAGON (Ph. S. G.).

1. Illustrations of Osteology. fol. Lond. 1839.

BOISSIER DE SAUVAGES ; *vide* SAUVAGES.

BOISSONNEAU (A.).

1. Classification méthodique d'Ornithologie européenne sur étiquettes. Par.—Rev. Zool. 1840, p. 350.

2. Nouvelles espèces d'Oiseaux-Mouches de Santa-Fé de Bogota.— Rev. Zool. 1839, p. 254.

3. Oiseaux nouveaux ou peu connus de Santa-Fé de Bogota.—Rev. Zool. 1840, pp. 2, 66.

4. Nouvelle espèce du genre Pic.—Rev. Zool. 1840, p. 36.

BOISSY (St. Ange de).

1. Description de quelques espèces d'Hélices fossiles provenant principalement des terrains d'eau douce du midi de la France.— Rev. Zool. 1839, p. 74.

2. *Helicina Ambicliana* des Antilles.—Mag. Zool. V. p. 68.

3. *Helix lanuginosa* de l'Ile Majorque.—Mag. Zool. V. p. 69.

BOITARD (M. P.).

1. Le Jardin des Plantes : description et mœurs des Mammifères de la Ménagerie et du Muséum d'Histoire naturelle. 8°. Paris, 1842.

2. Manuel d'Histoire naturelle, comprenant les trois Règnes de la Nature, ou Genera complet des Animaux, des Végétaux et des Minéraux. Paris, 1827, 18°. 2 vols.—*Féruss.* Bull. 1827, X. p. 219.

3. Le Cabinet d'Histoire naturelle. Paris, 1821, 18°. 2 vols.—Bibl. Ent. I. p. 38.

4. Histoire naturelle des Oiseaux d'Europe, avec la figure de chaque espèce et variété, dessinée et coloriée d'après nature. Paris, 6 livr. 4°. fig. 1825–1826.—*Féruss.* Bull. 1824, III. p. 75.

5. Manuel d'Entomologie, ou Histoire naturelle des Insectes. Paris, 1828, 2 vols. 18°. atl.—ed. 2, 1843.—*Féruss.* Bull. 1828, XIV. p. 148.

BOITARD (M.P.) et CANIVET (   ).

1. Manuel du Naturaliste préparateur, ou l'Art d'empailler les Animaux, de conserver les Végétaux et les Minéraux. Paris, 1828 ; 1832, 18°.—*Féruss.* Bull. XV. p. 347.—(Deutsch. übers. v. *F. Bauer.*) 8°. Quedlinb. 1835.

BOITARD (M.P.) et CORBÉE (   ).

1. Les Pigeons de volière et de colombier. 8°. Paris, 1824.

BOITET (P.).

1. Le Tableau des Merveilles du Monde. Paris, 1617, 8°.—*Böhm.* Bibl. I. 1, p. 238.

BOIVIN (   ).

1. Nouvelles Recherches sur la Mole vésiculaire. Paris, 1827, 8°. fig.

BOLIUS (Jac.).

1. De Fontibus et Fluviis, eorumque origine. Regiom. 1649, 4°.—
Böhm. Bibl. V. p. 63.

BOLL (Ant.).

1. Dissertationes philosophici Argumenti. Prag. 4°.—*Böhm.* Bibl.
IV. 1, p. 227 (Diss. II. de Lapidum Origine).

BOLLAERT ( ).

1. On the insulated masses of Silver found in the Mines of Huan-
taxaya, in the province of Tarapaca, Peru.—Proc. Geol. Soc. II.
p. 598.—Phil. Mag. ser. 3, XII. p. 578.

BOLLETI (G. G.).

1. Dell' Origine dell' Instit. di Bologna; *vide sup.* Pars I. p. 64.

BOLLEY (P.).

1. Die Lias-Formation bei Langenbrücken im Grossherzogthum
Baden geognostisch beschrieben, nebst chemischer Untersuchung
eines darin vorkommenden Schwefelwassers. Heidelb. 1837, 8°.

2. Ueber die feuerbeständigen Thon-Arten im Schweizerischen Jura.
—*L.* u. *Br.* N. Jahrb. 1840, p. 515.

3. Ueber das Vorkommen von Bittersalz im östlichen Jura der
Schweiz.—*L.* u. *Br.* N. Jahrb. 1841, p. 631.

BOLLEY (P.) et MÖLLINGER ( ).

1. Schweiz. Gewerbesblatt; *vide sup.* Pars I. p. 62.

BOLLMANN (G.).

1. Beschreibung des Pyrmontischen Sauerbrunnen. Rintel. 1661,
8°; 1670, 8°.—Marb. 1682, 8°.—*Böhm.* Bibl. V. p. 350.

BOLLMANN (J. Arnd.).

1. Dissert. de Loquelâ Animantium brutorum. Ups. 1708, 8°.—
Böhm. Bibl. II. 1, p. 105.

BOLTEN (J. Fr.).

1. Nachricht von einer neuen Thierpflanze. Hamb. 1770, 4°. fig.
—*Mod.* B. Helm.

2. Epist. ad *Linnæum* de novo quodam Zoophytorum genere (As-
cidiæ species). Hamb. 1771, 4°. fig.; 1776, 4°.—*Böhm.* Bibl. II.
2, p. 511.—*Mod.* Bibl. Helm.

3. Etwas von Ammonshörnern.—Berl. Beschäft. II. p. 499.—*Mod.* B. Helm.

4. Epistola de novo quodam Zoophytorum genere. Hamb. 1771, 4°. fig.

BOLTON (J. .

1. Harmonia ruralis, or Natural History of British Song Birds. 2 vols. 4°. Stannary, 1794–1796.

BOMARE (J. Chr. Valmont de).

1. Dictionnaire raisonné universel d'Histoire naturelle. Paris, 1746, 6 vols. 12°; 1765, 5 vols. 8°; 1768, 4 vols. 4°. Yverdun, 1768–1770, 6 vols. 8°. Paris, 1775, 6 vols. 4° et 9 vols. 8°. Lyon, 1791, 15 vols. 8°, et 1793, 8 vols. 4°.—Copenh. 1767–1770, 8°. (Trad. Holl.)—Bibl. Ent. I. p. 38.—*Dum.* et *Bib.* I. p. 342. *Vide sup.* Pars I. pp. 46, 53.

2. Catalogue de son Cabinet d'Histoire naturelle. Paris, 1758, 8°. —*Böhm.* Bibl. I. 1, pp. 397.

3. Observation d'une espèce d'Hermaphrodisme dans un individu de l'espèce du Daim.—J. de Phys. VI. p. 501.

4. Extrait Nomenclateur du Système complet de Minéralogie. Paris, 1759, 12°.

5. Minéralogie, ou nouvelle Exposition du Règne minéral, avec un Dictionnaire Nomenclateur et des Tables synoptiques. Par. 1762, 8°. 2 vols.; 1774, 8°.—(Germ.) Dresd. 1769, 2 vols. 8°.—*Böhm.* Bibl. IV. 1, p. 63.

BOMME (Leendert).

1. Natuurkundige Waarneeming, etc. (*Observ. sur un Nid de Guêpes singulier*).—Genootsch. te Vliss. VII. p. 213.—Bibl. Ent. I. p. 39.

2. Waarneemingen omtrent de gesteldheid en groeijing der Zeepolypen.—Genootsch. te Vliss. II. p. 277.

3. Bericht aangaande verscheiden Zonderlige Zee-Insecten.—Act. Vliss. I. p. 394; III. p. 283; VI. p. 357.

BOMPIEDE (Xav.).

1. De Sono et Auditu.—De Ciborum Canali.—De Temperamentis. —De Ape, Cantharide, Limace et Millepede, etc. (Theses pro Aggreg.) Taurini, 1749, 8°.

BON ( ).

1. Diss. sur l'utilité de la Soye des Araignées. 12°. Avignon, 1710; 8°. Avignon, 1748.

2. Dissert. sur l'Araignée, contenant la vertu et les propriétés de cet Insecte, etc., avec une Lettre du *P. Pouget.* Paris, 1710, 8º.—(Ital.) 12º. Siena, 1710.

BONACCIOLI (L.).

1. De Uteri partiumque ejus confectione, etc. Strasb. 1537, 8º. Bâle, 1566, 4º.—Biogr. Un. V. p. 78.

2. De Conceptionis indiciis, necnon maris femineique Partûs significatione, etc. Strasb. 1538, 8º. Lyon, 1639, 1641, 1650, 1660, 12º. Amst. 1663, 12º.—Biogr. Un. V. p. 78.

3. Enneas Muliebris. 1580, fol.—Biogr. Un. V. p. 78.

BONAFOUS (C.).

1. Note sur un bouleversement du Sol observé aux environs de Sassari.—C. R. V. p. 424.

2. Lettre relative à des observations tendant à prouver que la Muscardine est réellement contagieuse.—Rev. Zool. 1839, p. 119.

3. Osservazioni intorno ad alcune varietà di Bachi da Seta. Torino, 1825, 8º.

4. Description d'une nouvelle espèce de Puceron.—Ann. Soc. Entom. Fr. IV. p. 657.

5. Mémoire sur l'éducation des Vers à soie. Lyon, 1823, 8º; 3e éd. Par. 1827.—Bibl. Ent. I. p. 39.

6. *Aphis Zeæ.*—Ann. Soc. Entom. Fr. p. 657, fig.—*Burm.* in *Wiegm.* Arch. 1836, II. p. 327.

7. Introduzione delle Capre del Tibet in Piemonte. Tor. 1827, 8º. —Verz. Bibl. Schw. Naturf. Ges.

BONALD (Vict. de).

1. Moïse et les Géologues modernes, ou le Récit de la Genèse comparé aux Théories nouvelles des Savans sur l'origine de l'Univers, la formation de la Terre, ses Révolutions, l'état primitif des Etres divers qui l'habitent. Avign. 1835, I. 18º. Brux. 1837, 16º.—(Ital. *V. Alizari*) Genova, 1837, 16º.

BONAMY (C.).

1. Recherches sur la Structure du Placenta dans les Mammifères. Paris, 1839, 4º.

BONAMY (C.) et BEAU ( ).

1. Atlas d'Anatomie descriptive du Corps humain. I. Livr. Paris, 1841, 8º.

BONANNI (Phil.).

1. Rerum naturalium Historia, nempe Quadrupedum, Insectorum, Piscium, variorumque marinorum Corporum, Fossilium, Plantarum ac præsertim Testaceorum existentium in Museo Kircheriano. Rom. 1709, fol. fig.; 1773; 1782, 2 vols. fol. fig. (curâ *J. Ant. Batarra*).—Comment. Lips. XXII. 1, p. 141.—Tr. Linn. Soc. Lond. VII. p. 232.—*Klein*, Disp. Echinod. p. VIII.—*Lamx.* Pol. Cor. p. 522.—*Böhm.* Bibl. I. 1, p. 379.

2. Observationes circa Viventia quæ in rebus non viventibus reperiuntur, cum Micrographiâ curiosâ, seu Rerum minutissimarum Observatio ope Microscopii, etc.   Romæ, 1691, 4°. fig.—Biogr. Un. VI. p. 272.—*Böhm.* Bibl. I. 1, p. 358; II. 2, p. 129.

3. Ricreazione dell' occhio e della mente nell' osservazione delle Chiocciole.   Rom. 1681, 4°. fig.—*Recreatio mentis et oculi in observatione Animalium Testaceorum; italico sermone primùm proposita, nunc Latinè reddita et aucta.*   Romæ, 1684, 4°. fig.—Tr. Linn. Soc. Lond. VII. p. 217.—Act. Erudit. V. p. 108.—*Böhm.* Bibl. II. 2, p. 436.

4. Micrographia curiosa, sive Rerum minutissimarum Observationes. Romæ, 1703, 4°. fig.

5. Aliquot Animalium Testaceorum Imagines non antea in lucem editæ; annex. ejusd. Observationes circa Viventia quæ in non Viventibus reperiuntur.   Romæ, 1691, 4°.

BONAPARTE (C. L.).

1. American Ornithology; or the Natural History of Birds inhabiting the United States, not given by *Wilson*.   I. 1825, Philad. 4°. fig. col.; II. 1828, Philad. and Lond.—Isis, 1832, IX. p. 987.— *Féruss.* Bull. 1836, VII. p. 100.

2. The Genera of North American Birds, etc.—Ann. Lyc. N. Y. 1826, II. p. 1; 1827, p. 293.—*Féruss.* Bull. 1827, XI. p. 108; XII. p. 337; 1828, XIII. p. 122; XIV. p. 115; 1831, XXIV. p. 358.—Isis, 1832, XI. p. 1135.

3. On the Birds of the genus *Tetrao*.—Trans. Am. Phil. Soc. ser. 2, III. p. 383.

4. Observations on the Nomenclature of *Wilson's* Ornithology. Philad. 1826, 8°.—J. Ac. Philad. III. 2, p. 340; IV. pp. 25, 163, 251; V. p. 57.—*Féruss.* Bull. 1826, VII. pp. 244, 375; 1827, XI. p. 110.

5. Description of a new species of S. American *Fringilla*.—J. Acad. Philad. IV. p. 350.—*Féruss.* Bull. 1826, VII. p. 249.

6. Additions to the Ornithology of the United States.—J. Acad. Philad. V. 1, 1825, p. 28.—Zool. Journ. III. p. 49.—Isis, 1830, X. p. 1067.—*Féruss.* Bull. 1827, X. p. 399.

7. On the distinction of two species of *Icterus*, hitherto confounded under the specific name of *icterocephalus.*—J. Acad. Philad. V. p. 222.—*Féruss.* Bull. 1827, XII. p. 266.

8. Catalogue des Oiseaux des États-Unis.—Contr. Macl. Lyc. 1827, I. p. 8.—*Féruss.* Bull. 1829, XVII. p. 434.

9. Notice of a nondescript species of Grouse (*Tetrao*, L.) from N. Amer.—Zool. Journ. III. p. 212.—*Féruss.* Bull. 1828, XIV. p. 107.—Isis, 1830, XI. p. 1160.

10. Specchio comparativo delle Ornitologie di Roma e di Filadelfia. Pisa, 1827, 8°.—Supp. ib. 1832.—N. Giorn. Lett. XXXIII.—Isis, 1834, II. p. 150.

11. A Geographical and Comparative List of the Birds of Europe and North America. Lond. 1838, 8°.—Ann. of Nat. H. I. p. 318.

12. An account of four species of Stormy Petrel (*Procellaria*).—Journ. Acad. Philad. III. 2, 1824, p. 227, cum fig.

13. Supplement to " An account of four species of Stormy Petrel" (*Thalassidroma*, Vig.).—Zool. Journ. III. p. 89.

14. On a new species of Duck (*Anas rufitorques*), described by *Wilson* as the same with the *Anas fuligula* of Europe.—Journ. Acad. Philad. III. 2, 1824, p. 381.

15. Descriptions of two new species of Mexican Birds(*Corvus*, Linn. et *Cassicus*, Lacep.).—Journ. Acad. Philad. IV. 2, 1825, p. 387.

16. Descriptions of ten species of South American Birds.—Journ. Acad. Phil. IV. p. 370; V. p. 137.—*Féruss.* Bull. 1828, XIII. p. 240.

17. On new and interesting Birds from South America and Mexico. —Proc. Zool. Soc. V. p. 108.

18. Catalogue d'Oiseaux du Mexique et du Pérou.—N. Ann. Sc. n. Flor.—Rev. Zool. 1840, p. 19.

19. On the Long-tailed Trogon.—Mag. Nat. Hist. ser. 2, II. p. 229. —Proc. Zool. Soc. V. p. 101.

20. *Rhamphocelus Passerinii*, aus Cuba.—Isis, 1833, VIII. p. 755.

21. Nouvelle espèce d'Oiseau du genre *Rhamphocèle.*—Rev. Zool. 1838, p. 8.

22. Iconografia della Fauna Italica per le quatro classi degli Animali vertebrati. Rom. 1832–1842, 3 vols. 4°.—Isis, 1833, VIII. p. 742; 1834, IV. p. 412.—Ann. a. Mag. N. H. X. p. 127.—Mag. Zool. a. Bot. I. p. 82.

23. Zusätze und Berichtigungen zu den Säugethieren.—Isis, 1833, XII. p. 1218.

24. Zusätze zu den Vögeln.—Isis, 1833, XII. p. 1223.

25. Die Italienischen Spitzmäuse, nach den Angaben der Iconografia della Fauna Italica ; im Auszuge von *A. Wagner.*—*Wiegm.* Arch. 1841, I. p. 297.

26. Ueber *Testudo caspica,* Gmel.—Isis, 1833, XII. p. 1229.

27. Description d'une espèce inédite de Lacertide français, du genre *Psammodrome.*—Ann. Sc. n. (2ᵉ S.) XII. p. 60, fig. (*Psammodromus cinereus*).—Rome, 1839, 8°.—Giorn. Arcad. 1839.

28. On the species of the genus *Mustela.*—Mag. Nat. Hist. ser. 2, II. p. 37.

29. Descrizione di un nuovo Leuciscino, etc. 8°.

30. Note sur deux Oiseaux nouveaux du Musée de Marseille.— Rev. Zool. 1841, p. 145.

31. Tentamen Monographiæ Leuciscorum Europæ.—Rev. Zool. 1840, p. 27.

32. Cenni sopra le variazioni a cui vanno soggette le Farfalle del gruppo *Melitæa.*—Antologia, Mai 1831.—Ann. Soc. Ent. Fr. I. p. 244.

33. Osservazioni sulla seconda edizione del Regno Animale del *Barone Cuvier.* Bologna, 1830, 8°.—*Bemerkungen über die* 2ᵗᵉ *Auflage v.* Cuvier's *Thierreich.*—Ann. di St. Nat. Bol. X.—Isis, 1833, XI. p. 1041.

34. Saggio di una distribuzione metodica degli Animali vertebrati. Rom. 1831, 1832, 8°.—Giorn. Arcad. XLIX. p. 1.—Isis, 1832, III. p. 283; 1833, XII. p. 1183.—*Féruss.* Bull. XXV. p. 103.

35. Classification des Animaux vertébrés (Fragmens par *Is. Geoffroi-St.-Hilaire*).—C. R. VII. p. 656.

36. Fragmens de la nouvelle Classification des Animaux vertébrés.— Rev. Zool. 1838, p. 208.

37. Synopsis Vertebratorum Systematis. 1837, 8°.—Rev. Zool. 1839, p. 308.

38. Prodromus Systematis Vertebratorum. 12°. pam.

39. Prodromus Systematis Mastozoologiæ.—N. Ann. Sc. nat. Bol. 1840.

40. Prodromus Systematis Ornithologiæ. 12°. pam.

41. Systema Ornithologiæ.—N. Ann. Sc. n. Bol. 1840.—Rev. Zool. 1840, p. 306.

42. Prodromus Systematis Herpetologiæ.—N. Ann. Sc. n. Bol. 1840. —Rev. Zool. 1840, p. 307.

43. Saurorum Tabula analytica.—N. Ann. d. Sc. nat. 8°.—Rev. Zool. 1839, p. 238.

44. Cheloniorum Tabula analytica. Romæ, 1836, 8°.—Mag. Zool. and Bot. 1838, II. p. 58.—Rev. Zool. 1839, p. 237.

45. Selachorum Tabula analytica.   8°. Rome, 1839.

46. Prodromus Systematis Ichthyologiæ.—N. Ann. Sc. n. Bol. 1840.

47. Amphibia Europæa ad Systema nostrum Vertebratorum ordinata.—Atti R. Accad. di Torino, 1841.

48. Catalogo metodico degli Ucelli Europei.   Bol. 1842, 8°.

49. Osservazioni sullo stato della Zoologia in Europa in quanto ai Vertebrati, nell' anno 1840–41. Firenze, 1842, 8°.—(Angl. *H. E. Strickland.*) RAY SOCIETY, 8°. London, 1845.

BONARDUS (J. M.).

1. La Miniera del Mondo, nella quale si tratta delle cose più secrete e più rare de Corpi semplici nel Mondo elementare, etc. mandata in luce da *L. Grotto* cieco d'Adoja.   Venez. 1589, 1600, 1611, 8°.—*Böhm.* Bibl. I. 1, p. 235.

BOND (J. W.).

1. Notes on various Insects—Entom. Mag. IV. p. 221.

BOND (Th.).

1. De Verme in Hepate generato.—Med. Obs. a. Inquir. I. p. 68. —*Mod.* B. Helm.

BONDIOLI (P. Ant.).

1. Sulle Vaginali del Testiculo.   Vic. 1789.   Pad. 1790, 8°.—Biogr. Un. Suppl. LVIII. p. 542.

BONELLI (Fr. A.).

1. Observations sur un Hippopotame nouvellement acquis par le Musée de Turin.—Mém. Ac. Tur. XXIX. p. 243.—*Féruss.* Bull. 1826, VII. p. 369.—Isis, 1834, I. p. 76.

2. Catalogue des Oiseaux du Piémont.   1811, 4°.—Ann. Observat. Turin. 1811.—*Cuv.* R. An. III. p. 339.

3. Nouvelle espèce de Poisson de la Méditerranée (*Trachypterus cristatus*).—Mém. Ac. Tur. XXIV. p. 485.—Isis, 1834, IV. p. 432.

4. Specimen Faunæ subalpinæ.   1807.—Biogr. Un. Suppl. LVIII. p. 544.

5. Observations entomologiques.   Turin, 1809–13, 2 part. 4°.— Biogr. Un. Suppl. LVIII. p. 544.

6. Observations entomologiques sur le genre *Carabus.*—Mém. Ac. Tur. 1813.—Bibl. Ent. I. p. 40.

7. Mémoire sur l'Eurychile, nouveau genre d'insecte de la famille des Cicindèles.—Mém. Ac. Tur. XXIII. fig.
8. Descrizioni di sei nuove specie d' Insetti Lepidotteri diurni della Sardegna. Torino, 1824. 4°. fig.—Mém. Ac. Tur. XXX. p. 171. —*Féruss.* Bull. 1826, VII. p. 148; 1828, XIII. p. 163.—Isis, 1834, I. p. 82.

BONETUS (Theophr.).

1. Sepulcretum anatomicum, seu Anatomia practica in III. Tomos distributa. Genevæ, 1700, fol.—*Mod.* B. H.

BONGE (D.).

1. Diss. physica de Salmonum naturâ, eorumque apud Ostroboth-nienses Piscatione, etc. 8°.

BONGIOVANNI ( ) e BARBIERI ( ).

1. Illustrazioni delle Terme di Caldiero. Verona, 1795, 4°.

BONJOUR (Ph.).

1. Catalogue d'Oiseaux indigènes et étrangers de sa Collection. Paris, 1828, 8°.—*Féruss.* Bull. 1828, XIV. p. 257.

BONN (Andr.).

1. De Simplicitate Naturæ, anatomicorum admiratione, chirurgico-rum imitatione dignissimâ. Amst. 1772, 4°.—Biogr. Un. Suppl. LVIII. p. 551.
2. Dissert. inaug. de continuationibus Membranarum. Leyd. 1763, 4°.—Biogr. Un. Suppl. LVIII. p. 551.

BONN (A. C.).

1. Verhandeling over der Mastodonte of Mammouth van den Ohio. 1810, 8°. Haarlem.

BONNAFOUS (M.).

1. Cenni sull' introduzione delle Capre del Tibet in Piemonte.

BONNAIRE-MANSUY ( ).

1. Cosmogonie. Paris, 1824, 8°.

BONNARD (De).

1. Aperçu géognostique des Terrains. Paris, 1819, 8°.

2. Rapport sur un Mémoire de *M. Leymerie* concernant les Terrains secondaires inférieurs du Dépt. du Rhône.—C. R. VII. p. 878.

3. Essai géognostique sur l'Erzgebirge et les Montagnes métallifères de la Saxe.—Acad. Sc. 1816.—Isis, 1817, VI. p. 669.—Bull. Soc. Géol. Fr. II. p. 93.—J. d. Mines, XXXVIII. p. 261.—*Leonh.* Tasch. 1822, I. p. 94; II. p. 508 (*Geognostischer Versuch über das Erzgebirge Sachsens*).

4. Notice géognostique sur la partie occidentale du Palatinat.— Bull. Soc. Phil. 1821, p. 129.

5. Notices géognostiques sur le Hartz.—Bull. Soc. Phil. 1822, p. 10. —Ann. Min. VII. p. 41.—*Féruss.* Bull. 1823, I. p. 220.—*Leonh.* Tasch. 1824, I. p. 131 (*Geognostische Bemerkungen über den Harz*).

6. Géologie des environs de Freiberg.—Acad. Sc. 1816.—Isis, 1817, VI. p. 675.

7. Sur la constance des faits géognostiques qui accompagnent le Terrain d'Arkose dans l'est de la France.—Ann. Sc. n. 1827, XII. p. 298.—*Féruss.* Bull. 1829, XVIII. p. 156.—Bull. Soc. Géol. Fr. II. p. 93.

8. Notice sur une Formation métallifère observée récemment dans l'ouest de la France.—Bull. Soc. Phil. 1823, p. 57.—*Féruss.* Bull. 1823, III. p. 45.—*Leonh.* Zeitschr. 1825, I. p. 369.

9. Mémoire sur le gîte de Manganèse de Romanèche-la-Naine, Dép. de Saône et Loire.—Acad. Sc. 17 Déc. 1827.—*Féruss.* Bull. 1828, XIII. p. 185.—Ann. Sc. n. Mars 1829, p. 285.—*L. u. Br.* N. Jahrb. 1833, p. 562.

10. Notice géognostique sur quelques parties de la Bourgogne.— Ann. des Mines, 1825.—Bull. Soc. Géol. Fr. II. p. 93.

11. Knochen in den Höhlen von Arcy-sur-Cure.—Bull. Soc. Géol. Fr. III. p. 222.—*L. u. Br.* N. Jahrb. 1834, p. 366.

BONNATERRE (L'Abbé).

1. Tableau encyclopédique et méthodique des trois Règnes de la Nature. Paris, 1788–1790, 4°. fig.—*Dum.* et *Bib.* I. p. 308.

2. Texte des Reptiles et des Poissons dans l'Encyclopédie méthodique.—*Cuv.* R. An. III. p. 338.

BONNER (James).

1. Plan for speedily increasing the number of Bee-hives in Scotland. Lond. 1795, 8°.—Bibl. Ent. I. p. 40.

BONNES (   ).

1. Observations sur l'*Acarus* de la gale du Cheval.—C. R. V. p. 613.

BONNET (Ch.).

1. Compendium Observationum de Insectis.—Phil. Trans. No. 470, p. 458.—*Eis.* Insect. p. 162.—*Böhm.* Bibl. II. 2, p. 132.

2. Lettre sur les moyens de conserver diverses espèces d'Insectes et de Poissons dans les cabinets d'Histoire naturelle.—Journ. de Phys.'III. p. 296.—Op. V. p. 12.—Bibl. Ent. I. p. 41.

3. Disquisitiones circa Respirationem Erucarum.—Mém. Acad. Sc. Par. V. p. 276.—Phil. Tr. 1748, p. 300.

4. Observations sur une nouvelle partie propre à plusieurs Chenilles. —Mém. Sav. étr. II. p. 44.—Op. II.—Bibl. Ent. I. p. 41.

5. Sur la grande Chenille à queue fourchue du Saule.—Mém. Sav. étr. II. p. 276.—Bibl. Ent. I. p. 41.

6. Observations sur les Stigmates des Papillons.—Mém. Sav. étr. V. p. 294.—Bibl. Ent. I. p. 41.

7. Lettre et Mémoire sur les Abeilles (avec l'hist. de la Reine des Abeilles par *Schirach*).—Journ. de Phys. V. pp. 327, 418 ; VI. p. 23.—Op. V.—Bibl. Ent. I. p. 41.

8. Schreiben an Hrn. *Riem*, nebst des letzeren Anmerkungen über die Bienen.—Berl. Samml. VII. p. 245.—Op. V.—Bibl. Ent. I. p. 41.

9. Aphides variæ.—Op. I. p. 1.—Bibl. Ent. I. p. 41.

10. Expériences sur la régénération de la tête du Limaçon terrestre. —Journ. de Phys. 1775, X. p. 165.—Tr. Linn. Soc. Lond. VII. p. 239.

11. Lettere relative al soggetto degli Animali infusori. *Spallanz.* Op. fis. I.

12. Notice sur la Chenille du Pommier.—Ann. Provenç. d'Agr. XI. 1838.

13. Traité d'Insectologie, ou Observations sur les Pucerons et sur quelques espèces de Vers d'eau douce, qui, coupés par morceaux, deviennent autant d'Animaux complets.   Paris, 1745, 2 vols. 8°. fig.—(Germ. *Göze*): *Abhandlungen aus der Insectologie.*   Halle, 1773, 1774, 8°. fig.—*Mod.* Bibl. Helm.

14. Considérations sur les Corps organisés, etc. Amst. 1762, 2 vols. 8°.—(Ital.) *Considerazioni sopra i Corpi organizzati, recata dal Franc. dal P. F. N. N.* Firenze, 1768, 8°. 2 vols.—(Germ.) *Betrachtungen über die organisirten Körper, aus dem Frans. mit*

Zusätzen v. J. A. E. Götz. Lemgo, 1775, 8°. 2 vols.—*Böhm.* Bibl. I. 1, p. 292.—*Mod.* B. H.

15. Dissertation sur le Ver nommé *Tænia* en Latin, etc.—Mém. Math. et Phys. I. p. 478.—(Belg.) Uytg. Verh. III. p. 309.— *Mod.* B. H.

16. Some new Observations upon Insects.—Phil. Trans. XLII. p. 458. —*Mod.* B. H.

17. Palingénésie philosophique, ou Idées sur l'état passé et l'état futur des Etres vivans. Genève, 1769, 2 vols. 8°.—Laus. 1770, 2 vols. 12°.—(Germ.) *Philosophische Palingenesie.* Zürich, 1770, 2 vols.—*Mod.* B. H.

18. Œuvres d'Histoire naturelle et de Philosophie. Neuch. 1779–1783, 8°. 19 vols. et 4°.—(Germ.) *Werke der Naturgeschichte u. Philosophie.* Leipz. 1783, 3 vols. fig.—Bibl. Ent. I. p. 41.

19. Contemplation de la Nature. Amst. 1764–1765, 2 vols, 8°.—Berne, 1768, 12°.—Genève, 1770, 8°.—(Angl.) Lond. 1766 ; 1775, 2 vols. 12°.—(Ital.) *Contemplazione della Natura,* etc. *trad. dall Abbate Spallanzani.* Modena, 1769 et 1770, 8°. 2 vols.—Venez. 1773.—4 vols. 8°. Venez. 1818.—(Germ. v. *D. Titius*) : *Betrachtungen über die Natur.* Leipz. 1766, 8°. fig. ; 1772, id. ; 1774, id. ; 1783, id.—(Belg. door *G. Coopman*) : *Beschouwing de Natuur,* etc. Franck. 1775–77, 8°. 2 vols.—Biogr. Un. V. p. 130.—*Böhm.* Bibl. I. 1, p. 293.

20. Observations sur le Pipa ou Crapaud de Surinam.—Journ. de Phys. XIV. p. 425.

21. Mémoires sur la reproduction des Membres de la Salamandre aquatique.—Journ. de Phys. X. p. 385 (1er Mém.).—Ibid. XIII. p. 1 (2e Mém.).—Ibid. p. 340 (3e Mém.).

22. Lettre sur les Sangsues considérées non comme Baromètres, mais comme Thermomètres.—Journ. de Phys. V. p. 70.—*Mod.* B. Helm.

23. Nouvelles Recherches sur la structure du Ténia.—Journ. de Phys. IX. p. 243.

Bonnet (S.).

1. Diss. sur les Sangsues. 4°. Par. 1826.

Bonnet (Th.).

1. Sepulchretum, seu Anatomia practica ex cadaveribus morbo denatis ; observationibus illustravit ac auctiorem fecit *J. J. Mangetus.* Genève, 1679, fol. 2 vols. ; 1700, 3 vols. fol.—Biogr. Un. V. p. 132.

BONNYCASTLE (R. H.).

1. On the Transition Rocks of the Cataraqui.—*Sill.* Amer. Journ.
XVIII. p. 85; XX. p. 74; XXIV. p. 97; XXX. p. 233.

BONOLA (Gio. Batt.).

1. Della Bibliografia malacologica Italiana. Diss. inaug. Milano,
1839, 8°.—Rev. Zool. 1840, p. 146.

BONOMO (J. Cosme).

1. Epistola che contiene osservazioni intorno da Pellicelli del Corpo
humano. Fir. 1687, 4°.—(Lat. *Jos. Larzans*): *Observationes
circa humani Corporis Teredines.*—Misc. Nat. Cur. 1691, Dec. 2,
p. 180; App. p. 33.—Bibl. Ent. I. p. 42.—*Böhm.* Bibl. II. 2,
p. 388.—*Mod.* Bibl. Helm.

2. An Abstract of part of a letter containing some Observations
concerning the Worms of Human Bodies.—Phil. Trans. 1703,
p. 1296.—Bibl. Ent. I. p. 42.

BONSDORF (B. J.).

1. Disquisitio physiologica de Cavitatibus Organismi humani. Hel-
singf. 1837, 8°.

BONSDORF (Gabr.).

1. Dissert. Organa Insectorum sensoria generatim, Oculorumque
Fabricam et Differentias speciatim exponens. Aboæ, 1789, 4°.
—Tr. Linn. Soc. Lond. V. p. 278.

2. Diss. Fabricam, Usum et Differentias Antennarum in Insectis exponens. Aboæ, 1790, 4°.—Tr. Linn. Soc. Lond. V. p. 278.—
Bibl. Ent. I. p. 42.

3. Dissertatio: Differentiæ Capitis Insectorum præcipuæ, exemplis
illustratæ. Aboæ, 1789, 4°.—Tr. Linn. Soc. Lond. V. p. 278.

4. Fabrica, Usus et Differentia Palporum in Insectis. Aboæ, 1792,
4°.—Bibl. Ent. I. p. 42.

5. Historia naturalis Curculionum Sueciæ. Ups. 1785, 4°. fig.—
Bibl. Ent. I. p. 42.

6. Lucani genus, etc. (*Le genre Lucane, et descr. de deux nouv.
espèces suédoises.*)—Vet. Acad. 1785, p. 220.—Bibl. Ent. I. p. 42.
—N. Schw. Acad. 1785, p. 215.

BONSTETTEN (Ch. Vict. de).

1. L'Homme du Midi et l'Homme du Nord, ou l'Influence du Cli-
mat. Gen. 1824, 8°.—Biogr. Un. Suppl. LVIII. p. 585.

2. La Scandinavie et les Alpes.   Genève, 1826, 8°.—*Féruss.* Bull.
1826, IX. p. 146.—(Deutsche Uebersetz.) *Skandinavien und die
Alpen, mit einem Anhang über Island.* Kiel, 1827, 8°.—Cat. Bibl.
Tur.—Naumb. 1828, 8°.—Isis, 1831, VII. p. 677.

BONTÉ ( ).

1. Sur une espèce de Ver singulière.—Journ. de Méd. XIV. p. 32.

BONTEKOE (C.).

1. Metaphysica, ejusdemque de Motu liber singularis, necnon
Œconomia animalis.   Accedit *A. Gulinex* Physica vera.   Lugd.
Batav. 1688.—Act. Erudit. VII. p. 487.

BONTEMPI (Gius. Ant.).

1. De communibus universi Corporis Integumentis.—De Lapide
Turmalino.—De Vasis lymphaticis, eorumque usu ; etc. (Theses
pro aggreg.)   Taurini, 1792, 8°.
2. Sulle Chiocciole.—Mem. Soc. Ital. VII.—Verona, 1794, 4°.

BONTIUS (Jac.).

1. Historia Animalium (cum variis Opusculis).   Amst. 1688, fol.
—*Böhm.* Bibl. II. 1, p. 33.
2. Historiæ naturalis et medicæ Indiæ orientalis libri VI.—*Piso* :
De Indiæ utriusque re naturali et medicâ.—*Dum.* et *Bib.* I.
p. 308.

BONVICINI (Jor.).

1. Lettera sulla Voce della Testuggine.—Opusc. scelti, XVII.p. 212.
—*Dum.* et *Bib.* I. p. 437.

BONVICINO (C. B.).

1. De la Pierre hydrophane du Piémont.—Mem. Accad. Tor. VI.
p. 475.
2. Remarques sur la véritable nature de la Turquoise, etc.—Mem.
Accad. Tor. XI. p. 305.
3. Sur la Culture des produits du R ègne minéral en Piémont.—Mem.
Accad. Tor. XII. p. 224.
4. Sur les Mines de Plombagine des Dép⁸. de la Stura et du Po.—
Mem. Accad. Tor. XIV. p. 145.
5. De Concretis calcareis.—De insensibili Cutis Halitu et Sudore ;
etc. (Theses pro aggreg.)   Taur. 1778, 8°.

Boon Mesch (Van der).   *Vide* Van der Boon Mesch.

Boot (A. B. de).

1. Gemmarum et Lapidum Historia.   8°. Leyden, 1636 et 1647.

Booth (J. C.).

1. Memoir on the Geological Survey of the State of Delaware.
Dover, 1841, 8°.—Philad. id.

Booth (W. B.).

1. On the Habits of *Bulinus hæmastomus.*—Zool. Journ. V. p. 101
—Isis, 1831, VII. p. 729.

Boppe (C. A. F.).

1. Ueber die Schilddrüse.   Tübing. 1840, 8°.—*Val.* Repert. VI.
p. 29.

Bor (A. H.).

1. Dissertatio de Indole singulis hominibus optandâ.   Embr. 1828,
8°.—Bibl. med.-ch. p. 58.

Borch (M. J. de).

1. Lettres sur la Sicile et sur l'I. de Malte, écrites en 1777.   Suppl.
au Voyage de *Mr. Brydone.* Tur. 1782, 8°. fig.—(Germ.) *Briefe
über Sizilien u. Maltha,* etc.   Bern, 1783, 8°.
2. Litografia di Sicilia.   Nap. 1777, 8°.—*Böhm.* Bibl. IV. 1, p. 113.
—(Gall.) *Lithographie Sicilienne, ou Catalogue raisonné de toutes
les Pierres de la Sicile.*   Naples, 1777, 4°.
3. Litologia Siciliana, o Conoscenza della natura delle Pietre della
Sicilia ; seguita da un Discorso sopra la Calcare di Palermo.
Roma, 1778, 4°.—(Germ.) *Pfingst,* Bibl. ausländ. Chym. etc. I.
—*Böhm.* Bibl. IV. 1, p. 113.—(Gall.) *Lithologie Sicilienne, ou
Connaissance,* etc.   Rome, 1778 ; 1787, 4°.
4. Minéralogie Sicilienne, docimastique et métallurgique, avec les
détails des Mines et des Carrières et l'hist. des travaux anciens
et actuels de ce pays.   Turin, 1780, 8°.
5. Memoria sopra il Fosforo marino.—Atti Accad. Siena, VI. p. 317.

Borchardt (J.).

1. Nonnulla de Ligamentorum Columnæ spinalis comparatione inter
Aves et Mammalia.   Berol. 1833.

BORCHOLLT (Wern.).

1. De Differentiâ inter Homines et Bruta. Luneb. 1716, 4°.—*Böhm.* Bibl. II. 1, p. 83.

BORDEAUX (Th. de).

1. Recherches sur les Eaux minérales des Pyrénées. Paris, 1833, 8°.

BORDEL-DESHAUCHAMPS (A.).

1. Diss. sur les diff. espèces de Vers du canal intestinal de l'Homme. 4°. Paris, 1816.

BORDENAVE (Touss.).

1. Mémoire sur l'Ostéogénie.—Acad. de Chir.—Biogr. Un. V. p. 161.
2. Recherches Anatomiques et Expériences pour éclaircir la doctrine de *Haller* sur la distinction à établir entre la Sensibilité et l'Irritabilité.—Biogr. Un. V. p. 161.
3. Essais de Physiologie. Paris, 1756, 1764, 12°; 1787, 2 vols. 12°; 1837, id.—Biogr. Un. V. p. 161.
4. Remarques sur l'Insensibilité de quelques parties. 1757, 12°.—Biogr. Un. V. p. 161.
5. Mémoire sur la Respiration.—Acad. d. Sc.—Biogr. Un. V. p. 161.

BORDEU (Th. de).

1. Recherches sur le Tissu cellulaire. Paris, 1767.—Biogr. Un. V. p. 165.
2. Recherches sur les différentes positions des Glandes et sur leur action. Par. 1752, 12°.—Biogr. Un. V. p. 164.
3. Mémoire sur les articulations des Os de la Face.—Mém. Sav. étr.—Biogr. Un. V. p. 164.
4. De Sensu genericè considerato, Dissertatio physiologica. Montp. 1742, 4°.—Biogr. Un. V. p. 163.
5. Chilificationis Historia. 1743, 4°.—Biogr. Un. V. p. 163.
6. Recherches sur le Tissu muqueux. Par. 1751 et 1767, 12°.—*Böhm.* Bibl. V. p. 166.

BORDIN (Cr.).

1. Description géologique des environs de Madrid, dans le but d'examiner si la vallée du Tage est propre à la formation des

Fontaines ascendantes.—Gaz. Bayon. 1830, N° 166.—*Féruss.* Bull. 1830, XXII. p. 208.

BORELL (P.).

1. Observationum microscopicarum Centuria. Hag. 1655, 4°.— *Böhm.* Bibl. I. 1, p. 354.

2. Les Antiquités, Raretés, Plantes, Minéraux et autres choses de la ville et comté de Castres, et un Recueil des Antiquités du Languedoc. Castres, 1649, 8°.—*Böhm.* Bibl. I. 1, p. 376.

3. Observation sur une Fontaine salée de Franche-Comté, qui a un reflux.—Mém. de l'Acad. des Sciences de Paris, II. p. 42.

BORELLI (J. Alph.).

1. De Vi Percussionis et Motionibus naturalibus à Gravitate pendentibus, ad illustrandum Librum de Motu Animalium. Lugd. 1686, 4°.—*Böhm.* Bibl. II. 2, p. 111.

2. Philosophia de Motu Animalium ex principio mechanico-statico. Romæ, 1680–1682, 2 vols. 4°, fig.—Lugd. Bat. 1685; 1711, 4°. fig.—Neap. 1734, 4°. fig.—Hag. Com. 1742, id.—Act. Erudit. I. p. 351; II. pp. 32, 63.—*Cuv.* et *Val.* I. p. 66.—*Böhm.* Bibl. II. 1, p. 111.

3. Historia et Meteorologia Incendii Ætnæi anni 1669; accessit Responsio ad censuras *H. Fabri* contra Librum auctoris de Vi Percussionis. 4°. Reg. Jul. 1670.—Cat. R. Soc. Lond.

BORETIUS (M. Ern.).

1. Dissert. de Anatomiâ Plantarum et Animalium analogâ. Regiom. 1727, 4°.—*Böhm.* Bibl. II. 1, p. 83.

BORGARUCCI (Prosp.).

1. Della Contemplazione anatomica sopra tutte le parti del Corpo umano. Venise, 1564, 8°.—Biogr. Un. V. p. 173.

BORGHESE (Jul. Jos.).

1. De Hepate.—De Bile; etc. (Theses pro aggreg.) Taurini, 1797, 8°.

BORKE ( ).

1. De Sarcogenesi et Morbis musculorum organicis. Gron. 1834, 8°.—*Müll.* Arch. 1835, p. 241.

BORKHAUSEN (M. Balth.).

1. Versuch einer Erklärung der zoologischen Terminologie. Frankf. 1790, 8°.—Biogr. Un. V. p. 187.

2. Deutsche Fauna, etc. Frankf. 1797, 8°.—Biogr. Un. V. p. 187.

3. Entomologische Bemerkungen u. Berichtigungen.—Rhein. Mag. I. p. 625.—Bibl. Ent. I. p. 43.

4. Naturgeschichte der europäischen Schmetterlinge nach systematischer Ordnung. 5 Thl. 8°. Frankf. 1788-1794.—*Eis.* Insect. p. 204.—Cat. Bibl. Turic.

5. Ornithologie von Ober-Hessen.—Rhein. Mag.—Biogr. Un. V. p. 187.

BORKHAUSEN (M. B.) et BRAHM (   ).

1. Rheinisches Magazin ; *vide sup.* Pars I. p. 20.

BORKOWSKY (Stan. Dunin).

1. Geognostische Beobachtungen in der Gegend von Rom.—*Leonh.* Tasch. 1816, II. p. 352.

BORLASE (Will.).

1. The Natural History of Cornwall. Oxf. 1758, fol. fig.—Trans. Linn. Soc. Lond. VII. p. 226.

2. On the great alterations which the Islands of Sylley have undergone since the time of the ancients.—Phil. Trans. 1753, p. 55.

BORN (G.).

1. Ueber den innern Bau der Lamprete. (*Sur la structure interne de la grande Lamproie (Petromyzon marinus).*)—Zeitschr. f. organ. Phys. 1827, I. p. 170, fig.—*Féruss.* Bull. 1828, XIII. p. 360.—Ann. Sc. n. 1828, XIII. p. 22, fig.

2. Bemerkungen über den Zahnbau der Fische. (*Observations sur la structure des Dents chez les Poissons.*)—Zeitschr. f. die organ. Phys. 1827, I. p. 182, fig.—*Féruss.* Bull. 1828, XV. p. 167.

BORN (Ign. v.).

1. Briefe über mineralogische Gegenstände auf einer Reise durch das Temeswarer Bannat, Siebenbürgen und Ungarn ; herausg. v. *J. J. Ferber.* Fr. u. Leipz. 1774, 8°.—(Angl. transl. by *R. E. Raspe*): Lond. 1777, 8°.—(Gall. par *Monnet*): *Voyage minéralogique fait en Hongrie et en Transylvanie.* Paris, 1780, 8°. —*Böhm.* Bibl. I. 1, p. 645.

2. Lithophylacium Bornianum, s. Index Fossilium quæ collegit, in classes ac ordines digessit. Pragæ, 1772, 1775, 2 vols. 8°. fig.— *Böhm.* Bibl. IV. 1, p. 73.

3. Schreiben über einen ausgebrannten Vulkan bei der Stadt Eger in Böhmen. Prag. 1773, 4°.—*Böhm.* Bibl. IV. 2, p. 388.

4. Catalogue méthodique et raisonné de la Collection des Fossiles de *Mlle. É. de Raab.* Vienne, 1790, 2 vols. 8°. fig.—(Germ.) ibid. 1790, id.

5. Testacea Musei Cæsarei Vindobonensis. Vind. 1780, fol. fig. col.—Tr. Linn. Soc. Lond. VII. p. 233.—*Böhm.* Bibl. I. 1, p. 405.

6. Index Rerum naturalium Musei Cæsarei Vindobonensis. Viennæ, 1758, 8°.—*Böhm.* Bibl. I. 1, p. 404.—*Mod.* Bibl. Helm.

7. Index Fossilium quæ collegit et in classes et ordines disposuit. 8°. Pragæ, 1775.—*Mod.* Bibl. Helm.

8. Zufällige Gedanken über die Anwendung der Conchylien- u. Petrefactenkunde auf die physikalische Erdbeschreibung.—Prag. Abh. IV. p. 305.—*Mod.* Bibl. Helm.

9. Antwort auf das Schreiben des Grafen von K.....—Abh. einer Privatges. in Böhm. I. p. 253.

10. Mineralogische Bemerkungen aus den neuesten Reise beschreibungen. St. 1. Aus *P. S. Pallas* Reise durch verschiedene Provinzen des Russischen Reichs. Th. I. St. Petersb. 1771, 4°.— Abh. Privatges. Böhm. I. p. 264.

11. Versuch einer Mineral-Geschichte des Ober-Oesterreichischen Salzkammergutes.—Abh. Privatges. Böhm. III. p. 166.

12. Versuch über den Topas der Alten und den Chrysolith des *Plinius.*—Abh. Privatges. Böhm. II. p. 1.

13. Relatio de aurilegio Daciæ transalpinæ.—Nova Acta Acad. Nat. Cur. VIII. p. 97.

14. Nachricht vom gediegenen Spiessglaskönig in Siebenbürgen.— Abh. Privatges. Böhm. V. p. 383.

BORN (Ign. v.) u. BLUMENBACH (J. Fr.).

1. Ueber die Nutritionskraft. (2 gekr. Abhandl.) Petersb. 1789, 4°.

BORN (J. E. v.).

1. Abhandl. einer Privatgesellsch.; *vide sup.* Pars I. p. 32.

2. Physik. Arbeiten; *vide sup.* Pars I. p. 35.

BORN (J. H.).

1. Petino Theologle, oder Versuch die Menschen durch nähere

Betrachtung der Vögel, zur Bewunderung ihres Schöpfers auf-
zumuntern. Pappenh. 1742–43, 2 vols. 8°.

BORNEMANN (J. F. C.).

1. Diss. anatomico-physiologica de Ruminatione. 4°. Gottingen,
1812.

BÖRNER (J. C. H.).

1. Sammlungen aus der Naturgeschichte, Oeconomie, etc. Dresd.
1774, 8°.—Bibl. Ent. I. p. 35.
2. Beschreibung eines neuen Insekts, *Dermestes sexdentatus.*—
Nachr. d. Schles. Ges. IV. p. 78.—*Eis.* Insect. p. 190.
3. Beschreibung eines neuen Insekts, *Scarabæus biguttatus.*--Schles.
Ges. IV. p. 199.—Bibl. Ent. I. p. 36.
4. Beschreibung eines seltenen Insekts, *Meloë monoceros,* L.--Bibl.
Ent. I. p. 36.—Oekon. Nachr. Ges. Schles. IV. p. 380.
5. Beschreibung eines neuen Insekts, *Coccinella transversepunctata.*
—Schles. Gesellsch. IV. p. 250.—Bibl. Ent. I. p. 26.
6. Von *Ichneumon agricolator.*—N. Nachr. d. Schles. Ges. 1781,
p. 55.—Bibl. Ent. I. p. 36.
7. Beschreibung eines neuen Insekts, *Ichneumon murarius.*—N.
Nachr. d. Schles. Ges. III. p. 165.—Bibl. Ent. I. p. 36.
8. Beschreibung und Abbildung der schädlichen Gerstenfliege
(*Musca Tritici*).—Schles. Patr. Ges. 1781, p. 55.—*Eis.* Insect.
p. 235.
9. Beschreibung eines Fisches, *Gadus Fischeri* dipterygius, cirra-
tus, circis 3, maxilla superiore longiore.—Oekon. Nachr. Ges. in
Schles. VI. p. 75.
10. Zoologiæ Silesiacæ Prodromus. Mammalia; Aves; Amphibia;
Pisces.--Neue Oekon. Nachr. Ges. in Schles. II. pp. 3, 71, 131,
187.

BÖRNER (Nic.).

1. Traité rationnel des Sciences naturelles. (Allem.) Leipz. 1735,
8°; 1741, 8°.—Biogr. Un. Suppl. LVIII. p. 430.

BOROLES (W.).

1. Geschichte der spanischen Heuschrecken. Madrid, 1781.—*Eis.*
Insect. p. 243.

BOROTT (J.).

1. Acroama über *Dr. Gall's* Schädellehre, mit nützlichen unterhalt.

Reflexionen für gebildete Leser. Zittau, 1825, 8º.—Bibl. med.-ch. p. 59.

BOROWSKI (G. H.).

1. Systematische Tabellen über die allgemeine und besondere Naturgeschichte. Berl. 1775, 8º.—*Böhm.* Bibl. I. 1, p. 326.

2. Abriss einer Naturgeschichte des Elementarreichs. Mannh. u. Berl.—*Böhm.* Bibl. I. 1, p. 326.

3. Gemeinnützige Naturgeschichte des Thierreiches, darin die merkwürdigsten und nützlichsten Thiere in systemat. Ordnung beschrieben, und alle Geschlechter in Abbildungen nach der Natur vorgestellt werden. (Fortges. v. *J. F. W. Herbst.*) Berl. u. Strals. 1780–1790, 8º. fig. 10 Bde.—*Lamx.* Pol. Cor. p. 523.—*Eis.* Insect. p. 141.—*Böhm.* Bibl. II. 1, p. 53.

BORRAL ( ).

1. Mémoire sur l'Histoire naturelle de la Corse. Lond. et Par. 1783, 12º.—*Böhm.* Bibl. I. 1, p. 576.

BORRER (W.).

1. On *Emberiza hortulana.*—Ann. and Mag. N. H. VII. p. 524.

BORRICH (Chr.).

1. Dissert. de Fontibus arenosis. Hafn. 1733, 4º.—*Böhm.* Bibl. V. p. 66.

BORRICH (Olaus).

1. De Somno et Somniferis, maximè papavereis, Dissertatio. Hafn. et Francof. 1681, 4º.—Act. Erudit. I. p. 153.

2. Aci marini Anatome.—Act. Hafn. 1673, II. p. 149.—*Cuv.* et *Val.* I. p. 71 (*Anatomie de l'Orphie*).

3. De Generatione Lapidum in Macro- et Microcosmo.—Act. Hafn. V. p. 184.—(Cum Addit. *J. Lanzoni*): Ferrar. 1687, 12º.—(Germ.) *Cartheus.* Verm. Schr. II. p. 126.—*Böhm.* Bibl. IV. 1, p. 217.

4. An visus sit unquam Lumbricus Lumbrico Fœtus?—Act. Hafn. III. et IV. p. 157.

5. De Lumbricis latis et cucurbitinis.—Act. Hafn. Barth. 1673, p. 148.—*Mod.* B. Helm.

BORRO (Girol.).

1. Del Flusso e Reflusso del Mare e dell' Inondatione del Nilo. Firenze, 1567; 1577; 1583, 8º.—*Böhm.* Bibl. V. p. 84.

BORSONI (St.).

1. Observations microscopiques et dessins d'après nature du Ver qui a rongé le Bled en Piémont, et de celui qui a endommagé le Chanvre en 1815.—Mag. Encycl. V.

2. Note sur des Dents de grand Mastodonte trouvées en Piémont, et sur des Mâchoires et Dents fossiles prises dans la Mine de Houille de Cadibona.—Mem. Accad. Tor. XXVII.

3. Memoria per Appendice dell' Orittografia Pedemontana dell' *Allioni*.

4. Substances minérales exploitées dans les Dép^s. du Piémont, etc. Turin, 1806, 8°

5. Statistique minéralogique du Dép^t. du Pô.—Annuaire Statist. Turin, 1806.

6. Essai sur l'Oryctographie du Piémont (*Oryctographie von Piemont*).—Mém. Ac. Tur. XXV. p. 180; XXVI. p. 297; XXIX. p. 251, fig.—Isis, 1834, I. pp. 70, 76; IV. p. 434.—*Féruss*. Bull. 1823, II. p. 96.

7. Mémoire sur des Mâchoires et des Dents du Mastodonte dit Mammouth, trouvées fossiles en Piémont.—Mém. Ac. Tur. XXIV. p. 160. fig.—Isis, 1834, IV. p. 431 (*Ueber Kiefer und Zähne des Mastodon aus Piemont*).

8. Notice sur quelques Fossiles de la Tarentaise en Savoie (*Notiz über einige fossile Reste der Tarentaise in Savoyen*).—Mem. Accad. Tor. XXXIII. p. 174, fig.—*L. u. Br.* N. Jahrb. 1834, p. 726.—*Féruss*. Bull. XXVI. p. 153.

9. Mémoire sur quelques Ossemens fossiles trouvés en Piémont.—Mem. Accad. Tor. XXXVI. p. 33, fig.—*L. u. Br.* N. Jahrb. 1835, p. 120 (*Abh. über einige in Piemont gefundene fossile Knochen*).

10. Ossa fossili in Val d'Andona, credute falsamente di Scheletro umano.—Mem. Accad. Tor. XXIX. p. xxxiv.

11. Osservazioni intorno alle Sostanze minerali di cui sono formati i Monumenti del Regio Museo Egizio.—Mem. Accad. Tor. XXXI.

12. Ad Oryctographiam Pedemontanam Auctarium. 1798, 4°.—Mem. Accad. Tor. VI.

13. Catalogue raisonnée du Musée d'Histoire naturelle de l'Académie de Turin. Partie minéralog. selon le syst. d'*Al. Brongniart*. Tur. 1811, 8°.

BORY DE ST. VINCENT (J. G. B. M.).

1. Voyage dans les quatre principales Iles des Mers d'Afrique. Paris, 1804, 3 vols. 8°. avec Atlas.—Dict. Sc. n. LX. p. 536.—

(Germ.) *Reise nach den vier vornehmsten Inseln der Afrikani-schen Meere*, etc. I. Th.   Leipz. 1805, 8°. fig.

2. De la Matière, considérée sous les rapports de l'histoire naturelle. Paris, 1823, 8°.—Dict. class. 1824, II. p. 205.—*Féruss.* Bull. 1824, I. p. 230.—Isis, 1824, VII. Litt. A. p. 105.

3. L'Homme; essai zoologique sur le genre humain.—Dict. cl. d'H. nat. VIII.—Isis, 1825, XII. p. 1322.—2ᵉ éd. Paris, 1827.— 3ᵉ éd. Paris, 1836, 2 vols. 8°.—(Germ.) 12°. Weimar, 1837.— *Féruss.* Bull. 1827, XI. p. 281.—Isis, 1832, IV. p. 409.—*Val.* Repert. II. p. 20.

4. Résumé d'Erpétologie ou d'Histoire naturelle des Reptiles, ac-compagnée d'une Iconographie; faisant partie de l'Encyclopédie portative. 18°. avec Atlas 8°.—(Ital.) 18°. Milan, 1835.—*Féruss.* Bull. 1828, XV. p. 153.

5. Description d'une nouvelle espèce de Couleuvre—Ann. Sc. n 1824, I. p. 408, fig.—*Féruss.* Bull. p. 205.—Isis, 1830, IV. p. 392 (*Coluber Richardi*).

6. Note sur la naturalisation de la Cochenille en Espagne (*Ueber die Naturalisation der Cochenille in Spanien*).—Ann. Sc. n. VIII. p. 105.—Isis, 1834, VIII. p. 863.

7. Description de quelques Insectes nouveaux du Dép. de la Gironde.—Soc. d'Hist. nat. Bord III. p. 72.—Bibl. Ent. I. p. 43.

8. Sur un nouveau genre d'Acaridiens, sorti du corps d'une femme. —Ann. Sc. n. 1828, XV. p. 125, fig.—*Féruss.* Bull. 1829, XVIII. p. 135.—Isis, 1832, VIII. p. 904, fig. (*Neue Haut-Milbe*).

9. Mémoire sur l'établissement d'une nouvelle Famille dans la classe des Infusoires, sous le nom de Bacillariées.—Bull. Soc. Phil. 1823, p. 8.

10. Essai monographique sur les Oscillaires. Paris, 1827, 8°.—*Cuv.* R. An. III. p. 340.

11. Essai d'une Classification des Animaux microscopiques. Paris, 1826, 8°.—*Féruss.* Bull. 1826, VIII. pp. 303, 407, 444.—Isis, 1827, X. p. 878.

12. Voyage souterrain, ou Description du Plateau de St. Pierre de Maëstricht et de ses vastes Cryptes, etc.   Brux. 1819, 8°. fg.— Paris, 1821, 8°.—Isis, 1822, VIII. p. 866.

13. La Mer, considérée sous tous ses rapports.—Dict. class. d'H. n. X.—*Féruss.* Bull. 1826, VIII. p. 331.

14. Essai sur les Iles Fortunées et l'antique Atlantide. Paris, 1802, 4°. fig.—Dict. Sc. n. LX. p. 536.

15. Rapport sur un Livre intitulé: Histoire de la Génération de l'Homme, par *MM. Grimaud de Caux* et *Martin St.-Ange.*—C. R. V. p. 329.

16. Sur une fouille faite dans le Terrain primitif de l'Ile de Santorin.—C. R. V. p. 585.

17. Expédition scientifique en Morée, entreprise et publiée par ordre du Gouvernement Francais. Paris, 1835, 3 vols. fol. atl.

18. Beiträge zur Naturgeschichte der Maskaren-Inseln, geordnet v. *Biedermann.* Wiem. 1805.

19. Relation du Voyage de la Commission scientifique de Morée, dans le Péloponnèse, les Cyclades et l'Attique. Paris, 3 vols. 8°. atl. fol. 1836–38.

20. Sur l'existence du Guacharo (*Steatornis*) à l'île de la Trinité. —C. R. VII. p. 474.

21. Note sur une espèce d'Acaride que vit sur le Corps humain.— Acad. Sc. Par. 1823, p. 42.—*Féruss.* Bull. 1824, II. p. 305.

22. Notice sur les premiers travaux de la Commission scientifique de l'Algérie.—Rev. Zool. 1840, p. 153.—C. R. X. p. 138.

23. Remarques à l'occasion d'une Note de *M. Milne Edwards,* concernant des Observations de *M. Nordmann* sur les Polypiers du genre Campanulaire.—C. R. IX. p. 717.

BORY DE ST. VINCENT (J. G. B. M.), DRAPIEZ (   ), et VAN MONS (J. B.).

1. Ann. des Sc. physiques ; *vide sup.* Pars I. p. 37.

Bosc (L. A. G.).

1. Description d'un nouveau *Calopus.*—Bull. Soc. Phil. I. p. 12.— Bibl. Ent. I. p. 44.

2. Description de trois Lépidoptères.—Soc. Phil. II. p. 102.—Bibl. Ent. I. p. 44.

3. Description de trois Lépidoptères de la Caroline.—Bull. Soc. Phil. II. No. 39. p. 113, fig.—Bibl. Ent. I. p. 44.

4. Mémoire pour servir à l'histoire de la Chenille qui a ravagé les vignes d'Argenteuil en 1786.—Soc. Roy. d'Agric. 1786, Trim. d'été, p. 22.—Bibl. Ent. I. p. 44.

5. Descriptions of two new species of *Phalæna* (*Pyralis tuberculosa* et *P. Sparmannella*).—Tr. Linn. Soc. Lond. I. p. 196, fig.—Bibl. Ent. I. p. 44 (*Descriptio duarum Phalænarum*).

6. Supplément à la Cynipédologie du Chêne (Descr. du *C. Quercus Tozæ*).—J. d'Hist. nat. II. p. 154.—Bibl. Ent. I. p. 44.

7. Observation sur une nouvelle espèce de Tenthrède.—N. Bull.

Soc. Phil. 1818, p. 111.—Bibl. Ent. I. p. 45.—J. de Phys. LXXXVI. p. 476.

8. Rapport sur l'ouvrage de *Huber* sur les mœurs des Fourmis. Paris, 1813, 12°.—Bibl. Ent. I. p. 45.

9. Sur une nouvelle espèce de Cécidomye (*C. Poæ*).—N. Bull. Soc. Phil. 1817, p. 133.—Bibl. Ent. I. p. 45.—J. de Phys. LXXXV. p. 161.

10. *Keroplatus.*—Soc. d'Hist. nat. Par. I. p. 44.—Bibl. Ent. I. p. 45.

11. Description de deux Mouches.—J. d'Hist. nat. II. p. 54.—Bibl. Ent. I. p. 44.

12. Gallen von Zweiflüglern.—J. de Phys. 1817.—Isis, 1818, IX. p. 1559.

13. *Pulex fasciatus.*—Bull. d. Sc.—*Wiedem.* Arch. III. 1. p. 24.

14. Description d'une espèce de Puce.—Bull. Soc. Phil. No. 44, II. p. 156.—Bibl. Ent. I. p. 44.

15. Description du *Dorthesia Characias.*—J. de Phys. XXIV. p. 171. —Bibl. Ent. I. p. 44.

16. Histoire naturelle des Crustacés, 2 vols. 18°; faisant suite au Buffon de *Castel.* Paris, an X. 1802.—Par. 1828, 2 vols. 12°. (corr. par *A. G. Desmarest*).—Ann. Sc. n. II.

17. Histoire naturelle des Vers. Paris, 1802, 18°. 3 vols. fig.— (Ital. *Farini*): 8°. Leghorn, 1835.—Biogr. un. Suppl. LIX. p. 30.

18. Histoire naturelle des Coquilles, contenant leur description, etc. Paris, 1801, 18°. 5 vols.—Paris, 1836, 5 vols. 18°.—Biogr. un. Suppl. LIX. p. 30.

19. Rapport concernant les nouvelles Observations de *M. Gaillou,* sur la cause de la coloration des Huitres, et sur les Animalcules qui servent à les nourrir. Instit. 1823, 8 Déc.—*Féruss.* Bull. II. p. 319.

20. Articles d'Erpétologie du N. Dict. d'Hist. nat. fig.—*Dum.* et *Bib.* I. p. 309.

21. *Lacerta exanthematica.*—Soc. d'Hist. nat. Par. I. p. 25.—Dict. Sc. n. XV. p. 250.

22. Histoire naturelle des Insectes. (Contin. de *Buffon.*)

23. Moyen simple de dessécher les Larves pour les conserver dans les Collections.—Journ. de Phys. XXVI. 1780, p. 241.—*Licht.* Mag. III. 2, p. 81.—Bibl. Ent. I. p. 44.

24. Observations sur le *Serropalpus* —Soc. d'Hist. nat. Par. I. p. 40. —Bibl. Ent. I. p. 45.

25. *Bostrichus furcatus.*—J. d'Hist. nat. II. p. 259.—Bull. Soc. Phil. I. p. 6.—Bibl. Ent. I. p. 44.

26. *Ripiphorus subdipterus.*—J. d'Hist. nat. II. p. 293.—Bibl. Ent. I. p. 44.

27. Notice sur deux Insectes du genre *Cerceris*, ennemis des Charancons.—Soc. d'Agric. Seine et Oise, 1812.

28. Note sur l'Animal du Madrépore.—Journ. de Phys. LXII. 1806, p. 435.

29. Note sur l'Écureuil capistrate (*Sciurus capistratus*) de la Caroline.—Ann. du Muséum, I. 1802, p. 281.

30. Rapport sur un nouveau genre de Vers intestinaux (*Thelazia*) découvert par *T. B. Rhodes.*—Journ. de Phys. LXXXVIII. 1819, p. 214.

31. Observation sur *Acheta sylvestris.*—Soc. d'Hist. nat. Par. I. p. 44.

32. Observation sur *Locusta punctatissima.*—Soc. d'Hist. nat. Par. I. p. 45.

33. Beytrag zur Kenntniss der edlen Opal-Arten.—Schr. Berl. Ges. Naturf. Fr. XI. p. 152.

34. Observation sur *Sepia rugosa.*—Actes de la Soc. d'Hist. nat. Par. I. p. 24.

35. Observation sur l'*Ardea gularis.*—Actes de la Soc. d'Hist. nat. de Paris, I. p. 4.

36. Description des objets nouveaux d'Histoire naturelle trouvés dans une traversée de Bordeaux à Charlestown.—Bull. Soc. Phil. 1797, p. 9.

Bosc (L. A. G.), Tessier (H. A.) et Bosc ( ).

1. Annales d'Agriculture française.

Boscius (J. Lon.).

1. Descriptio Balneorum Wembdingensium, quæ in Bavariâ superiori maximo ægrotorum emolumento frequentantur. Ingolst.—*Böhm.* Bibl.

Bose (A. J.).

1. De differentia Fibræ in corporibus trium Naturæ Regnorum. 4°. Wittenberg, 1768.

Bose (E. G.).

1. Progr. de Fabricâ vasculosâ vegetabili et animali. Lips. 1783, 4°.—*Böhm.* Bibl. II. 1, p. 118.

2. Mém. sur la Génération spontanée dans les trois regnes. 8°. Montpellier, 1831.

3. Progr. de Generatione hybridâ. Lips. 1777, 4°.—Tr. Linn. Soc. Lond. V. p. 278.

4. Decas Librorum anatomicorum variorum. Lips. 1761.—Biogr. un. V. p. 218.

Bose (G. Matth.).

1. Otia Wittebergensia. Wittenb. 1739, 4°.—Bibl. Ent. I. p. 45.—
Böhm. Bibl. II. 2, p. 231.

2. Dissert. de naturâ et origine Nebularum. Wittenb. 1756, 4°.—
Böhm. Bibl. V. p. 50.

3. Anatome Ranæ in vacuo extinctæ et vivæ, 4°. Wittenb. 1739.

Bösius (D. Salv.).

1. Septem Miracula Delphinatûs : 1° de Fonte ardente ; 5° de Fonte vinoso. Gratian. 1638 et 1650, 8°.—Lugd. 1661, 8°.—*Böhm.* Bibl. V. p. 180.

Bosman (W.).

1. Naauwkeurige beschryving van de Guinese Goud-Tand-en Slave-kust. 4°. Utrecht, 1704.—(Angl.) *New Description of the Coast of Guinea*, etc. 8°. London, 1705, 1721.—(Gall.) *Voyage en Guinée.* Utrecht, 1705, 1 vol. 8°.—*Dum. et Bib.* I. p. 309.

Bossart (J. J.).

1. Kurze Anweisung Naturalien zu sammeln. Barby, 1774, 8°.—
Böhm. Bibl. I. 1, p. 371.—*Mod.* B. H.

Bosse ( ).

1. De Vermibus in pustulis Cutis indentatis.—Journ. de Méd. XXXII. p. 336.—*Mod.* B. Helm.

Bosset (De).

1. Notice sur la Carinaire de la Méditerranée.—Mém. Soc. Sc. nat. Neuch. II. p.   fig.

2. Notice sur la présence temporaire de l'*Ophidium imberbe* dans la cavité du corps d'une Holothurie orangée.—Mém. Soc. Sc. nat. Neuch. II. p.

Bossi (Giac.).

1. Compendio di Storia naturale. 12°. Turin, 1842.

Bossi (L.).

1. Dizionario portatile di Geologia, Litologia e Mineralogia. Milano, 1819, 8°.

2. Osservazioni orittologiche sulle colline.—Scelta de Opusc. interess. XVI.

3. Observations sur l'Or natif en paillettes que l'on trouve dans les Sables.—Mem. Accad. Tor. XIV. p. 270.

4. Dei Basilischi, Dragoni, ed altri animali creduti favolosi. 8°. Milan, 1792.

Bossuet (   ).

1. De naturâ Aquatilium Carmen. Lugd. 1558, 4°. fig.—*Mod.* B. H.

Bost (L. v.).

1. Oude nieuws der ontdekte Weerelt, vervattende een duidelijke Beschrijving van uytstekende hedendaagshe en alveide Steden, Gebouwen, Bergen, Wateren, Fonteinen, Vruchten, Vogelen, Beesten en Menschen. Amst. 1667, 12°. fig.—*Eis.* Insect. p. 147.

Bostock (J.).

1. On the Analysis of a Mineral Water from the Island of St. Paul, in lat. 38° 45' S. and long. 77° 53' E.—Trans. Geol. Soc. ser. 2, V. 261.—Proc. Geol. Soc. II. p. 112.

2. On the Pebbles in the Bed of Clay which covers the New Red Sandstone in the S.W. of Lancashire.—Trans. Geol. Soc. ser. 2, II. p. 138.

3. On the Domestic Habits of a minute species of Ant.—Trans. Entom. Soc. II. p. 65.

4. Versuch einer Lehre über das Athemholen. Aus d. Engl. v. *Nolde.* 2te Aufl. Erfurt, 1817, 8°. fig.—Bibl. med.-ch. p. 59.

5. An elementary System of Physiology. 3 vols. 8°. Lond. 1824–27; Lond. 1836.—Cat. R. Soc. L.

Bostock (J. A.).

1. On a remarkable Flight of Locusts.—Linn. Soc. Lond. 1844, Jan.—Ann. and Mag. N. Hist. XIII. p. 517.

Boswell (Jam.).

1. An Account of Corsica. Glasg. 1768, 8º.—Lond. 1768, 8º;
1769, 8º.—(Gall. *J. P. Dubois*): La Haye, 1769.—Par. 1769, 8º.
—(Germ. *A. E. Klausing*): Leipz. 1768, 8º; 1769, 8º.—(Ital.)
Lond. 1769, 4º.—*Böhm.* Bibl. I. 1, p. *575.*

Boswell (P.).

1. Art of Taxidermy. 8º. London, 1841.

Botal (Leonh.).

1. De viâ Sanguinis à dextro ad sinistrum cordis ventriculum, etc.
—Biogr. un. V. p. 249.

Bothen (Barth.).

1. Gute Bothschaft von denen Gungelsbrunnen, oder Beschreibung
des Gnadenbrunnen so nicht weit von dem Kloster Lühne bey
Lüneburg quillt. Lüneb. 1647, 4º.—*Böhm.* Bibl. V. p. 364.

Böthlingk (W.).

1. Ueber die Diluvial- u. Alluvial-Gebilde im südlichen Finnland.—
Bull. Ac. Pétersb. V. p. 270.—*L.* u. *Br.* N. Jahrb. 1839, p. 722.

2. Bericht einer Reise durch Finnland und Lappland: 1te Hälfte,
von Petersburg bis Kola.—Bull. Acad. Pétersb. 1840, VIII. p. 107.
—*L.* u. *Br.* N. Jahrb. 1840, p. 613.—2te Hälfte: Reise längs
den Küsten des Eis-Meeres u. Weissen Meeres.—Bull. Ac. Pét.
1840, VII. p. 191.—*L.* u. *Br.* l. c. p. 717.

Botta (C.).

1. Storia naturale e medica dell' Isola di Corfu. Milano, 1774;
1799, 2 vols. 12º.—12º. Milan, 1823.

Botta (P. E.).

1. Sur la Structure géognostique du Liban et de l'Anti-Liban.—
Bull. Soc. Géol. Fr. I. pp. 212, 225, 234.—Mém. I. p. 134, fig.—
L. u. *Br.* N. Jahrb. 1834, p. 464 (*Beobachtungen über den Liba-
non u. Antilibanon*).

2. Note sur un fait observé sur les Chameaux.—Ann. d'Anat. I.
p. 141.

3. Description du *Saurothera californiana.*—N. Ann. Mus. IV.
p. 121, fig.—*Wiegm.* Arch. 1836, II. p. 269.

BOTTA-DESMORTIERS (   ).

1. Examen comparatif de l'influence de chacun des deux Nerfs de la Face sur la production de la Sensibilité et du Mouvement de cette partie. Par. 1834, 8°.

BÖTTGER (C. H.).

1. Beschreibung der Gesundbrunnen und Bäder bey Hofgeismar. Cassel, 1772, 8°. fig.—*Böhm.* Bibl. V. p. 331.

BOTTIN DESYLES (   ).

1. Notice sur la Chenille de l'*Urapterix sambucata*, et sur la manière dont elle construit sa coque (Essai sur les Lépidoptères du Dép. de la Manche).—Ann. Soc. Ent. Fr. VI. p. 401.

BOTTIS (G. de).

1. Ragionamento istorico intorno all' Eruzione del Vesuvio, che cominicio il di 29 Juglio 1779, etc. Nap. 1779, 4°. fig.

2. Ragionamento istorico intorno a nuovi Vulcani comparsi nella fine del anno 1760, etc. Nap. 1761, 4°.

3. Ragionamento istorico del Incendio del Vesuvio accadute nel mese d'Ottobr. Nap. 1768, 4°.

BOTTO (Hier.).

1. De Sensibilitate. Pisa, 1811.

2. De Humano Fœtu. Genuæ, 1817.

BOTTONI (Dom.).

1. Pyrologia topographica. Nap. 1692, 4°.—Bibl. Ent. I. p. 45.

2. De immani Trinacriæ Terræmotu ; idea historico-physica, in quâ non solùm Telluris concussiones transactæ recensentur, sed novissimæ anni 1717. Messana, 1718, 4°.—Cat. R. Soc. L.

BOUBEE (Nérée).

1. Tableau mnémonique des Terrains primitifs, destiné au Géologue voyageur, avec son explication.—Ann. Sc. n. 1831, XXIV.—Rev. Bibl. p. 81.

2. Bassin de Toulouse.—Ann. Sc. n. 1831, XXIII.—Rev. Bibl. p. 52.—Bull. Soc. Géol. I. p. 146.

3. Deux Promenades au Mont Dore pour l'étude de la question des Cratères de Soulèvement. 12°. Paris, 1833.—Cat. R. Soc. L.

4. La Géologie dans ses rapports avec l'Agriculture et l'Economie politique. Paris, 1840, 18º.

5. Carte géologique, minéralogique, agricole et industrielle de la France, en 4 grandes cartes coloriées. Par. 1840.

6. Nouveaux moyens propres à faciliter la détermination des Fossiles.—Bull. Soc. Géol. I. p. 230.

7. Note sur les Sources thermales de Bagnères de Luchon.—C. R. II. p. 534.

8. Lettre sur un Tremblement de terre ressenti à St. Bertrand de Comminges.—C. R. I. p. 322.

9. Tableau de l'état du Globe à ses différens âges ; basé sur l'examen des faits. fol. col.

10. Relation des Expériences physiques et géologiques, faits au lac Doo. Paris, 1832, 18º. fig.

11. Géologie populaire élémentaire, à la portée de tout le monde. Paris, 1833, 8º. fig.

12. Sur le Terrain houiller de la France centrale.—C. R. VIII. p. 133.

13. Recueil d'Itinéraires pour servir de Guide au Minéralogiste, au Conchyliologiste et au Géologue dans toute la France, accompagné d'un Bulletin de nouveaux Gisemens pour toutes les parties de l'Hist. nat. Par. 1832, 8º.

14. Cours abrégé de Géologie. Par. 1834, 8º.

15. Abh. über die Aushöhlung der Treppen-Thäler.—Instit. 1833, I. p. 94.—L. u. Br. N. Jahrb. 1835, p. 593.

16. Fossiles d'un Terrain de Calcaire d'eau douce au Sud-est de Toulouse.—Bull. Soc. Géol. Fr. I. p. 212.

17. Considérations générales sur les Animaux qui vivaient aux diverses époques géologiques.—Ann. Sc. n. 1830, XIX.—Rev. Bibl. p. 42.

18. Géologie élémentaire appliquée à l'Agriculture et a l'Industrie. 18º. Par. 1833.—Cat. R. Soc. Lond.—(2e éd.) 1836, 18º fig.— (Trad. Allem.) *Elemente der Geologie, oder Grundzüge,* etc. Weim. 1837, 8º.

BOUCHARD-CHANTEREAUX (   ).

1. Animaux sans vertèbres observés dans le Boulonnais. Boulogne, 1829, 8º.—*Féruss.* Bull. 1830, XX. p. 165.

2. Catalogue des Mollusques marins et des Crustacés du Boulonnais. Boulogne, 1835, 8º

3. Observations sur les mœurs de divers Mollusques terrestres et fluviatiles observés dans le Dépt. du Pas-de-Calais.—Ann. Sc. n. (2ᵉ S.) XI. p. 295.

4. Note sur le genre *Productus*.—Ann. Sc. n. (2ᵉ S.) XVIII. p. 158, fig.

5. Catalogue des Mollusques terrestres et fluviatiles du Dépt. du Pas-de-Calais. Boul. 1838, 8°. fig.

BOUCHARDAT (A.).

1. Cours des Sciences physiques, à l'usage des Élèves de philosophie. Paris, 1844, 12°. fig (Hist. nat. 2 vols.)

BOUCHARDAT (A.) et SANDRAS (   ).

1. Recherches sur la Digestion.—Ann. Sc. n. (2ᵉ S.) XVIII. p. 225.

2. Recherches sur la Digestion et l'Assimilation des Corps gras, suivies de Considérations sur le rôle de la Bile et de l'Appareil chylifère.—Acad. Sc. 14 Août 1843.—Ann. Sc. n. (2ᵉ Sér.) XX. p. 169.

BOUCHE (Hon.).

1. Chorographie de Provence. Aix, 1674, fol.—*Böhm.* Bibl. I. 1, p. 512.

BOUCHÉ (P. F.).

1. Die Körpertheile der zweiflügeligen Insekten; ein terminologischer Versuch.—Berl. Mag. VI. p. 36.—Isis, 1818, IX. p. 1473.

2. Naturgeschichte der Insekten besonders in Hinsicht ihrer ersten Zustände als Larven u. Puppen. 1ᵗᵉ Lief. Berlin, 1834, 8°. fig. —*Wiegm.* Arch. 1835, II. p. 9.

3. Naturgeschichte der schädlichen u. nützlichen Garten-Insekten, etc. Berl. 1833, 8°.—Bibl. Ent. I. p. 45.

4. Beiträge zur Insektenkunde.—N. Act. Nat. Cur. XVII. p. 493.

5. Bemerkungen über die Larven der zweiflügligen Insecten.—N. Act. Nat. Cur. XVII.

6. Bemerkungen über die Gattung *Pulex*.—N. Act. Nat. Cur. XVII. fig.

BOUCHER D'ABBEVILLE (   ).

1. De Filaria in Bupresti reperta.—Rapp. gén. Soc. Philom. III. p. 72.

Boucher (   ).

1. Catalogue raisonné des Tableaux, etc., Minéraux, Cristallisations, Madrepores, Coquilles, etc. qui composent son Cabinet. Par. 1771, 8°.—*Böhm.* Bibl. I. 1, p. 404.

2. Observations sur un Squelette d'Auerochs trouvé à Picquigny. —Mag. Encycl. IV. p. 24.

3. Conchyologie. 80 pl. 4°.

Boucheron (P. P.).

1. Traité anatomique, physiologique et pathologique du Système pileux, et en particulier des Cheveux et de la Barbe.   Paris, 1837, 8°.—*Val.* Repert. III. p. 16.

Boudet (   ).

1. Essai critique et expérimental sur le Sang.   Par.

Boudier (H. Ph.).

1. Observations sur divers Insectes parasites.—Ann. Soc. Ent. Fr. III. p. 327, fig.— *Wiegm.* Arch. 1835, II. p. 46.

2. Description du genre *Psammœcus.*—Ann. Soc. Ent. Fr. III. p. 367.— *Wiegm.* Arch. 1835, II. p. 39.

3. Description d'une nouvelle espèce de *Lema* (*L. brunnea,* Fabr.). —Ann. Soc. Linn. Par. IV. p. 239, fig.—*Féruss.* Bull. 1827, XI. p. 395.—Bibl. Ent. I. p. 46.

4. Observations sur les habitudes des Larves d'Ichneumons vivant aux dépens de la Chenille du Bombyx du chêne.—Ann. Soc. Ent. Fr. V. p. 357, fig.— *Wiegm.* Arch. 1837, VI. p. 316 (*Cryptus Bombycis*).

Boué (A.).

1. On Serpentine and Diallage Rocks.—Edinb. New Phil. Journ. II. p. 265.—*Féruss.* Bull. 1828, XIV. p. 7.

2. On the Geological Structure of the Archduchy of Austria, and of the South of Bavaria.—Proc. Geol. Soc. I. p. 223.—Phil. Mag. ser. 2, VIII. p. 64.—*Féruss.* Bull. 1830, XX. p. 235.—Bull. Soc. Géol. Fr. I. p. 144.

3. On *M. Beudant's* Opinions regarding the Crystalline Rocks of the Red Sandstone Formation.—Edinb. Phil. Journ. X. p. 67 — *Féruss.* Bull. 1824, I. p. 310.

4. On Alluvial Rocks; on Formations; and on the Changes that appear to have taken place during the different periods of the Earth s formation on the Climate of our Globe, and in the nature and the physical and the geographical distribution of its Animals and Plants.—Edinb. New Phil. Journ. I. p. 82.—*Féruss.* Bull. 1826, IX. p. 261.

5. Geological distribution of the Fossil Organic Remains enumerated by *Baron von Schlotheim.*—Edinb. Phil. Journ. XII. pp. 142, 270.—*Féruss.* Bull. 1826, IX. p. 2.—Isis, 1833, VIII. p. 706 (*Geologische Vertheilung der von Schlotheim aufgeführten Versteinerungen*).—*Leonh.* Zeitschr. 1826, II. p. 129.

6. Observations in Answer to a Memoir by *Messrs. Sedgwick* and *Murchison* on the Austrian Alps.—Edinb. New Phil. Journ. X. p. 14.—Ann. Sc. n. 1831, XXII.—Rev. Bibl. p. 18.—Bull. Soc. Géol. I. p. 40.

7. Short Comparison of the Volcanic Rocks of France with those of a similar nature found in Scotland.—Edinb. Phil. Journ. II. p. 326.

8. Critical Observations on the Ideas of *M. Alex. Brongniart,* relating to the Classification and probable Origin of Tertiary Deposits.—Edinb. New Phil. Journ. XII. pp. 159, 340.

9 On the Geognosy of Germany, with Observations on the Igneous origin of Trap.—Mem. Wern. Soc. IV. p. 91.

10. Mémoire géologique sur le Sud-Ouest de la France, suivi d'observations comparatives sur le Nord du même royaume, et en particulier sur les bords du Rhin.—Ann. Sc. n. 1824, II. p. 387; III. p. 299; 1825, IV. p. 125.

11. Mémoire sur les Terrains secondaires du versant nord des Alpes allemandes.—Ann. des Mines, 1824, IX. p. 477.—*Féruss.* Bull. 1824, III. p. 136.

12. Observations générales sur la distribution géographique, la nature et l'origine des Terrains de l'Europe.—Zeitschr. f. Mineral. Jul. 1827, p. 18.—*Féruss.* Bull. 1828, XIV. p. 307.

13. Hippuriten-Kalk.—*Leonh.* u. *Br.* Jahrb. 1834, p. 614.—*Bronn,* Leth. II. p. 551.

14. Résultats d'Observations géognostiques faites en Allemagne en 1821 et 1822.—Bull. Soc. Phil. 1822, p. 38.

15. Note sur l'âge du Dépôt de Terre-Nègre à Bordeaux.—Ann. Sc. n. Oct. 1826.—*Féruss.* Bull. 1827, X. p. 433.

16. Mémoires géologiques et paléontologiques. T. I. Paris, 1832, 8°. fig.—Isis, 1833, IX. p. 823.

17. Catalogue des Cartes et des Coupes géologiques.—Zeitschr. f. Mineral. Avr. 1828, p. 283.—*Féruss.* Bull. 1829, XVI. p. 203.

18. Supplément au Catalogue des Cartes géognostiques.—Ann. Sc. n. 1829, XVI.—Rev. Bibl. p. 21.

19. On the Theory of the Elevation of Mountain Chains, as advocated by *M. Elie de Beaumont.*—Edinb. New Phil. Journ. XVII. p. 123.—*L.* u. *Br.* N. Jahrb. 1836, p. 426.

20. Réponse au Cte *de Sternberg,* par rapport à l'existence des Variolaires dans le grès du Lias d'Amberg.—*Féruss.* Bull. 1826, IX. p. 287.

21. Bemerkungen zu *Bronn's* Bestimmung Salzburger Petrefakten. —*Leonh.* u. *Br.* Jahrb. 1833, p. 63.

22. Ossemens humains présumés fossiles.—Ann. Sc. n. 1829, XVIII. —Rev. Bibl. p. 150.

23. Notice sur la Chaussée des Géans en Irlande.—Bibl. Univ. Gen. III. p. 278.

24. Mémoire sur des Coupes géologiques aux environs d'Edimbourg et dans le Fifeshire.—Journ. de Phys. 1818 (ou 1819).

25. Guide du Géologue voyageur, sur le modèle de l'*Agenda geognostica* de *M. Leonhard.* Paris, 1835–36, 2 vols. 12°. fig.

26. Exposé de la Géologie faite par zones et régions, et non par pays.—Rapport à la Soc. Géol. Fr. 1832, III. p. LXXXI.—Guide du Géol. voy. II. p. 354.

27. Extraits des Observations géologiques faites en Turquie.—Bull. Soc. Géol. 1837–1839.—*Leonh.* et *Br.* Jahrb. f. Min.—Edinb. Philos. Journ. 1837 & 1839.

28. Zweite geognostische Reise in die Türkey.—*L.* u. *Br.* N. Jahrb. 1838, p. 44 ; 1839, p. 553.

29. Esquisse géologique de la Turquie d'Europe. Par. 1840, 8°.

30. Gegen *Bielz*, wegen Siebenbürgen.—*L.* u. *Br.* N. Jahrb. 1833, p. 181.

31. Cratères de Soulèvement dans les Terrains non volcaniques, et appliquées particulièrement au Sol de la Carinthie.—Bull. Soc. Géol. Fr. VI. p. 29.

32. Aperçu sur le Sol tertiaire de la Gallicie.—Bull. Soc. Géol. Fr. I. pp. 15, 90.

33. Sur le Mercure dans le Calcaire des Alpes.—Journ. Géol. V.— Bull. Soc. Géol. Fr. I. p. 91.

34. Du Thermomètre géologique de *M. Daubeny.*—Zeitschr. f. Mineral. 1828, p. 617.—*Féruss.* Bull. 1829, XVII. p. 163.

35. Tableau synoptique des Formations de la Croûte du Globe et de ses masses subordonnées principales.—Ed. Phil. Journ. XIII. p. 180; 1825, p. 371.—Mém. Soc. Linn. Norm. 2e Sér. I. pp. 1, 165.—*Féruss.* Bull. 1826, VIII. p. 314; IX. p. 258 ; 1828, XIV.

p. 305; 1831, XXIV. p. 131.—*Leonh.* Zeitschr. 1827, II. p. 1
(*Synoptische Darstellung der die Erdrinde ausmachenden Formazionen*).

36. Résumé de nos connaissances géologiques principales sur la
Turquie Européenne et l'Asie Mineure.—Zeitschr. f. Mineral.
Avril 1828, p. 270.—*Féruss.* Bull. 1828, XV. p. 49.

37. Essai géologique sur l'Écosse. Paris, 1820, 8°. fig. et cartes.—
Isis, 1822, X. p. 1072.

38. Mémoire sur les Terrains tertiaires et basaltiques du Sud-Ouest
de l'Allemagne, au nord du Danube.—Ann. Sc. n. II. p. 1.—
Féruss. Bull. 1824, II. p. 125.—*Leonh.* Zeitschr. 1825, II. pp. 252,
427 (*Die älteren und neueren Felsgebilde im südwestl. Deutsch-
lande, nordwärts der Donau*).

39. Geognostisches Gemählde von Deutschland, mit Rücksicht auf
die Gebirgs-Beschaffenheit nachbarlicher Staaten. Herausg. v.
C. C. v. Leonhard. (*Tableau géognostique de l'*Allemagne, *mis
en rapport avec la géologie des états voisins, et publié par* C. de
Leonhard.) Frankf. 1829, 8°. fig.—*Féruss.* Bull. 1829, XIX.
p. 180.—Bull. Soc. Géol. Fr. VI. p. 103.

40. Mémoire géologique sur les Terrains anciens et secondaires du
S.O. de l'Allemagne au N. du Danube.—Ann. Sc. n. Juin 1824,
p. 173.—*Féruss.* Bull. 1824, III. p. 268.

41. Observations sur le Sud de l'Allemagne.—Zeitschr. f. Miner.
1829, VII. p. 513.—*Féruss.* Bull. 1830, XX. p. 236.

42. On the Secondary Rocks in the Alps and the Carpathians.—
Edinb. New Phil. Journ. VIII. p. 176.—*Féruss.* Bull. 1830, XX.
p. 228.

43. Carte géologique de l'Europe.—Bull. Soc. Géol. Fr. I. p. 90.

44. Mémoire sur le Sol tertiaire de la Bavière, de l'Autriche et de
la Hongrie.—Journ. Géol. VIII.—Bull. Soc. Géol. Fr. I. p. 170.

45. Description de divers Gisemens intéressans de Fossiles dans les
Alpes autrichiennes.—Bull. Soc. Géol. Fr. I. p. 128.

46. Nachrichten über die Gegenden von Narbonne, Pézenas, La
Corniche u. das Vicentinische.—Bull. Soc. Géol. Fr. III. p. 324.
—*L.* u. *Br.* N. Jahrb. 1834, p. 689.

47. Ueber das Zusammenvorkommen von Orthoceratiten mit Ammoniten und Belemniten.—*L.* u. *Br.* N. Jahrb. 1844, p. 328.

48. Outlines of a Geological Comparative View of the South-west
and North of France and the South of Germany.—Edinb. Phil.
Journ. IX. p. 128.—*Féruss.* Bull. 1824, I. p. 10.—*Leonh.* Zeitschr.
1825, I. p. 374.

49. Note sur les idées de *M de Beaumont,* relativement au Sou-

lèvement successif des diverses Chaînes du Globe.—Journ. Géol. XII. p. 338.—*Féruss.* Bull. XXVI. p. 116.

50. Sur les environs de Kandern, dans le pays de Bade.—Journ. Géol. Sept. 1830, p. 107.—*Féruss.* Bull. XXVI. p. 149.

51. Allgemeine geologische Beobachtungen über die Entstehung der Gebirge Schottlands (Essai géolog. s. l'Écosse).—*Leonh.* Tasch. 1823, II. p. 239.

52. Geognostische Beschaffenheit von Siebenburgen.—*Leonh.* Zeitschr. 1825, I. p. 508.

53. Ueber die Alluvial-Gebilde.—Ed. N. Phil. J. 1826, Apr. p. 82. —*Leonh.* Zeitschr. 1827, I. p. 316.

54. Ueber die Aenderungen, welche während der verschiedenen Perioden der Erd-Bildung in den Klimaten auf unserer Erde Statt gehabt.—Ed. N. Phil. J. 1826, Apr. p. 88.—*Leonh.* Zeitschr. 1827, I. p. 184.

55. Ueber die Formazionen.—Ed. N. Phil. J. 1826, Apr. p. 84.— *Leonh.* Zeitschr. 1827, I. p. 261.

56. La Turquie d'Europe, ou Observations sur la Géographie, la Géologie, l'Histoire naturelle, la Statistique, etc. de cet Empire. Paris, 1840, 4 vols. 8°.

57. On the Geology of Moravia and the West of Hungary.—Proc. Geol. Soc. I. 239.—Phil. Mag. ser. 2, IX. p. 50.

58. On the Geology of Transylvania.—Proc. Geol. Soc. I. 242.— Phil. Mag. ser. 2, IX. p. 134.

59. On the Geography and Geology of Northern and Central Turkey.—Edinb. New Phil. Journ. XXII. pp. 47, 253 ; XXIII. p. 54.

60. Geographical and Geological Observations on some parts of European Turkey, namely, Mæsia, Bulgaria, Romelia, Albania, and Bosnia.—Edinb. New Phil. Journ. XXV. p. 174.

61. Essai pour apprécier les avantages de la Paléontologie appliquée à la Géognosie et à la Géologie.—Bull. Soc. Géol. Fr. II. pp. 81, 87.

62. Geognostische Ergebnisse in der Türkei.—*L.* u. *Br.* N. Jahrb. 1836, p. 700.

63. Mémoire géologique sur l'Allemagne. Paris, 1822, 4°.—*Féruss.* Bull. I. p. 219 ; II. p. 46.

64. Aperçu sur la constitution géologique des provinces Illyriennes. 4°. 1 pl.

65. Coup-d'œil d'ensemble sur les Carpathes, le Marmarosh, la Transylvanie, et certaines parties de la Hongrie. 4°. 1 pl.

66. Résumé des Progrès des Sciences géologiques pendant l'année 1833. 8°.

67. Détails sur une Exploration de la Turquie d'Europe par une Société de Naturalistes.—C. R. III. p. 444.

BOUÉ (A.), JOBERT (A. C. G.), et ROZET ( ).

1. Journal de Géologie; *vide sup.* Pars I. p. 52.

BOUESNEL ( ).

1. Notice sur un gisement de Calamine, dans les environs de Philippeville, province de Namur.—Ann. d. Mines, 1826, p. 243.—*Féruss.* Bull. 1827, X. p. 22.

BOUGAINVILLE (L. A.).

1. Voyage autour du Monde par la frégate la Boudeuse et l'Étoile en 1766–69. 4°. Paris, 1771 ; 2 vols. 8°. 1772.—(Angl. *J. R. Forster*): 4°. London, 1772.

BOUGAINVILLE (Baron de).

1. Journal de la Navigation autour du Globe de la frégate la Thétis et de la corvette l'Espérance pendant 1824, 25, et 26. 2 vols. 4°. Paris, 1837.

BOUGEANT ( ).

1. Observ. Curieuses ; *vide sup.* Pars I. p. 51.

BOUI (Rocco).

1. Dissert. ital. e franc. sopra la produzione de Coralle, e Riflessioni critiche sopra i Polypi creduti costruttori dei medesimi Coralli. Firenze, 1769, 8°.—(Germ.) Mag. d. Ital. Literat,—*Böhm.* Bibl. II. 2, p. 498.

BOUILLAUD ( ).

1. Recherches expérimentales sur les Fonctions du Cerveau en général et sur celles de sa partie antérieure en particulier. Paris, 1830.

BOUILLET (J. B.).

1. Catalogue des espèces et variétés des Mollusques terrestres et fluviatiles observés dans la Haute-Auvergne. Clerm. 1836, 8°.

2. Topographie minéralogique du Dép. du Puy-de-Dôme, suivie d'un Dictionnaire oryctognostique et d'un Tableau synoptique des hauteurs d'un grand nombre de montagnes, villes et villages du même département. Clerm. 1829, 8°. fig.—1 vol. 8°. fig.—Ann. Sc. n. 1829, XVIII.—Rev. Bibl. p. 104.

3. Catalogue des espèces de Coquilles vivantes du Dép. du Puy-de-Dôme.—Ann. d'Auv. 1832.

4. Lettre sur les Coquilles fossiles.—Bull. Soc. Géol. Fr. VI. p. 99.

5. Description historique et scientifique de la Haute-Auvergne, suivie d'un Tableau alphabét. des Roches et Minéraux du même pays, etc. Par. 1835, 2 vols. 8°. Atl.—Bull. Soc. Géol. Fr. VI. p. 83.

6. Notice sur les Coquilles fossiles du Calcaire d'eau douce du Cantal. Clerm. 1834, 8°. fig.—Bull. Soc. Géol. Fr. VI. p. 61.

7. Essai géologique et minéralogique sur les environs d'Issoire (Puy-de-Dôme), et principalement sur la Montagne de Boulade. Paris, fol.

8. Catalogue des espèces et variétés de Mollusques terrestres et fluviatiles observées jusqu'à ce jour à l'état vivant dans la Haute et Basse Auvergne. 8°. 1837.

BOUILLET (J. B.) et LECOQ (   ).

1. Vues et Coupes des principales formations géologiques du Dép. du Puy-de-Dôme, accompagnées de la description et des échantillons des Roches qui les composent. Clerm. 1829, 1830 (par livr.).—*Féruss.* Bull. 1829, XVII. p. 164; 1831, XXIV. p. 255.

BOUILLET (   ).

1. Sur la grotte de la Balme en Bugey.—Mém. Acad. Dijon, I. p. LXIII.

BOUILLIER (E.).

1. Sur une espèce de Polypier fossile, rapporté au genre Favosite, *Lamk.*—Ann. Soc. Linn. Par. Sept. 1826, p. 428.—*Féruss.* Bull. 1829, XVII. p. 474.

BOUISSON (   ).

1. Tableau des progrès de l'Anatomie dans l'École de Montpellier. Montp. 1837, 8°.

BOULLAND (   ).

1. Observations sur les Ichthyosarcolites.   8°. Angoulême, 1819.

BOULLEMIER.(   ).

1. Observatio de Insecto, quod *Myrmeleon formicarium* dicitur.— Mém. Acad. Dijon, I. p. 403.—*Eis.* Insect. p. 230.

BOURASSÉ (J. J.).

1. Esquisses entomologiques. 12°. Tours, 1842 ; 1844.

2. Hist. nat. des Oiseaux, des Reptiles et des Poissons. 12º. Tours, 1843.

BOURCHIER (R. J.).

1. On *Vultur Kolbii.*—Proc. Zool. Soc. Lond. III. p. 81.

BOURCIER (Jules).

1. Description de l'adulte de l'*Ornismya Bonapartei.*—Rev. Zool. 1841, p. 177.
2. Oiseau-Mouche nouveau.—Rev. Zool. 1840, p. 275.
3. Description de quelques espèces nouvelles d'Oiseaux-Mouches. —Rev. Zool. 1839, p. 294.

BOURDELIN (C. L.).

1. Sur le Succin.—Mém. Ac. Sc. Paris, 1742, Hist. p. 47, Mém. p. 143.

BOURDELIN (L. Ch.).

1. Dissertatio : An ut Insectis sic et Fœtui sua Metamorphosis ? —Paris, 1773, 4º.—*Eis.* Insect. p. 162.

BOURDELOT (P. Mich.).

1. Recherches et Observations sur les Vipères ; réponse à une Lettre de *M. Rédi.* Paris, 1671, 12º.—Dict. Sc. n. XV. p. 254.

BOURDET (P. F. M.).

1. Mémoire sur le gisement des Ossemens fossiles du Mont de la Molière.—Ann. Soc. Linn. Par. Sept. 1825, p. 361.—*Féruss.* Bull. 1826, VII. p. 297.
2. Notice sur les Brèches osseuses de l'île de Corse.—J. de Phys. 1822, Août, p. 143.—Ann. Soc. Linn. Par. Mars 1825, p. 52. —*Féruss.* Bull. 1823, I. p. 219 ; 1826, VII. p. 300.
3. Note sur un nouveau gisement de la Strontiane sulfatée.—Bull. Soc. Phil. 1823, p. 37.
4. Histoire naturelle des Ichthyodontes, ou Dents fossiles qui ont appartenu à la famille des Poissons, sous les rapports zoologiques et géologiques. 4º. fig.—*Féruss.* Bull. 1824, III. p. 231.
5. Mémoire sur deux Tortues fossiles du genre Chélonée et du genre Emyde.—Bull. Soc. Phil. 1822, p. 99, fig.
6. Notice sur des Fossiles inconnus qui semblent appartenir à des plaques maxillaires de Poissons dont les analogues vivans sont perdus, et que j'ai nommés *Ichthyosiagônes.* Genève et Par. 1822, 4º. fig.—*Féruss.* Bull. 1824, II. p. 100.

7. Le Naturaliste Voyageur et la Taxidermie.   Berne, 1820, 8°.—
Cat. Bibl. Schw. Naturf. Ges.

8. Mémoire sur l'Histoire naturelle.   Lettre à MM. les Prof. du
Muséum.   1820, 8°.

9. On the Position of the Fossil Bones of Mont de la Molière
(Switzerland).—Phil. Mag. ser. 2, II. p. 8.

BOURDON (Aimé).

1. Nouvelle Description anatomique de toutes les parties du Corps
humain et de leurs usages. Paris, 1674, 1679, 1683, 12°.—Biogr.
Un. V. p. 367.

2. Nouvelles Tables anatomiques, où sont représentées toutes les
parties du Corps humain. Paris, 1678, fol.—Biogr. Un. V. p. 367.

BOURDON (Isid.).

1. Lettres à *Camille* sur la Physiologie de l'Homme.   Paris, 1830,
18°.—Ann. Sc. n. 1830, XIX.—Rev. Bibl. p. 8.

2. Principes de Physiologie comparée.   8°. Paris, 1830.

3. Considérations sur les Animaux en général.   8°. Paris, 1822.

4. Guide aux Eaux minérales de la France et de l'Allemagne. Par.
1834, 12° ; 1837, 18°.

5. Recherches sur le Mécanisme de la Respiration et sur la Circulation du Sang. 8°. Paris, 1820.—Cat. R. Soc. L.

BOURDON (M. H.).

1. Rapport sur une Collection d'échantillons de Vers à soie malades,
présentée à l'Académie, avec un mémoire explicatif.—Rev. Zool.
1838, p. 35.

BOURGELAT (Cl.).

1. De Vermibus in Sinubus frontalibus, Ventriculo et externâ superficie Equorum.—Mém. prés. à l'Acad. III. p. 409.—*Mod.* B.
Helm.

BOURGERY (J. M.).

1. Traité complet de l'Anatomie de l'Homme, comprenant la Médecine opératoire.   Paris, fol.—Ann. Sc. n. 1832, XXVI. p. 428.

2. Recherches microscopiques sur la Structure intime de la Rate
dans l'Homme et les Mammifères.—Compt. Rend. Juin 1842.—
Ann. Sc. n. (2ᵉ Sér.) XVII. p. 380.

3. Note sur la Structure des Poumons.—C. R. II. p. 496.

4. Recherches microscopiques sur la Structure intime et les Fonctions de la Rate dans l'Homme et les Animaux (Extrait).—Ann. Sc. n. (2ᵉ Sér.) XIX. p. 218.

5. Collection de Mémoires sur la Structure intime et les Fonctions des Organes et des Tissus. Paris, 1843, 4°. (1ᵉʳ Fascic.)

BOURGERY (J. M.) et JACOB ( ).

1. Anatomie élémentaire en 20 planches. Paris, 1836, 8°.—(Deutsch übers. v. *Wilhelmi*): *Anfangsgründe der Anatomie.* Leipz. 1837, 1838.

BOURGUET (L.).

1. Lettres philosophiques sur la formation des Sels et des Crystaux, sur la génération des Plantes et des Animaux, sur la génération et le méchanisme organique des Plantes et des Animaux, à l'occasion de la pierre Bélemnite et de la pierre Lenticulaire. Amst. 1729, 12°. fig.—(2ᵉ éd. avec un Mémoire sur la Théorie de la Terre) Amst. 1762, 8°. fig.—*Böhm.* Bibl. IV. 1, p. 308.

2. Traité des Pétrifications, ou Mémoires pour servir à l'histoire naturelle des Pétrifications dans les quatre parties du Monde. Paris et La Haye, 1742, 4°. fig.—8°. Par. 1778.—*Klein*, Disp. Echinod. p. VIII.—*Cuv.* R. An. III. p. 340.—*Böhm.* Bibl. IV. 2, p. 238.

BOURJOT ST.-HILAIRE (A.).

1. Considérations sur le Nerf facial et sur son influence dans l'acte de la Respiration chez le Marsouin. (Rapport de *Duméril* à l'Acad. d. Sc.)—Ann. Sc. n. 1834, II. p. 255.

2. Collection de Perroquets, pour faire suite à la publication de *Levaillant*, contenant les espèces laissées par cet auteur, ou récemment découvertes ; destinée à compléter une Monographie figurée de la famille des Psittacidés. Paris, 1835 et suiv. fol. et 4°.—Mag. Zool. and Bot. 1837, I. p. 282.

BOURLET (L'Abbé).

1. Mém. sur les Podurelles. 8°. Douai, 1843.

BOURNE (A.).

1. Freshwater and Land Shells from the neighbourhood of Chillicothe, Ohio, presented to Yale College.—*Sill.* Am. J. XXXIX. p. 195.

2. On the Prairies and Barrens of the West.—*Sill.* Am. J. II. 1, p. 30.

BOURNE (W. Ol.).

1. Notice of a locality of Zeolites, &c. at Bergen, Bergen County,
N. Jersey.—*Sill.* Am. J. XL. p. 69.

BOURNON (J. L. de).

6 1. Catalogue de sa Collection minéralogique.   Lond. 1813, 8°, fig.
4°.

2. Essai sur la Lithologie des environs de St. Etienne en Forez, et
sur l'origine de ses Charbons de pierre.   Par. 1785, 12°.—Lyon,
1787, 8°.—J. d. Mines, III.—Biogr. Un. Suppl. LIX. p. 131.

3. Traité de Minéralogie.   Lond. 1808, 4°. 3 vols.—Tr. Linn. Soc.
Lond. X. p. 408.

4. On Bardiglione or Sulphate of Lime, containing a Sketch of the
true nature of Plaster, as well as of its properties, in order to de-
termine the differences that exist between it and Bardiglione.—
Trans. Geol. Soc. ser. 1, I. p. 355.

5. On the Laumonite.—Trans. Geol. Soc. ser. 1, I. p. 77.

BOUROS (Joh.).

1. Περὶ τριῶν Ἰχθύων.   8°.   Ἀθήναις. 1840.

2. Sulle Acque termali della Grecia.—Mem. letta in Pisa al 1°
Congr. dei Natural. Ital.   Milano, 1839.

BOURRIT (M. Th.).

1. Voyage pittoresque aux Glaciers de Savoie.   Gen. 1773, 12°.—
(Germ.) *Mahlerische Reise nach den Eisbergen u. Gletschern von
Savoyen.*   Nürnb. 1775, 8°.—*Schilderung seiner Reise nach den
Savoyischen Eisgebirgen; aus d. Franz. mit Anmerk.*   Gotha,
1775, 8°. 2 Th.—(Belg.) Amst. 1779, 8°. fig.—(Angl.) Lond.
1775, 8°.—*Böhm.* Bibl. I. 1, p. 562.—Biogr. Un. Suppl. LIX.
p. 139.

2. Description des Glacières, Glaciers et amas de Glace du Duché
de Savoie.   Gen. 1774, 8°. fig.—Biogr. Un. Suppl. LIX. p. 139.
—(Germ.) *Beschreibung der Savoyischen Eisgebirge.*   Zürich,
1786, 8°.—Cat. Bibl. Turic.

3. Nouvelle description des Glacières et Glaciers de la Savoie.   Gen.
1785, 8°.—Biogr. Un. Suppl. LIX. p. 139.

4. Nouvelle description des Alpes pennines et rhétiennes, etc.   Gen.
1781-85, 8°. 2 vols. fig.; 1789, 8°. 3 vols.—(Germ.) *Beschreibung der Penninischen u. Rhätischen Alpen.*   Zür. 1782, 8°. fig.;
1812, id.—Biogr. Un. Suppl. LIX. p. 139.—*Böhm.* Bibl. I. 1,
p. 548.

5. Description des aspects du Mont-Blanc, du côté du Val d'Aost, des Glaciers qui en déscendent, et de la découverte de la Mortine. Lausanne, 1776, 8°.—Biogr. Un. Suppl. LIX. p. 139.

6. Itinéraire de Genève à Chamouni, Lausanne, etc. Gen. 1791, 12°; 1792, 8°; 1818, 12°.—Biogr. Un. Suppl. LIX. p. 140.

7. Description des Cols ou passages des Alpes. Gen. 1803, 2 vols. 8°. fig.—*Salis*, Alpina, 1817, II.—*Eis*. Insect. p. 158.

8. Schreiben an die *Miss Craven* über zwei Reisen auf dem Gipfel des Mont-Blanc, etc.; nebst der Beschreibung einer Reise welche er über das berühmte Eismeer des Montanverts nach Piemont gemacht hat. A. d. Franz. mit Anm. v. *A. T. v. Gersdorf.* Dresd. 1787, 8°.

BOURSSE-WILS (H.).

1. Obs. anatomicæ comparatæ de *Squatina lævi*. 8°. Lugd. Bat. 1844.

BOURY ( ).

1. Notices sur les Concrétions des Grottes de Baume et de Loisia. 8°. Lons-le-Saunier, 1835.

BOUSQUET ( ).

1. De Vermibus in Sanguine.—*Van der M.* Rec. VII. p. 65.—*Mod.* B. Helm.

BOUSSINGAULT (J. B.).

1. Versuch einer Ersteigung des Chimborazo, unternommen am 16 Dez. 1831.—*Pogg.* Ann. XXXIV. p. 193.—*L.* u. *Br.* N. Jahrb. 1836, p. 74 ; (Nachtrag) p. 402.

2. Coquilles et Echinodermes fossiles de la Colombie, rec. par *A. D'Orbigny.* 4°. Paris, 1842.

3. Notice sur les Tremblemens de terre des Andes.—Bull. Soc. Géol. Fr. VI. p. 52.—*L.* u. *Br.* N. Jahrb. 1836, p. 712 (*Ueber die Erdbeben in den Anden*).

4. Abh. über die Tiefe des Bodens, wo man zwischen den Wendkreisen die Temperatur unveränderlich findet.—*L.* u. *Br.* N. Jahrb. 1835, p. 478.—Ann. Chim. et Phys. LIII. p. 295.

5. Considérations sur l'Alimentation des Animaux. (*Bouss.* Écon. rur.)—Ann. Sc. n. (3ᵉ S.) I. p. 229.

BOUSSINGAULT (J. B.) et MARIANO DE RIVERO ( ).

1. Sur les Eaux chaudes de la Cordillère de Venezuela.—Ann.

Chim. et Phys. Juill. 1823, p. 272.—*Féruss.* Bull. 1823, IV. p. 204.

BOUSSINGAULT (J. B.) et D'ORBIGNY ( ).

1. Lettres sur un Organe vasculaire découvert dans les Cétacés. Paris, 1836.

BOUSSUET (Fr.).

1. De naturâ Aquatilium carmen, in universam *G. Rondeletii,* quam de Piscibus marinis scripsit Historiam ; cum vivis eorum imaginibus. Lugd. 1558, 4°.—Biogr. Un. V. p. 399.

BOUTEILLE (J. A. M.).

1. Beobachtung einer chronischen durch den Abgang zweier grossen im linken Ohre enthaltenen Würmer geheilten Cephalalgie.— Journ. de Méd. XIII.

BOUTEILLE (Hipp.) et LABATIE (De).

1. Ornithologie du Dauphiné. 2 vols. Grenoble, 1844, 8°. fig.

BOUTON (Louis).

1. Dixième Rapport annuel sur les Travaux de la Société d'Histoire naturelle de l'île Maurice.—Rev. Zool. 1840, p. 296.

2. Onzième Rapport annuel sur les Travaux de la Société d'Histoire naturelle de l'île Maurice. Maurice, 1841.

BOUVÉ ( ).

1. Report upon some Fossil Shells from the Tertiary of Europe, and some Remarks upon the prevalence of the different strata of this class in Great Britain, as also of their Paleontological contents.—Proc. Bost. Soc. N. H. 1843, p. 156.

2. Report on some Fossils from the Greensand of New Jersey, presented by *Prof. Rogers.*—Proc. Bost. Soc. N. H. 1844, p. 171.

BOUVIER ( ).

1. Analyse de la Coralline (*Corallina officinalis,* Linn.).—Ann. de Chimie, VIII. p. 308.—*Lamx.* Polyp. Corall. p. 523.

2. Note sur le mode de réunion des Racines antérieures et postérieures des Nerfs spinaux.—Bull. Soc. Phil. 1823, p. 28.—*Féruss.* Bull. IV. p. 233.

BOVI (R.).

1. Diss. sopra la produzione de' Coralli. 8°. Firenze, 1769.

BOWDICH (T. E.).

1. Anecdotes of a Diana Monkey.—Mag. Nat. Hist. ser. 1, II. p. 9.

2. On the Geognosy of Madeira and Porto Santo.—Edinb. Phil. Journ. IX. p. 315.—*Féruss.* Bull. 1823, IV. p. 325.

3. Voyage to Ashantee. Lond. 1819, 4°.—*Cuv.* R. An. III. p. 320. —Biogr. Un. Suppl. I.IX. p. 170.

4. Natur- und Kunstprodukte des Königreichs Ashantea.—Phil. Mag. LIV. p. 26.—Isis, 1834, V. p. 453.

5. Analysis of the Classification of Mammalia. 8°. Paris, 1821.

6. Introduction to Ornithology. 8°. Paris, 1821.

7. Elements of Conchology. 2 vols. 8°. Paris, 1822.

8. Ueber die Aggry Beads Africas.—Phil. Mag. LIII. p. 445.— Isis, 1834, V. p. 453.

9. Excursions in Madeira and Porto Santo. 4°. London, 1825.— (Gall.) *Excursions à Madère et à Porto Santo* ; avec un appendice zoologique par Mme *Bowdich*, et avec des notes de *MM. Cuvier* et *de Humboldt.* Paris, 1826, 8°. fig.--*Féruss.* Bull. 1826, VIII. p. 281 ; 1827, X. p. 126.

10. Freshwater Fishes of Great Britain. Lond. 1828, 4°.--Lond. Liter. Gaz. 15 Mars 1828.--Athen. 18 Mars 1828.--*Féruss.* Bull. 1828, XV. p. 165.

BOWERBANK (Jas. Scott).

1. Observations on the Circulation of the Blood in Insects.—Ent. Mag. I. p. 239.

2. Observations on the Circulation of the Blood and the Distribution of the Tracheæ in the Wing of *Chrysopa Perla.*—Entom. Mag. IV. p. 179.— *Wiegm.* Arch. VI. p. 314.

3. On the Structure of the Scales on the Wings of Lepidopterous Insects.—Entom. Mag. V. p. 300.

4. Microscopical Observations on the Structure of the Bones of *Pterodactylus giganteus* and other Fossil Animals.—Quart. Journ. Geol. Soc. IV. p. 2.

5. A History of the Fossil Fruits and Seeds of the London Clay. Lond. 1840, 8°. fig.—Ann. of Nat. H. V. p. 410.—*L. u. Br.* N. Jahrb. 1844, p. 767.

6. On a new variety of Vascular Tissue found in a Fossil Wood from the London Clay.—Trans. Microsc. Soc. I. p. 16.—Ann. and Mag. N. H. VI. p. 312.

7. On three new species of Sponges (*Halichondria Johnstoniana,*

Dyseideia Kirkii, D. fragilis).—Trans. Microsc. Soc. I. p. 63.—Ann. and Mag. N. H. VIII. p. 393.

8. Description of a new genus of Calcareous Sponge (*Dunstervillia*).—Ann. and Mag. N. H. XV. p. 297, fig.

9. On the Structure of the Cocoon of a Leech.—Ann. and Mag. N. H. XV. p. 301, fig. pl. 18.

10. On the Keratose or Horny Sponges of Commerce.—Trans. Microsc. Soc. I. p. 32.—Ann. and Mag. N. H. VII. p. 72.

11. Observations on a Keratose Sponge from Australia.—Ann. and Mag. N. H. VII. p. 129, fig.

12. On a Deposit containing Land-Shells at Gore Cliff, Isle of Wight.—Proc. Geol. Soc. II. 449.—Phil. Mag. ser. 3, XI. p. 103.

13. On the Organic Tissues in the Bony Structure of the *Corallidæ*.—Phil. Trans. CXXXII. p. 215.

14. On the Lower Freshwater Formation in the Isle of Wight.—Mag Nat. Hist. ser. 2, II. p. 674.

15. On Moss Agates and other Siliceous Bodies.—Proc. Geol. Soc. III. 431.—Phil. Mag. ser. 3, XIX. p. 542.—Ann. and Mag. N. H. VIII. p. 460; X. pp. 9, 84.

16. On the London Clay Formation at Bracklesham Bay, Sussex.—Mag. Nat. Hist. ser. 2, IV. p. 23.

17. On a new species of Pterodactyl found in the Upper Chalk of Kent.—Journ. Geol. Soc. II. 7.

18. On the Siliceous Bodies of the Chalk, Greensands and Oolites.—Trans. Geol. Soc. ser. 2, VI. p. 181.—Proc. Geol. Soc. III. 278.—Phil. Mag. ser. 3, XVIII. p. 220.—Ann. and Mag. N. H. VII. p. 223.

19. On the London and Plastic Clay Formations of the Isle of Wight.—Trans. Geol. Soc. ser. 2, VI. p. 169.—Proc. Geol. Soc. III. 125.—Phil. Mag. ser. 3, XV. p. 405.

20. On the Structure of the Shells of Molluscous and Conchiferous Animals.—Trans. Microsc. Soc. I. p. 123.

BOWLES (Don G.).

1. Introduction a la Historia natural y a la Geografia fisica de Espanna. Madr. 1775, 4º; 1782, id.; 1789, id.; (Trad. en Franç. par le Vic. de *Flavigny*): *Introduction à l'H. nat. et à la Géogr. phys. de l'Espagne.* Paris, 1776, 8º.—Biogr. Un. V. p. 415.—*Böhm.* Bibl. I. 1, p. 500.—(Germ. *Dieze*): *Einleitung in die Naturgesch. u. physikal. Erdbeschreib. v. Spanien.* Leipz. 1778, 8º.—(Ital. *F. Milizia*): *Introduzione alla Storia nat. ed alla Geografia fisica di Spagna.* Parma, 1785, 2 vols. 8º.

2. Historia da Langosta de Espanna. Madr. 1781.—Bibl. Ent. I. p. 46.

BOWMAN (J. E.).

1. On the Gossamer Spider.—Mag. Nat. Hist.—*Sill.* Amer. Journ. XVI. 2, p. 399.

2. On the natural Terraces on the Eildon Hills being formed by the action of ancient Glaciers.—Ann. and Mag. N. H. VI. p. 346.

3. On some Objections to the Theory of attributing the natural Terraces on the Eildon Hills to the action of Water.—Ann. and Mag. N. H. VI. p. 207.

4. On the Characters of the Fossil Trees lately discovered near Manchester, etc.; and on the Formation of Coal by gradual subsidence.—Geol. Soc. 1841, Febr.—Ann. and Mag. N. H. VI. p. 388.

5. Fossil Infusoria in England.—*Sill.* Am. J. XL. p. 174.

6. On the question, Whether there are any evidences of the former existence of Glaciers in North Wales?—Phil. Mag. ser. 3, XIX. p. 469.

7. On the Characters of the Fossil Trees lately discovered near Manchester, on the line of the Manchester and Bolton Railway; and on the Formation of Coal by gradual subsidence.—Proc. Geol. Soc. III. 270.—Phil. Mag. ser. 3, XVIII. p. 212.

8. On the Fossil Trees found on the line of the Bolton Railway, at Dixon Fold, near Manchester; and the light they throw on several points still undecided among Geologists.—Edinb. New Phil. Journ. XXXI. p. 154.

9. On the Luminosity of the Sea—Mag. Nat. Hist. ser. I, V. p. 1.

10. On the Upper Silurian Rocks of Denbighshire.—Rep. Brit. Assoc. 1841, Sect. p. 59.

11. On the Upper Silurian Formation in the Vale of Llangollen, North Wales.—Rep. Brit. Assoc. 1840, Sect. p. 100.

12. On a small patch of Silurian Rocks to the West of Abergele, on the Northern Coast of Denbighshire, visited 18th and 19th July 1837.—Trans. Geol. Soc. ser. 2, VI. p. 195.—Proc. Geol. Soc. II. 666.—Phil. Mag. ser. 3, XIII. p. 225.

13. On the Bone-Cave of Cefn in Denbighshire.—Rep. Brit. Assoc. 1836, Sect. p. 88.

BOWMAN (W.).

1. On the Minute Structure and Movements of Voluntary Muscle. —Phil. Trans. CXXX. p. 457; CXXXI. p. 69.

2. Additional Notes on the Contraction of Voluntary Muscle in the Living Body. Lond. 1841, 4°.

3. On the Structure and Use of the Malpighian Bodies of the Kidney.—Phil. Trans. 1842, I. p. 57.—Ann. Sc. n. (sér. 2) XIX. p. 108.

BOYER DE FONSCOLOMBE (E. L. J. N.).

1. Description de la *Psyche Febretta*, nouvelle espèce de Bombycites.—Ann. Soc. Ent. Fr. IV. p. 107, fig.—*Wiegm.* Arch. 1836, II. p. 318.

2. Notice sur deux Teignes qui attaquent l'Olivier.—Ann. Soc. Ent. Fr. VI. p. 179.

3. Monographia Chalciditum Gallo-Provinciæ, circa Aquas-Sextias degentium.—Ann. Sc. n. XXVI. p. 273 ; sér. 2, XIII. p. 186.—Bibl. Ent. I. p. 115.—Isis, 1835, VI. p. 512.

4. Description de la *Megachile sericea.*—*Guér.* Mag. de Zool. 1832, No. 50.—Bibl. Ent. I. p. 116.

5. Description des Insectes de la famille des Diplolépaires qui se trouvent aux environs d'Aix.—Ann. Sc. n. 1832, XXVI. p. 184.—Isis, 1835, VI. p. 505 (*Beschreibung der Diplolepariæ um Aix*).

6. Notice sur les genres d'Hyménoptères *Lithurgus* et *Phyloxera.*—Ann. Soc. Ent. Fr. III. p. 219, fig.—*Wiegm.* Arch. 1835, II. pp. 50, 72.

7. Description du *Ceramius Fonscolombii*, Latr.—Ann. Soc. Ent. Fr. IV. p. 421, fig.—*Burm.* in *Wiegm.* Arch. 1836, II. p. 314.

8. Monographie des Libellulines des environs d'Aix.—Ann. Soc. Ent. Fr. VI. p. 129 ; VII. p. 75, fig.

9. Description des Kermes qu'on trouve aux environs d'Aix.—Ann. Soc. Ent. Fr. III. p. 201, fig.—*Wiegm.* Arch. 1835, II. p. 72 (*Coccina*).

10. Des Insectes nuisibles à l'Agriculture, principalement dans les Dép⁸. du midi de la France. Aix, 1840, 8°. fig.—Mém. Ac. d'Aix, IV.

BOYER (Al.).

1. Traité complet d'Anatomie. Paris, 1815, 4 vols. 8°.—Biogr. Un. Suppl. LIX. p. 176.

BOYER ( ).

1. Observations sur un Ver qui se trouve dans l'intérieur des pepins de la Pomme d'Apis.—Bull. Soc. Phil. III. p. 241.—Bibl. Ent. I. p. 47.

BOYLE (Rob.).

1. Heads for the Natural History of a Country, etc. (*Capita gene-ralia Historiæ naturalis Regionum tam magnarum quam parva-rum, in usum Peregrinatorum et Navigantium quondam con-signata.*) Lond. 1692, 12°. fig.— Act. Erudit. XII. p. 256.—Phil. Trans. I. p. 186.—*Böhm.* Bibl. I. 1, p. 436.

2. Memoirs for the Natural History of Human Blood. Lond. 1683, 8°.—*Apparatus ad Historiam naturalem Sanguinis humani ac spiritus præcipuè ejusdem liquoris. Ex Anglico sermone in Latinum translatus.* Lond. 1684, 8°.—Act. Erudit. III. p. 434.

3. Tentamen porologicum. Lond. 1684, 8°.—Act. Erudit. IV. p. 422.

4. Of some Phænomena afforded by Shell-Fishes.—Phil. Trans. 1670, V. p. 2023.—Tr. Linn. Soc. Lond. VII. p. 237.

5. Memoirs for the Natural History of Human Blood. 8°. Lond. 1683, 1684.—Cat. R. Soc. L.

6. Usefulness of Experimental Philosophy. Oxf. 1663-64, 2 vols.; 1671, 4°.—(Lat.) *Exercitatio de utilitate Philosophiæ naturalis.* Lind. 1692, 4°.—*Böhm.* Bibl. I. 1, p. 245.

7. Tentamina quædam physiologica. Amst. 1667, 8°.—Cat. Bibl. Turic.

8. Observat. de Generatione Metallorum. Lond. 1676, 12°; 1678, 8°.

9. Tractatus de Temperie subterranearum Regionum. Geneva, 1680, 4°.

10. An Essay about the Origine and Virtues of Gems. Lond. 1672, 12°.—(Lat.) Hamb. 1673, 12°.

Boys (H.).

1. Account of the *Flustra arenosa*, and some other marine produc-tions.—Tr. Linn. Soc. Lond. V. p. 230, fig.

Boys (W.).

1. History of Sandwich. Canterb. 1792, 4°.—Bibl. Ent. I. p. 47.

2. A List of the Beasts and some of the Insects, etc.—Hist. of Sandw. p. 847.—Bibl. Ent. I. p. 47.

Boys (W.) et WALKER (G.).

1. Testacea minuta rariora nuperrimè detecta in arenâ Littoris Sandvicensis. Lond. 1784, fig. 4°.—Tr. Linn. Soc. Lond. VII. p. 234.

Boys (W. J. E.).

1. On the genus *Paussus.*—Journ. As. Soc. Beng. XII. p. 421.

Bozza (Vinc.).

1. Catalogo sistematico dei piu rari Ittioliti del Monte Bolca che si conservano nel gabinetto privato.—Trans. R. I. Acad. V. p. 312.
2. Delle Acque mazziali di Rovere di Velo.   Verona, 1767, 8°.

Bozzelli (J. B.).

1. De Hepate, etc.   Taur. 1776, 8°.

Braam-Houckgeest (A. E. van).

1. Bericht van een Chamelion aan de Kaap de Goede Hoope.—Verhand. Maatsch. Haarl. IX. 3, p. 637.—Dict. Sc. n. XV. p. 251.
2. Bericht wegens den *Lophius histrio.*—Verhand. Maatsch. Haarl. XV. p. 20.—Dict. Sc. n. XXII. p. 528.

Braccini (J. C.).

1. Relazione del Incendio fattosi nel Vesuvio 16 Dec. 1631.   Nap. 1631, 8°; 1632, 4°.

Braccio (Ign.).

1. Remoræ pisciculi effigies. Romæ, 1634, fol.—Dict. Sc. n. XXII. p. 535.

Brace (J. P.).

1. Description of the *Phalæna devastator.*—*Sill.* Amer. Journ. I. 2, p. 154.
2. Observations on the Minerals connected with the Gneiss range of Litchfield County.—*Sill.* Amer. Journ. I. 4, p. 350.

Brachet (J. L.).

1. Réflexions sur quelques points de Physiologie relatifs au Système nerveux-ganglionaire au sujet de quelques opinions professées par MM. les Prof. *Medici* et *Berruti.*   Lyon, 1842.
2. Recherches expérimentales sur les Fonctions du Système nerveux-ganglionaire, et sur leur application à la Pathologie. 8°. Paris, 1830, 1837.—Brux. 1834, 12°.—*Praktische Untersuchungen über die Verrichtungen des Gangliennervensystemes, und über ihre Anwendung auf die Pathologie.*   Aus dem Franz. v. *H. E. Flies.* Quedlinb. 1836, 8°.—Bibl. med.-ch. p. 60.

3. Mémoire sur les Fonctions du Système nerveux-ganglionaire. Paris, 1823, 8°.—*Féruss.* Bull. II. p. 108.

4. Recherches expérimentales sur les Fonctions du Système nerveux-ganglionaire, etc. Paris, 1837, 8°.

BRACHIUS (Jac.).

1. De Ovis Ostreorum.—·Misc. Nat. Cur. Dec. II. 1689, p. 506.— *Mod.* B. Helm.—Tr. Linn. Soc. Lond. VII. p. 237.

2. Pensieri fisico-medici circà gli Animali che muciono, etc. Cogitata physico-medica circa Animalia quæ moriuntur in Recipientibus, et Aëre vacuis, et Aëre factitio à diversis mixtis fermentationis ope elevato repletis. Venet. 1685, 8°.—Act. Erudit. V. p. 502.—*Böhm.* Bibl. II. 1, p. 111.

BRACHMANUS (J.).

1. Oratio an cruentæ Pluviorum Guttæ naturales. Lignit. 1619, 4°. —*Böhm.* Bibl. V. p. 45.

BRACKENHAUSEN ( ).

1. Beobachtungen über die Processions-Raupen.—Hall. Naturf. I. p. 203.—Bibl. Ent. I. p. 47.

BRACY-CLARK (F. L. P.).

1. Observations on the genus *Œstrus.*—Tr. Linn. Soc. Lond. 1797, III. p. 289, fig.—Ann. of Phil. 1628, Avr. p. 283.—*Féruss.* Bull. 1829, XVII. p. 149 (*De l'Insecte appelé Œstrus par les Anciens, et sur la véritable espèce à laquelle ils donnaient ce nom*).

2. An Essay on the Bots of Horses and other animals. Lond. 1815, 4°. fig.

BRADBURY (John).

1. Travels in the Interior of North America in the years 1809–1811. Lond. 1817, 8°.—Isis, 1818, V. p. 826.

BRADLEY (Rich.).

1. A Philosophical Account of the Works of Nature, etc. Lond. 1721, 4°. and fol.; 1739, 8°. fig.—(Belg.) Amst. 1744, 8°.—Tr. Linn. Soc. Lond. VII. p. 217.—Biogr. Un. V. p. 462.—*Böhm.* Bibl. I. 1, p. 261.

BRADLEY (T.).

1. On the Habits of the Electric Eel.—Mag. Nat. Hist. ser. 2, II. p. 668.

BRADY (T.).

1. An Account of some remarkable Insects of the Polype kind, found in the waters near Brussels in Flanders.—Phil. Trans. XLIX. p. 248, fig.—(Belg.) Holl. Mag. III. p. 89.—*Mod.* B. Helm.

BRAGUIER (B.).

1. Histoire naturelle des Élémens de la Faune française, Mammifères. Paris, 1839–1840, 12º.—Niort, 1844, 8º.—Rev. Zool. 1840, p. 336 ; 1841, p. 280.

BRAHM (Nic. J.).

1. Handbuch der ökonomischen Insectengeschichte, in Form eines Kalenders bearbeitet. Mainz, 1791, 2 vols. 8º.

2. Entwurf einer Fauna entomologica der Wetterau.—Ann. Wetter. Gesellsch. 1809, I. pp. 59, 229 ; 1811, II. p. 189.

3. Versuch einer Fauna entomologica der Gegend um Mainz. Giesen, 1793, 8º.—*Eis.* Insect. p. 157.

4. Insectenkalender für Sammler und Oekonomen. 2 Th. 8º. Mainz, 1790, 1791.—*Eis.* Insect. p. 127.

5. Entomologische Nebenstunden.—Journ. f. die Entom. I. pp. 1, 193.—Bibl. Ent. I. p. 47.

6. Vergeblicher Insectenregen.—Naturf. XXII. 1802.—*Eis.* Insect. p. 171.

7. Verzeichniss in Form eines Kalenders der im Jahr 1786 um Mainz gesammelten Schmetterlinge u. Raupen.—*Fuessl.* N. Mag. Entom. III. p. 141.—Bibl. Ent. I. p. 47.

8. Arten der *Coccinella* und *Cassida* bei Mainz.—Naturf. XXIX. 1802.—*Eis.* Insect. p. 177.

BRAID (J.).

1. On the Formation of the various Lead-Spars.—Mem. Wern. Soc. IV. p. 508.

BRAMBILLA (J. A. v.).

1. Geschichte der von den berühmtesten Männern Italiens gemachten Entdeckungen in der Physik, Medicin, Anatomie u. Chirurgie. 4º. Wien, 1798.

BRAMBILLA ( ).

1. Enumeratio Caraborum Ticinensium. 1830.

BRAMSON (M.).

1. De Diverticulo quodam a casibus descriptis abhorrente. Bero..
1841, 8°.—*Val.* Repert. VII. p. 46.

BRANCHE (Ph.).

1. Anatomie du Cœur, suivie de Recherches sur la mesure des parties qui le constituent. Strasb. 1838, 4°.

BRAND ( ).

1. On the Latin Terms used in Natural History.—Trans. Linn. Soc
III. 1797, p. 70.

BRAND (Ad.).

1. Beschreibung der Chinesischen Reise, welche vermittelst einer
Zaarischen Gesandschaft verrichtet worden. Hamb. 1698, 12°.
—*Böhm.* Bibl. I. 1, p. 692.

BRAND ( ).

1. Note sur la prétendue Mine d'étain de Ségur (Corrèze).—Ann.
Sc. n. 1826, VIII. p. 111.

BRAND (J.).

1. Anfangsgründe der Naturwissenschaft. Frankf. 1820, 8°. fig.
(3e Aufl.).

BRAND (J. Arn. v.).

1. Reisen durch Brandenburg, Preussen, Plescovien, Grossnaugar-
dien, Tiwerien u. Muscovien, nebst *Alb. Dobbin's* Beschreibung
von Sibirien ; mit *H. C. v. Hennin's* Anmerkungen. Wesel, 1702,
8°. fig.—(Belg.) Utrecht, 1703, 8°. fig.—*Böhm.* Bibl. I. 1, p. 453.

BRAND (Paul).

1. Disputatio de Ovo humano. Hafn. 1677, 4°.—*Böhm.* Bibl. II.
1, p. 137.

BRANDE (W. T.).

1. Outlines of Geology. 8°. Lond. 1817 ; 12°. Lond. 1829 ; 8°. 1837.
—Journ. Roy. Inst. XIX. pp. 63, 184 ; XX. pp. 24, 235 ; XXI.
p. 15 ; XXII. pp. 51, 249.—*Féruss.* Bull. 1826, VII. p. 273 ;
VIII. p. 313 ; IX. p. 1, 1827 ; XII. p. 305.

2. A Descriptive Catalogue of the British specimens deposited in
the Geological Collection of the Royal Institution. Lond. 1816,
4°.—Cat. Roy. Soc. Lond.

BRÄNDEL (Matth.).

1. Beschreibung des heiligen Thiergartens, sonderlich derer Hirsche. Augsb. 1692, 12°.—*Böhm.* Bibl. II. 1, p. 74.

BRANDENSTEIN (W. F. von).

1. Getreu aufgenommene Gebirgsgegenden und Höhlen um u. bei Muggendorf, etc. Nürnb. 1811–1816, 4 Hefte.
2. Fauna œconomica, oder Beschr. u. Abb. der wichtigsten Insekten u. Würmer, etc. Nürnb. 1811, 4°. fig.

BRANDER (Fr. Reg.).

1. De regiâ Piscaturâ Cumoënsi. Aboæ, 1751, 4°. fig.—*Mod.* B. Helm.

BRANDER (Gust.).

1. Fossilia Hantoniensia collecta et in Museo Britannico deposita. Lond. 1766, 4°. fig.—4°. Lond. 1829 (by *Wood*).—Phil. Trans. XLIX. p. 295, fig.—*Klein,* Disp. Echinod. p. IX.—Biogr. Un. Suppl. LIX. p. 188.—*Mod.* B. Helm.
2. An Account of a remarkable *Echinus.*—Phil. Trans. XLIX. p. 295, fig.—*Mod.* B. Helm.
3. Dissert. on the Belemnites.—Phil. Trans. XLVIII. p. 803, fig.— *Mod.* B. Helm.

BRANDES (R.).

1. Die Mineralquellen und Schwefelschlammbäder zu Meinberg, nebst Beiträgen zur Kenntniss der Vegetation und der klimatischen und mineralogisch-geognostischen Beschaffenheit des Fürstenth. Lippe-Detmold. Lemgo, 1832, 4°.—Isis, 1833, III. p. 184.

BRANDES (R.) u. KRÜGER (F.).

1. Neue physikalische chemische Beschreibung der Mineralquellen zu Pyrmont, etc. (*Nouvelle Description physico-chimique des Eaux de Pyrmont, avec une description naturelle des environs.*) Pyrmont, 1826, 8°. carte.—*Féruss.* Bull. 1827, XII. p. 31 ; 1828, XV. p. 74.
2. Mineralogisch-geognostische Bemerkungen über die Umgebungen von Pyrmont. (Neue phys. chem. Beschreib. etc.)—*Leonh.* Zeitschr. 1827, I. p. 560.

BRANDES (W.).

1. Bemerkungen über die spiegelnden Flächen des Sandsteins.—

Ann. d. Pharm. 1832, I. p. 90.—*L.* u. *Br.* N. Jahrb. 1833,
p. 447.

BRANDIS (J. Dietr.).

1. Ueber Leben und Polarität. 8°.—*Val.* Rep. II. p. 18.
2. Einige Beiträge zum Studio der Alten in der Insectengeschichte.
   —*Licht.* Götting. Mag. IV. p. 129.—*Eis.* Insect. p. 124.
3. Ueber die Lebenskraft. Hannov. 1795, 8°.

BRANDSTEN (P. W.).

1. Bombi Scandinaviæ, monographice tractati et Iconibus illustrati.
   Lond. Goth. 1832.—Ann. Soc. Ent. Fr. I. p. 244.

BRANDT (J. F.).

1. Versuch einer kurzen Uebersicht der Fortschritte, welche die
   Kenntniss der thierischen Körper den Schriften der K. Ac. d.
   Wissensch. zu Petersburg verdankt.—Act. Ac. Pet. 1831.—Isis,
   1834, p. 317.
2. Prodromus Descriptionis Animalium ab *H. Mertensio* in orbis
   terrarum circumnavigatione observatorum. Petrop. 1835, 4°.—
   Wiegm. Arch. 1836, II. p. 188.
3. Descriptiones et Icones Animalium rossicorum novorum vel minùs rectè cognitorum. Petrop. 1835, 4°. fig.
4. Bemerkungen über die Mundmagen, oder Eingeweidenerven der
   Evertebraten. Leipz. 1836, 4°. fig.—*Müll.* Arch. 1836, V. p. c ;
   1837, III. p. 324.—Mém. Ac. Pét. Sc. nat. II. p. 597, fig.—Ann.
   Sc. n. 1836, V. pp. 81, 138 (*Remarques sur les Nerfs stomato-
   gastriques ou intestinaux, dans les Animaux invertébrés*).
5. Beobachtungen über die Systeme der Eingeweidenerven der Insekten.—Isis, 1831, X. p. 1103.
6. Observationes anatomicæ de Instrumento Vocis Animalium, in
   Museo zootomico Berolinensi factæ. Berol. 1826, 4°. fig.—Isis,
   1827, XI. p. 965.
7. Beiträge zur Kenntniss des inneren Baues von *Glomeris margi-
   nata.*—Mém. Ac. Pét. VI. p. 11.—*Müll.* Arch. 1837, III. p. 320,
   fig.
8. Bemerkungen über den inneren Bau des Wuychuchol (*Myogale
   moschata*), im Vergleich mit dem des Maulwurfs und der Spitzmaus.
   —*Wiegm.* Arch. 1836, I. p. 176.—*Müll.* Arch. 1837, p. LVIII.
9. Anatomie et Physiologie de la Sangsue.—Mém. Ac. Pét. II. 1834.
   —*Wiegm.* Arch. 1835, I. p. 342.
10. Ueber den eigenthümlichen Bau der Spitze der Stacheln bei
    den amerikanischen Stachelschweinen.—*Müll.* Arch. 1835, p. 551.

11. Bemerkungen über die Differenzen im Schädel und Zahnbau zwischen den Stachelschweinen der alten und neuen Welt.—Ac. Pét. Déc. 1834.—*Müll.* Arch. 1835, p. 548, fig.

12. Observations sur les espèces qui composent le genre *Scolopendra.* 1840, 8º.

13. Rapport relatif aux Recherches ultérieures sur l'Histoire, l'Anatomie et la Physiologie des Gloméridés.—Bull. Acad. Pétersb. VI. 1839.—Rev. Zool. 1840, p. 303.

14. Remarques sur les espèces du genre *Glomeris* et sur leur Distribution géograph.—Bull. Ac. Pétersb. 1840.—Rev. Zool. 1840, p. 304.

15. Sur le prétendu nouveau Cartilage du Larynx de *M. E. Rousseau.*

16. Conspectus Monographiæ Crustaceorum Oniscodorum *Latreillii.*

17. Tentaminum quorundam monographicorum Insecta Myriapoda Chilognatha *Latreillii* spectantium. Tab. II. 8º. Mosquæ, 1833.

18. Versuch einer Anatomie des medizinischen Blutegels.—Mém. Acad. Pét. 1833.

19. Ausführliche Beschreibung der von *C. H. Mertens* beobachteten Schirmquallen, etc. Pétersb. 1838, 4º. fig. col.—Mém. Acad. Pét. sér. 6, II.—Rev. Zool. 1840, p. 249.

20. Note sur une nouvelle espèce du genre *Catarhactes* de Brisson. —Bull. Ac. Pétersb. 1837.—Rev. Zool. 1838, p. 114.

21. Rapport sur une Monographie de la famille des Alcadées.— Bull. Pétersb. 30 Juin 1837.—Rev. Zool. 1840, p. 302.

22. Mammalium exoticorum novorum, vel minus ritè cognitorum Musei academici zoologici Descriptiones et Icones. Petrop. 1835, 4º. fig. col.—*Müll.* Arch. 1836, p. XLVIII.

23. *Cricetus nigricans.*—Bull. Ac. Pétersb.—*Wiegm.* Arch. 1837, II. p. 167.

24. *Erinaceus Hypomelas.*—Bull. Acad. Pétersb.—*Wiegm.* Arch. 1837, II. p. 154.

25. Naturhistorische Bemerkungen über die Wurzelcochenille, im Vergleich mit der mexikanischen.—Isis, 1835, I. p. 66.

26. Ueber eine neue Ordnung der Myriapoden.—Bull. Ac. Pétersb. I. No. 23, p. 182.— *Wiegm.* Arch. 1837, I. p. 238.—Ann. Sc. n. 1837, VIII. p. 376 (*Note sur un ordre nouveau de la classe des Myriapodes*).

27. Einige Bemerkungen über die Stellung der Cirripeden im System.—*Müll.* Arch. 1835, p. 304.

28. Mittheilung einiger Modifikationen seines frühern Systems der

Scheibenquallen.—Bull. Sc. Acad. Pétersb. I. p. 185.— *Wiegm.* Arch. 1837, II. p. 277.—Rev. Zool. 1839, p. 272.

29. Beiträge zur Kenntniss des Baues der innern Weichtheile des Lama (*Auchenia Lama*). Petersb. 1841, 4°. fig.—Mém. Acad. Pét. (6e Sér.) IV. fig.—Rev. Zool. 1840, p. 301.

30. Spicilegia ornithologica exotica. Fasc. I. Petrop. 1839, 4°. fig.—Mém. Acad. Pét. (6e Sér.) V. 2, fig.—Rev. Zool. 1840, p. 306.

31. Recueil de Mémoires relatifs à l'ordre des Insectes Myriapodes. —Bull. Ac. Pét. V.-IX.

32. Recherches sur l'Anatomie des Araignées.—Ann. Sc. n. (2e S.) XIII. p. 180, fig.

33. Sur la Classification des Gerboises.—Bull. Phys. Ac. Pét. II. No. 14, 15.

34. Beiträge zur Kenntniss der Naturgeschichte der Vögel, mit besonderer Beziehung auf Skelettbau u. vergleichende Zoologie. I. Lief. Pétersb. 1839, 4°. fig.—Rev. Zool. 1840, p. 278.

35. On certain species of Siberian Birds described by *Latham,* but which have not hitherto been sufficiently determined.—Ann. and Mag. N. Hist. XI. p. 113.

36. Bruchstücke zu einer Fauna der Berberei.— *Wagner,* Reise in Algier. III.

37. De Solenodonte, novo Mammalium insectivororum genere.— Mém. Ac. Pétersb. II. 1833.

38. Ueber den Zahnbau der *Rytina Stelleri,* etc. (*Sur la Structure des Dents du* Rytina Stelleri, *avec des Observations sur la division en deux familles des Cétacés herbivores*). 4°. fig.

39. Bemerkungen über die Klassifikation der Springer, hauptsächlich hinsichtlich der in Russland vorkommenden Arten, etc.— Bull. Phys. Ac. Pétersb. 1844, II. p. 209.— *Wiegm.* Arch. 1845, II. p. 31.

40. Bemerkungen über die verschiedenen Arten Fasane der russischen Fauna.—Bull. Phys. Ac. Pétersb. III. p. 49.

41. Observations sur plusieurs espèces nouvelles du genre *Carbo* ou *Phalacrocorax,* qui se trouvent dans le Muséum de l'Académie des Sciences de St. Pétersbourg.—Bull. Pétersb. 6 Oct. 1837.— Rev. Zool. 1840, p. 302.

42. Ueber den Bau der sogenannten Moschusdrüsen des *Sorex moschatus.*—Pall. Nov. Act. Ac. Leop. Car. XVIII.

43. Mémoire sur le Squelette du genre *Rhynchops,* comparé à celui des *Larus, Lestris* et *Sterna,* etc.—Bull. Pétersb. 28 Sept. 1838. —Rev. Zool. 1840, p. 302.

44. Rapport sur les *Oniscides* et les *Myriapodes* de la régence d'Alger.—Bull. Pétersb. 21 Févr. 1840.—Rev. Zool. 1840, p. 304.

45. Observations sur les espèces qui composent le genre *Scolopendra.*—Bull. Pétersb. 1840, 13 Mars.—Rev. Zool. 1840, p. 304.

46. Notice sur une nouvelle espèce du genre Cormoran (*Carbo nudigula*, Brandt).—Bull. Pétersb. 29 Nov. 1839.—Rev. Zool. 1840, p. 302.

47. Note sur quatre nouvelles espèces de Serpens de la cote occidentale de la mer Caspienne et de la Perse septentrionale, découvertes par *M. Kareline.*—Bull. Pétersb. 22 Sept. 1837.—Rev. Zool. 1840, p. 302.

48. Rapports sur les acquisitions des Musées zoologiques et zootomiques et les travaux qui y ont été exécutés en 1837 et 1838.—Bull. Pétersb.—Rev. Zool. 1840, p. 301.

49. Sur l'Organisation de ce que l'on appelle les Glandes musquées (Glandes anales) des Desmans.—Rev. Zool. 1840, p. 361.

50. Sur trois espèces nouvelles d'Oiseaux chanteurs.—Bull. Phys. Ac. Pétersb. II. No. 9, 10.

51. De *Cetotherio*, novo Balænarum familiæ genere in Rossia meridionali effosso.—Bull. Phys. Ac. Pétersb. I. No. 10–12.

52. Sur les Perdrix-géants du Caucase et de l'Altai, comme types d'un sousgenre des Perdrix —Bull. Phys. Ac. Pétersb. I. No. 17, 18; III. No. 12.

53. Sur trois nouvelles espèces d'Oiseaux chanteurs de Sibérie; sur une nouv. esp. de Souslik; sur une nouv. esp. de Perdrix; sur une nouv. esp. d'*Accentor*; sur le *Passer pusillus*, Pall.—Bull. Phys. Ac. Pétersb. I. No. 23.

54. Sur les différentes espèces des Sousliks de Russie.—Bull. Phys. Ac. Pétersb. II. No. 23, 24.

55. Ueber die bis jetzt bekannten Wirbelthiere West-Sibiriens.—Bull. Phys. Ac. Pétersb. III. No. 2.

56. Sur deux Oiseaux nouveaux pour la Faune de Russie.—Bull. Phys. Ac. Pétersb. III. No. 1.

57. Sur le *Cervus pygargus* de Pallas.—Bull. Phys. Ac. Pétersb. III No. 18.

58. Observationes ad structuram cranii *Rytinæ Stelleri* spectantes.—Bull. Phys. Ac. Pétersb. IV. No. 8, 9; V. No. 6.

59. Sur la récolte zoologique du voyage de *M. Kolenati.*—Bull. Phys. Ac. Pétersb. IV. No. 10, 11.

60. Ueber die Weichtheile u. aüssern Organe des *Rhinoceros tichorhinus.*—Bull. Phys. Ac. Pétersb. V. No. 6.

61. Ueber das Vorkommen eines zweifachen Haarkleides beim Songarischen Hamster.—Bull. Phys. Ac. Pétersb. V. No. 7, 8.

62. Ueber den gleichzeitig mit der Ausrottung der Pflegemutter bewerkstelligten, geschichtlich nachweisbaren Untergang einer kleinen parasitischen Krebsart u. eines Eingeweidewurmes der Jetztwelt.—Bull. Phys. Ac. Pétersb. V. No. 12.

63. Acquisitions du Musée zoologique dues au voyage du préparateur *Voznessensky.*—Bull. Phys. Ac. Pétersb. V. No. 23, 24.

64. Fragments du squelette de la Rytine de *Steller.*—Bull. Phys. Ac. Pétersb. VI. No. 2, 3.

65. Note relative à la Classification des espèces qui composent le genre *Polydesmus,* et suivie d'une caractéristique de dix espèces nouvelles, ainsi que de quelques remarques sur la distribution géographique des espèces en général.—Bull. Pétersb. 1839, V. No. 20.
—Rev. Zool. 1839, p. 270 ; 1840, p. 303.

BRANDT (J. F.) u. ERICHSON ( ).

1. Monographia generis Meloë.—N. Act. Nat. Cur. XVI. p. 101, fig.

BRANDT (J. F.) et HAMEL ( ).

1. Cochenille d'Arménie.—Mém. Acad. Pétersb. Sc. phys. III. 2, p. 9.—*Wiegm.* Arch. 1835, II. p. 73.

BRANDT (J. F.) u. RATZEBURG ( ).

1. Medizinische Zoologie, oder getreue Darstellung und Beschreibung der Thiere, die in der Arzneimittellehre in Betracht kommen. Berl. 1827-1833. 2 vols. 4°. fig.—N. Act. Nat. Cur. XVI. p. xvi.—Isis, 1827, XI. p. 964 ; 1828, XI. p. 1168 ; 1834, III. p. 322 ; IV. p. 400.—Ann. Sc. n. 1831, XXII.—Rev. Bibl. p. 26 (*Description et représentation des Animaux employés dans la Thérapeutique*).

2. Hauptresultate aus der Untersuchung über Naturgeschichte der Störarten.—Isis, 1831, X. p. 2002 (1102).

3. Beiträge zur Anatomie der Spinnen.—Isis, 1831, X. p. 1105.

4. Einige Bemerkungen zum Maywurm.—Isis, 1831, X. p. 1104.

BRANDTS (H. A. A.).

1. De Medullæ spinalis in Secretiones et Nutritionem actione. Berol. 1841, 8°.

BRANTS (A.).

1. Dissert. zoologica inauguralis de Tardigradis. Leyd. 1828, 4°. fig.—Isis, 1834, II. p. 144.

2. Bijdrage tot de kennis van de eenvondige oogen der gelede dieren (*Articulata*, Cuv.).— *Van der Hoev*. Tijdsch. IV. p. 135. (*Observations sur les Yeux simples des Animaux articulés*.)—Ann. Sc. n. 1838, IX. p. 308.

3. Ontleedkundige Beschouwing van de Schorpioenvlieg (*Panorpa communis*).— *Van der Hoev*. Tijdsch. VI. p. 173.

4. Bijdrage tot de kennis der monddeelen van eenige vliesvleugelige gekorvenen (*Insecta hymenoptera*).— *Van der Hoev*. Tijdsch. VIII. p. 71.

5. Over het gezigtswerktuig der gelede dieren.— *Van der Hoev*. Tijdsch. X. p. 12.

6. Over de luchtbuizen in het zamengestelde oog der gelede dieren. — *Van der Hoev*. Tijdsch. XII. p. 233.

7. Waarnemingen omtrent een schadelijk insekt op *Pinus Larix*. — *Van der Hoev*. Tijdsch. VI. p. 321.

8. Het geslacht der Muizen door *Linnæus* opgesteld, in familien, geslachten en soorten verdeeld. 8°. Berlin, 1827.

BRANTZ ( ).

1. Mémoire sur l'*Euriotis* (Holl.).— *Cuv*. R. An. III. p. 341.

BRARD (C. P.).

1. Minéralogie appliquée aux Arts, ou Histoire des Minéraux qui sont employés dans l'Agriculture et l'Économie domestique, etc. 3 vols. 4°. av. 15 pl. Paris.

2. Éléments pratiques d'exploitation des Mines, contenant tout ce qui est relatif à l'art d'explorer la surface des terrains, d'y faire des travaux de recherche et d'y établir des exploitations réglées, etc. 8°. av. atl. de 22 pl.

3. Description historique de sa Collection de Minéralogie appliquée aux Arts. Par. 1833, 8°.

4. Minéralogie populaire, ou Avis aux Cultivateurs et aux Artisans sur les terres, les pierres, les sables, les métaux et les sels qu'ils emploient journellement, etc. Par. 1826, 12°.—Bull. Soc. Géol. Fr. VI. p. 102.

5. Histoire des Coquilles terrest. et fluviat. qui vivent aux environs de Paris. Paris, 1815, 8°.

6. Traité des Pierres précieuses. Par. 1808, 2 vols. 8°.

BRARD (P.).

1. Mémoire sur la Natrolithe et sur le Gisement de cette substance. --Ann. Mus. XIV. p. 367.

2. Mémoire sur les Coquilles fossiles du genre Lymnée qui se trouvent aux environs de Paris, sur les autres Coquilles qui les accompagnent, et sur la nature des Pierres qui renferment ces fossiles.—Ann. Mus. XIV. p. 426 ; XV. p. 406, fig.

3. Nouveaux Élémens de Minéralogie, ou Manuel du Minéralogiste voyageur (3e éd. rev. par *Guillebot*). Par. 1838, 8°.—(1re éd.) Manuel du Minéralogiste et du Géologue voyageur. Par. 1805, 8°.—(2e éd.) Par. 1824, 8°.

BRAUER (A.).

1. Von den Rebenstichern ; *vide* BREUCHEL.

BRAUN (Alex.).

1. Fossile Pflanzen von Oeningen.—*L.* u. *Br.* N. Jahrb. 1838, p. 310.

2. Ueber den Nutzen der Naturwissenschaft. 8°. Carlsruhe, 1833.

3. Ueber die Naturgeschichte als Bildungsmittel. 8°. Carlsruhe, 1839.

BRAUN (C. F.).

1. Versuchte Beantwortung der von der Herzogl. Oldenburgischen Regierung im Jahr 1822 aufgestellten Preisfrage über das gelbe Fieber. Marburg, 1827, 8°.—N. Act. Nat. Cur. XIII. p. xxvi.

2. Zur Geschichte des Vorkommens von fossilem Braunstoffe. Bayr. 1839.

3. Beiträge zur Urgeschichte der Pflanzen. 4°. Bayr. 1843.

BRAUN (J. Ad.).

1. De Vermium intestinalium primâ Origine. Jenæ, 1806, 8°.

2. Oratio de insignioribus Telluris mutationibus. 4°. Petrop. 1756.—*Böhm.* Bibl. IV. 2, p. 245.

3. Beiträge zur Geschichte der Eingeweidewürmer.—Schrift. Berl. Naturf. Fr. VIII. p. 236 ; X. p. 57.

BRAUN (J. Ern.).

1. Amœnitates subterraneæ, i. e. Breviarium sufficiens quod agit de Metalli fodinarum Harcicarum, cùm inferiorum tùm superiorum primâ origine, progressu et præstantiâ. Goslar, 1726, 4°.—*Böhm.* Bibl. IV. 1, p. 131.

BRAUN (J. J. Ph.).

1. Systematische Beschreibung einiger Egelarten. Berl. 1805, 4°.

Braun (Max.).

1. Die Pyrenæen; alte Gletscher und Moränen daselbst.—*L.* u. *Br.*
N. Jahrb. 1843, p. 80.

2. Ueber eine neue Art von *Strophostoma* und ein neues genus,
Scoliostoma, mit ähnlicher Bildung des Gehäuses.—*L.* u. *Br.* N.
Jahrb. 1838, p. 291, fig.—Rev. Zool. 1838, p. 323 (*Sur une nouv.
esp. de Strophostoma, et un nouv. genre de Coquille, Scoliostoma*).

3. Maynzer Tertiär-Schichten an der Hard.—*L.* u. *Br.* N. Jahrb.
1838, p. 316.

Braun (Ph.).

1. Beiträge zu der Lehre von den Fels-Spiegelflächen.—*L.* u. *Br.*
N. Jahrb. 1842, p. 656.

2. Versuch einer allgemeinen Theorie der Fels-Spiegelflächen.—
L. u. *Br.* N. Jahrb. 1842, p. 757.

3. Ueber *Trematosaurus.*—*L.* u. *Br.* N. Jahrb. 1844, p. 569.

4. Spiegel u. Schichtung des Bunt-Sandsteins; Kohlensandstein
Hessens.—*L.* u. *Br.* N. Jahrb. 1842, p. 89.

5. Harmotome im Dolerit, und Relief-Figuren am Sandstein bei
Marburg.—*L.* u. *Br.* N. Jahrb. 1841, p. 666.

Braune (K. H.).

1. Galerie der interessantesten Erscheinungen in der Naturkunde.
8°. Naumberg, 1837.

Braune (Sal.).

1. Teutscher Jordan, oder Biberacher Bad, dabey nicht allein eine
schöne Vergleichung der Kleinen Welt mit der grossen, der Ursprung aller Flüste und Brunnen, mineralischen Wassern u. Bä-
dern, sammt derselben Nutzen, sondern auch eine Kurze Beschreibung der Stadt Biberach u. des dabey gelegenen mineralischen
Gesundbades, der Jordan genannt. Augsb. 1673, 8°. 2 Th.—
Böhm. Bibl. V. p. 312.

Braunhofer (A. G.).

1. Naturwissenschaftliche Vorbegriffe für Naturgeschichte, nebst
dem präparativen Theile der oryktognostischen Mineralogie. Wien,
1816, 8°. fig.

2. Lehrbuch der Naturgeschichte. 8°. Wien, 1830.

Brauss (R.).

1. De Caloris in Organismum actione observationes et experimenta
nonnulla. 8°. Berol. 1841.—*Val.* Repert. VII. p. 29.

BRAVAIS (A.).

1. Sur les Lignes d'ancien Niveau de la Mer dans le Finmark.—
Voy. de la Comm. Scient. du Nord.—Journ. Geol. Soc. I. p. 534.
—C. R. X. p. 691.

BRAVARD (Aug.).

1. Considérations sur la distribution des Mammifères terr. fossiles
dans le Dép. du Puy-de-Dôme. 8°. Clermont, 1845.

2. Monographie de la Montagne de Perrier, près d'Issoire (Puy-de-
Dôme), et de deux espèces fossiles du genre *Felis*, découvertes
dans l'une de ses couches d'alluvion. Paris, 1828, 8°. fig.—
Féruss. Bull. 1829, XIX. p. 166.

BRAVARD (Aug.), CROIZET ( ) et JOBERT (A. C. G.).

1. Recherches sur les Ossemens fossiles du Département du Puy-de-
Dôme. Paris, 1825, 2 vols. 4°. fig.—*Féruss.* Bull. 1827, XI.
pp. 92, 379.

BRAVENDER (J.).

1. Indications of the Fertility or Barrenness of Soils.—Journ. Agr.
Soc. Engl. V. p. 559.

BRAY (De).

1. Mémoire sur la Livonie.—Münch. Acad. 1814.—Isis, 1818, IV.
p. 612.

BRAYLEY (E. W.).

1. On the Odour exhaled from certain Organic Remains in the Di
luvium of the Arctic Circle.—Phil. Mag. ser. 2, IX. p. 411.—*L.
u. Br.* N. Jahrb. 1833, p. 370.

2. On the frequent deficiency of the Ungueal phalanx in the Hallux
of the Orang Outang.—Phil. Mag. ser. 3, VII. p. 72.

3. On the probable Connection of Rock-Basins, in Form and Situation, with an internal concretionary Structure in the Rocks on
which they occur.—Phil. Mag. ser. 2, VIII. p. 331.—*Féruss.* Bull.
XXV. p. 146.

4. On certain Organs of the *Helicidæ* usually regarded as their
Eyes.—Zool. Journ. II. p. 497.—Isis, 1830, X. p. 1066.—*Féruss.*
Bull. 1827, XII. p. 165.

BRÉBISSON (L. A. de).

1. Sur un nouveau genre d'Insecte de l'ordre des Hyménoptères

(*Pinicola*).—N. Bull. Soc. Phil. 1818, Août, p. 116.—Bibl. Ent.
I. p. 48.

2. Catalogue des Arachnides, des Myriapodes et des Insectes aptères
que l'on trouve dans le Département du Calvados.—Mém. Soc.
Linn. Calvad. III. p. 254.—*Féruss.* Bull. 1828, XIV. p. 285.

3. Catalogue méthodique des Crustacés terrestres, fluviatiles et
marins, recueillis dans le Département du Calvados.—Mém. Soc.
Linn. Calvad. II. p. 225.—*Féruss.* Bull. 1826, VIII. p. 403.

4. Observations sur les Diatomées.—C. R. III. p. 577.

BREDA (Van).

1. Betrachtungen über die beim Brunnenbohren zu Zeist herauf-
gebrachten Erdarten u. deren geologisches Alter.—Algem. K. en
Lett. Bode, 1835, X. XI.—*L.* u. *Br.* N. Jahrb. 1837, p. 71.

2. Notice sur un Cétacé échoué près d'Ostende le 5 Nov. 1827.—
Allgem. Konst en Letterb. Nov. 1827, p. 341.—*Féruss.* Bull.
1828, XV. p. 298.

BREDA (Van) et HEES (Van).

1. Notice sur les Dents de Ruminans, de Pachydermes et de Carnassiers trouvés dans la Formation crayeuse de la Montagne de
St. Pierre de Maestricht.—Ann. Sc. n. XVII. p. 446.

BREDE (F. J.).

1. Beschaffenheit der Erde. 8°. Altona, 1837.

BREDECZKY (S.).

1. Beiträge zur Topographie des Königreichs Ungarn. Wien,
1803–5, 4 vols. 8°. fig.—(Neue Beitr.) Wien u. Triest, 1807, 8°.

2. Sur les Lignites d'Orenburg.—Ann. Soc. Minér. Jéna, 1825, VI.
p. 101.—*Féruss.* Bull. 1826, IX. p. 143.

BREDIN (C. J.).

1. Découverte d'Ossemens d'Éléphans près de Lyon.—*Féruss.* Bull.
1824, III. p. 166.

2. Notice sur des Os fossiles de grands Mammifères, trouvés à la
Croix-Rousse, près de Lyon, en Avril 1824.—Arch. hist. et stat.
Rhône.—*Féruss.* Bull. 1827, X. p. 319.

BREDOW (Von).

1. Auszüge aus den Schreiben des reisenden Naturforschers *C.
Moritz* in Süd-Amerika.—*Wiegm.* Arch. 1837, I. p. 408.

BREDSDORF (J. H.).

1. De Mappis geognosticis. Copenh. 1828, 8º.—*Féruss.* Bull. 1831, XXIII. p. 1.

2. Aperçu sur les systèmes de Montagnes du continent d'Europe. —Tidskr. f. Naturw. 1827, XIII. p. 1.—*Féruss.* Bull. 1828, XIV. p. 171.

3. Craie régénérée en Sélande.—Tidskr. f. Naturw. 1827, XIII. p. 64.—*Féruss.* Bull. 1828, XIV. p. 193.

4. De l'Argile bleue d'Olden en Danemark.—Tidskr. f. Naturw. 1826, XII. p. 378.—*Féruss.* Bull. 1827, X. p. 335.

5. Observations géognostiques et minéralogiques faites dans un Voyage dans le Jutland septentrional.—Tidskr. f. Naturw. IX. p. 243.—*Féruss.* Bull. 1826, VIII. p. 417.

BREE (W. T.).

1. Caractère servant à distinguer les deux sexes de la Bécasse ordinaire (*Scolopax rusticola*).—Mag. of Nat. Hist. 1830, XII. —*Féruss.* Bull. 1831, XXIV. p. 368.

2. Fall, wo eine Kröte in einem soliden Sandstein eingeschlossen gefunden wurde.—*Loud.* Mag. IX. p. 316.—*Wiegm.* Arch. 1837, II. p. 234.

3. On the mode in which the common Frog takes its prey.—Mag. Nat. Hist. ser. 1, III. p. 326.

4. List of *Papilionidæ* in the vicinity of Dover.—Mag. Nat. Hist. ser. 1, V. p. 330.

5. On a species of *Acarus* which infests Butterflies.—Mag. Nat. Hist. ser. 1, V. p. 336.

6. On the Arrival and Retreat of the British Hirundines from 1800 to 1828.—Mag. Nat. Hist. ser. 1, II. p. 16.—*Féruss.* Bull. 1830, XXII. p. 118.

7. Instances of singular Nidification in Birds.—Mag. Nat. Hist. ser. 1, VI. p. 32.

8. On the fall of an aged Ash-tree.—Mag. Nat. Hist. ser. 1, VI. p. 327.

9. Attempt to naturalise the *Arctia phæorrhæa.*—Mag. Nat. Hist. ser. 1, II. p. 66.

10. On the capture of *Argynnis aphrodite* in Warwickshire.—Mag. Nat. Hist. ser. 2, IV. p. 131.

11. On *Trochilium crabroniforme,* the Lunar Hornet Sphinx.— Mag. Nat. Hist. ser. 2, I. p. 19.

12. On some singular varieties of *Papilionidæ* in *Weaver's* Museum, Birmingham.—Mag. Nat. Hist. ser. 1, V. p. 749 ; VI. p. 175.

13. On Stones found in the Stomachs of Pikes.—Mag. Nat. Hist. ser. 1, III. p. 241.

14. On *Pontia chariclea* and *metra*.—Mag. Nat. Hist. ser. 1, III. p. 242.

BREFELD (F.).

1. Der Stockfisch-Leberthran in naturhistorisch-chemisch-pharmazeutischer Hinsicht. Hamm, 1835, 8°.—N. Act. Nat. Cur. XVII. p. XXVI.

BREHDAHL (Fr.).

1. Comment. anat.-patholog. de Testiculorum in scrotum descensu, adjectâ novâ de Cryptorchide observatione. Leipz. 1824, 4°. fig. —Bibl. med.-ch. p. 63.

BREHM (Chr. L.).

1. Das Studium der Naturgeschichte in der Natur.—Isis, 1835, III. p. 241.

2. Einige merkwürdige Beobachtungen über die Fledermäuse.— Ornis, III. 1827, p. 17.—Isis, 1827, XI. p. 960.—*Féruss.* Bull. 1828, XIV. p. 250 (*Quelques observations sur les Chauves-Souris, et Description de cinq nouv. espèces*).

3. Zusammenwohnen der weiblichen Fledermäuse, etc.—Isis, 1829, VI. p. 640.—*Féruss.* Bull. XXIII. p. 115 (*Observ. sur les Chauves-Souris, et Descr. d'une nouv. esp.*, Vespertilio rufescens).

4. Ein noch unbekannter, gefährlicher Feind der Fische (*Sorex fodiens*).—Isis, 1830, XI. p. 1126.

5. Der Löwe keine Katze.—Isis, 1829, VI. p. 636.

6. Abstammung der Hauskatze.—Isis, 1829, VI. p. 639.

7. Zoologische, vorzüglich ornithologische Bemerkungen auf einer Reise von Renthendorf nach Berlin.—Isis, 1832, VII. p. 734; VIII. p. 836; 1833, VIII. p. 791; 1834, I. p. 38.

8. Ornis, oder das Neueste und Wichtigste der Vögelkunde. (In Verbind. mit mehreren Naturforschern.) 8°. Jena, 1824–1827; Hft. I.-III.—Isis, 1826, VI. p. 629; 1827, XI. p. 957.—*Féruss.* Bull. 1827, X. p. 150; 1828, XIV. p. 255 (*Ornis*, etc.).

9. Das Ausstopfen der Vögel.—Isis, 1827, II. p. 147.—*Féruss.* Bull. 1828, XIV. p. 261 (*La Taxidermie des Oiseaux*).[sr]

10. Die Hauptbewegungen der Vögel.—Isis, 1831, II. p. 145. [sl]

11. Der Zug der Vögel.—Isis, 1828, IX. p. 912.—*Féruss.* Bull. 1830, XXI. p. 132 (*De la Migration des Oiseaux*).

12. Beobachtungen über die Ehen und Pflegmutterschaft der Vögel.—Isis, 1835, II. p. 127 ; III. p. 233.—*Wiegm.* Arch. 1836, II. p. 262.

13. Lehrbuch der Naturgeschichte aller europäischen Vögel. Jena, I. 1823 ; II. 1824, 8°. fig.—Isis, 1825, I. p. 104.—*Féruss.* Bull. 1826, VII. p. 241.

14. Beiträge zur Vögelkunde, in vollständiger Beschreibung mehrerer neu entdeckter oder nicht gehörig beobachteter deutscher Vögel.—Cat. Bibl. Turic. Neustadt, 1820–1822. 3 vols. 8°. fig.

15. Observation ornithologique.—Ann. Sc. n. 1829, XVIII.—Rev. Bibl. p. 104.

16. Eine Vergleichung verwandter Vögelarten, und zugleich eine Erwiederung auf *Faber*'s Bemerkungen über meine neue Arten hochnordischer Schwimmvögel.—Isis, 1826, IX. p. 927 ; X. p. 983. —*Féruss.* Bull. 1828, XIV. p. 258.

17. Handbuch der Naturgeschichte aller Vögel Deutschlands. 8°. Ilmenau, 1831.

18. Etwas über den Borkenkäfer (*Bostrychus typographus*).—Isis, 1829, VIII. p. 877.

19. Die Kunst, Vögel als Bälge zubereiten, auszustopfen, aufzustellen und aufzubewahren. Nebst einer kurzen Anleitung, Schmetterlinge und Käfer zufangen, zu präpariren, etc. Weimar, 1842, 8°.

20. On some of the Domestic Instincts of Birds.—Mag. Nat. Hist. ser. 2, II. p. 399.

21. Monographie der Papageien. 3 Hft. Jena, 1842–45, fol. fig.

22. Ueber seine neuen Vögelarten.—Isis, 1826, II. p. 190.—*Féruss.* Bull. 1827, XI. p. 113.

23. Noch etwas über seine neuen Vögelarten, etc.—Isis, 1828, I. pp. 23, 39.—*Féruss.* Bull. 1828, XIV. p. 259.

24. Ornithologischer Ausflug nach Thüringen, im Juni 1827.—Isis, 1830, XI. p. 1113.

25. Uebersicht der deutschen Vögelarten.—Isis, 1828, XII. p. 1268 ; 1830, X. p. 985.

26. Mittheilungen von Beobachtungen der Herren *v. Seyffertitz* u. *Homeyer* über Schneeeulen, deren sich mehrere im Winter 1832 und 1833 im nördlichen Deutschland, selbst in Sachsen und Thüringen sehen liessen.—Isis, 1834, III. p. 240.—*Wiegm.* Arch. 1835, II. p. 302.

27. Etwas über das Sommerkleid der Entenmannchen.—Isis, 1835, III. p. 238.—*Wiegm.* Arch. 1836, II. p. 274.

28. Wann werden die Krähenarten brutfähig?—Isis, 1835, IV. p. 317.

29. Die deutschen Goldhähnchen.—Isis, 1832, I. p. 19.

30. Ueber *Columba domestica, livia* et *amaliæ.*—Isis, 1828, II. p. 136.—*Féruss.* Bull. 1828, XIV. p. 260 (*Sur les* Columba domestica, etc.).

31. Ueber die Sumpf- und Waldschnepfen.—Isis, 1835, II. pp. 116, 126.—*Wiegm.* Arch. 1836, II. p. 273.

32. *Telmatias platyura,* s. *Gallinago heterura.*—Isis, 1835, II.— *Wiegm.* Arch. 1837, II. p. 218.

33. Ueber die doppelte Mauser der zur Sippe Taucher (*Colymbus*) gehörigen Vögel.—Isis, 1830, X. p. 979.

34. Die kleinen europäischen Rohrhühner.—Isis, 1834, VII. p. 705.

35. Der deutsche Baumfalke in seinem Betragen.—Isis, 1830, I. p. 107.

36. Die grossen Adler mit befiederten Füssen.—Isis, 1830, I. p. 69.

37. Betragen des rauchfüssigen Kauzes (*Strix dasypus,* Bechst.).— Isis, 1830, XI. p. 1128.

38. Etwas über die Kreuzschnäbel.—Isis, 1827, VIII. p. 704.— *Féruss.* Bull. XIV. p. 259 (*Sur les Becs-croisés*).

39. Ueber *Aquila borealis, Procellaria hyemalis, Platypus Leisleri* u. *Carbo subcormoranus.*—Ornis, I. p. 1.

40. Beurtheilung von *Naumann*'s Vögelwerk.—Ornis, I. p. 133.

41. Kurze Uebersicht der europäischen Vögel.—Ornis, II. p. 1; III. p. 1.—*Féruss.* Bull. 1827, X. p. 288.

42. Die einheimische Wasserspitzmäuse.—Ornis, II. p. 25.—*Féruss.* Bull. 1827, XI. p. 287.

43. Eine neue Art Kreuzschnäbel mitten in Deutschland.—Ornis, III. p. 77.—*Féruss.* Bull. 1828, XIV. p. 260.—Isis, 1830, I. p. 110.

BREHM (C. L.) et GOURCY-DROITAUMONT (F. von).

1. Handbuch für den Liebhaber der Stuben-Haus- u. Zähmung werthen Vögel. 8°. Ilmenau, 1832.

BREHM (C. L.) et RICHTER ( ).

1. Erwiederung auf die Bemerkungen des Hrn. *Dr. Carus* über die Haare im Kuckucksmagen.—Isis, 1823, XI. p. 1249.

BREHM (C. L.) et SCHILLING (W.).

1. Mémoires pour servir à la connaissance des Oiseaux, ou Descrip-

tion détaillée de plusieurs Oiseaux nouvellement découverts, et d'un grand nombre d'Oiseaux rares d'Allemagne. Neustadt, 1823, 3 vols. 8°. fig.—*Féruss.* Bull. 1824, I. p. 377.

BREHM (C. L.) et THIENEMANN (F. A. L.).

1. Exposé systématique de la Propagation des Oiseaux d'Europe, avec les figures de leurs Œufs. 3ᵉ section : Granivores. Leipzig, 1829, 4°. fig.—*Féruss.* Bull. 1830, XXIII. p. 269.

BREIBISIUS (J. G.).

1. Neueste Beschreibung der Sauerbrunnen zu Jebenhausen, eine Stunde v. Göppingen, darinnen von des Orts Beschaffenheit, der Quellen Ursprung, langjährigen Ruhme, etc., gehandelt wird. Rotenb. 1723, 8°.—*Böhm.* Bibl. V. p. 317.

BREIDENSTEIN (J. Ph.).

1. Naturgeschichte des Sperlings deutscher Nation, nebst vielen Mitteln dessen Anzahl zu vermindern. Giessen, 1779, 8°.—*Böhm.* Bibl. II. 1, p. 597.

BREISLAK (Scip.).

1. On the Gypsum of Monte Seano.—Trans. Geol. Soc. ser. 2, I. p. 169.—Isis, 1823, IV. Litt. Λ. p. 223.—*Féruss.* Bull. II. p. 44.

2. Introduzione alla Geologia. Milano, 1811, 8°. 2 vols.—Biogr. Un. Suppl. LIX. p. 313.—(Gall. *J. J. B. Bernard*): *Introduction à la Géologie, ou à l'Hist. nat. de la Terre.* Par. 1812, 8°.

3. Instituzioni geologiche. Milano, 1818, 8°. 3 vols. atl. 4°.— (Trad. Fr.) *Traité sur la structure extérieure du Globe, ou Institutions géologiques.* Paris, 1822.—Biogr. Un. Suppl. LIX. p. 213. —(Germ. *Strombeck*): *Lehrbuch der Geologie.* Braunschw. 1819, 3 vols. 8°. atl. 4°.

4. Sulla Giacitura di alcune Rocce porfiritiche e granitose osservate nel Tirolo dal Cte. *Marzari-Pencati.* 8°. Milano, 1821. (*Sur le gisement de quelques Roches porphyritiques et granitoïdes, observées en Tyrol par le* Comte Marzari-Pencati.)—Mem. I. R. Istit. Lombardo-Ven. III. p. 3.—*Féruss.* Bull. 1826, VIII. p. 324.

5. Memorie sulle osservazioni fate da cel. Geologi, etc. intorno alla giacitura di Graniti del Tirolo merid. Milano, 1824, 8°.—Biogr. Un. Suppl. LIX. p. 213.

6. Topografia fisica della Campania, etc. Firenze, 1797, 2 vols. 8°.

7. Sur des Observations faites par quelques Géologues, postérieurement à celles déjà signalées au Comte *Marzari-Pencati,* relatives

au gisement du Granite dans le Tyrol méridional.—Mem. I. R. Istit. Lomb.-Ven. III. p. 357.—*Féruss.* Bull. 1826, VIII. p. 324.

8. Ueber die Vulkane. (Introd. alla Geologia, C. VIII.)—*Leonh.* Tasch. 1819, II. p. 497.

9. Anwendung geogonischer Hypothesen auf die geognostische Klassifikazion der Felsarten.—*Féruss.* Bull. Janv. 1826, p. 22.— *Leonh.* Zeitschr. 1826, II. p. 440.

10. Essais minéralogiques sur la Solfatare de Pouzzole, trad. du mscr. italien par *Pommereul.* Naples, 1792, 8°.

11. Viaggi litologici nella Campania. Firenze, 1798.—(Gall. *Pommereul*): *Voyages physiques et lithologiques dans la Campanie,* etc. Par. 1801, 2 vols. 8°.—(Germ. *F. A. Reuss*): *Physikalische u. lithologische Reisen in Kampanien,* etc. Leipz. 1802, 2 vols. 8°. fig.

12. Descrizione geologica della Provincia di Milano. Mil. 1822, 8°.—Cat. Bibl. Turic.

13. Atlas géologique, ou Vues d'amas de Colonnes basaltiques, faisant suite aux Institutions géologiques. Mil. 1818, fol.—Cat. Bibl. Turic.

BREISLAK (Scip.) e WINSPEARE (Ant.).

1. Memoria sull' Eruzione del Vesuvio accaduta la sera da 15 Giugno 1794. Nap. 1794, 8°.

BREITHAUPT (Aug.).

1. Observations sur l'Anthracite.—Zeitschr. f. Miner. 1827, p. 47. —*Féruss.* Bull. 1827, XII. p. 7.

2. Ueber eine eigene Art von Palmen-Versteinerung, den Röhrenstein.—Isis, 1820, Litt. A. p. 440.

3. Vollständiges Handbuch der Mineralogie. Dresd. u. Leipz. 1836–41, 2 vols. 8°. fig.

4. Uebersicht des Mineral-Systems. Freiberg, 1830, 8°.

5. Vollständige Charakteristik des Mineral-Systems. (2te Aufl.) Dresd. 1823, 8°. (3te Aufl.) Dresd. u. Leipz. 1832, 8°.

6. Die Erzgebirgische Lager-Formation in Böhmen, etc.—*L.* u. *Br.* N. Jahrb. 1835, p. 676.

7. Geognostische und mineralogische Bemerkungen über den Nord-Amerikanischen Freistaat Nord-Karolina.—*Leonh.* Zeitschr. 1827, II. p. 349.

BREMBATI (Ottav.).

1. La Mineralogia, divisa in IV Libbri, etc. Bergamo, 1663, 12°.

BREME (F. De).

1. Nouvelle espèce européenne du genre *Larus*.—Rev. Zool. 1839, p. 321.

2. Monographie de quelques genres Coléoptères, Hétéromères, appartenant à la tribu des Blapsides.   Paris, 1842, 12°. fig.

3. Essai monographique et iconographique de la tribu des Cossyphides.   Paris, 1842, 1ᵉ part. 1 vol. 8°. 7 pl. col.

4. Note sur une nouvelle espèce du genre *Saperda*, Fab.—Rev. Zool. 1840, p. 277.

5. Description de deux nouveaux *Mélasomes* du genre *Adesmia* de *Fischer*.—Rev. Zool. 1840, p. 112.

BRÉMOND (C. P. de).

1. De l'Asphalte et de la Mine du Val de Travers, dans la Princip. de Neuchâtel.   Neuch. 1838, 8°.

BRÉMOND (C. P. de) et DEMOURS (   ).

1. Transact. Philos.; *vide sup.* Pars I. p. 54.

BREMSER (J. G.).

1. Ueber lebende Würmer im lebenden Menschen, etc. Wien, 1819, 4°.—Isis, 1819, VII. p. 1169.

2. Traité zoologique et physiologique sur les Vers intestinaux de l'Homme. Trad. de l'Allem. par *Gründler*.   Paris, 1824, 4°. fig. —2ᵉ éd. revue et augmentée de notes par *De Blainville* et d'un nouvel Atlas avec texte par *C. Leblond*.   Par. 1837, 8°. atl. 4°.

3. Icones Helminthum Systema *Rudolphii* entozoologicum illustrantes. Viennæ, 1824 et seqq. (Heftw.) fol.—Isis, 1824, I. p. III; IX. p. 991; XII. Beil. p. 5; 1825, VIII. p. 906.—*Féruss.* Bull. 1823, IV. p. 212; III. 1824, p. 351.

4. Notitia insignis Vermium intestinalium collectionis Vindobonensis.   Vienna, 1811.

BRENTA (L.).

1. Fenomeni della Visione, etc.   Milano, 1838.

BRERA (V. L.).

1. Lezioni medico-pratiche sopra i principali Vermi del Corpo umano vivente, etc.   Crema, 1802, 4°.—(Germ. *Weber*): *Ueber die vornehmsten Eingeweidewürmer des menschl. Körpers*.   Leipz. 1803, 4°. fig.—(Gall. *Barroli* et *Calvet*): 8°. Par. 1804; 1807.— —(Angl. *Coffin*): 8°. Boston, 1817.

2. Singolare Monstruosità di un Feto umano, e Congetture sul primitivo Sviluppo dell' Embrione. Verona, 1816, 4º.—Soc. Ital. d. Sc. XVII. fig.—Isis, 1818, XII. p. 1976.

3. Anatomisch-tabellarische Uebersicht der Knochen- Gefäss- und Nervenlehre. Aus dem Ital. Wien, 1800, fol.—Bibl. med.-ch. p. 64.

4. Memorie fisico-mediche sopra i principali Vermi del Corpo umano, etc. Crema, 1811, 4º.

5. Antologia medica ; *vide sup.* Pars I. p. 73.

6. Giornale di Medicina; *vide sup.* Pars I. p. 69.

BRÈS (J. P.).

1. Observations sur la forme arrondie considérée dans les Corps organisés et principalement dans le corps de l'Homme. 1813.—Biogr. Un. Suppl. LIX. p. 219.

BRESCHET (Gilb.).

1. Recherches anatomico-physiologiques et chimiques sur la Matière colorante du Placenta de quelques animaux.—Ann. Sc. n. 1830, XIX. p. 379.—Isis, 1834, XI. p. 1108.

2. Sur le développement de l'Œuf humain.—Ann. Sc. n. 1832, XXVII. p. 208.—Isis, 1835, VII. p. 632.

3. Quelques Recherches sur la Structure des Membranes de l'Œuf des Mammifères.—Ann. Sc. n. 1827, VIII. p. 224.

4. Du Périone ou Membrane caduque, de l'Hydropérione ou liquide contenu dans cette membrane, et de la Nutrition du Fœtus pendant les premières périodes de la Gestation.—Ann. Sc. n. 1832, XXVI. p. 160.—Isis, 1835, VI. p. 501.

5. Études anatomiques et physiologiques de l'organe de l'Ouïe et de l'Audition, dans l'Homme et les Animaux vertébrés. Paris, 1833, 4º. fig.—(2e éd.) Paris, 1836.—Ann. Sc. n. 1833, XXIX. pp. 89 et 304, fig.—*Müll.* Arch. 1837, p. xLvii.

6. Recherches anatomiques et physiologiques sur l'organe de l'Ouïe dans les Oiseaux. Par. 1836, 8º. atl. 4º.—Ann. Sc. n. Janv. 1836. —*Müll.* Arch. 1837, p. Lxiv.—Mag. Zool. and Bot. I. p. 101.

7. Ouïe des Poissons.—Ann. Sc. n. 1829, XVIII.—Rev. Bibl. p. 145.

8. Le Système lymphatique, considéré sous les rapports anatomique, physiologique et pathologique. Paris, 1836, 8º. fig.—(Germ. *E. Martiny*): *Das Lymphsystem in Hinsicht auf Anatomie, Physiologie u. Pathologie.* Quedl. 8º. fig.— *Val.* Repert. II. p. 23.

9. Anatomisch-physiologische Untersuchungen über einige neuentdeckte Theile des Nervensystems. Erste Abhandlung: Von den Venen der Knochen überhaupt und den Blutadercanälen der

schwammigen Substanz der Schädelknochen insbesondere. fig.—
N. Act. Nat. Cur. XIII. p. 359.

10. Description d'un Organe vasculaire découvert dans les Cétacés ;
suivie de quelques Considérations sur la Respiration chez ces
animaux et chez les Amphibies. (Rapport de *Duméril* à l'Acad.
des Sc.)—Ann. Sc. n. 1834, II. p. 376.

11. Histoire anatomique et physiologique d'un Organe de nature
vasculaire découvert dans les Cétacés. Paris, 1836, 4°. fig.—*Val.*
Repert. II. p. 23.

12. Recherches sur différentes pièces du squelette des Animaux
vertébrés, encore peu connues, et sur plusieurs vices de conformation des Os.—C. R. VII. p. 912.—Ann. Sc. n. (2ᵉ S.) X. p. 91 ;
(3ᵉ S.) I. p. 25, fig.—Rev. Zool. 1838, p. 275.

13. Aperçu descriptif de l'Organe auditif du Marsouin (*Delphinus
Phocæna*, L.).—Ann. Sc. n. (2ᵉ S.) X. p. 221, fig.

14. Abhandlung über den Nervenplexus der Paukenhöhle im
Menschen und den Thieren, nebst einem doppelte Anhange.
I. Ueber den eigenthümlichen Nerven des Musculus tensor tym-
pani. II. Ueber das Gehörwerkzeug des Braunfisches, *Delphinus Phocæna.*—*Heusinger's* Zeitschrift, III. 1833, p. 581.

15. Ueber die Gesichtsnerven des Pferdes.—*Heusinger's* Zeitschrift,
I. 1827, p. 462, cum fig.

16. Rapport sur un Mémoire de M. le Dr. *Gerdy*, ayant pour titre :
De la Structure des Os.—Ann. Sc. n. (2ᵉ S.) XI. p. 33.—Rev.
Zool. 1839, p. 7.—C. R. VIII. p. 120.

17. Études anatomiques, physiologiques et pathologiques de l'Œuf
dans l'espèce humaine et dans quelques-unes des principales familles des Animaux vertébrés. 4°. Paris, 1833, fig.—Cat. R. Soc.
Lond.

18. Mémoire sur un Vice de conformation congéniale des enveloppes
du Cœur. 4°. Paris, 1826.—Cat. R. Soc. L.

19. Mémoire sur l'ectopie de l'Appareil de la Circulation, et particulièrement sur celle du Cœur. 4°. Paris, 1826.—Cat. R. Soc. L.

20. Recherches anatomiques et physiologiques sur l'organe de l'Ouïe
et sur l'Audition dans l'Homme et les Animaux vertébrés. Paris,
1836, 4°. avec 13 pl. gravées.

21. Essai sur les Veines du Rachis, la formation du Cal, la Hernie
fémorale ; de la Désiccation et des autres moyens de conservation
des pièces anatomiques. Paris, 1819, 4°. fig.

22. Remarques sur la communication faite par M. Serres, concernant le développement de l'Amnios chez l'Homme.—Rev. Zool.
1838, p. 308.—C. R. VII. p. 1031.

23. Recherches anatomiques, physiologiques et pathologiques sur le Système veineux. Paris, 1829, fol. fig. col.

24. Rapport sur un Mémoire de *M. Milne Edwards* relatif à la Circulation du Sang dans les Annelides.—C. R. VII. 1838, p. 633.—Rev. Zool. 1838, p. 207.

25. Nouv. Recherches anatomiques sur la structure de la Peau. Paris, 1835, 8°. fig.

26. Recherches anatomiques et physiologiques sur l'organe de l'Ouïe des Poissons. Paris, 1838, 4°. fig.— *Val.* Repert. IV. p. 18.

BRESCHET (G.) et GLUGE (   ).

1. Recherches sur la structure des membranes de l'Œuf des Mammifères.—Rev. Zool. 1838, p. 6.—C. R. V. p. 79.

BRESCHET (G.) et MILNE EDWARDS (H.).

1. Recherches expérimentales sur l'Exhalation pulmonaire.—Ann. Sc. n. 1826, IX. p. 5.—Isis, 1834, IX. p. 895 (*Ueber die Lungen-Exhalation*).

2. Mémoire sur le mode d'action des Nerfs pneumogastriques dans la production des phénomènes de la Digestion.—Ann. Sc. n. 1825, IV. p. 257.—N. Act. Nat. Cur. XIII. p. xxiv.

BRESCHET (G.), MILNE EDWARDS (H.) et VAVASSEUR (   ).

1. De l'influence du Système nerveux sur la Digestion stomachale. —Arch. Gén. Méd. Août 1823, p. 479.—*Féruss.* Bull. IV. p. 339.

BRESCHET (G.) et RASPAIL (F. V.).

1. Anatomie microscopique des Nerfs, pour démontrer leur structure intime, etc. 4°.—Cat. R. Soc. L.

2. Anatomie microscopique des flocons du Chorion de l'Œuf humain. 4°. Paris, 1827.—Cat. R. Soc. L.

BRESCHET (G) et ROUSSEL DE VAUZÈME (   ).

1. Analyse d'un premier Mémoire sur la structure et les fonctions de la Peau. 8°. Paris, 1834.—Cat. R. Soc. L.

2. Nouvelles Recherches sur la structure de la Peau. 8°. Paris, 1835.—Cat. R. Soc. L.

3. Recherches anatomiques et physiologiques sur les Appareils tégumentaires des Animaux.—Ann. Sc. n. 1834, II. pp. 167 et 321. —*Müll.* Arch. 1835, p. 18 (*Untersuchungen über die Structur der Haut*).

BRESCIUS (Zach.).

1. Dissertatio de Lumbricis. Lugd. Bat. 1699, 4°.—*Böhm.* Bibl.
II. 2, p. 376.

BRESMAL (J. Fr.).

1. La Circulation des Eaux, ou l'Hydrographie des Eaux minérales
d'Aix et de Spa. Liege, 1690, 12° ; 1699, 8° ; 1716 et 1718, 8°.
—*Böhm.* Bibl. V. p. 345.

BRESSEY (De).

1. Nouvelles observations sur le volcan de Drevin.—Nouv. Mém. de
Dijon, 1785, Sem. 2, p. 105.

BRETON (P.).

1. On the Topography, Animals and Reptiles of some districts of
India.—*Brewst.* Journ. Science, ser. 1, VII. p. 316.
2. On the Worm found within the Eye of the Horse.—Trans. Med.
and Phys. Soc. Calcutta, I. p. 337.
3. Observations sur les variétés produites par le Changement de
Peau dans les Lizardes et dans les Vipères.—Journ. de Phys.
LXXXIV. 1819, p. 456.

BRETON (Lieut.).

1. Account of the Habits of a specimen of *Echidna*, which survived
during a considerable part of its voyage to Europe.—Proc. Zool.
Soc. Lond. II. p. 23.—*Wiegm.* Arch. 1835, II. p. 334.
2. On the Habits of the Musk Duck of New Holland (*Hydrobates
lobatus,* Temm.).—Proc. Zool. Soc. II. p. 19.
3. On a mode of preserving Bird-skins, in the absence of the ordi-
nary means.—Proc. Zool. Soc. Lond. IV. p. 21.
4. Rapport sur les Mines de Diamant de Sumbhulpore.—Edinb.
Journ. of Sc. Juill. 1827, p. 134.—*Féruss.* Bull. 1828, XIV. p. 39.

BRETONNEAU (   ).

1. Blasenziehende Eigenschaft einiger Canthariden.—Ann. Sc. n.
XIII. p. 75. (Rapp. de *Dum.* et *Latr.*)—Isis, 1834, X. p. 1011.

BRETSCHNEIDER (K. B.).

1. Auch ein Beitrag zur Kenntniss der Fichtenraupen. Weimar,
1798, 8°.—*Eis.* Insect. p. 214.

BREUCHEL (J. P.), WALTHER (J.), BRAUER (A.), u. ein Un-
genannter.

1. Von den Rebenstiehern (einem noch nicht ganz bekannten In-
sekte aus dem Geschlechte der Käfer), vier Preisschriften der
Churpfälz. Akademie. Mannh. 1767, 8°.—Bibl. Ent. I. p. 49.—
Böhm. Bibl. II. 2, p. 183.

BREUER (F.).

1. De speciebus Graviditatis variis. Berol. 1839, 8°.— *Val.* Repert.
VI. p. 35.

BREUGEL (H. G. van).

1. Naspooringen aangaande de Oöulogie of Eijerkunde en de oor-
spronkelyke voortteling van menschen en beesten. 8°. Dordrecht,
1794.

2. Aanhangzel of physiologische Verhandelingen over het leerstel-
zel der ontwikkeling en dat der bygeboorte. 8°. Dordrecht, 1796.

BREVAL (J.).

1. Remarks on several parts of Europe, relating chiefly to the History, Antiquities and Geography, etc. Lond. 1726, 2 vols. fol.
fig.—*Böhm.* Bibl. I. 1, p. 459.

BREVENTANI (U.).

1. Del Moto vibratorio rinvenuto in varie Membrane degli Animali.
—N. Ann. Sc. n. di Bol. 1840.

2. Sui . . . . del Cuore.—Bull. Soc. Med. Bol. IV.

3. Sul Coloramento del Sangue.—Bull. Soc. Med. Bol. I.

4. Esperienze elettro-fisiologiche istituite nel Gabinetta di Fisica
della Università di Bologna.—Bull. Soc. Med. Bol. XII.

BREWER (J.).

1. On Beds of Oyster-Shells found near Reading in Barkshire.—
Phil. Trans. XXII. p. 484.

BREWER (T. M.).

1. On the Habits of certain Birds.—Bost. Soc. Oct. 1839.—*Sill.*
Am. J. XXXVIII. p. 392.

BREWSTER (Sir D.).

1. Report on the Phænomena and Cause of *Muscæ volitantes.*—Rep.
Brit. Assoc. Sept. 1840.—*Sill.* Am. J. XL. p. 333.

2. On the anatomical and optical Structure of the Crystalline Lenses of Animals.—Roy. Soc. Lond. Jan. 1836.—Mag. Zool and Bot. I. p. 206.—L. and Ed. Phil. Mag. VIII. p. 193, fig.

3. On the Structure and Functions of the Human Eye, &c.—Edinb. Roy. Soc.—*Féruss.* Bull. II. p. 115 (*De la structure et des fonctions de l'Œil, et particulièrement du Cristallin chez l'Homme*).

4. On the Crystalline Lens in Fishes, Birds, Reptiles and Quadrupeds.—Rep. Brit. Assoc. 1831, p. 81.

5. On the Structure and Origin of the Diamond.—Trans. Geol. Soc. ser. 2, III. p. 455.—Proc. Geol. Soc. I. 466.—Phil. Mag. ser. 3, III. p. 219.

6. On the Form of the Integrant Molecule of Carbonate of Lime.—Trans. Geol. Soc. ser. 1, V. p. 83.

7. On the Structure of the Fossil Teeth of the Sauroid Fishes.—Rep. Brit. Assoc. 1838, Sect. p. 90.

8. Observations sur la Température moyenne du Globe.—Trans. Roy. Soc. Edinb. I. p. 301.—*Féruss.* Bull. 1824, I. p. 3.—N. Ed. J. of Sc. 1831, VIII. p. 300.—*L. u. Br.* N. Jahrb. 1833, p. 379.

9. A Treatise of the Microscope. Lond. 1837, 8°.—*Val.* Repert. III. p. 11.

10. Edinburgh Encyclopædia ; *vide sup.* Pars I. p. 75.

11. Edinb. Journ. of Science ; *vide sup.* Pars I. p. 75.

12. Note sur le Mémoire de *Home* : Production et formation des Perles.—Edinb. Journ. of Sc. Avril 1827, p. 275.—*Féruss.* Bull. 1827, XII. p. 154.

BREWSTER (D.) and JAMESON (R.).

1. Edinb. Philos. Journal ; *vide sup.* Pars I. p. 75.

BREWSTER (P.).

1. On a Fossil Tree found in a Quarry at Nites-hill.—Trans. Roy. Soc. Edinb. IX. p. 103.

BREY (Gaët.).

1. Sulle utili Applicazioni del nuovo Systema di Perforamento denominato *Hauts Sondages.* Milano, 1835, 8°.—*L. u. Br.* N. Jahrb. 1835, p. 683.

BREYER (F. G.).

1. Obs. anatomicæ circa fabricam *Ranæ pipæ.* 4°. Berl. 1811.

BREYN (J. Ph.).

1. Addenda et corrigenda in Historiâ Cocci seorsin editâ.—Act.

Erudit. 1733, p. 167.—Commerc. Norimberg. 1733, p. 11.—Phil. Trans. No. 426, p. 364.—*Eis.* Insect. p. 250.

2. Epistola varias Observationes continens in Itinere per Italiam suscepto, anno 1703.—Phil. Trans. XXVII. p. 447.—Tr. Linn, Soc. Lond. VII. p. 228.

3. De quibusdam Conchis minùs notis.—Mem. di Fis. e d' Ist. nat. Lucc. 1743, I. p. 175.—Tr. Linn. Soc. Lond. VII. p. 217.

4. Dissertatio de Polythalamiis novâ Testaceorum classe, cui quædam præmittuntur de Methodo Testacea in classes et genera distribuendi; acc. Commentatiuncula de Belemnitis Prussicis et Schediasma de Echinis methodice disponendis. Gedani, 1732, 4°. fig.—*Böhm,* Bibl. II. 2, p. 438.—*Mod.* B. Helm.

5. Epist. de Melonibus petrefactis Montis Carmel vulgò creditis. Lips. 1722, 4°. fig.—Cat. Bibl. Turic.

6. Tentaminis de Lithozois ac Lithophytis olim marinis, jam verò subterraneis, Prodromus.—De Coronâ Serpentum.—De Alcyonio Britannici Maris.—Misc. Nat. Cur. Cent. VIII. App. p. 153.—*Mod.* B. Helm.

7. De Belemnitis Prussicis.—De Echinis et Echinitis. Gedani, 1732, 4°. fig.

8. A Description of some Mammouts Bones dug up in Siberia, etc. —Phil. Tr. XL. p. 124.

9. De Insectis quibusdam (et Vermibus) rarioribus, in Hispaniâ observatis.—Phil. Tr. XXIV. No. 301, p. 2050.—Ephem. Nat. Cur. Cent. V. et VI. App. p. 101.—Bibl. Ent. I. p. 49.

10. Historia naturalis Cocci tinctorii, quod Polonicum vulgò audit, præmisis quibusdam Coccum in genere et in specie Coccum ex Ilice, quod Grana Kermes, et alterum Americanum, quod Cochinilla Hispanis dicitur, spectantibus. Gedani, 1731, 4°. fig. col.— Biogr. Un. V. p. 573.—*Böhm.* Bibl. II. 2, p. 229.

11. De Echinis et Echinitis, sive methodica Echinorum distributione Schediasma. 8°. Gedani, 1732.

BREZ (Jac.).

1. La Flore des Insectophiles, précédée d'un Discours sur l'utilité des Insectes et de l'étude de l'Insectologie. Utrecht, 1791, 8°. fig. —*Eis.* Insect. p. 129.

2. Observation entomologique sur une Larve de Staphylin.—Mém. Laus. III. p. 18.—Bibl. Ent. I. p. 49.

BRICKELL (J.).

1. The Natural History of N. Carolina. Dublin, 1739, 8°.—*Böhm.* Bibl. I. 1, p. 744.

BRIDELLE DE NEUILLAN ( ).

1. Nouvelle Découverte concernant la Punaise des Jardins.—Journ. de Phys. XX. 1782, p. 154.—*Licht.* Mag. II. 1, p. 87 (*Eine neue Entdeckung die Baumwanze betreffend*).

BRIDGES (Jerem.).

1. No Foot no Horse; or an Essay on the Anatomy of the Foot of a Horse. Lond. 1751, 8°.—*Schreb.* N. Cameralschr. IV. p. 126.—*Böhm.* Bibl. II. 1, p. 448.

BRIDGES (T.).

1. On South American Ornithology.—Proc. Zool. Soc. XV. p. 28.

2. On Bolivian Mammals and Birds.—Proc. Zool. Soc. XIV. p. 7.

3. On various Birds and Mammals from Chile.—Proc. Zool. Soc. IX. p. 93.

4. On the Habits of some of the smaller species of Chilian Rodents. —Proc. Zool. Soc. XI. p. 129.—Ann. and Mag. N. Hist. XIV. p. 53.

BRIÈRE (J. de la).

1. Récit véritable du monstrueux et effroyable Dragon, occis en une Montagne de la Haute-Auvergne. Par. 1632, 8°.—*Böhm.* Bibl. II. 2. p. 534.

BRIÈRE DE BOISMONT (A.).

1. Anthropotomie, ou Traité élémentaire d'Anatomie, contenant les préparations, l'Anatomie descriptive, et les principales Régions du Corps humain. Paris, 1831, 8°.

2. De la Menstruation considérée dans ses rapports physiologiques et pathologiques. Paris, 1842, 8°.—(Germ. *T. C. Krafft*): *Die Menstruation in ihren physiologischen, patholog. u. therapeut. Beziehungen.* Mit Zusätzen von *A. Moser.* Berl. 1842, 8°.

BRIGANTI (Jos.).

1. An Essay on the Method of carrying to perfection the E. India raw Silk. Lond. 1779, 8°.—Bibl. Ent. I. p. 50.

BRIGANTI (Vinc.).

1. Description de la structure, de la métamorphose, de la manière de vivre et des mœurs de la Mouche qui perce les Olives.—Atti Instit. Sc. nat. di Napoli, III. 1822, p. 97, fig.—*Féruss.* Bull. 1826, VII. p. 150.

2. Descrizione delle Ligule che abitano nell' addome de' Ciprini del lago di Palo in provincia di Principato citeriore.—Giorn. enc. 1820; Genn. p. 59.—*Féruss*. Bull. 1828, XIII. p. 167 (*Descr. des Ligules qui habitent dans l'abdomen d'une espèce de Cyprin*).

BRIGELIUS (J. Matth.).

1. Beschreibung des Gesundbrunnens zu Zeisenhaussen. Stuttg. 1715 et 1746, 8°.—*Böhm*. Bibl. V. p. 322.

BRIGGS (G.).

1. Ophthalmographia et nova Visionis Theoria. Lugd. Bat. 1686, 12°.—Act. Erudit. II. p. 454; V. p. 318.

BRIGGS (J. J.).

1. Occurrence of rare British Birds in the county of Derby.—Zoologist, p. 178; Capture and Appearance of certain British Birds, p. 311; Capture of Large Fishes in the Trent, p. 323; Migration of Birds in Derbyshire, p. 440; Seeds sown by Animals, p. 442; Song of Birds, p. 442; Wood-Wren, p. 451; British Birds in Derbyshire, pp. 553, 644; Summer Birds at Melbourne, p. 652; Carrion-Crow, p. 656; On the Rook, p. 656; Redwings and Fieldfares, p. 656; Song Thrush, p. 657; Skylark, p. 657; Tree Pipit, p. 658; On the Otter, p. 714; On the Stoat, p. 714; Habits of the Hedgehog, p. 714; Departure of the Redwing and Fieldfare from Melbourne, p. 724; On the Starling, p. 724; Birds at Spurn Head, p. 820.

BRIGHT (R.).

1. On the Hills of Badacson, Szigliget, etc. in Hungary.—Trans. Geol. Soc. ser. 1, V. p. 4.

2. On the Strata in the neighbourhood of Bristol.—Trans. Geol. Soc. ser. 1. IV. p. 193.

BRIGHT ( ).

1. Observations on Abdominal Tumours and Intumescence, illustrated by some cases of acephalocyst Hydatids. 1837.— *Wiegm.* Arch. 1838, II. p. 307.

BRIGHTWELL (Th.).

1. On the Food and Habits of certain Insects.—Zool. Journ. V. p. 396.

2. Characters of *Cebrionidæ*.—Zool. Journ. I. p. 282.

3. On the Young of the common Lobster.—Mag. Nat. Hist. ser. 1, VIII. p. 482.

Brignoli de Brunnhoff (Giov.).

1. Relazione accademica dell' ultima Eruzione accaduta nel vulca-
netto Arreo, o cosi salsa di Sassuolo nel Modanese, e Considerazioni geognostiche intorno alle Salse e alle loro cause. Reggio,
1836, 8°.—N. Act. Nat. Cur. XVIII. p. x.

Brilli (Hipp.).

1. Opusc. de Vermibus in Corpore humano genitis.   Venet. 1540,
8°.—*Böhm*. Bibl. II. 2, p. 374.

Brinckerhoff (   ).

1. Shells and Minerals presented to the N. York Lyceum of Nat.
Hist. (Reports).—*Sill*. Am. J. XXXVIII. p. 198.

Bring (Sven).

1. Dissertatio de Piscaturis in Oceano Boreali; resp. *C. Estenberg.*
Lond. Goth. 1750, 4°.—Dict. Sc. n. XXII. p. 514.

Bringier (L.).

1. Notices of the Geology, Mineralogy, Topography, Productions,
and aboriginal Inhabitants of the regions around the Mississippi
and its confluent waters.—*Sill*. Am. J. III. 1, p. 15.

2. Observations sur les Alluvions du Mississippi.—Gentl. Mag.1823,
Nov. p. 452.—Amer. Philos. Journ.—*Féruss*. Bull. 1824, I. p. 112.

Brink (Dietr.).

1. Prodromus è Norwegiâ, s. Descriptio Loufoudiæ.   Amst. 1676,
8°.—*Böhm*. Bibl. I. 1, p. 615.

Brinkmann (H. Ph.).

1. Sententia *Leoreti* de Partus Mechanismo.   Kiliæ, 1839, 4°.

Brisseau (   ).

1. Traité des Mouvemens sympathiques.   Mons, 1692, 12°.

Brisson (M. J.).

1. Dictionnaire raisonné de Physique.   Paris, 1781, 4°. 2 vols.—
Bibl. Ent. I. p. 50.

2. Regnum animale in Classes IX. distributum, sive, Synopsis methodica sistens generalem Animalium distributionem in Classes
IX., etc. (Le Règne animal, divisé en IX. Classes, ou Méthode
contenant la division générale des Animaux en IX. Classes, et la

division particulière des deux premières classes, savoir de celle des Quadrupèdes et de celle des Cétacées en ordres, sections, genres et espèces, etc. (Gall. et Lat.). Lyon, 1742, 8º.—Par. 1756, 4º. fig.—*Böhm.* Bibl. II. 2, p. 40.

3. Ornithologia, sive Synopsis methodica, etc. (Gall. et Lat.). Paris, 1760, 6 vols. 4º. fig.; 2 vols. 8º. Lugd. Bat. 1763. Paris, 1788. —*Cuv.* R. An. III. p. 331.—*Böhm.* Bibl. II. 2, p. 41.

4. Observations sur une espèce de Limaçon terrestre dont le sommet de la coquille se trouve cassé sans que l'Animal en souffre. —Mém. Acad. Sc. 1759, p. 99, fig.—Tr. Linn. Soc. Lond. VII. p. 223.

5. Principes élémentaires de l'Histoire naturelle et chimique des Substances minérales. Par. 1797, 8º.—(Germ. *F. C. Drechsler* mit Anmerk. v. *J. B. Trommsdorff*): *Anfangsgründe der Naturgeschichte u. Chemie der Mineralien.* Mainz, 1798, 8º.—(Angl.) *Elements of the Chemical and Natural History of Mineral Substances.* Lond. 1800, 8º.

BRIVET (V.).

1. Nouveau Traité des Robes ou Nuances chez le Cheval, l'Ane et le Mulet, chez l'espèce Bovine et les petites espèces domestiques. Paris, 1844, 8º. fig.

BROC (D. F.).

1. Essai sur les Races humaines considérées sous les rapports anatomique et philosophique. Paris, 1836, 8º.— *Val.* Repert. II. p. 20.

2. Traité complet d'Anatomie descriptive et raisonnée. Paris, 1835, 2 vols. 8º.

3. Introduction à l'étude de l'Anatomie, ou l'Homme considéré en grand, sous le rapport des Appareils et des Fonctions. Paris, 1835, 8º. atl. 4º.

BROCCHI (G. B.).

1. Osservazioni naturali fatte all' Isola de' Ciclopi, e nella contigua spiaggia di Catania.—Bibl. Ital. No. 59, p. 217.—Biogr. Un. Suppl. LIX. p. 281.

2. Osservazioni naturali fatte in alcune parti degli Appennini nell' Abruzzo ulteriore.—Bibl. Ital. No. 42, p. 363.

3. Osservazioni naturali fatte al Promontorio Argentaro ed all' Isola del Giglio.—Bibl. Ital. Nos. 31-33.—Biogr. Un. Suppl. LIX. p. 280.

4. Observations naturelles faites sur la montagne de Sila dans la Calabre intérieure.—Mem. Istit. Lomb.-Ven. III. p. 133.—*Féruss.* Bull. 1826, IX. p. 148.—*Leonh.* Tasch. 1827, I. p. 338 (*Beobachtungen über den Sila-Berg*).

5. Viaggio al Capo Circeo, ed Osservazioni naturali in quei contorni. —Bibl. Ital. N° 20, 21.—Biogr. Un. Suppl. LIX. p. 280.

6. Lettre et fragment sur l'Arabie et la Syrie.—Giorn. di Fis. etc. Mars et Avril 1824, p. 136.—*Feruss.* Bull. 1824, III. p. 325.

7. Catalogo di una serie di Conchiglie raccolte presso la Costa Affricana del Golfo Arabico, etc.—Bibl. Ital. N° 70 et 71.—Biogr. Un. Suppl. LIX. p. 281.

8. Intorno a delle Conchiglie mar. rinvenute nel peperino di Albano. —Bibl. Ital. N° 30, p. 424.—Biogr. Un. Suppl. LIX. p. 280.

9. Ragguaglio di alcuni Molluschi e Zoof. del mar. Tirreno presso la costa Romana, comun. al Sign. *Renieri,* etc.—Bibl. Ital. N° 39, 40.—Biogr. Un. Suppl. LIX. p. 280.

10. Descrizione di una nuova Conchiglia bivalve della costa del Brasile, con osserv. di alcuni alteri Testacei.—Bibl. Ital. N° 23, p. 276.—Biogr. Un. Suppl. LIX. p. 280.

11. Catalogo ragionato di una Raccolta di Rocce, disposto con ordine geografico per servire alla Geognosia d'Italia. Milano, 1817, 8°.—Biogr. Un. Suppl. LIX. p. 280.

12. Sulle diverse formazioni di Rocce della Sicilia.—Bibl. Ital. N° 69, p. 357.—Biogr. Un. Suppl. LIX. p. 281.

13. Sulle geognostiche relazioni delle Rocce calcarie e vulcaniche in Val di Noto, nella Sicilia.—Bibl. Ital. N° 79, p. 53.—Biogr. Un. Suppl. LIX. p. 281.

14. Osservazioni geologiche su'i contorni di Reggio in Calabria, etc.—Bibl. Ital. N° 55, p. 69.—Biogr. Un. Suppl. LIX. p. 281.

15. Osservazioni geologiche fatte nella Terra di Otranto.—Bibl. Ital. N° 52, p. 52.—Biogr. Un. Suppl. LIX. p. 281.

16. Osservazioni sulle Montagne metallifere della Tolfa.—Bibl. Ital. N° 26, p. 192.—Biogr. Un. Suppl. LIX. p. 280.

17. De' colli Iblei in Sicilia.—Bibl. Ital. N° 75, p. 75.—Biogr. Un. Suppl. LIX. p. 281.

18. Sull' eruzione del Vesuvio del 1812.—Bibl. Ital. N° 17, p. 275. —Biogr. Un. Suppl. LIX. p. 280.

19. Descrizione del Monte Soratte.—Bibl. Ital. N° 73, p. 74.—Biogr. Un. Suppl. LIX. p. 281.

20. Osservazioni naturali sulle Spelonche di Adelsberg in Carniola. —Bibl. Ital. N° 74 e 75.—Biogr. Un. Suppl. LIX. p. 281.—*Féruss.* Bull. 1823, I. p. 305 (*Observations sur les Grottes d'Adelsberg en Carniole*).

21. Osservazioni sulla corrente di Lava di Capo di Bove, presso Roma, etc.—Bibl. Ital. N° 19, p. 102.—Biogr. Un. Suppl. LIX. p. 280.

22. Dello stato fisico del Suolo di Roma. (*De l'état physique du Sol de Rome ; Mémoire pour servir d'explication à la carte géognostique de cette ville.*) Rome, 1820, 8°. fig.—*Féruss.* Bull. 1824, III. p. 4.—*Leonh.* Zeitschr. 1826, I. p. 542 (*Ueber den Boden der Gegend um Rom*).

23. Intorno ad uno scavo interessante la Geognosia, fatto in Roma a Campo Vaccino.—Bibl. Ital. N° 37, p. 114.—Biogr. Un. Suppl. LIX. p. 280.

24. Conchiologia fossile subapennina, con Osservazioni geologiche sugli Apennini e sul suolo adjacente. Milano, 1814 ; 1843, 2 vols. 4°. fig.— *Cuv.* R. An. III. p. 342.

25. Intorno alle Conchiglie fossili del Piemonte; lettere in rip. a quella del *Deluc.*—Bibl. Ital. N° 40, p. 282.—Biogr. Un. Suppl. LIX. p. 281.

26. Trattato mineralogico e chimico sulle Miniere di Ferro del dipartimento del Mella, con l'esposizione della costituzione fisica delle Montagne metallifere della Val-Trompia. 2 vols. 8°. Brescia, 1807 e 1808.—Cat. R. Soc. L.

27. Memoria mineralogica sulla Valle di Fassa in Tirolo. Milano, 1811, 8°.—*Leonh.* Tasch. 1823, II. p. 439 (*Allgemeine geologische Ansichten,* etc.).

28. Sulla Lignite di Valgandino.—Giorn. Soc. d'Incorrag. Milano, 1809.

29. Viaggi litologici nella Campania. Firenze, 1798.

30. Ragguaglio de' Manoscritti e della Raccolta di Minerali e di Piante lasciati dal defunto *Brocchi,* da *G. Acerbi.*—Bibl. Ital. Avr. 1828, p. 80 ; Mai, p. 208.—*Féruss.* Bull. XVI. p. 210.

BROCH (J. K.).

1. Correspondance entomologique. 8°. Mühlhausen, 1823.

2. Entomologische Briefe. Französisch und Deutsch. Mühlh. 1823, 8°. fig.—*Eis.* Insect. p. 145.

BROCHANT DE VILLIERS (A. J. M.).

1. Mémoire sur les Montagnes primitives.—Acad. Sc. 1816.—Isis, 1817, VI. p. 682.

2. Considérations sur la place que les Granits du Mont-Blanc, etc. doivent occuper dans l'ordre des Roches primitives.—Acad. Sc. 1816.—Isis, 1817, VI. p. 708.

3. Ueber die mineralogischen und geologischen Verschiedenheiten der granitartigen Felsen des Mont-Blanc, etc.—Bull. Sc. 1816, p. 86.—Isis, 1818, VII. p. 1160.

4. Ueber zwei Abhandlungen von *De Buch* : Ueber die Ursachen, welche die Blöcke von Urgebirgsarten auf den Jura geführt, und über den Trapp-Phosphor.—Par. Verh. 1817.—Isis, 1818, VIII. p. 1285.

5. Traité élémentaire de Minéralogie suivant les principes du *Prof. Werner,* etc. 2 vols. 8°. Paris, 1801–1803 ; 1808, 8°. fig. —Cat. R. Soc. L.

6. Notice sur la Carte géologique générale de la France.—C. R. I. p. 423.

7. Uebergangsgyps der Alpen.—Bull. Soc. Philom. 1816, Avr. p. 61. —*Leonh.* Tasch. 1817, I. p. 220.

8. Ueber den Gyps von Val Canaria von *Lardy.*—Ann. des Mines, 1817, I. p. 55 ; III. p. 257.—*Leonh.* Tasch. 1819, II. p. 572.

9. Rapport sur la partie géologique et minéralogique de l'Histoire physique des Antilles, de *M. Moreau de Jonnès.*—Globe, 5 Janv. 1826.—*Féruss.* Bull. VII. p. 185.

10. Geognostische Beobachtungen über das Uebergangs-Gebirge in Tarentaise und in andern Theilen der Alpen-Kette.—Journ. des Mines, No. 137, p. 321.—*Leonh.* Tasch. 1817, I. p. 59.

11. Alters-Verhältnisse der granitischen Felsarten des Mont-Blanc und anderer erhabenen Alpen-Gipfel. — Ann. des Mines, IV. p. 282.—*Leonh.* Tasch. 1823, III. p. 607.

BROCHANT DE VILLIERS ( ), DUFRÉNOY ( ) et BEAU-
MONT (L. Elie de).

1. Sur les Mines de Plomb du Cumberland et du Derbyshire. 1e partie: Gisement des Minérais.—Ann. d. Mines, 1826, II. p. 339. —*Féruss.* Bull. 1827, X. p. 22.—*Leonh.* Zeitschr. 1827, I. p. 270 (*Ueber die Blei-Gruben in Cumberland u. Derbyshire*).

BROCHMAND (Er. J.).

1. De causis rerum naturalium intrinsecis. Hafn. 1644, 4°.—*Böhm.* Bibl. I. 1, p. 242.

BROCKÆUS (Marc.).

1. Dissert. de Animâ Brutorum. Erf. 1715, 4°.—*Böhm.* Bibl. II. 1, p. 93.

BROCKEDON (W.).

1. On a Method of forming the Dust of Graphite into a Solid Mass. —Journ. Geol. Soc. I. 31.

Brockhoff (E.).

1. De Circuitu Sanguinis in Fœtu humano.—Berol. 1839, 8°.

Brockmann (H.).

1. De Pancreate Piscium.   4°. Paris, 1846.

Brockwell (Ch.).

1. Natural and Political History of Portugal.   London, 1726, 8°.— *Böhm*. Bibl. I. 1, p. 497.

Broderip (W. J.).

1. Characters of new species of Mollusca and Conchifera.—Proc. Com. Zool. Soc. II. pp. 25, 50, 104, 124, 173, 194.—Proc. Zool. Soc. I. pp. 4, 52, 82 ; II. pp. 2, 13, 35, 47, 114, 148 ; III. pp. 41, 192 ; VIII. pp. 83, 94, 119, 155, 180 ; IX. pp. 22, 34, 36, 44 ; X. p. 53.—Ann. and Mag. N. H. VII. pp. 226, 335, 546 ; VIII. pp. 62, 148, 381, 466, 527, 536 ; XI. p. 311.—*Müller*, Synopsis Testaceorum, 1836, p. 238.— *Wiegm*. Arch. 1836, II. p. 211 ; 1837, II. p. 271.—Isis, 1835, IV. pp. 371, 375.—Revue Zool. 1838, p. 154.

2. Descriptions of some new species of *Cuvier's* Family of Brachiopoda.—Trans. Zool. Soc. I. p. 141.—Proc. Zool. Soc. I. p. 124. —Ann. Sc. n. III. p. 26.—Isis, 1835, II. p. 143.— *Wiegm*. Arch. 1835, I. p. 317.

3. On a new genus of *Gasteropoda* (*Scutella*).—Proc. Zool. Soc. II. p. 47.— *Wiegm*. Arch. 1835, I. p. 322.

4. On *Clavagella*, with characters of new species.—Proc. Zool. Soc. Lond. II. p. 115.—Tr. Zool. Soc. I. p. 261, fig.— *Wiegm*. Arch. 1835, I. p. 312.—Rev. Zool. 1838, p. 154.

5. Descriptions of new species of *Calyptraeidae*.—Proc. Zool. Soc. II. pp. 13, 35.—Tr. Zool. Soc. I. p. 195, fig.— *Wiegm*. Arch. 1835, I. p. 323 ; 1836, II. p. 213.—Rev. Zool. 1838, p. 149.

6. Observations on the Habits, etc. of a male Chimpanzee (*Troglodytes niger*, Geoff.) living in the Menagerie of the Zoological Society of London.—Zool. Proc. Part III. p. 160.—Lond. and Ed. Phil. Mag.— *Wiegm*. Arch. 1836, II. p. 276.

7. On some new species of Shells belonging to the genera *Spondylus*, *Voluta*, *Conus*, *Purpura* and *Bulinus*.—Proc. Zool. Soc. IV. p. 43.

8. On some species of *Bulinus*.—Proc. Zool. Soc. IX. p. 14.

9. Hints for collecting Animals and their Products.   8°.

10. The Family of the Cowries (*Cypraeidae*)

11. The Family of the Strombs (*Strombidæ*).

12. On some Fossil Crustacea and Radiata found at Lyme Regis, in Dorsetshire.—Trans. Geol. Soc. ser. 2, V. p. 171.—Proc. Geol. Soc. II. p. 201.—L. and Ed. Phil. Mag. VII. p. 517.—*L.* u. *Br.* N. Jabrb. 1836, p. 739.

13. On a new Cowry and other Testacea.—Zool. Journ. V. p. 330.

14. Observations upon *Volvox Globator*.—Zool. Journ. V. p. 51.

15. On the Manners of a live Toucan exhibited in this country.—Zool. Journ. I. pp. 484, 591.—Isis, 1828, IX. p. 939.

16. On the Utility of preserving Facts relative to the Habits of Animals, with additions to two Memoirs in "White's Natural History of Selborne."—Zool. Journ. II. p. 14.—Isis, 1830, VIII. p. 830.—*Féruss.* Bull. 1826, VIII. p. 243.

17. On some new and rare Volutes.—Zool. Journ. II. p. 27.—*Féruss.* Bull. 1826, VII. p. 447.

18. On two new species of Shells from the Mauritius.—Zool. Journ. II. p. 198.—*Féruss.* Bull. 1826, VIII. p. 134.

19. On the mode in which the Boa Constrictor takes its Prey, and on the adaptation of its organization to its habits.—Zool. Journ. II. p. 215.—Isis, 1828, IX. p. 941.

20. On some new and rare Shells.—Zool. Journ. III. p. 81.

21. On a Fossil Volute from the Montagne de St. Pierre near Maestricht.—Zool. Journ. III. p. 234.—*Féruss.* Bull. 1828, XV. p. 175.

22. On the Jaw of a Fossil Mammiferous Animal found in the Stonesfield Slate.—Zool. Journ. III. p. 408.—*Féruss.* Bull. 1828, XIV. p. 109.

23. On the Animals hitherto found in the Shells of the genus *Argonauta*.—Zool. Journ. IV. pp. 57, 224.—Isis, 1830, XII. p. 1247.—*Féruss.* Bull. 1829, XVII. p. 130.

24. On a new species of *Cypræa*.—Zool. Journ. IV. p. 163.—*Féruss.* Bull. 1829, XIX. p. 390.

25. On the Habits and Structure of *Paguri* and other Crustacea.—Zool. Journ. IV. p. 200.—Isis, 1830, XII. p. 1256.—*Féruss.* Bull. 1830, XX. p. 342.

26. On a new Land-Shell from South America.—Zool. Journ. IV. p. 222.—*Féruss.* Bull. 1829, XIX. p. 389.

27. On two new species of *Buccinum* from the English and Irish seas.—Zool. Journ. V. p. 44.

28. On the genus *Chama*, with descriptions of some species apparently not hitherto characterized.—Trans. Zool. Soc. I. p. 301.

29. Introduction to the Monographs of the genera *Cymba, Melo* and *Voluta*; and Monograph of *Cymba.—Sowerby,* Species Conchyliorum, etc. London, folio.

30. Table of the Situations and Depths at which recent genera of Marine and Estuary Shells have been observed.—Appendix to "Researches in Theoretical Geology:" by *H. T. De la Beche,* F.R.S. etc. London, 12°.

31. Notice on *Mus messorius,* Shaw (Less Long-tailed Field Mouse of *Pennant*).—Zool. Journ. V. p. 496.

32. Note on *Caryophyllia.*—Zool. Journ. III. p. 485.

33. Account of the Manners of a tame Beaver.—"The Gardens and Menagerie of the Zoological Society." London, 8°. I. p. 167.

34. Zoological articles in the " Penny Cyclopædia " from A s t to the end, including the whole of those relating to Mammals, Birds, Reptiles, Crustacea, Mollusca, Conchifera, Cirrigrada, Pulmograda, etc., *Buffon, Brisson,* etc., and Zoology.

35. Zoological Recreations. 8°. London, 1847.

BRODERIP (W. J.) and KING (Capt.).

1. Descriptions of the Cirripeda, Conchifera and Mollusca in a collection formed by the Officers of H.M.S. Adventure and Beagle, 1826–30, in surveying the Southern Coast of South America, etc. —Zool. Journ. V. 1832–34, p. 332.

BRODERIP (W. J.) and SOWERBY (G. B.).

1. On new or interesting Mollusca.—Zool. Journ. IV. p. 359; V. p. 46.—Isis, 1831, II. pp. 105, 220.—*Féruss.* Bull. 1829, XIX. p. 378.

BRODIE (Benj.).

1. Observations sur l'action de la Bile dans la Digestion.—Bull. Soc. Phil. 1823, p. 58.—*Féruss.* Bull. I. p. 408.

BRODIE (P. B.).

1. History of the Fossil Insects in the Secondary Rocks of England, accompanied by a particular account of the strata in which they occur. ·8°. London, 1845.—Journ. Geol. Soc. I. p. 399.

2. On the existence of Purbeck strata with remains of Insects and other Fossils, at Swindon, Wilts.—Journ. Geol. Soc. III. p. 53.

3. On the occurrence of Plants in the Plastic Clay of the Hampshire Coast.—Proc. Geol. Soc. III. p. 592.—Phil. Mag. ser. 3, XXI. p. 546.

4. On the Discovery of the Remains of Insects, and a new genus of Isopodous Crustacea belonging to the family *Cymothoidæ*, in the Wealden formation in the Vale of Wardour, Wilts.—Proc. Geol. Soc. III. p. 134.—Phil. Mag. ser. 3, XV. p. 534.—*L.* u. *Br.* N. Jahrb. 1843, p. 238.

5. Notice on the Discovery of Insects in the Wealden of the Vale of Aylesbury, Bucks, with some Observations on the Distribution of these and other Fossils in the Vale of Wardour, Wiltshire.— Proc. Geol. Soc. III. p. 780.—Ann. and Mag. N. H. XI. p. 480. —Phil. Mag. ser. 3, XXIII. p. 512.

6. Notice of the Discovery of the Remains of Insects in the Lias of Gloucestershire, with some remarks on the lower members of this formation.—Proc. Geol. Soc. IV. p. 14.—Ann. and Mag. N. H. XI. p. 509.—*L.* u. *Br.* N. Jahrb. 1844, p. 127.—Phil. Mag. ser. 3, XXIII. p. 529.—Rep. Brit. Assoc. 1842, Sect. p. 58.

BRODIE (P. B.) and BUCKMAN (J.).

1. On the Stonesfield Slate of the Cotteswold Hills.—Journ. Geol. Soc. I. p. 220.

BRŒCKHUYSEN (Benj.).

1. Œconomia Corporis animalis, sive Cogitationes succinctæ de Mente, Corpore et utriusque conjunctione, juxta methodum Cartesianam. Amst. 1683, 4°.—Act. Erudit. II. p. 341.

BROERS (J. C.).

1. Observationes anatomico-pathologicæ. Lugd. Bat. 1839, fol.

BROGANTI (Jos.).

1. Instruction familière sur l'éducation des Vers à soie. Gen. 1775, 8°.—*Böhm.* Bibl. II. 2, p. 263.

BROGIANI (Dom.).

1. De veneno Animantium naturali et acquisito Tractatus. Florent. 1755, 4°.—*Böhm.* Bibl. II. 2, p. 147.

BROMEL (Ol.).

1. Catalogus generalis, seu Prodromus Indicus specialior rerum curiosarum, tam artificialium quam naturalium, quæ servantur in Pinacothecâ *Olai Bromellii.* Gothenb. 1698, 4°.—Biogr. Un. VI. p. 21.

2. De Lumbricis terrestribus, illorumque in Medicinâ proprietatibus atque recto usu. Lugd. Bat. 1673, 4°.—Biogr. Un. VI. p. 21.

3. *Cloris gothica.*   Gothob. 1794, 8°.—Tr. Linn. Soc. Lond. V.
p. 279.

BROMELL (Magn. v.).

1. Mineralogia, eller Inledning til nödig Kundskap om Bergarter,
Mineralier, Metaller, samt Fossilier, etc. Stockh. 1730, 8°; 1739,
8°.—*Böhm.* Bibl IV. 1, p. 47.—(Germ. *Mikfandern*): Mineralogia et Lithographia Suecana; d. i. Abhandlung derer im Königr.
Schweden befindlichen Mineralien u. Steine. Aus d. Schw. übers.
durch *Mikfandern*; etc.   Stockh. u. Leipz. 1740, 8°. fig.—*Böhm.*
Bibl. IV. 1, p. 152.—*Mod.* B. Helm.

2. Lithographiæ Suecanæ Specimen I. et II.   Ups. 1726–1727, 4°.
fig.—8°. Holm. 1740.—*Böhm.* Bibl. IV. 1, p. 151.

3. Commentatio de Petrificatis Suecanis.—Act. Upsal. 1729–1730.

BROMHEAD (Sir E. F.).

1. On Zoological Classification.—Mag. Nat. Hist. ser. 2, II. pp. 412,
484.

BROMME (T.).

1. Zonengemälde.   4°. Stuttg. 1846.

BRONGNIART (Ad.).

1. Zeugung und Entwickelung des Embryo bei den Phanerogamen.
—Ann. Sc. n. XII. pp. 14, 145, 225.—Isis, 1834, IX. pp. 944,
947, 954.

2. Neue Untersuchungen über den Blüthenstaub und die Befruch-
tungskörnchen der Gewächse.—Ann. Sc. n. XV. p. 381.—Isis,
1834, X. p. 1036.

3. Note sur la composition de l'Atmosphère à diverses époques de
la formation de la Terre, et sur l'opinion de M. le professeur *Parrot* relative à ce sujet.—Ann. Sc. n. 1830, XX. p. 427.

4. Sur la classification et la distribution des Végétaux fossiles en
général, et sur ceux des Terrains de Sédiment supérieur en particulier.— Bull. Soc. Phil. 1822, p. 25.—Mém. Mus. VIII. pp. 203,
297.—*Féruss.* Bull. 1823, I. p. 252.

5. Considérations générales sur la nature de la Végétation qui cou-
vrait la surface de la Terre aux diverses périodes de la formation
de son écorce.—Ann. Sc. n. Nov. 1828.—*Féruss.* Bull. 1829,
XVII. p. 1.—Bull. Soc. Géol. VI. p. 96.—C. R. V. p. 403.—*Sill.*
Am. Journ. XXXIV. 2, p. 215.—Mag. Nat. Hist. ser. 2, II. p. 1.

6. Histoire des Végétaux fossiles, ou Recherches botaniques et géologiques sur les Végétaux renfermés dans les diverses couches du Globe. 4º. Par. 1828–1843, 2 vols. 4º. fig.—Ann. Sc. n. 1829, XVIII.—Rev. Bibl. p. 129.

7. Observations sur les Arborisations dans les Calcédoines dendritiques.—Ann. Sc. n. 1829, XVI.—Rev. Bibl. p. 7.

8. Essai d'une Flore du Grès-bigarré.—Ann. Sc. n. 1828, X. p. 435. —*Féruss.* Bull. XVIII. p. 193.—Isis, 1834, X. p. 1038 (*Versuch einer Flora des bunten Sandsteins*).

9. Sur les Plantes fossiles d'Armissan, près Narbonne (*Fossile Pflanzen im Mergel bei Armissan in der Gegend von Narbonne*). —Ann. Sc. n. XV. p. 43.—Isis, 1834, X. p. 1031.—*Féruss.* Bull. XXV. p. 10.

10. Note sur les Végétaux fossiles de l'Oolite à Fougères de Mamers.—Ann. Sc. n. 1825, IV. p. 417.—*Féruss.* Bull. 1826, VII. p. 237.

11. Notice sur un Conifère fossile du terrain d'eau douce de l'île d'Iliodroma.—Ann. Sc. n. 1833, XXV. p. 138.—*L.* u. *Br.* N. Jahrb. 1834, p. 240 (*Note über eine fossile Conifere des Süsswassergebildes der Insel Iliodroma*).

12. Note sur la présence du *Pecopteris reticulata* dans des Couches de formation contemporaine en Angleterre et en France.—Ann. Sc. n. 1828, XIII. p. 335.—*Féruss.* Bull. XIV. p. 437.

13. Notice sur des Végétaux fossiles traversant les couches du Terrain houiller. Paris, 1821.—Isis, 1822, V. Litt. Anz. p. 145.— Sill. Am. Journ. IV. 2, p. 266 (*Notice on Vegetable Fossils which traverse the Layers of Coal formation*).—Bull. Soc. Géol. Fr. VI. p. 95.

14. Observations sur les Végétaux fossiles des terrains d'Anthracite des Alpes.—Ann. Sc. n. XIV. p. 127.

15. Observations sur les Végétaux fossiles renfermés dans les Grès de Hoer en Scanie.—Ann. Sc. n. 1825, IV. p. 200.—Bull. Soc. Géol. VI. p. 96.—*Féruss.* Bull. 1826, VII. p. 236.

16. On Fossil Organic Remains as a geognostic character.—Edinb. Phil. Journ. VIII. p. 226.

17. Résumé des travaux de la Société d'Hist. nat. de Paris pendant l'année 1821. 8º. Paris, 1823.

18. Prodrome d'une Histoire des Végétaux fossiles. 8º. Paris, 1828. —Cat. R. Soc. L.

19. Notice sur un Gisement de Végétaux fossiles et de Belemnites, situé à Petit-Cœur, près Moutiers. 1828, 8º.

20. On some Fossil Vegetables of the Coal Formation, and on their

relations to living Vegetables.—Ann. des Sci. nat. Jan. 1825, p. 23.
—Edinb. New Phil. Journ. I. p. 282.

BRONGNIART (Ad.) and GÖPPERT (H. R.).

1. On the Distribution of Fossil Plants in the different Geological
Formations.—C. R. 1845.—Ediñb. New Phil. Journ. XL. p. 285.

BRONGNIART (Alex.).

1. Sur les Terrains calcaréo-trappéens du pied méridional des Alpes
Lombardes.—Bull. Soc. Phil. 1821, p. 87.—*Leonh.* Zeitschr. 1825,
II. p. 530 (*Ueber das Kalkige Trapp-Gebilde am südlichen Fusse
der Lombardischen Alpen*).

2. Sur quelques Terrains d'eau douce de la Suisse et de l'Italie,
propres à établir la théorie de ces Terrains.—Bull. Soc. Phil.
1822, p. 17.

3. Sur le gisement ou position relative des Ophiolithes, Euphotides,
Jaspes, etc., dans quelques parties des Apennins. Paris, 1821, 8°.
—Bull. Soc. Philom. 1820, p. 174.—Isis, 1826, VII. p. 696.—
Bull. Soc. Géol. Fr. VI. p. 95.

4. Notes sur les Coquilles qui se trouvent dans les Terrains décrits
par *M. Studer*, sur les époques géognostiques qu'elles indiquent,
et sur la Montagne des Diablerets au N.E. de Bex.—Ann. Sc. n.
Juillet 1827, p. 266.—*Féruss.* Bull. 1828, XIII. p. 383.

5. Sur l'existence des Fossiles microscopiques dans les Roches en
apparence homogènes.—Ann. Sc. n. 1836, VI. p. 56.

6. *Strombites denticulatus.*—Ann. d. Mines, IV. fig.

7. On the Lignite or Plastic Clay Formation of Mount Meissner.—
Phil. Mag. ser. 2, II. p. 107.

8. Essai d'une classification naturelle des Reptiles. Paris, 1805, 4°.
—*Cuv.* R. An. III. p. 342.—Bull. Soc. Phil. 3ᵉ ann. II. pp. 81, 89.

9. Sur un nouveau genre d'Insecte des environs de Paris (*Dasyce-
rus*).—Bull. Soc. Phil. 4ᵉ ann. II. p. 115.—Bibl. Entom. I. p. 50.
—*Wiedem.* Arch. III. 1, 1802, p. 190.

10. Description d'une nouvelle espèce de *Lamia.*—Bull. Soc. Phil.
7ᵉ ann. p. 34, fig.—Bibl. Ent. I. p. 51.

11. Mémoire sur le *Limnadia*, nouveau genre de Crustacés.—Mém.
Mus. VI. p. 83.—Isis, 1822, II. p. 212.

12. Tableau des Terrains qui composent l'écorce du Globe, ou Essai
sur la Structure de la partie connue de la Terre. Paris, 1829, 8°.
—*Féruss.* Bull. 1829, XIX. p. 153 ; 1830, XXI. p. 3.—N. Act.
Nat. Cur. XIV. p. xiv.—(Germ. *Kleinschrod*): *Die Gebirgsfor-*

mationen der Erdrinde, oder Versuch über d. Struktur des bekann-ten Theils unsers Erdbodens. Strasb. 1830, 8°.

13. Description géologique des Couches des environs de Paris. (*Cuvier*, Recherches sur les Ossemens fossiles, II.)—*Bronn*, Le-thæa, II. p. 547.

14. Classification et caractères minéralogiques des Roches homo-gènes et hétérogènes. Paris, 1827, 8°.—*Féruss.* Bull. 1817, XII. p. 201.—Bull. Soc. Géol. Fr. VI. p. 95.

15. Notice sur les blocs de Roches des Terrains de transport en Suède—Ann. Sc. n. 1828, XIV. p. 5.—*Féruss.* Bull. 1829, XVIII. p. 151.

16. Rapport fait à l'Académie des Sciences sur le Mémoire de *M. Constant Prévost,* ayant pour titre : Géologie des Falaises de la Normandie.—Ann. Sc. n. 1824, I. p. 195.—*Féruss.* Bull. 1824, II. p. 2.

17. Notice sur les Brèches osseuses et les Minérais de fer pisiforme de même position géognostique.—Ann. Sc. n. 1828, XIV. p. 410. —N. Act. Nat. Cur. XIV. p. xiv.—Bull. Soc. Géol. Fr. VI. p. 96.

18. Observations additionelles à la Notice sur les Minérais de fer pisiforme de position analogue à celle des Brèches osseuses.— —Ann. Sc. n. 1828, XVI. p. 89.—Bull. Soc. Géol. Fr. VI. p. 96.

19. De l'Arkose ; caractères minéralogiques et histoire géognostique de cette Roche.—Ann. Sc. n. 1826, VIII. p. 113.—*Féruss.* Bull. 1827, XII. p. 2.—N. Act. Nat. Cur. XIII. p. xxvi.—Isis, 1834, VIII. p. 864 (*Ueber die Arkose*).

20. Rapport sur deux Mémoires de *M. Virlet,* relatifs à la Géologie de la Messénie, et notamment à celle des environs de Modon et de Navarin.—Ann. Sc. n. 1830, XIX. p. 259.

21. Rapport fait à l'Académie royale des Sciences, sur les travaux géologiques de *M. Gay.*—Ann. Sc. n. 1832, XXVIII. p. 394.— *L. u. Br.* N. Jahrb. 1834, p. 238 (*Bericht an die K. Ak. d. Wis-sensch. über* Gay's *geologische Arbeiten*).

22. Rapport sur un Mémoire relatif à la Formation des Montagnes. —Ann. Sc. n. 1829, XVIII.—Rev. Bibl. p. 128.

23. Note sur la présence de la Webstérite dans l'Argile plastique d'Auteuil près Paris.—Ann. Sc. n. 1828, XIII. p. 225.

24. Des caractères particuliers que présente le Terrain de Craie dans le Sud de la France et sur les pentes des Pyrénées.—Ann. Sc. n. 1831, XXIII.—Rev. Bibl. p. 54.

25. Mémoire sur des Terrains qui paraissent avoir été formés sous l'eau douce.—Ann. Mus. XV. p. 357, fig.—Bull. Soc. Géol. Fr. VI. p. 95.

26. Sur les caracteres zoologiques des Formations, avec l'application de ces caractères à la détermination de quelques terrains de Craie.—Bull. Soc. Phil. 1822, p. 7.—*Leonh.* Tasch. 1824, III, p. 667 (*Ueber die zoologischen Merkmale der Formazionen*).

27. Miscellaneous Observations relating to Geology, Mineralogy, and some connected topics.—*Sill.* Am. Journ. III. 2, p. 216.

28. Géologie de la Morée.—Ann. Sc. n. XXII.—Rev. Bibl. p. 16.

29. Notice concerning the Method of collecting, labelling, and transmitting Specimens of Fossil Organized Bodies, and of the accompanying Rocks.—*Sill.* Am. Journ. I. 1, p. 71.

30. On the Freshwater Formations of Italy, posterior to the Coarse Limestone.—Edinb. Phil. Journ. VIII. p. 92.

31. On the Freshwater Formation of the Environs of Rome.—Phil. Mag. ser. 2, II. p. 172.

32. Premier Rapport des Expériences faits d'après l'*Abbé Spallanzani* sur la Génération des Grenouilles. 1792, 8°.

33. Note sur la présence de quelques Métaux dans les Grès supérieurs du Terrain de Paris.—C. R. II. p. 221.

34. Tableau théorique de la succession et de la disposition des Terrains et Roches qui composent l'écorce de la Terre.

35. Introduction à la Minéralogie, ou Exposé des Principes de cette Science et de certaines Propriétés des Minéraux. 8°. Paris, 1825.

36. Mémoires sur les Terrains de sédiment supérieurs calcaréo-trappéens du Vicentin, et sur quelques Terrains d'Italie, de France et d'Allemagne qui peuvent se rapporter à la même époque. Paris, 1823, 4°. fig.—*Feruss.* Bull. 1823, I. p. 218.

37. Traité élémentaire de Minéralogie, avec des applications aux Aits. 8°. 2 vols. Paris, 1807.

38. Premier Mémoire sur les Kaolins ou Argiles à Porcelaine, sur la nature, le gisement, l'origine et l'emploi de cette sorte d'Argile. —C. R. VII. p. 1085.—Par. 1839, 4°.—Arch. Mus.—*L. u. Br.* N. Jahrb. 1839, p. 484 (*Ueber Zusammensetzung u. Charaktere der Porzellan-Thone*).

39. Essai sur les Orbicules siliceux et sur les formes à surfaces courbés qu'affectent les Agates et les autres Silex. Paris, 1831, fig.—Bull. Soc. Géol. Fr. II. p. 65.

40. Notice pour servir à l'Histoire géognostique du Cotentin, suivie de quelques Considérations sur la Classification géologique des Terrains.—Bull. Soc. Géol. Fr. VI. p. 95.—*Leonh.* Tasch. 1816, II. p. 534 (*Geognostische Nachrichten über den Cotentin, etc.*).

41. Des Volcans et des Terrains volcaniques.—Dict. Sc. nat. 1829.

42. Rapport sur l'Essai sur la Constitution géognostique du Bassin de Vienne en Autriche, par *M. Const. Prévost.* 4°. fig.

43. Notice sur la Magnésie du Bassin de Paris. Par. 1822, 8°. fig.

44. Tableau de la Distribution méthodique des Espèces minérales. Paris, 1828 ; 1833, 8°.

45. Lettre sur l'existence de Dépouilles fossiles très-abondantes d'Animaux Infusoires dans les Tripolis provenant de localités différentes.—C. R. III. p. 31.

46. Catalogue des Mammifères, envoyés de Cayenne par le Blond. —Actes de la Soc. d'Hist. nat. de Paris, I. p. 115.

47. Lagerungs-Verhältnisse von Serpentin, Gabbro, Jaspis, u. s. w. in einigen Theilen der Apenninen.—Ann. des Mines, 1821.— *Leonh.* Tasch. 1822, III. p. 731.

48. Rapport sur un Mémoire de *M. Provana de Collegno,* intitulé : Sur les Terrains tertiaires du Nord-Ouest de l'Italie.—C. R. VII. p. 232.

49. Rapport sur un travail de *M. Fournet,* concernant les Filons métallifères et le Terrain des environs de l'Arbresle.—C. R. V. p. 51.

50. Versuch einer mineralogischen Klassifikation der gemengten Gebirgsarten.—J. d. Mines, No. 199, p. 5.—*Leonh.* Tasch. 1815, II. p. 378.

BRONGNIART (Al.) et BEAUMONT (Elie de).

1. Voyages en Scandinavie, etc. Observations sur le Phénomène diluvien dans le Nord de l'Europe.—Paris, 1840, 8°.

BRONGNIART (Al.) et COQUEBERT ( ).

1. De la formation de la Coquille du *Strombus Fissurella,* et sur des espèces analogues.—Bull. Soc. Phil. No. XXV.

BRONGNIART (Al.) et CUVIER (Fr.).

1. Dent fossile d'Hippopotame.—Ann. Sc. n. 1829, XVIII.—Rev. Bibl. p. 134.—*Müll.* Arch. 1835, p. 45 (*Ueber den gefundenen Unterkiefer des fossilen* Hippopotamus medius).

2. Rapport fait à l'Acad. des Sciences sur un Mémoire de *M. de Christol* ayant pour objet de ramener au genre Dugong les débris fossiles que *M. G. Cuvier* avait rapprochés des Hippopotames. —Ann. Sc. n. 1834, I. p. 282.—*L. u. Br.* N. Jahrb. 1835, p. 369

(*Bericht über* Christol's *Abh. über die Zurückführung der fossilen Reste, etc. zum Geschlechte Dugong*).

BRONGNIART (Al.) et DE LA BÈCHE (H. T.).

1. Collection de Mémoires géologiques tirés des Annales des Mines, avec une Table synoptique des Formations équivalentes et la Classification des Roches mélangées. Lond. 1828, 8°. fig.—*Féruss.* Bull. 1830, XX. p. 10.

BRONGNIART (Al.) et DESMAREST (A. G.).

1. Histoire naturelle des Trilobites et des Crustacés fossiles. Par. 1822, 4°. fig.—Bull. Soc. Phil. 1822, pp. 59, 73.

BRONGNIART (Al.) et TIGNY (F. M. G. de).

1. Hist. Nat. des Insectes, dans l'éd. de *Buffon* par *Castel.* Paris, 1799–1802, 18°. 10 vols.—Bibl. Ent. I. p. 51.

BRONN (H. G.).

1. Observations ajoutées à un extrait du Tableau des Corps organisés de *M. Defrance.*—Zeitschr. f. Mineral. 1826, I. p. 41.— *Féruss.* Bull. 1827, XI. p. 3.

2. Briefe aus der Schweiz, etc. (*Lettres sur la Suisse, l'Italie et la France méridionale, écrites pendant l'été de* 1824.) Heidelb. et Leipz. 1826, 8°. fig.—*Féruss.* Bull. 1827, X. p. 322.

3. *Gæa Heidelbergensis,* oder Mineralogische Beschreibung der Gegend von Heidelberg. Heidelb. 1830, 12°.—Isis, 1833, VIII. p. 721.—Bull. Soc. Géol. Fr. I. p. 190.

4. Sur la Constitution géologique des Apennins de l'Italie supérieure.—Zeitschr. f. Mineral. III. 1828, p. 214.—Giorn. Ligust. I. et III. 1827, p. 122.—Antolog. 1827, III. p. 146.—*Féruss.* Bull. 1829, XVI. p. 37.

5. Essai géognostique sur les environs du Necker inférieur autour de Heidelberg, avec une carte géol. col.—Bad. Arch. 1827, II.— *Féruss.* Bull. 1829, XIX. p. 190.—Bull. Soc. Géol. Fr. II. p. 109.

6. Ueber gewisse Kreidebildungen an den Pyrenäen und im südlichen Frankreich.—Jahrb. 1837, III.—*Bronn,* Lethæa, II. p. 553.

7. *Lethæa geognostica,* oder Abbildungen und Beschreibungen der für die Gebirgs-Formationen bezeichnendsten Versteinerungen. Stuttg. 1835–1837, 8°. 2 vols. Atl. 4°.—2te Aufl. 1838.—3e Aufl. 1846.—*L. u. Br.* N. Jahrb. 1835, p. 238.

8. Ueber die Versteinerungen im Hippuriten-Kalke.—Jahrb. f. Mineral. 1837, p. 43.—Lethæa, II. p. 552.

9. Sur quelques collections de Pétrifications d'Italie.—Zeitschr. f. Mineral. VI. 1828, p. 417.—*Féruss.* Bull. 1829, XVII. p. 198.

10. *Testudo antiqua*, eine im Süsswasser-Gypse von Hohenhöwen untergegangene Art.—N. Act. Nat. Cur. XV. 2, p. 201, fig.

11. Sur les impressions de Poissons dans les Nids ferrifères des Terrains houillers du Palatinat du Rhin, et sur le *Palæoniscum macropterum*.—Zeitschr. f. Mineral. VII. 1829, p. 477.—*Féruss.* Bull. 1830, XX. p. 410.

12. Sur deux nouveaux Poissons fossiles du Calcaire à Gryphées de Donau-Eschingen.—Jahrb. f. Mineral. I. p. 14.—*Féruss.* Bull. 1830, XXII. p. 140.

13. Sur les Restes fossiles du Lignite feuilleté (*Papier-Kohle*) de Geislinger-Busch, dans les Sept-Montagnes.—Zeitschr. f. Mineral. V. 1828, p. 374.—*Féruss.* Bull. 1829, XVI. p. 193.

14. Notice sur les Fossiles du Calcaire lithographique de Pappenheim.—Zeitschr. f. Mineral. 1828, II. p. 608.—*Féruss.* Bull. 1829, XVII. p. 174.

15. Sur les Fossiles des Tubicolées de *Lamarck*.—Zeitschr. f. Miner. 1828, I. p. 1 ; V. p. 321.—*Féruss.* Bull. 1829, XVII. p. 311 ; 1831, XXV. p. 361.

16. System der Urweltlichen Pflanzenthiere durch Diagnose, Analyse und Abbildung der Geschlechter erläutert (*Système des Zoophytes du Monde primitif, éclairci par le diagnostic, l'analyse, et les figures des différens genres*). Heidelb. 1825, fol. fig.—*Féruss.* Bull. 1827, X. p. 422.

17. Ueber die Krinoideen-Reste im Muschelkalk.—*L. u. Br.* N. Jahrb. 1837, p. 30, fig.

18. Ueber das geologische Alter und die organischen Ueberreste der tertiären Gesteine des Maynzer-Beckens.—*L. u. Br.* N. Jahrb. 1837, p. 153.

19. Notizen über das Vorkommen der Tegel-Formation u. ihrer Fossil-Reste in Siebenbürgen u. Galizien.—*L. u. Br.* N. Jahrb. 1837, p. 653.

20. Note über die mit *Homalonotus* verwandten Trilobiten-Genera.—*L. u. Br.* N. Jahrb. 1840, p. 445.

21. *Ctenocrinus*, ein neues Krinoiden-Geschlecht der Grauwacke. *L. u. Br.* N. Jahrb. 1840, p. 542, fig.

22. Die Gletscher-Theorie u. Eiszeit-Hypothese des Herrn L. *Agassiz*, aus dem physikalisch-geologischen Gesichtspunkte beleuchtet.—*L. u. Br.* N. Jahrb. 1842, p. 56.—Edinb. New Phil. Journ. XXXIII. p. 36.

23. Gedrängte Anleitung zum Sammeln, Zubereiten u. Verpacken v. Thieren, Pflanzen u. Mineralien für naturhistorische Museen. Heidelb. 1838, 8°.

24. Uebersicht und Abbildungen der bis jetzt bekannten *Nerinea*-Arten.—*L.* u. *Br.* N. Jahrb. 1836, p. 544.

25. Description des Dents fossiles d'un nouveau genre de Pachyderme appelé *Cœlodonta.*—Bull. Soc. Géol. Fr. I. p. 171.

26. Ueber die fossilen Gaviale der Lias-Formation u. der Oolithe. —*Wiegm.* Arch. 1842, I. p. 77.

27. Ergebnisse meiner naturhist. ökonomischen Reisen. Heidelb. 1826, 1831, 2 vols. 8° fig.

28. Handbuch der Geschichte der Natur. Stuttg. 1841–1843, 2 vols. 8°.

29. System der urweltlichen Konchylien. Heidelb. 1824, fol. fig. —*Leonh.* Tasch. 1824, III. p. 702.

30. Geschichte der Natur. Stuttg. 1842, I. u. II. Bd. 8°.

31. Ueber Ichthyosauren in den Lias-Schiefern der Gegend von Boll in Würtemberg.—*L.* u. *Br.* N. Jahrb. 1844, p. 385, fig.— (Nachtrag) ibid. p. 676.

32. Recherches sur des Epis et d'autres parties végétales pétrifiées appartenant au *Cupressus Ullmanni* et provenant du Minerai de Frankenberg.—Zeitschr. f. Min. VII. p. 509.—*Féruss.* Bull. XVII. p. 42.

33. Sur les Coquilles pétrifiées du Steinsalzgebirge, qui ont été jusqu'a ce jour réunies sous le nom de *Pectinites salinarius.*—Jahrb. I. p. 279, fig.—*Féruss.* Bull. XXVI. p. 94.

34. Italiens Tertiär-Gebilde u. deren organische Einschlüsse. 8°. Heidelb. 1831.

35. Ueber zwei neue Trilobiten-Arten zum Calymene-Geschlechte gehörig (*C. latifrons* u. *Schlotheimii*).—*Leonh.* Zeitschr. 1825, I. p. 317, fig.

36. Ueber die Versteinerungen und über verschiedene Felsarten in Piemont.—*Leonh.* Zeitschr. 1825, I. p. 55.

37. Zur angewandten Naturgeschichte u. Physiologie. 8°. Heidelberg, 1824.

38. Paläontologische Collectaneen ; *vide sup.* Pars I. p. 34.

BRONN (H. G.) u. KAUP (J. J.).

1. Abhandlungen über die Gavialartigen Reptilien der Lias-Formation. Stuttg. 1841–44, fol. fig.—*L.* u. *Br.* N. Jahrb. 1842, p. 374.

BROOKE (H. J.).

1. On Conchology regarded as a distinct branch of Science.—Zool.
Journ. V. p. 207.

BROOKE (James).

1. On the Habits and points of distinction in the Orangs of Borneo.
—Proc. Zool. Soc. IX. p. 55.—Ann. and Mag. N. H. IX. p. 54.

BROOKES (Josh.).

1. On a new genus of the Order Rodentia (*Lagostomus trichodac-
tylus*). (Lagostomus, *eine neue Sippe von Nagthieren.*)—Trans.
Linn. Soc. XVI. 1, p. 95.—Isis, 1838, IX. p. 905.—*Féruss.* Bull.
1831, XXIV. p. 71 (Lagostomus, *nouveau genre de Rongeurs*).

2. On the remarkable formation of the Trachea in the *Egyptian
Tantalus.*—Trans. Linn. Soc. XVI. 1833, p. 499.

BROOKES (Rich.).

1. A new and accurate System of Natural History, etc. Lond.
1763, 12°. 6 vols. fig.; 2d ed. Lond. 1773, 8°. fig.—*Böhm.* Bibl.
I. 1, p. 296.—*Eis.* Ins. p. 139.

BROOKES (S.).

1. Introduction to the Study of Conchology. Lond. 1815, 4°.—Tr.
Linn. Soc. Lond. XI. 2, p. 422.—(Deutsch) *Einleitung zu dem
Studium der Conchylienlehre, mit einer Tafel über die Anatomie
der Flussmuscheln, vermehrt von* Carus. Leipz. 1825, 4°. fig. col.

BROSCHE (J. N. J.).

1. Beiträge für eine allgemeine Naturlehre der Pflanzen, Thier-
körper u. des Menschen überhaupt. 8°. Wien, 1817.

BROSSE (De).

1. Histoire des Navigations aux Terres australes, contenant ce que
l'on sait des mœurs et des productions des Contrées découvertes
jusqu'à ce jour. Par. 1756, 4°. 2 vols. fig.—(Angl) *Terra
Australis cognita, or Voyages to the Terra Australis or Southern
Hemisphere,* etc. Lond. 1767-68, 3 vols. 8°.—(Germ.) *Vollständige Geschichte der Schiffarthen nach den, noch grösstentheils
unbekannten Sudländern,* etc. *übers. von* J. Chr. Adelung. Halle,
1767, 4°. 5 vols.—*Böhm.* Bibl. I. 1, pp. 468, 473.

Brotz (J.).

1. Einleitung in die Geschichte der Naturwissenschaft. 8°. Heidelb. 1842.

Brotz (J.) et Wagemann (C. A.).

1. De Amphibiorum Hepate, Liene ac Pancreati observationes zootomicæ. Diss. inaug. Friburg, 1838, 4°.

Brougham (H.).

1. Discourse of Natural Theology, showing the nature of the evidence, and the advantages of the study. 8°. London, 1835.—Brussels, 1835, 12°.

Broughton (Arth.).

1. Dissert. de Vermibus Intestinorum. Edinb. 1779, 8°.—*Böhm.* Bibl. II. 2, p. 383.

Broughton (S. D.).

1. Experiments and Observations upon the fifth, seventh and eighth pairs of Nerves, etc.—Lond. Med. and Phys. J. 1823, p. 463.—*Féruss.* Bull. 1823, III. p. 292.

2. Sur l'usage des Moustaches dans le Chat et chez plusieurs autres animaux.—Lond. Med. and Phys. J. May 1823, p. 397.—*Féruss.* Bull. 1823, IV. p. 72.

3. On the Progress of Physiological Research.—Rep. Brit. Assoc. 1832, p. 589.

Broughton (Th. D.).

1. Letters written in a Mahratta Camp in 1809, descriptive of the Manners, etc. Lond. 1813, 4°.—Bibl. Ent. I. p. 51.

Broughton (   ).

1. Voyage to the N. Pacific Ocean. 4°. 1802.

Broussais (F. Jos. Vict.).

1. Traité de Physiologie appliquée à la Pathologie. 8°. 2 vols. Paris, 1822 et 1823.—Cat. R. Soc. L.

Brousse (De la).

1. Des Mûriers et de l'éducation des Vers à soie.—Mél. d'Agriculture, Nismes, 1789.—Bibl. Ent. I. p. 51.

Broussonnet (P. M. A.).

1. Mémoire sur les Dents.—Biogr. Un. VI. p. 46.

2. Observations sur les Vaisseaux spermatiques des Poissons épineux.—Mém. Acad. Sc. Par. 1785, p. 170.—Dict. Sc. n. XXII. p. 509.

3. Mémoire sur la Régénération de quelques parties du corps des Poissons.—Mém. Acad. Sc. Par. 1786, p. 684.—J. de Phys. XXXV. p. 62.—Dict. Sc. n. XXII. p. 509.

4. Mémoire pour servir à l'Histoire de la Respiration des Poissons. —Mém. Acad. Sc. Par. 1785, p. 174.—J. de Phys. XXXI. p. 289. —*Cuv.* et *Val.* I. p. 165.—Dict. Sc. n. XXII. p. 508.

5. Account of the *Ophidium barbatum*, Linn.—Philos. Trans. LXXI. p. 436.—Trans. Linn. Soc. Lond. V. p. 279.

6. Observations sur les Écailles de plusieurs espèces de Poissons, qu'on croit communément dépourvues de ces parties.—J. de Phys. XXXI. p. 12.

7. Ichthyologia, sistens Piscium descriptiones et icones. Lond. 1782, 4°.—Par. 1792, 4°. fig.—Tr. Linn. Soc. Lond. V. p. 279.

8. Mémoire sur le Voilier.—Acad. Sc. Par. 1786, p. 450.—*Cuv.* et *Val.* I. p. 138.

9. Mémoire sur le Trembleur, espèce peu connue de Poisson électrique.—Acad. Sc. Par. 1782, p. 692.—J. de Phys. XXVII. p. 189.

10. Observations sur le Loup marin.—Acad. Sc. Par. 1785, p. 161. —Dict. Sc. n. XXII. p. 532.

11. Mémoire sur différentes espèces de Chiens de mer.—Acad. Sc. Par. 1788.—J. de Phys. XXVI. pp. 51, 120.

12. Essai de comparaison entre les mouvemens des Animaux et ceux des Plantes.—Mém. Acad. Sc. Par. 1784, p. 609.

Broussuet (Aug.).

1. Ichthyologia, sistens Piscium descriptiones et icones. Lond. et Viennæ, 1782, fol. fig.— *Böhm.* Bibl. II. 2, p. 59.

Brovardi (Nic. J.).

1. De Sanguinis Circulatione in Fœtu natoque Homine.—De Fœtûs origine et incremento (Diss. inaug.). Taur. 1743, 8°.

Browallius (J.).

1. Dissert. de Scientiâ naturali ejusque Methodo. Ups. 1737, 4°. —*Böhm.* Bibl. I. 1, p. 182.

2. Betänkande om Vattuminskningen. Stockh. 1755, 8°.

Brown (Col.).

1. On the Streams of Sea-Water which flow into the Land in the Island of Cephalonia.—Proc. Geol. Soc. II. p. 393.—Phil. Mag. ser. 3, VIII. p. 573.

Brown (Edward).

1. A brief Account of some Travels in Hungaria, Servia, Bulgaria, Macedonia, Thessaly, Austria, Styria, etc. Lond. 1673 and 1677, 4°; 1685, fol. fig.—(Gall.) *Relation des Voyages*, etc. Par. 1674 et 1684, 4°.—(Belg. *J. Leew*): Amsterd. 1682, 4°. fig.—(Germ.) *Itinerarium ; oder sonderbare Reisen durch Nieland, Deutschland, Ungarn, Servien*, etc. Nürnb. 1686, 1711, 1750, 4°. fig.—*Böhm.* Bibl. I. 1, p. 447.

2. De Re Metallicâ et Fodinis in Hungariâ et vicinis visis. Lond. 1673, 4°.—*Böhm.* Bibl. IV. 1, p. 158.

3. De nonnullis Hungariæ Mineralibus.—Philos. Trans. LVIII. p. 1189.—*Böhm.* Bibl. IV. 1, p. 158.

4. An Account of several Travels through a great part of Germany, etc.; wherein the Mines, Baths, etc. of those parts are treated of. 4°. Lond. 1677.—Cat. R. Soc. L.

Brown (Edwin).

1. A Specimen of *Locusta migratoria* captured at Mickleover, near Derby.—Ann. and Mag. N. H. X. p. 157.

2. Occurrence of Coleopterous Insects during Floods.—Zoologist, p. 177 ; Piscivorous Habits of the Brown Rat, p 212 ; Keen Scent of the Weasel tribe, p. 213 ; More frequent Occurrence of the Woodcock, p. 249 ; Multiplication of *Dreissena polymorpha*, p. 255.

Brown (Henr.).

1. *Testudo antiqua*, etc.—N. Act. Nat. Cur. XV. p. 201, fig.—*Dum.* et *Bib.* I. p. 435.

Brown (John).

1. Nova Animalium Generatio, necnon Corporis humani Delineatio anatomica. Lugd. Bat. 1686, 12°.—*Böhm.* Bibl. II. 1, p. 128.

2. Epistola de Hepate quodam manifestè et ad oculum glanduloso. —Phil. Trans. 1685, No. 178, p. 1166, fig.—Act. Erudit. VI. p. 26.

Brown (John, II.).

1. On some Pleistocene Deposits near Copford, Essex.—Proc. Geol.

Soc. IV. p. 164.—Phil. Mag. ser. 3, XXIV. p. 63.—Ann. and Mag. N. H. XII. p. 476.—*L. u. Br.* N. Jahrb. 1844, p. 375 (*Einige pleistocene Ablayerungen bei Copford in Essex*).

2. Table of Mineral Substances and Organic Remains in the Gravel at Stanway, Essex.—Mag. Nat. Hist. ser. 1, VIII. p. 349.

3. On a new locality of Fossil Bones of Elephant and Deer.—Mag. Nat. Hist. ser. 1, VIII. p. 353.

4. Geological conditions of the Chalk and Argillaceous beds at Ballingdon Hill, Essex.—Mag. Nat. Hist. ser. 1, IX. p. 42.

5. On the contents of the Freshwater Formation at Copford, Essex. —Mag. Nat. Hist. ser. 1, VII. p. 436; IX. p. 429.

6. On a Fluvio-marine Deposit containing Mammalian Remains, occurring in the Parish of Little Clacton, on the Essex Coast.— Mag. Nat. Hist. ser. 2, IV. p. 197.

7. On certain Conditions and Appearances of the Strata on the Coast of Essex near Walton.—Journ. Geol. Soc. I. p. 341.

8. On the Boulders of Trap Rocks, etc. which occur in the Diluvium of Essex.—Mag. Nat. Hist. ser. 2, I. p. 145.

9. On Fossil Remains in Essex.—Ann. and Mag. Nat. Hist. XI. p. 325.

10. Observations on the Common Toad, and on its long Abstinence from Food.—Ann. and Mag. N. H. X. p. 180.

11. Capture of some rare Birds on the Cotswold Hills —Ann. and Mag. N. H. VI. p. 395.

12. On the Analysis of the Nodules usually regarded as Coprolitic in the Crag and London Clay.—Lond. Geol. Journ. I. p. 17.

BROWN (Littleton).

1. The same sort of Insects found in Kent, with a Remark by *Mortimer.*—Phil. Trans. No. 447, p. 153.—Badd. X. p. 341, fig.—Bibl. Ent. I. p. 51.

BROWN (Patr.).

1. The Civil and Natural History of Jamaica, etc. Lond. 1756, fol. fig.—Ibid. 1789, fol.—*Böhm.* Bibl. I. 1, p. 766.—*Mod.* B. Helm.

BROWN (Peter).

1. New Illustrations of Zoology (*Nouvelles Illustrations de Zoologie*). Lond. 1776, 4°. fig. col.—*Dum.* et *Bib.* I. p. 309.—*Böhm.* Bibl. II. 1, p. 50.

452 BRO

Brown (P. J.).

1. List of Diurnal Lepidoptera in Switzerland.—Mag. Nat. Hist. ser. 1, VIII. p. 205.
2. List of Crepuscular Lepidopterous Insects in Switzerland.—Mag. Nat. Hist. ser. 1, VIII. p. 553.

Brown (R.).

1. On the Geology of Cape Breton.—Journ. Geol. Soc. I. pp. 23, 207 ; III. 1, p. 257.
2. On a Group of Erect Fossil Trees in the Sydney Coal-field of Cape Breton.—Journ. Geol. Soc. II. p. 393.

Brown (Sam.).

1. On a curious Substance which accompanies the native Nitre of Kentucky and of Africa.—*Sill.* Am. Journ. I. 2, p. 146.

Brown (S.).

1. On a Cave on Crooked Creek.—Trans. Am. Phil. Soc. VI. p. 235.

Brown (Sir Thos.).

1. Enquiries into Vulgar Errors. fol. Lond. 1646 ; 1650 ; 4°. 1664 ; 1666 ; 1672.—(Gall. *Souchay*) 3 vols. 12°. Paris, 1733 ; 1738.
2. Works, edited by *Wilkin.* 4 vols. 8°. London, 1836.

Brown (Thos. ii.).

1. On the *Exocœtus volitans*, or Flying Fish.—Phil. Trans. LXVIII. p. 791.—*Cuv.* et *Val.* I. p. 138.
2. Observations on the Zoonomia of *E. Darwin.* 8°. Edinb. 1798.

Brown (Thos. iii.).

1. An Account of the Whidah Bird.—Edinb. Journ. Nat. Geogr. Sc. I. pp. 341, 425.—Isis, 1832, VI. p. 583.
2. On a new species of British Fish.—Edinb. Journ. Nat. Geogr. Sc. II. p. 99.—*Féruss.* Bull. 1830, XXIII. p. 272.
3. On the Irish Testacea.—Mem. Wern. Soc. II. p. 501.
4. Description of five new British species of Shells. — Edinb. Journ. Nat. Geogr. Sc. I. p. 1.—*Féruss.* Bull. 1830, XXIII. p. 274.
5. Observations on *Mr. Kenyon's* paper on British Land and Fresh-water Shells.—Edinb. Journ. Nat. Geogr. Sc. I. p. 65.

6. A Monograph on the *Pisidium*, a new genus of British Fresh-water Testaeea.—Edinb. Journ. Nat. Geogr. Sc. I. p. 411.

7. Illustrations of the American Ornithology of *Alex. Wilson* and *Charles Lucien Bonaparte*, with all the new discoveries, and the addition of the whole Forest Sylva. fol. Edinb. 1835.

8. The Elements of Fossil Conchology. Royal 18mo. 10 plates, 1842.

9. Book of Butterflies, Sphinges and Moths. 3 vols. 12°. Lond. 1832–34. *[Lonsh mac]*

10. Zoologist's Text-book. 2 vols. 18°. Glasgow, 1833.

11. Elements of Conchology. Lond. 1816, 8°.—Tr. Linn. Soc. Lond. XII. 2, p. 590.—Dict. Sc. n. X.

12. Illustrations of the Conchology of Great Britain and Ireland. 4°. Edinb. 1827. 2nd ed. Lond. 4°. 1842.—*Féruss.* Bull. 1827, XII. p. 282.

13. Illustrations of the Fossil Conchology of Great Britain and Ireland. Edinb. 1827 ; 1837–45, 4°.

14. Ueber Reste einer Eiche, aus einem Torfe bei Lanfyne, Ayr-shire, gezogen.—Edinb. Soc. Apr. 1834.—Instit. 1834, p. 335.—*L. u. Br.* N. Jahrb. 1834, p. 728.

15. Description of some new species of the genus *Pachyodon.*—Ann. and Mag. N. H. XII. p. 390.—*L. u. Br.* N. Jahrb. 1844, p. 240 (*Beschreibung einiger neuen Pachyodon-Arten*).

16. Taxidermist's Manual. 12°. Glasgow, 1833.

17. Conchologist's Text-book. Glasg. 1833, 8°.—6th ed. 1841.

18. Indication d'une espèce de Ver non décrite (*Ascaris pellucidus*), trouvée aux Indes orientales dans les yeux des chevaux.—Tr. Soc. Roy. Edinb. 1821, I. p. 107.—*Féruss.* Bull. 1824, I. p. 300.

BROWN (W. G.).

1. Travels in Africa, Egypt and Syria, ann. 1792–1798. Lond. 1799, 4°.—Biogr. Un. Suppl. LIX. p. 328.

BRUAND (T.).

1. Entomologie. 8°. Besancon, 1844.

BRUCE (Arth.).

1. A curious Fact in the Natural History of the [Common Mole (*Talpa europæa*, Linn.).—Tr. Linn. Soc. Lond. III. p. 5.—Phil. Mag. ser. 1, II. p. 36.

BRUCE (Jam.).

1. Travels to discover the Sources of the Nile, in the years 1768–

1772. Edinb. 1790, 4°. *5* vols. fig.—(2nd ed.) Lond. 8°. 7 vols.
atl.—(3rd ed.) 7 vols. 8°. Edinb. 1813.—(Trad. Franç.) Paris,
1790 : *Voyage aux Sources du Nil, en Nubie et en Abyssinie,* etc.
—Biogr. Un. VI. p. 79.—*Dum.* et *Bib.* I. p. 310.

BRUCE (P. H.).

1. Memoir containing an Account of his Travels in Germany,
Russia, Tartary, Turky, etc. Lond. 1782, 8°.—(Germ.) Leipz.
1784, 8°.—*Böhm.* Bibl. I. 1, p. 494.

BRUCH (C.).

1. Untersuchungen zur Kenntniss des körnigen Pigments der Wirbelthiere in physiologischer und pathologischer Hinsicht. Zürich,
1844, 4°.

BRUCH ( ).

1. Ornithologische Mittheilungen.—Isis, 1832, X. p. 1105.

2. Verzeichn. der zu Mainz aufgestellten Sammlungen. 4°. Mainz,
1843.

3. Bemerkungen über einige Artkennzeichen der Vögel.—Isis, 1829,
VI. p. 629.—*Féruss.* Bull. 1830, XXII. p. 121 (*Observ. sur
quelques Caractères spécifiques des Oiseaux*).

4. Ornithologische Beiträge. 5ᵗᵉ Lief.—Isis, 1828, VII. p. 718,
fig.—*Féruss.* Bull. 1829, XVII. p. 293 (*Matériaux ornithologiques*).

5. Ornithologische Bemerkungen. 2ᵗᵉ Lief.—Isis, 1825, V. p. 577.

6. Ein Beitrag zur Beschreibung des Geyeradlers (*Gypaëtos bar-
batus*).—Isis, 1831, IV. p. 404.

7. Zoologische Bemerkungen (Ueber Mäuse u. mehr. Vögel).—
Isis, 1824, VI. p. 674.

BRUCH (H.).

1. Allgemeine Naturgeschichte des Thier-Pflanzen und Mineral-
reiches. Nürnberg, in fol. cum tab. col.

BRUCHHAUSEN (A.).

1. Institutionum Physicæ Pars I. Münst. 1775, 8°. fig. Pars II.
1777.—*Böhm.* Bibl. I. 1, p. 326.

BRÜCK (A. T.).

1. Ueber die Wiederholung geschlechtiger Emotionen in dem
Kopfe.—Isis, 1828, VIII. p. 845.

BRÜCKE (Ern.).

1. Ueber den innern Bau des Glaskörpers.—*Müll.* Arch. 1842, p. 345.

2. Ueber die physiologische Bedeutung der stabförmigen Körper und der Zwillingszapfen in den Augen der Wirbelthiere.—*Müll.* Arch. 1844, p. 444.

3. Ueber die Ursache der Todtenstarre.—*Müll.* Arch. 1842, p. 178.

4. De Diffusione Humorum per Septa mortua et viva. Berol. 1842, 8°.

5. Ueber die Stereoskopischen Erscheinungen und *Wheatstone*s Angriff auf die Lehre von den identischen Stellen der Netzhäute. —*Müll.* Arch. 1841, p. 459, fig.

BRÜCKMANN (C. Ph.).

1. Neue verbesserte undvollständige Beschreibung der gesunden warmen Brunnen und Bäder zu Ems. Frankf. 1772, 8°. fig.— *Böhm.* Bibl. V. p. 329.

BRÜCKMANN (Ern. L.).

1. Epist. de Petrificationis fiendi Modo. Jenæ, 1750, 4°.—(Germ.) *Grund.* Nat. u. Kunstgesch. II. p. 579.—*Böhm.* Bibl. IV. 2, p. 242.

BRÜCKMANN (Fr. Ern.).

1. De figurâ Lapidum ex Pisce quodam Indico, cum effigie S^{ti} Petri, St Peterstein genannt.—Commerc. Norib. 1739, No. 27, p. 209, fig.—*Böhm.* Bibl. IV. 2, p. 293.

2. Historia naturalis Lapidis nummalis Transylvaniæ. Wolfenb. 1727, 4°. fig.—*Böhm.* Bibl. IV. 2, p. 308.

3. Specimen physicum exhibens Historiam naturalem Oolithi, etc. Helmst. 1721, 4°. fig.

4. De fabulosissimæ originis Lapide, Arachneolitho dicto, Epistola. Wolfenb. 1722, 4°. fig.

5. Lapides fungiformes Maris Rubri et Dendrites Abachiensis.— Act. phys.-med. VIII. p. 217, fig.—*Mod.* B. Helm.

6. Petrefactum singulare dentem seu palatum Piscis Ostracionis referens.—Act. Nat. Cur. IX. p. 116.

7. Verzeichniss der Schriften, so von Thieren und deren Theilen handeln, nebst Fortsetzung. Wolfenb. 1743.

8. Observatio curiosa de Vermis nunquam antea excreti excretione. Wolfenb. 1723, 4°. fig.—*Böhm.* Bibl. II. 2, p. 388.

9. *Magnalia Dei in locis subterraneis*, oder Unterirrdische Schatz-kammer aller Königreiche und Länder, in Beschreibung aller Bergwerke. Braunschw. 1727-30, 2 vols. 4º. fig.—Erstes Suppl. die Schwedischen Bergwerke betreffend. Wolfenb. 1734, fol. fig. —*Böhm.* Bibl. IV. 1, p. 94.

10. *Thesaurus subterraneus Ducatûs Brunsvigii*, i. e. Braunschweig mit seinen unterirrdischen Schätzen u. Seltenheiten der Natur. Braunschw. 1728, 4º. fig.—*Böhm.* Bibl. IV. 1, p. 135.

11. De Lapide violaceo Sylvæ Hercyniæ. Guelph. 1725, 4º.—*Böhm.* Bibl. IV. 1, p. 227.

12. Historia naturalis curiosa Lapidis τοῦ ἀσβέστου ejusque præpa-rationum, etc. Brunsw. 1727, 4º.—*Böhm.* Bibl. IV. 1, p. 267.

13. Von einem zu Osterode ausgegrabenen Skelete eines Elephanten. —Hamb. Berichte, 1744, p. 497.—*Böhm.* Bibl. IV. 2, p. 280.

14. Bibliotheca animalis, oder Verzeichniss der meisten Schriften so von Thieren und deren Theilen handeln. Wolfenb. 1743, 1747, 8º. 2 Th.—Biogr. Un. VI. p. 83.—*Böhm.* Bibl. II. 1, p. 3.

15. Epistolæ itinerariæ. Cent. I.-III. Wolfenb. 1742-50, 4º. fig.— *Klein*, Disp. Echinod. p. ix.—*Mod.* B. Helm.

16. Von der Gackerlake.—Epist. itin. Cent. I. Ep. 23.—Bibl. Ent. I. p. 52.

17. De Vermibus quibus Helgolandi ad Piscatum hamatilem utun-tur.—Comm. Norimb. 1742, p. 381, fig.—Bibl. Ent. I. p. 52.— *Mod.* B. Helm.

18. Relatio brevis physica de curiosissimis duabus Conchis marinis, Vulvâ marinâ et Conchâ venereâ. Brunsw. 1722, 4º. fig.—Tr. Linn. Soc. Lond. VII. p. 221.—*Böhm.* Bibl. II. 2, p. 475.—*Mod.* B. Helm.

19. Serpentes et Viperæ sylvæ Hercynicæ.—Epist. itin. Cent. II. p. 137.—Dict. Sc. n. XV. p. 245.

20. Observationes de Insectis.—Epist. itin. Cent. II. Ep. 5.—Bibl. Ent. I. p. 52.

21. De Cervo volante et ejus hybernaculo.—Epist. itin. Cent. I. Ep. 78, fig.—*Eis.* Insect. p. 194.

22. Verzeichniss der vornehmsten Stücke welche in dem nunmehr zertheilten Curiositäten u. Naturalienkabinet *J. Chr. Olearii* be-findlich gewesen sind. Jena, 1750, 4º.—*Böhm.* Bibl. I. 1, p. 396.

23. Petrefacta Havelbergensia.—Comm. litt. Norimb. 1763, p. 391.

BRÜCKMANN (J. A. v. und A. E.).

1. Vollständige Anleitung zur Anlage, Fertigung u. neuern Nut-zanwendung artesischer Brunnen. Heilbr. 1833, 8º. fig.

Brückmann (U. Fr. B.).

1. Ueber den so genannten Stahrenstein.—Schr. Berl. Ges. Naturf. Fr. VI. p. 416.
2. Vom krystallisirten Sandstein.—Schr. Berl. Ges. Naturf. Fr. I. p. 393.
3. Ueber den Serpentinstein mit schielenden Flecken.—Schr. Berl. Ges. Naturf. Fr. IX. p. 201.
4. Abhandlung von Edelsteinen, etc. Braunschw. 1757 ; 1773, 8°. —(Gesammelte Beiträge) : Braunschw. 1778–83, 8°.
5. Beschreibung eines besondern Encriniten.—Schr. Berl. Ges. Naturf. Fr. VI. p. 410.

Brückner (Em. Dan.).

1. Versuch einer Beschreibung historischer und natürlicher Merkwürdigkeiten der Landschaft Basel. Bas. 1748–63, 8°. fig.— *Böhm.* Bibl. I. 1, p. 550.—*Mod.* B. Helm.

Brückner (G. A.).

1. Wie ist der Grund und Boden Mecklenburgs? etc. (*Quelles sont les Couches minérales du sol du Mecklenbourg, et quelle est leur origine?*)—Fragment géologique sur le Mecklenbourg, etc. Neu-Strel. 1825, 8°.—*Féruss.* Bull. 1827, XI. p. 9.

Bruderamon (J. B.).

1. Coleccion de laminas que representan los animales y monstruos del real cabinete de historia natural de Madrid.   Madrid, 1784–1786, 2 vols. fol. avec 71 pl. col.

Brüggemann (L. W.).

1. Ausführliche Beschreibung des gegenwärtigen Zustandes des Kön. Pr. Herzogthums Vor- u. Hinterpommern. Stett. 1779–84, 2 vols. 4°.—*Böhm.* Bibl. I. 1, p. 605.

Brugmann (Sch. Just.).

1. Lithologia Groningiana juxta ordinem *Wallerii* digesta cum Synonymiis aliorum imprimis *Linnæi* et *Cronstedii.*   Gröningen, 1781, 1787, 8°.—*Böhm.* Bibl. IV. 1, p. 107.

Brugnatelli (Gasp.).

1. Trattato delle Cose naturali e dei loro Ordini conservatori. Pavia, 1837, 4 vols. 8°.

2. Della coltura sociale e del contribuire delle cognizioni naturali alla medisima. 8°. Pavia, 1841.

3. Elementi di Storia naturale generale. (*Elémens d'Histoire naturelle générale.*) 8°. Pavie, 1825 (1er vol.); 1826 (2e vol.); (2e éd.) 1830.—*Féruss.* Bull. 1826, IX. p. 23; 1827, XII. p. 42.

BRUGNATELLI (L.).

1. Lettera sulla maniera di conservare varii Insetti.—Opusc. scelt. VII. p. 226.—Bibl. Ent. I. p. 52.

2. Biblioteca Fisica; *vide sup.* Pars I. p. 70.

BRUGNONE (C. G.).

1. Ippometria ossia della conformazione externa del Cavallo, del Asino e del Mulo. 8°. Torino, 1802.

2. Bometria, ossia delle conformazione externa delle Bestie bovine. 8°. Torino, 1802.

3. Sur les Vésicules séminales.—Mém. Acad. Tur. VIII. p. 609.

4. De Ovariis eorumque Corpore luteo.—Mém. Acad. Tur. IX. p. 393.

5. Observations sur l'Origine de la Membrane du Tympan et de celle de la Caisse.—Mém. Acad. Tur. XII. p. 1.

6. Observations myologiques.—Mém. Acad. Tur. XII. p. 157.

7. Sur le Labyrinthe de l'Oreille.—Mém. Acad. Tur. XVI. p. 167.

8. Sur la Digestion dans les Oiseaux.—Mém. Acad. Tur. XVI. p. 306.

9. Des Animaux ruminans et de la Rumination.—Mém. Acad. Tur. XVIII. 1, p. 309.

10. Observations anatomiques sur l'Origine de la Membrane du Tympan et celle de la Caisse. 4°.

11. De Testium in Fœtu positu; de eorum in Scrotum descensu; de Tunicarum quibus hi continentur, numero et origine.—Mém. Acad. Tur. VII. p. 13.

BRUGUIÈRE (J. G.).

1. Sur un Ver singulier rendu par les Selles.—Journ. de Phys. XXXI. 1787, p. 109.

2. Sur la formation de la Coquille des Porcellaines, et sur la faculté qu ont leurs Animaux de s'en détacher et de les quitter à différentes époques.—J. d'Hist. nat. 1792, I. pp. 307 et 321.—Tr. Linn. Soc. Lond. VII. p. 239.

3. *Chiton spinosus.*—J. d'Hist. nat. 1792, I. p. 20, fig.—Tr. Linn. Soc. Lond. VII. p. 224.

4. *Unio granosa.*—J. d'Hist. nat. 1792, I. p. 103, fig.—Tr. Linn. Soc. Lond. VII. p. 224.

5. *Purpura tubifera.*—J. d'Hist. nat. 1792, I. p. 20, fig.—Tr. Linn. Soc. Lond. VII. p. 224.

6. *Bulimus sinamarinus.*—J. d'Hist. nat. 1792, I. p. 339, fig.—Tr. Linn. Soc. Lond. VII. p. 224.

7. Tableau encyclopédique des trois Règnes de la Nature. Paris, 1791, 4°. fig.—*Lamx.* Polyp. Corall. p. 524.

8. Description d'une espèce particulière de Serpent à Madagascar. —J. de Phys. XXIV. p. 132.—Dict. Sc. n. XV. p. 252.

9. Description d'une nouvelle espèce de Tortue de Cayenne.—J. d'Hist. nat. 1792, VII. p. 253, fig.—*Dum.* et *Bib.* I. p. 426.

10. Catalogue des Coquilles, envoyées de Cayenne, par *Le Blond.* —Mém. de l'Acad. des Sc. de Paris, I. p. 120.

BRUGUIÈRE (J. G.) et LAMARCK (J. B. P. A.).

1. Histoire naturelle des Vers, continuée par *Deshayes.*—Encycl. Méthod. Par. 1789–1792; 1830, 2 vols.—Tr. Linn. Soc. Lond. VII. p. 219.—*Féruss.* Bull. XXV. p. 123.

BRÜHL (B. C.).

1. Zur Kenntniss des Wirbelthier-skelettes. 8°. Wien, 1845.

BRÜHL (J. U. Chr.).

1. Dissert. de Pabulo Vitæ, seu de Materià cui, cùm Animalia, tùm Vegetabilia, vitam debent ac nutritionem. Marb. 1781, 4°.— *Böhm.* Bibl. II. 1, p. 125.

BRUINSMA (J. J.).

1. Buitengewone afwijkingen waargenomen bij de gedaante verwisseling des Zijdeworms (*Bombyx mori*).— *Van der Hoev.* Tijdsch. VII. p. 257.

BRUINT (Corn. de).

1. Reizen over Moskovie, door Persie en Indie. Amst. 1711, fol.— *Fisch.* Cat.

BRULLÉ (Aug.).

1. Coup-d'œil sur l'Entomologie de la Morée. Par. 1831, 8°.— Ann. Sc. n. XXIII. p. 244.—Isis, 1835, III. p. 285.

2. Résumé des travaux de la Société Entomologique de France pendant l'année 1832.—Ann. Soc. Ent. Fr. II. p. 321.

3. Histoire naturelle des Insectes. (Coléoptères, Paris, 1834, fig.)— Bibl. Ent. I. p. 53.

4. Observations critiques sur la Synonymie des Carabiques.—*Silberm.* Rev. Entom. II. p. 89 ; III. p. 271.— *Wiegm.* Arch. 1835, II. p. 23.

5. Description du *Procerus Duponchelii.*— *Guér.* Mag. de Zool. 1832, Ins. No. 9.—Bibl. Ent. I. p. 52.

6. Examen des genres *Brachinus* et *Ditomus*, de la tribu des Carabiques.—Ann. Soc. Ent. Fr. IV. p. 621.—*Burm.* in *Wiegm.* Arch. 1836, II. p. 305.

7. Sur les Transformations du *Cladius difformis,* hyménoptère de la famille des Tenthrédines.—Ann. Soc. Ent. Fr. I. p. 308, fig.

8. Mémoire sur un Insecte hyménoptère parasite et voisin du genre Alison.—Ann. Soc. Ent. Fr. II. p. 403.

9. Mémoire sur un genre nouveau de Diptères, de la famille des Tipulaires (*Xiphura*).—Ann. Soc. Ent. Fr. II. p. 205, fig.

10. Note sur le genre *Xiphura* formé aux dépens de celui de *Stenophora* de *Meigen.*—Ann. Soc. Ent. Fr. II. p. 398.

11. Observations sur la Bouche des Libellulines.—Ann. Soc. Ent. Fr. II. p. 343, fig.

12. Recherches sur la Classification des Animaux en Séries parallèles.—Ann. Sc. n. (2e S.) XVIII. pp. 50, 298.

13. Recherches sur les Transformations des Appendices dans les Articulés.—Ann. Sc. n. (3e S.) II. p. 271, fig. (*Researches upon the Transformations of the Appendages of the Articulata.*)—Ann. and Mag. N. Hist. XIII. p. 484.

14. Thèse sur le Gisement des Insectes fossiles, etc. Paris, 1839, 4°.

15. Idées nouvelles sur la Classification des Insectes.—C. R. 6 Dec. 1841.—Rev. Zool. 1841, p. 387.

16. Sur quelques points de la Méthode en Histoire naturelle, et en particulier sur les Limites du Genre et de l'Espèce. Par. 1839, 4°.

17. Observations sur l'absence des Tarses dans quelques Insectes.— Ann. Sc. n. 1837, VIII. p. 246.

Brullé (Aug.) et Guérin Méneville (F. E.).

1. Expédition scientifique en Morée: Entomologie. 4°. atl. fol.— Ann. Soc. Ent. Fr. I. p. 116.

BRULLEY (C. A.).

1. Essai sur la Cochenille. 8°. Paris, 1794.

BRUMATI (Leon.).

1. Catalogo sistematico delle Conchiglie terrestri e fluviatili osservate nel territorio di Monfalcone. Gorizia, 1838, 8°.—Rev. Zool. 1839, p. 143.

BRUN (T.).

1. Formation soudaine d'une Ile nouvelle sur les Côtes de l'Italie, à la suite d'éruptions volcaniques.—Bull. Soc. Géogr. XVI. p. 87. —*Féruss.* Bull. XXVI. p. 235.

BRUN DES BEAUNES (De).

1. Tableau méthodique de tous les genres des Productions naturelles qui se trouvent en France. Par. 1811, 8°.

BRUNCRONA (   ).

1. Observations relatives à l'Abaissement présumé du niveau de la Mer Baltique, suivies de remarques sur le même sujet par *C. P. Haelström.*—Vet. Akad. Handl. 1824, I. p. 21.—*Féruss.* Bull. II. 1824, p. 133.

BRUNELLI (Gabr.).

1. De Reptilium Organo Auditus.—Comm.Instit. Bonon.VII. p. 301. —Dict. Sc. n. XV. p. 244.

2. De Locustarum Anatome.—Comm. Instit. Bonon. VII. p. 24.— Bibl. Ent. I. p. 53.

BRUNET (Cl.).

1. Traité raisonné sur la Structure des Organes des deux sexes destinés à la Génération. 1696.—Biogr. Un. VI. p. 115.

BRUNET DE LAGRANGE (   ).

1. Vers à soie: Tableau synoptique publié sous les auspices du Ministre du Commerce et de l'Agriculture. Paris, 1838.—Rev. Zool. 1838, p. 73.

BRUNI (L.).

1. Sulle duplici funzioni del Muscolo sterno-cleido-mastoideo. Napoli, 1840.

BRUNN (Fr. Leop.).

1. Anmerkungen und Zusätze zu *Ahrens* Verzeichniss der Schmetterlinge, welche zu Schloss-Ballenstädt gefunden und beobachtet worden.—*Fuess.* N. Ent. Mag. II. pp. 64, 80.—Bibl. Ent. I. p. 53.

BRÜNNER (E.).

1. De Vesicularum Sanguinis naturâ Observationes microscopicæ et chemicæ. 1835, 8°.—*Val.* Repert. I. p. 8.

2. Disquisitio experimentis illustrata de singularum Nervi sympathici partium vi in cor. Berol. 1836, 4°.—*Val.* Repert. II. p. 21.

BRUNNER (Jos.).

1. Neue Hypothese von Entstehung der Gänge. Leipz. 1801, 4°. fig.

2. Handbuch der Gebirgskunde für angehende Geognosten. Leipz. 1803, 8°. fig.

3. Handbuch der mineralogischen Diagnosis. Leipz. 1804, 8°.

4. Versuch eines neuen Systems der Mineralogie. Leipz. 1800, 8°.

5. Ueber das Bairische Waldgebirge.—*Leonh.* Tasch. 1817, p. 357,

BRUNNER (J. Conr.).

1. Dissertatio de Glandulâ pituitariâ. Heidelb. 1688, 4°.—Biogr. Un. VI. p. 123.

2. Glandulæ Duodeni, seu Pancreas secundarium detectum.—Francf. et Heidelb. 1715, 4°.—Biogr. Un. VI. p. 123.

3. Experimenta nova circa Pancreas; accedit Diatriba de Lymphâ et genuino Pancreatis usu. Amst. 1682, 8°. Leyde, 1709, 1722, 8°.—Act. Erudit. II. p. 198.—Biogr. Un. VI. p. 123.

4. De Animalium et Insectorum variorum Excretione per Os.—Eph. Nat. Cur. Dec. III. Ann. 5 et 6, p. 661.

BRÜNNICH (M. Thr.).

1. Literatura Danica Scientiarum naturalium. Hafn. 1783, 8°.—*Böhm.* Bibl. I. 1, p. 175.

2. Introductio ad Historiam Progressuum quos fecit Historia naturalis eique analogæ Scientiæ. Copenh. 1782, fol.—(Gall. *Yanss. des Campeaux*). Copenh. 1783, 8°.—*Böhm.* Bibl. I. 1, p. 609.

3. Dyrenes Historie udi Universitetets Natur-Theater. 4°. Copenh. 1782.

4. Forsög til Mineralogie for Norge. Nidros. 1777, 8°.—*Böhm.* Bibl. IV. 1, p. 78.

5. Mineralogie afhandlende Egenskaber og Brug af Jord-og Steenarter, etc. Copenh. 1777, 8°.—(Germ.) .... *aus dem Dän. übers. mit Zusätzen des Verfassers, u. einer Anzeige der bisher bekannten russischen Mineralien vermehrt.* Petersb. u. Leipz. 1781, 8°. (v. *G. Georgi*).—*Böhm.* Bibl. IV. 1, p. 78.

6. Les Progrès de l'Histoire naturelle en Danemark et en Norwège. Copenh. 1789, 8°.

7. Beschreibung einer Seltenen Dünn- oder Tellmuschel.—Berl. Beschäft. III. p. 313, fig.—*Mod.* B. H.

8. Sur les Trilobites.—Danske Vidensk. etc. (N. Samml.) I. p. 384.

9. Catalogus Bibliothecæ Historiæ naturalis. Hafniæ, 1793, 8°.— Tr. Linn. Soc. Lond. V. p. 279.

10. Zoologiæ Fundamenta prælectionibus acad. accommodata (Lat. et Dan.). Hafn. et Lips. 1772, 8°.—Tr. Linn. Soc. Lond. VII. p. 241.—*Mod.* B. H.

11. Ichthyologia Massiliensis, etc. Hafn. et Lips. 1768, 8°.—*Cuv.* R. An. III. p. 343.—*Böhm.* Bibl. II. 2, p. 62.

12. Spolia e Mari Adriatico reportata. (Impr. avec son Ichthyologie de Marseille.) Hafn. et Lips. 1768, 8°.—*Dum.* et *Bib.* I. p. 310.

13. Om den Islandske Fisk Vogmeren.—Danske Vidensk. etc. III. p. 408.—Dict. Sc. n. XXII. p. 535.

14. Den barbudgede Pampelfisk (*Coryphæna Apus*), en nye Art. —Danske Vidensk. etc. II. p. 319; III. p. 406.—Dict. Sc. n. XXII. p. 535.

15. Descriptio *Gadi ranini.*—Act. Hafn. XII.—Dict. Sc. n. XXII. p. 529.

16. Prodromus Insectologiæ Siællandicæ. Hafniæ, 1761, 8°.—*Eis.* Insect. p. 153.

17. Entomologia, sistens Insectorum tabulas systematicas, cum Introductione et Iconibus. Copenh. 1764, 8°. fig. (Lat. et Dan.) —*Eis.* Insect. p. 139.—*Böhm.* Bibl. II. 2, p. 142.

18. Lettre sur quelques Plantes et Insectes rares observés en Espagne.—Phil. Tr. XXIV. No. 301, p. 2015.—Bibl. Ent. I. p. 54.

19. Ornithologia borealis, sistens collectionem Avium in omnibus, Imperio Danico subjectis, Insulis borealibus, factam. Hafn. 1764, 8°. fig.—*Böhm.* Bibl. II. 1, p. 512.

20. Ederfuglens beskrivelse. Copenh. 1763, 8°. fig.—*Böhm.* Bibl. II. 1, p. 567.

21. Om Sild-tuften; *Regalecus remipes.*—Skr. Kiöbenh. Selsk. N. Saml. III. p. 414.

22. Om en ny fiskeart den draabe plettede pladefisk, **fanget ved** Helsignoër i Nordsöen 1786. *Zeus guttatus.*—Skr. Kiöb. Selsk. N. Saml. III. p. 398.

23. Velsens beskrivelse.—Danske Vidensk. etc. XII. p. 291.

BRUNNWISER (M.).

1. Lithologische Beobachtungen.   Abh. Baiersch. Akad. IX. p. 153.

BRUNO (Giov. Dom.).

1. Illustrazione di un nuovo Cetaceo fossile.

BRUNO (J.).

1. De Origine Fontium.   Hamb. 1666, 4°.—*Böhm.* Bibl. V. p. 63.

BRUNS (V.).

1. De Nervis Cetaceorum cerebralibus.   Tübing. 1836. 8°.—*Müll.* Arch. 1837, IV. p. LVI.

2. Lehrbuch der allgemeinen Anatomie des Menschen, etc. Braunschw. 1841, 8°.— *Val.* Repert. VII. pp. 8, 31.

BRUNSMANN (J.).

1. Diss. de Ceto Jonæ, quâ eum verum fuisse Cetum ostenditur adversus *S. Bochart.*   Jenæ, 1687, 8°.—*Böhm.* Bibl. II. 2, p. 77.

BRUSCH (C.).

1. Beschreibung des Fichtelbergs.   1542.—Witteb. 1612, 4°.— Nürnb. 1683, 4°. (*Z. Theobald* auct.)

BRUSCOLI (   ).

1. Mémoire sur les habitudes d'un Boa qui a vécu pendant 18 mois au Musée de Florence.—Congr. sc. de Pise, 11 Oct. 1839.—Rev. Zool. 1840, p. 30.

BRUSENIUS (D.).

1. Forsök til en beskrifning öfver Löt och Alböke församlingar på Öland.—Vet. Ac. 1776, p. 38.—Schwed. Acad. Abh. 1776, p. 43.

BRUUN-NEERGAARD (T. C.).

1. Journal du dernier Voyage du *Cit. Dolomieu* dans les Alpes. Paris, 1802, 8°.—(Dan. *P. H. Mönster*): *Reise giennem Alperne med Dolomieu.* Kiöb. 1802, 8°.—(Germ. *Karsten.*) Berl. 1802, 8°.—(Germ. ab alio.) Hamb. u. Mainz, 1802, 8°.

2. Voyage pittoresque au Nord de l'Italie. Par. 1812–15, fol. fig. (Livr. I.–VI.).

BRUYN (Abr. de).

1. De Zee-worm beschouwt in zyn eigen Aart en Natuur, waar door zy die schadelyke, etc. Rotterd. 1735, 8°.—*Böhm.* Bibl. II. 2, p. 489.—*Mod.* B. H.

BRUYN (Corn. de).

1. Relatio de Ostreis petrificatis, illustrata per *J. Ch. Klein.*—Phil. Trans. XLI. p. 568.—*Mod.* B. H.

2. Reysen door der Levant, of Klein Asien, etc. Delft, 1698, fol. fig.—(Gall.) *Voyage au Levant,* etc. Delft, 1700.—Par. 1714, fol. fig. et 1725, 4°.—Rouen, 1725, 4°. 5 vols. (avec le Voyage en Russie).—*Böhm.* Bibl. X. 1, p. 663.—*Mod.* B. H.

3. Reisen door Moscovien, door Persien en Indien. Amst. 1711, fol. fig.—Ibid. 1714, fol.—(Gall.) *Voyages par la Moscovie, en Perse et aux Indes Orientales, ouvrage enrichi de plus de 300 pl. représentant les Animaux, les Oiseaux, les Poissons et les Plantes, et quelques remarques contre MM. Chardin et Kæmpfer.* Amst. 1700 et 1718, fol. fig.—*Böhm.* Bibl. I. 1, p. 631.

BRUYN (S. J.).

1. On the Retreat of Swallows and the torpid state of certain Animals in Winter.—Mem. Amer. Acad. II. p. 96.

BRUZELIUS (N.).

1. Diss. sistens species cognitas Asteriarum. 4°. Lund. 1805.

BRYAM (Fr.) and POND (Arth.).

1. An Account of the Impression on a Stone dug up in the Island of Antigua.—Phil. Trans. XLIX. p. 295.

BRYANT (H. J.).

1. Report upon a Volume of the Naturalist's Library, on Gallinaceous Birds.—Proc. Bost. Soc. N. H. 1841, p. 19.

2. Report on a Parrot recently presented by *M. Teschemacher.*—Proc. Bost. Soc. N. H. 1842, p. 60.

BRYANT (Will.).

1. Account of an Electrical Eel, or the Torpedo of Surinam.—Tr. Amer. Soc. II. p. 166.—Dict. Sc. n. XXII. p. 546.

BRYCE (J.).

1. On the Discovery of the Plesiosaurus in Ireland.—Phil. Mag. ser. 2, IX. p. 331.—L. u. Br. N. Jahrb. 1833, p. 708.

2. On the Geological Structure of the North-eastern part of the county of Antrim.—Trans. Geol. Soc. ser. 2, V. p. 69.—Proc. Geol. Soc. I. 397.—Phil. Mag. ser. 3, I. p. 228.

3. Ueber Diluvial-Wirkungen im nördlichen Ireland.—J. Geol. Soc. Dubl. I. 1, p. 34.—L. u. Br. N. Jahrb. 1836, p. 400.

4. On the Discovery of some Remains of the Ichthyosaurus in Ireland.—Phil. Mag. ser. 3, XX. p. 83.

5. Manual of Zoology. 8°. Belfast, 1834.

6. Tables of simple Minerals, Rocks and Shells, with local Catalogue of Species. Belfast, 1831, 8°.

7. On Caverns containing Bones near the Giant's Causeway.—Rep. Brit. Assoc. 1834, p. 658.

BRYDONE (Patr.).

1. A Tour through Sicily and Malta, in a series of Letters to W. Beckford. Lond. 1773, 2 vols. 8°.—(Germ.) Leipz. 1774, 2 vols. 8°.—Ibid. 1777.—(Gall. de Meunier.) Amst. and Par. 1775–76, 12°. 2 vols. ; 1780, id.

BRYLLI (Hipp.).

1. Opusculum de Vermibus in Corpore humano genitis. Venet. 1540, 8°.—Mod. B. Helm.

BSCHERER (Dan.).

1. De Gallis Quercuum larvatis.—Misc. Nat. Cur. 1689, Dec. p. 73. —Bibl. Ent. I. p. 54.

BUCH (L. v.).

1. Geognostische Uebersicht der Gegend von Rom.—N. Schr. Ges. Naturf. Fr. Berl. III. pp. 478, 535.

2. Ueber die geognostische Beschaffenheit der Gegend von Pergine.—Neue Schrift. Gesellsch. Naturforsch. Freunde Berlin, III. p. 233.

3. Pétrifications recueillies en Amérique par M. *Al. de Humboldt* et par M. *Ch. Degenhard*, décrites. Berl. 1839, fol. fig.

4. Ueber die Muscheln im Granaten-Lager von Trziblitz.—*Karst.* Arch. 1838, XI. p. 315.—*L.* u. *Br.* N. Jahrb. 1839, p. 100.

5. Ueber den Jura in Deutschland.—Abh. Berl. Akad. 1839, p. 49, Cart. et fig.—*L.* u. *Br.* N. Jahrb. 1839, p. 339.—Ann. of Nat. H. I. p. 332 (*On the Jura in Germany*).

6. Ueber Goniatiten und Clymenien in Schlesien.—Berl. Akad. März, 1838.—Berl. 1839, 4°. fig.

7. Ueber Sphæroniten u. einige andre Geschlechte, aus welchen Krinoiden entstehen, u. über einige Brachiopoden der Gegend von Petersburg.—Ber. Berl. Ak. 1840, März, fig.—*L.* u. *Br.* N. Jahrb. 1840, p. 732.—Ann. and Mag. N. H. VI. p. 12 (*On Sphæronites and some other genera from which Crinoidea originate*).

8. Beiträge zur Bestimmung der Gebirgs-Formationen in Russland, nebst 3 Taf. u. 1 Karte. Berl. 1840, 8°.—Arch. f. Min. XV.

9. Metamorphismus und Glättung der Gesteine Schwedens.—*L.* u. *Br.* N. Jahrb. 1842, p. 282.

10. Ueber Produkten u. Terebrateln.—*L.* u. *Br.* N. Jahrb. 1842, p. 230.

11. Ueber Productus oder *Leptæna.* 8°. Berl. 1842.—Abhand. Berl. Ak. 1843, p. 1.—Bull. Ac. Berl. 1841.—*L.* u. *Br.* N. Jahrb. 1842, p. 369.

12. Ueber Terebrateln, mit einem Versuche sie zu classificiren. 4°. Berl. 1834.—*L.* u. *Br.* N. Jahrb. 1833, p. 257, fig.—*Wiegm.* Arch. 1835, p. 319.

13. Salzburger u. andere Ammoniten, etc.—*L.* u. *Br.* N. Jahrb. 1833, p. 186.

14. *Zieten's* Versteinerungen Würtembergs.—*L.* u. *Br.* N. Jahrb. 1833, p. 322.

15. Erscheinungen bei Meissen, etc.—*L.* u. *Br.* N. Jahrb. 1834, p. 532.

16. Carte physique de l'Ile de Ténériffe.—Bull. Soc. Géol. Fr. II. p. 43.

17. Lettre à M. *Al. Brongniart* sur le Gisement des Couches calcaires à empreintes de Poissons, et sur les Dolomies de la Franconie.—Bull. Soc. Géol. Fr. VI. p. 97.—J. de Phys. Oct. 1822.—*Féruss.* Bull. 1823, II. p. 47.—*Leonh.* Tasch. 1824, II. p. 239 (*Ueber die Lagerungs-Verhältnisse der Kalkschichten mit Fisch-Abdrücker u. über den Dolomit im Frankenlande*).

18. Einige Bemerkungen über die Alpen in Baiern.—Abh. Berl. Akad. 1831, p. 73, fig.—*L.* u. *Br.* N. Jahrb. 1834, p. 612.

19. Ueber zwei neue Arten von Cassidarien in den Tertiär-Schichten v. Mecklenburg.—Abh. Berl. Akad. 1831, p. 43.

20. Ueber das genus *Delthyris.* 4°. Berlin, 1838.—*L.* u. *Br.* N. Jahrb. 1836, p. 175.

21. Recueil de planches de pétrifications remarquables. fol. Berl. 1831.

22. Versuch einer mineralogischen Beschreibung von Landeck. Bresl. 1797, 4°.

23. Reise durch Norwegen und Lappland. Berl. 1810, 2 vols. 8°. fig.

24. Ueber die Zusammensetzung der basaltischen Inseln und über Erhebungs-Cratere.—Abh. Berl. Ak. 1818–19, p. 51.—*Leonh.* Tasch. 1821, II. p. 391.

25. Ueber einen vulcanischen Ausbruch auf den Insel Lanzerote.— Abh. Berl. Ak. 1818–19, p. 69, fig.—*Leonh.* Tasch. 1821, II. p. 428.

26. Ueber einige Berge der Trappformation in der Gegend von Grätz.—Abh. Berl. Ak. 1818–19, p. 111.—*Leonh.* Tasch. 1821, II. p. 457.

27. Ueber Granit und Gneiss, vorzüglich in Hinsicht der äussern Form, etc.—Abh. Berl. Akad. 1844, p. 57.—*Pogg.* Annal. 1843, p. 289.—Edinb. New Phil. Journ. XXXV. p. 316.—Journ. Geol. Soc. I. p. 126.

28. Einige Bemerkungen über Quellen-Temperatur.—Abh. Berl. Akad. IX. 1828, p. 93.—Isis, 1834, IV. p. 355.—*Féruss.* Bull. 1830, XXI. p. 18 (*Quelques observations sur la Température des Sources*).

29. Reise zwischen Glarus und Chiavenna.—Berl. Mag. III. p. 102. —Isis, 1818, XI. p. 1819.

30. Physikalische Beschreibung der Canarischen Inseln. Berl. 1825, 4°. Atl. fol.—Isis, 1829, VII. p. 695.—*Féruss.* Bull. 1826, IX. p. 388 (*Description physique des Canaries*).—(Trad. fr. par *C. Boulanger*): *Description physique des Iles Canaries, suivie d une Indication des principaux Volcans du Globe.* Paris, 1836, 8°. Atl. fol.

31. Ueber die Silification organischer Körper, nebst einigen andern Bemerkungen über wenig bekannte Versteinerungen.—Abh. Berl. Ak. XII. 1831, p. 43, fig.—Isis, 1834, IV. p. 363.

32. Ueber locale und allgemeine Gebirgsformationen.—Berl. Mag. IV. p. 69.—Isis, 1818, XI. p. 1821.—*Leonh.* Tasch. 1814, II. p. 518.

33. Sur les formes et les relations des Volcans ; d'après sa Description physique des îles Canaries.—Ann. Sc. n. 1830, XIX. p. 390.

34. Ueber den Pic von Teneriffa.—Abh. Berl. Ak. 1820-21, p. 93.
—Min. Taschenb. 1823, IV. p. 813.—*Féruss.* Bull. 1824, p. 222.
(*Sur le Pic de Ténériffe.*)

35. Das Berninagebirge in Graubündten.—Abh. Berl. Akad. 1818,
p. 105, fig.—*Leonh.* Tasch. 1822, I. p. 31.

36. Sur la dispersion des Blocs alpins.—*Poggend.* Ann. 1827, IX.
4, p. 575.—*Féruss.* Bull. 1828, XIV. p. 5.

37. Ueber einige geognostische Erscheinungen in der Umgebung
des Luganer Sees in der Schweiz.—Abh. Berl. Ak. 1830, XI.
p. 193, fig.—Isis, 1834, IV. p. 361.—Unterhalt. Blätt. 1826, No.
823-834.— *Féruss.* Bull. 1827, X. p. 43.—Ann. Sc. n. 1827,
X. p. 195.—*Leonh.* Zeitschr. 1827, I. p. 289.

38. Ueber den Gabbro bei Plimouth.—Berl. Mag. IV. p. 128 ; VII.
p. 234.—Isis, 1818, XI. p. 1822.—*Leonh.* Tasch. 1815, II. p. 467.

39. Ueber Dolomit als Gebirgsart (*Sur la Dolomie considérée
comme roche*).—Abh. Berl. Akad. 1823.—*Féruss.* Bull. 1824,
III. p. 280.

40. Note sur l'Ile de Madère.—Ann. Sc. n. 1825, IV. p. 14.

41. Ueber die geognostische Konstitution von Diemensland, nach
der Pariser Sammlung.—Berl. Mag. VI. p. 234.—Isis, 1818, XI.
p. 1826.

42. Ueber die Lagerung von Melaphir und Granit in den Alpen
von Mailand, nebst Charte.—Abh. Berl. Akad. 1830, XI. p. 205.
—Isis, 1834, IV. p. 362.—*L.* u. *Br.* N. Jahrb. 1834, p. 421.

43. Ueber die Steinkohlen von Entrévernes in Savoyen.—Berl.
Mag. I. p. 25.—Isis, 1818, XI. p. 1813.

44. Von den geognostischen Verhältnissen des Trapp-Porphyrs.—
Abh. Berl. Akad. 1816, p. 129.—*Leonh.* Tasch. 1819, I. p. 200.

45. Ueber das Vorkommen des Tremolits im Norden.—Berl. Mag.
III. p. 172.—Isis, 1818, XI. p. 1820.

46. Ueber die Eisenerzlager in Schweden.—Berl. Mag. IV. p. 46.—
Isis, 1818, XI. p. 1821.

47. Carte géologique du Terrain entre le lac d'Orta et celui de
Lugano.—Ann. Sc. n. Nov. 1829.—*Féruss.* Bull. 1830, XXI. p. 43.

48. Ueber die Verbreitung des Hippuriten-Kalkes.—Min. Zeitschr.
1829, p. 376.—*Bronn* Lethæa II. p. 551.

49. Ueber Ammoniten, etc. 4°. Berlin, 1832.—Ann. Sc. n. 1829.
XVII. p. 267 ; XVIII. p. 417.—*Féruss.* Bull. 1829, XIX. p. 120 ;
1830, XXI. p. 325.—Isis, 1834, XI. p. 1078 et 1101.

50. Geognostische Briefe an Hen. *Al. v. Humboldt,* herausg. von
Geheimr. *v. Leonhard.* Hanau, 1824, 8°.—Cat. Bibl. Turic.

51. Geognostische Beobachtungen auf Reisen. Berl. 1802, 1809, 2 vols. 8°.—Cat. Bibl. Turic.

52. Ueber einige neue Versteinerungen aus Moskau.—*L.* u. *Br.* N. Jahrb. 1844, p. 536, fig.

53. Ueber die Cystideen, eingeleitet durch die Entwickelung der Eigenthümligkeiten des *Caryocrinus ornatus.*—Ber. Berl. Akad. 1844, März, p. 120.—*L.* u. *Br.* N. Jahrb. 1844, p. 507.—Journ. Geol. Soc. I. pt. 2. p. 20 ; II. p. 11.

54. Ueber die Uebergangsgebirgsarten.—*Leonh.* Tasch. VI. p. 335.

55. Ueber die Ursachen der Verbreitung grosser Alpengeschiebe. —Abh. Berl. Akad. 1815, p. 161.—*Leonh.* Tasch. 1818, II. p. 458.

56. Ueber das Uebergangs-Gebirge bei Fangsbierg (Reise durch Norw. u. Lappl. I. p. 169).—*Leonh.* Tasch. 1816, II. p. 471.— Ueber die Gebirgsarten in Guldbrandsdalen (*Ibid.* p. 188).— *Ibid.* p. 473.

57. Mineralogische Nachrichten über Finnmarken (Reise d. Norw. u. Lappl. p. 3).—*Leonh.* Tasch. 1816, II. p. 485.

58. Ueber den Dolomit in Tyrol.—Tyrol. Bot. 1822, July.—*Leonh.* Tasch. 1824, II. p. 272.

59. Geognostisches Gemälde von Süd-Tyrol.—Ann. Chim. XXIII. p. 276.—*Leonh.* Tasch. 1824, II. p. 288.

60. Ueber das Vorkommen des Dolomites in der Nähe der vulka- nischen Gebilde der Eifel.—*Nögg.* Gebirge in Rh. Westph. III. p. 280.—*Leonh.* Tasch. 1824, II. p. 331.

61. Ueber geognostische Erscheinungen im Fassathale (Brief an *Leonhard*).—*Leonh.* Tasch. 1824, II. p. 343.

62. Ueber die geognostischen Systeme von Deutschland (Brief an *Leonhard*).—*Leonh.* Tasch. 1824, II. p. 501.

63. Ueber *Terebratula Mantzelii* im Tarnowitzen Muschelkalke.— *L.* u. *Br.* N. Jahrb. 1843, p. 253.

64. Ueber die Karnischen Alpen.—*Leonh.* Tasch. 1824, II. p. 396.

65. Ueber den Thüringen Wald.—*Leonh.* Tasch. 1824, II. p. 437.

66. Ueber den Harz.—*Leonh.* Tasch. 1824, II. p. 471.

67. Ueber *Delthyris* oder *Spirifer* und *Orthis.*—Abh. Berl. Akad. 1838, p. 1, fig.—*L.* u. *Br.* N. Jahrb. 1838, p. 221.

68. Ueber *Thurmann's* Soulèvemens jurassiques.—*L.* u. *Br.* N. Jahrb. 1838, p. 385.

69. Ueber den zoologischen Charakter der Secundär-Formationen in Süd-Amerika.—Berl. Akad. Apr. 1838, p. 54.—*L.* u. *Br.* N. Jahrb. 1838, p. 607.

70. *Terebratula hastata* u. *T. Sacculus.—L.* u. *Br.* N. Jahrb. 1839, p. 431.

71. Die Formationen bei St.-Triphon.—*Bunsen's* Beobachtung v. Erdöl-Quellen bei Peina u. Celle.—*L.* u. *Br.* N. Jahrb. 1839, p. 696.

72. Sur les Ammonites et leur distribution en familles; sur les espèces qui appartiennent aux Terrains les plus anciens, et sur les Goniatites en particulier.—Ann. Sc. n. 1833, XXIX. p. 5.—Abh. Berl. Akad.—*L.* u. *Br.* N. Jahrb. 1833, p. 231 (*Ueber Ammoniten über ihre Sönderung in Familien, über die Arten, etc.*).

73. Ueber die Hippuriten bei Reichenhall.—Naturf. in Münch. 1827.—Isis, 1828, V. p. 438.—*Féruss.* Bull. 1829, XVIII. p. 335 (*Sur les Hippurites*).

74. Note sur les Huitres, les Gryphées et les Exogyres.—Ann. Sc. n. 1835, III. p. 296.—*Quenst.* in *Wiegm.* Arch. 1836, II. p. 335 (*Note über Austern, Gryphäen und Exogyren*).—*L.* u. *Br.* N. Jahrb. 1836, p. 250.

75. Nature des Phénomènes volcaniques dans les Iles Canaries.— Mém. Soc. Linn. Norm. 1829, I. p. 76.—Bull. Soc. Géol. Fr. I. p. 240.—*Féruss.* Bull. XXV. p. 9.

76. On Volcanos and Craters of Elevation.—*Poggend.* Annalen, XXXVII. p. 169.—Edinb. New Phil. Journ. XXI. p. 206.

77. Ueber *Chabrier's* Abhandlung von der Sündfluth.—*Leonh.* Zeitschr. 1825, I. p. 482.

78. Die Bären-Insel nach *B. M. Keilhau* geognostisch beschrieben; und, Ueber *Spirifer Keilhavii*, über dessen Fundort und Verhältniss zu ähnlichen Formen. — Abh. Berl. Akad. 1846.— Journ. Geol. Soc. III. pt. 2. p. 48.

BUCH (Lud. Theo.).

1. Diss. de *Tænia solio*. Kiliæ, 1841.

BUCHAN ( ).

1. Storia d'un Uomo cui mancava la Vesica urinaria.—Ann. Med. Stran. VI.

BUCHANAN (C.).

1. Note sur de l'Eau fraiche trouvée en mer loin de la Terre.— Edinb. Phil. Journ. Janv. 1827, p. 369.—*Féruss.* Bull. 1828, XV. p. 247.

BUCHANAN (Fr. Ham.).

1. Description of the *Vespertilio plicatus.*—Tr. Linn. Soc. Lond. V. p. 261, fig.—Phil. Mag. VII. p. 145.

2. An Account of the *Onchidium*, a new Genus of the Class of Vermes, found in Bengal.—Tr. Linn. Soc. Lond. V. p. 132. fig.

3. An Account of the Fishes found in the River Ganges and its Branches, etc. Edinb. 1822, 4°. fig.—*Féruss.* Bull. 1823, III. p. 253 (*Description des Poissons du Gange, etc.*).

4. A Journey from Madras through Mysore, Canara and Malabar. Lond. 1807, 3 vols. 4°.—Cat. Bibl. Turic.

BUCHANAN (G.).

1. Diss. inaug. de causis Respirationis ejusdemque effectibus. 8°. Philad. 1789.—Cat. R. Soc. L.

2. Physiological Illustrations of the Organ of Hearing. Lond. 1828, 8°. fig. (*Illustrazioni fisiologiche sull Organo dell' Udito*). —*Omod.* Ann. Univ. Med. XXXVII. e XLVIII.

BUCHANAN (S.).

1. A Lecture introductory to a Course of Anatomy, delivered to the Students of *Anderson's* University. Glasgow, 1842, 8°.

BUCHANAN (T.).

1. Sketches of the Comparative Anatomy of the Organ of Hearing, founded chiefly on the Ear of the *Squalus.*—Mem. Wern. Soc. VI. p. 144.

BUCHARD (E. F.).

1. Epistola de Cocco Polonico.—Act. Ups. 1742, p. 53.—N. Hamb. Mag. IV. p. 481.—Bibl. Ent. I. p. 54.

BUCHER (P.).

1. Histoire véritable et naturelle des Mœurs et Productions du pays de la Nouvelle France vulgairement dite le Canada. Paris, 1664, 12°.—*Böhm.* Bibl. I. 1, p. 748.

BUCHER (Sam. Fr.).

1. Dissert. de variis Corporibus petrefactis. Witteb. 1715, 4°.— *Böhm.* Bibl. IV. 2, p. 234 (Falso *Büchner*).—*Mod.* B. H.

BUCHER (Urb. G.).

1. Sachsenlandes Naturhistorie. Dresd. 1723, 8°.—*Böhm.* Bibl.
I. 1, p. 601.—Biogr. Un. VI. p. 201.

2. Ursprung der Donau und Beschaffenheit der Landgrafschaft
Fürstenberg. Nürnb. 1720, 8°. fig.—*Böhm.* Bibl. I. 1, p. 589.

BUCHET (J. P. A.).

1. Extrait d'un Mémoire sur une Caverne à Ossemens fossiles, découverte à l'est de St.-Jean-du-Gard.—Mém. Soc. Phys. et H.
n. VI. p. 369 ; VII. p. 1.—*L.* u. *Br.* N. Jahrb. 1835, p. 242.

BÜCHNER (Andr. El.).

1. Falsò credita Metamorphosis summa miraculosa Insecti cujusdam
Americani.—N. Act. Nat. Cur. III. 1767, p. 437.—Bibl. Ent. I.
p. 55.

2. Acad. Leop. Car. Historia ; *vide sup.* Pars I. p. 22.

BÜCHNER ( ).

1. Repertorium der Pharmacie ; *vide sup.* Pars I. p. 32.

BÜCHNER (G.).

1. Mémoire sur le Système nerveux du Barbeau (*Cyprinus Bar-
bus,* Linn.). 4°. Strasb. 1835.—Mém. Soc. Strasb. II.

BÜCHNER (J. G.).

1. De rarioribus quibusdam animalibus in Voigtlandia quondam
natis ac degentibus.—Acta Acad. Nat. Cur. IV. p. 261.

2. De Equo Cornuto et Hippolitho.—Acta Acad. Nat. Cur. VII.
p. 289.

3. De memorabilibus Voigtlandiæ Subterraneis. Plaviæ, 1743, 4°.
(In *Brückmann* Centuria Epistolarum, 4°. 1742.)—Cat. R. Soc.
L.—*Böhm.* Bibl. IV. 1, p. 149.

BÜCHNER (J. God. II.).

1. Observatio de Lapidibus figuratis in Voigtlandia occurrentibus.
—Act. Phys. Med. VII. p. 281.—*Mod.* B. H.

BUCHOLZ (Fr. H.).

1. Von Bereitung des Ameisenäthers.—*Crell,* Chem. Entd. VI. p. 55.
—Bibl. Ent. I. p. 54.

2. Insectorum species novæ aut parùm saltem descriptæ.   Argent.
1778, 4°.—Bibl. Ent. I. p. 54.

3. Reise auf die Karpatischen Gebirge.—Ungrisches Magazin, IV.
1787.

BUCHOZ (P. JOS.).

1. Lettres périodiques sur les avantages que la Société peut retirer
de la connaissance des Animaux. Paris, 1769 et 1770, 8°.—
Dict. Sc. n. XXII. p. 511.

2. Histoire naturelle de la France. Paris, 1776, 8°. 14 vols.—
Biogr. Un. VI. p. 205.

3. *Aldrovandus Lotharingiæ*, ou Catalogue des Animaux qui habitent la Lorraine et les Trois Evêchés. Par. 1771, 8°.—Bibl.
Ent. I. p. 54.

4. Dons merveilleux et diversement coloriés de la Nature dans le
règne animal. Par. 1781, 1797, fol. 2 vols. fig. col.—Bibl. Ent.
I. p. 55.

5. Histoire des Insectes nuisibles à l'homme, aux bestiaux, à l'agriculture et au jardinage, avec les moyens qu'on peut employer
pour les détruire ou s'en garantir, ou remédier aux maux qu'ils
ont pu occasionner. Par. 1781, 12°. Par. an. 7.—*Eis.* Insect.
p. 170.—*Böhm.* Bibl. II. 2, p. 181.

6. Hist. nat. physique et médicinale de l'Homme.   4 vols. 8°. Paris,
1784.

7. Planches enluminées et non enluminées pour servir d'intelligence
à l'Hist. gen. des trois Règnes de la Nature. Paris, 2 vols. fol.
1775–79.—*Mod.* Bibl. Helm.

8. Hist. gén. des Animaux, des Végétaux, et des Minéraux qui se
trouvent dans le Royaume. Par. 1778, fol.

9. Hist. gén. des Animaux, des Végétaux, et des Minéraux qui se
trouvent hors du Royaume. Par. 1778, fol. fig.

10. Les Richesses de la France dans ses productions minéralogiques
et hydrologiques. Par. 1782, 8°.

11. Diss. sur l'Hist. nat. des environs de Pont à Mousson en Lorraine. Par. 1790, fol.

12. Hist. générale et économique des trois Règnes de la Nature.
fol. Paris, 1777 ; 2 vols. 8°. Paris, 1781–83.

13. Hist. Nat. de la Taupe et de la Taupe-Willon. 8°. Paris, 1806.

14. Traité économique et physique des Oiseaux de Basse-cour, etc.
Par. 1775, 2 vols. 12°.—(Germ.) *Œkonomisch-physische Ab-*

handlung vom Federvieh, etc. übers. v. J. W. Consbruch. Münst. 1777, 8°.—*Böhm.* Bibl. II. 1, p. 504.

15. Dictionnaire minéralogique et hydrologique de France, etc. Paris, 1773–76, 4 vols. 8°.—Ibid. 1785, id.—*Böhm.* Bibl. IV. 1, p. 98.

16. *Wallerius Lotharingiæ,* ou Catalogue des Mines, Terres, Fossiles, Sables et Cailloux qu'on trouve dans la Lorraine et les Trois-Evêchés, etc. Nancy, 1768, 8°; 1769, 12°.—*Böhm.* Bibl. IV. 1, p. 103.

17. Traité de la Pêche. 12°. Paris, 1786.

18. Dissert. sur les Vosges, de même que sur leurs productions naturelles et économiques. Par. 1798, fol.

19. Traité physique et économique du Règne Animal. 2 vols. fol. Paris, 1790.

20. La Nature considérée, etc.; *vide sup.* Pars I. p. 51.

21. Dissert. sur le Cantal, montagne fameuse de la Haute-Auvergne. Par. 1792, fol.

BUCHWALD (B. J. v.).

1. Insectologiæ Danicæ specimen. Hafn. 1760, 4°.—Bibl. Ent. I. p. 55.

BUCK (J. F.).

1. Von einigen in der Erde befindlichen denkwürdigen Höhlen, etc. 1768, 4°.

BUCKINGHAM (Duke of).

1. Extrait d'une Lettre contenant le récit de certaines Phénomènes qui ont accompagné la dernière Eruption du Vésuve.—Ann. Sc. n. 1829, XVI.—Rev. Bibl. p. 34.

2. On the geological structure of the Island of Pantellaria.—Rep. Brit. Assoc. 1832, p. 584.—*L.* u. *Br.* N. Jahrb. 1834, p. 89.

BUCKLAND (W.).

1. Account of an assemblage of Fossil Teeth and Bones of Elephant, Rhinoceros, Hippopotamus, Bear, Tiger, and Hyæna, and sixteen other animals, discovered in a cave at Kirkdale, Yorkshire, in the year 1821.—Phil. Trans. CXII. p. 171.—Isis, 1825, VII. Litt. Anz. p. 35, 59.

2. *Reliquiæ Diluvianæ,* or Observations on the Organic Remains attesting the action of an universal Deluge. Lond. 1820, 4°. fig. —*Féruss.* Bull. 1823, III. p. 281.

3. On an Insulated Group of Rocks of Slate and Greenstone in Cumberland and Westmoreland, on the east side of Appleby, between Melmerby and Murton.—Trans. Geol. Soc. Lond. ser. 1, IV. p. 105.

4. On a series of Specimens from the Plastic Clay near Reading, Berks : with Observations on the Formation to which those Beds belong.—Trans. Geol. Soc. Lond. ser. 1, IV. p. 277.

5. On the Paramoudra, a singular fossil body that is found in the Chalk of the North of Ireland ; with some general observations upon Flints in Chalk, tending to illustrate the History of their formation.—Trans. Geol. Soc. ser. 1, IV. p. 413.

6. On the Geological Structure of a part of the Island of Madagascar, founded on a Collection transmitted by Governor Farquhar, in the year 1819 ; with Observations on some Specimens from the Interior of New South Wales, collected during Mr. Oxley's Expedition to the River Macquarie, in the year 1818.— Trans. Geol. Soc. ser. 1, V. p. 476.—Isis, 1823, IV. Litt. Anz. p. 204.—Bull. Soc. Philom. 1820, p. 95.

7. On the Quartz Rock of the Lickey Hill in Worcestershire, and of the Strata immediately surrounding it ; with considerations on the evidences of a Recent Deluge afforded by the gravel beds of Warwickshire and Oxfordshire, and the valley of the Thames from Oxford downwards to London.—Trans. Geol. Soc. ser. 1, V. p. 506.—Isis, 1823, IV. Litt. Anz. p. 188.

8. On the Megalosaurus, or great Fossil Lizard of Stonesfield.— Trans. Geol. Soc. ser. 2, I. p. 390.—Isis, 1825. X. Litt. Anz. p. 113.—*Féruss.* Bull. II. p. 20.

9. On the Excavation of Valleys by diluvial action, as illustrated by a succession of Valleys which intersect the South Coast of Dorsetshire and Devonshire.—Trans. Geol. Soc. ser. 2, I. p. 95. — *Féruss.* Bull. I. 1823, p. 51. — Isis, 1823, IV. Litt. Anz. p. 235.

10. On the Formation of the Valley of Kingsclere and other Valleys, by the Elevation of the Strata that inclose them ; and on the Evidences of the original Continuity of the Basins of London and Hampshire.—Trans. Geol. Soc. ser. 2, II. p. 119.—Phil. Mag. LXV. p. 214.—Mag. Nat. Hist. ser. 1, I. p. 249.—*Féruss.* Bull. 1827, X. p. 26.

11. On a collection of Vegetable and Animal Remains, and Rocks, from the Burmese Country, presented to the Geological Society by J. Crawford, Esq.—Proc. Geol. Soc. I. 71.—Phil. Mag. ser. 2, III. p. 446.—Trans. Geol. Soc. ser. 2, II. p. 377.—Ann. Sc. n. XVI. p. 238.—Isis, 1834, X. p. 1027.

12. On the Cycadeoideæ, a family of Plants found in the Oolite

Quarries of the Isle of Portland.—Trans. Geol. Soc. Lond. ser. 2, II. p. 395.—Ann. Sc. n. 1829, XVI.—Rev. Bibl. p. 36.—Proc. Geol. Soc. I. 80.—Phil. Mag. ser. 2, IV. p. 225.

13. Supplementary Remarks on the supposed Power of the Waters of the Irawadi to convert Wood to Stone.—Trans. Geol. Soc. Lond. ser. 2, II. p. 403.—N. Ed. Phil. J. Jan. 1829, p. 67.—*Féruss.* Bull. XXV. p. 162.

14. On the Secondary Formations between Nice and the Col di Tendi.—Trans. Geol. Soc. ser. 2, III. p. 187.—Proc. Geol. Soc. Lond. I. p. 94.—Phil. Mag. ser. 2, V. p. 384.—*Féruss.* Bull. 1829, XVIII. p. 342.

15. On the Discovery of a new species of Pterodactyle in the Lias at Lyme Regis.—Trans. Geol. Soc. Lond. ser. 2, III. p. 217.— Proc. Geol. Soc. I. p. 96.—Edinb. Phil. Journ. XXII. p. 21.— *Féruss.* Bull. 1829, XVIII. p. 345.—Isis, 1832, VIII. p. 822 (*Neuer Pterodactylus im Lias bei Lyme Regis.*).—Bull. Soc. Géol. Fr. VI. p. 89.—*L.* u. *Br.* N. Jahrb. 1834, p. 369.

16. On the Discovery of Coprolites, or Fossil Fæces, in the Lias at Lyme Regis, and in other Formations.—Trans. Geol. Soc. Lond. ser. 2, III. p. 223.—Proc. Geol. Soc. I. pp. 97, 142.—Phil. Mag. ser. 2, VI. p. 60.—Edinb. Phil. Journ. XXII. p. 23.—*L.* u. *Br.* N. Jahrb. 1833, p. 704.—Bull. Soc. Géol. Fr. VI. p. 89.

17. On the Occurrence of Agates in Dolomitic Strata of the New Red Sandstone Formation in the Mendip Hills.—Trans. Geol. Soc. ser. 2, III. p. 421.—Proc. Geol. Soc. Lond. I. p. 149.

18. On the Discovery of Fossil Bones of the Iguanodon in the Iron Sand of the Wealden Formation in the Isle of Wight, and in the Isle of Purbeck.—Trans. Geol. Soc. ser. 2, III. p. 425.—Proc. Geol. Soc. I. p. 159.—Phil. Mag. ser. 2, VII. p. 54.—*Féruss.* Bull. 1830, XXIII. p. 49.—*L.* u. *Br.* N. Jahrb. 1836, p. 730.

19. On the Bones of Hyænas and other animals in the cavern of Lunel near Montpelier, and in the adjacent strata of Marine Formation.—Proc. Geol. Soc. Lond. I. p. 3.—Phil. Mag. ser. 2, I. p. 66.—Edinb. Journ. Avr. 1827, p. 242.—*Féruss.* Bull. 1828, XIV. p. 21.—*Leonh.* Zeitschr. 1827, II. p. 392.—*Brewst.* Journ. Science, ser. 1, VI. p. 242.

20. On the Discovery of a number of Fossil Bones of Bears in the Grotto of Osselles or Quingey, near Besancon in France.—Proc. Geol. Soc. Lond. I. p. 21.—Phil. Mag. ser. 2, II. p. 147.—Ann. Sc. n. Mars 1827.—*Féruss.* Bull. 1827, XII. p. 13.

21. On the Discovery of a Black Substance resembling Sepia, or Indian Ink, in the Lias at Lyme Regis.—Proc. Geol. Soc. Lond. I. p. 97.—Phil. Mag. ser. 2, V. p. 387.—Edinb. New Phil. Journ. VII. p. 22.

22. On a newly-discovered Gigantic Reptile.—Proc. Geol. Soc. II. p. 190.—Phil. Mag. ser. 3, VII. p. 327.

23. On the Fossil Beaks of four extinct species of Fishes, referable to the genus Chimæra, which occur in the oolitic and cretaceous formations of England.—Proc. Geol. Soc. Lond. II. p. 205.—Phil. Mag. ser. 3, VIII. p. 4.—*L. u. Br.* N. Jahrb. 1836, p. 625.

24. On the Occurrence of Silicified Trunks of large trees in the New Red Sandstone or Poikilitic series at Allesley near Coventry.—Proc. Geol. Soc. Lond. II. p. 439.—Phil. Mag. ser. 3, X. p. 475.

25. On the Discovery of Fossil Fishes in the Bagshot Sands at Goldworthy Hill, four miles north of Guildford.—Proc. Geol. Soc. Lond. II. p. 687.—Phil. Mag. ser. 3, XIII. p. 387.—Rev. Zool. 1839, p. 25.

26. On the Discovery of a Fossil Wing of a Neuropterous Insect in the Stonesfield Slate.—Proc. Geol. Soc. Lond. II. p. 688.—Phil. Mag. ser. 3, XIII. p. 388.—Rev. Zool. 1839, p. 29.

27. On the Occurrence of Keuper-Sandstone in the upper region of the New Red Sandstone formation, or *Poikilitic system*, in England and Wales.—Proc. Geol. Soc. Lond. II. p. 453.—Phil. Mag. ser. 3, XI. p. 106.

28. Address to the Geological Society, 1840.—Proc. Geol. Soc. Lond. III. p. 210.—Phil. Mag. ser. 3, XVII. pp. 307, 387, 508. —Edinb. New Phil. Journ. XXX. p. 57.

29. On the Evidence of Glaciers in Scotland and the North of England.—Proc. Geol. Soc. Lond. III. pp. 332, 345.—Phil. Mag. ser. 3, XVIII. pp. 574, 587.—Edinb. New Phil. Journ. XXX. pp. 194, 202.—Ann. and Mag. N. H. VII. p. 326.

30. On the Agency of Land Snails in corroding and making deep excavations in compact Limestone Rocks.—Proc. Geol. Soc. Lond. III. p. 430.—Phil. Mag. ser. 3, XIX. p. 541.—Ann. and Mag. N. H. VIII. p. 459.—Rep. Brit. Assoc. 1845, Sect. p. 48.

31. Address to the Geological Society, 1841.—Proc. Geol. Soc. Lond. III. p. 469.—Phil. Mag. ser. 3, XX. pp. 418, 512.—Ann. and Mag. N. H. IX. p. 156.

32. On the Glacio-Diluvial phænomena in Snowdonia and the adjacent parts of North Wales.—Proc. Geol. Soc. Lond. III. p. 579.

33. On *Ichthyopatolites*, or Petrified Trackways of Ambulatory Fishes upon Sandstone of the Coal formation.—Proc. Geol. Soc. IV. p. 204.—Ann. and Mag. N. H. XIII. p. 152.—*L. u. Br.* N. Jahrb. 1844, p. 511 (*Ueber Ichthyopatoliten, oder versteinerte Flossen-Spuren wandelnder Fische auf Kohlen-Sandstein*).

34. On the Occurrence of Nodules (called Petrified Potatoes) found

on the shores of Lough Neagh in Ireland.—Journ. Geol. Soc. Lond. II. p. 103.

35. An Account of the Footsteps of the Cheirotherium and five or six smaller animals in the Stone Quarries of Storeton Hill, near Liverpool.—Rep. Brit. Assoc. 1838, Sect. p. 85.—*Sill.* Am. Journ. XXXV. 2, p. 307.

36. On the Action of Acidulated Waters on the surface of the Chalk near Gravesend.—Rep. Brit. Assoc. 1839, Sect. p. 76.

37. On Recent and Fossil Semicircular Cavities caused by air-bubbles on the surface of soft clay, and resembling impressions of rain-drops.—Rep. Brit. Assoc. 1842, Sect. p. 57.

38. On the Mechanical Action of Animals on Hard and Soft Substances during the progress of stratification.—Rep. Brit. Assoc. 1845, Sect. p. 52.

39. Reply to some Observations in Dr. Fleming's Remarks on the Distribution of British Animals.—Edinb. Phil. Journ. XII. p. 304.

40. On the Vitality of Toads enclosed in Stone and Wood.—Zool. Journ. V. p. 314.—Ed. N. Phil. J. XIII. p. 26.—*Sill.* Am. J. XXIII. 2, p. 272.—Isis, 1834, X. p. 988.

41. On Artesian Wells.—Edinb. New Phil. Journ. XXXVII. p. 318.

42. On the remarkable Accumulation of the Exuviæ of Bears, in a Cave at Kühloch in Franconia.—Phil. Mag. LXII. p. 112.

43. On the Discovery of Coprolites in North America.—Phil. Mag. ser. 2, VII. p. 321.

44. Lettre et Envoi de Coprolithes.—Bull. Soc. Géol. Fr. I. p. 227.

45. Ueber die fossilen Megatherium-Reste, welche neuerlich von Südamerika nach England gebracht worden sind.—Rep. 1. a. II. Meet. Brit. Assoc. p. 104.—*L.* u. *Br.* N. Jahrb. 1834, p. 112.

46. Ueber den Bau u. die mechanische Kraft des Unterkiefers des Dinotherium.—*L.* u. *Br.* N. Jahrb. 1835, p. 516.

47. Notiz über die hydraulische Wirkung des Siphons bei den Nautilen, Ammoniten u. andern Polythalamien.—*L.* u. *Br.* N. Jahrb. 1835, p. 631.

48. Bemerkungen über das genus *Belemnosepia* und über den fossilen Dintensack in dem vorderen Kegel der Belemniten.—*L.* u. *Br.* N. Jahrb. 1836, p. 36.

49. On the occurrence of the Remains of Elephants and other Quadrupeds in the Cliffs of Frozen Mud, in Eschscholtz Bay, in Behring's Strait, and in other distant parts of the Shores of the Arctic Seas.— Beechey's Voyage to the Pacific, p. 593.

50. Sur la découverte de l'*Anoplotherium* commun, dans l'île de

Wight.—Ann. of Philos. Nov. 1825, p. 360.—*Féruss.* Bull. 1827, XII. p. 112.

51. Lettre à *Mr. Jameson* sur les Antres habités par des Hyènes.— Ed. N. Phil. J. Jan. 1827, p. 377.—*Féruss.* Bull. XV. p. 59.

52. Ueber die Knochen-Brekzien bei Gibraltar, etc.—Reliqu. Diluv. p. 148.—*Leonh.* Zeitschr. 1825, II. p. 373.

53. Ueber die Struktur der Alpen und der diese Gebirge begren- zenden Theile des Kontinentes, sowie über ihre Beziehung zu den Flöz- u. Uebergangs-Gebilden Englands.—Ann. of Phil. (new ser.) I. p. 450.—*Leonh.* Zeitschr. 1826, I. p. 468.

54. Note sur des traces de Tortues observées dans le Grès rouge.— Ann. Sc. n. 1828, XIII. p. 85.

55. Instructions for conducting Geological Investigations and col- lecting Specimens.—*Sill.* Amer. J. III. 2, p. 249.

56. Geology and Mineralogy considered with reference to Natural Theology. 2 vols. 8°. London, 1836.—Bridgewater Treatises, VI.—*Sill.* Am. J. XXXI. 2, p. 419.—(Trad. fr. p. *Doyère*) *De la Géologie et de la Minéralogie, considérées dans leurs rapports avec la Théologie naturelle.* Par. 1838, 2 vols. 8°. fig.—(Deutsch.) *Ge- ologie u. Mineralogie in Beziehung zur natürlichen Theologie,* a. d. Engl. nach der 2ten Ausg. übersetzt u. mit Anmerkungen u. Zu- sätzer versehen v. *L. Agassiz.* Neuch. 1838–39, 2 vols. 8°.— *L. u. Br.* N. Jahrb. 1839, p. 443.

57. Die Urwelt und ihre Wunder (ein Theil der Bridgewater Bü- cher), a. d. Engl. übersetzt von *Fr. Werner.* Stuttg. 1837, 8°. fig.

58. Supplementary Notes to the first and second edition of the Bridgewater Treatise, with a Plate of the fossil Head and restored Figure of the *Dinotherium.* Lond. 1837, 8°.

59. On the Adaptation of the Structure of the Sloths to their pecu- liar mode of life.—Tr. Linn. Soc. Lond. XVII. p. 17.—*Wiegm.* Arch. 1835, II. p. 332.

60 *Vindiciæ geologicæ* ; or the Connexion of Geology with Religion explained, in an inaugural Lecture, etc. 4°. Oxf. 1820.—Cat. R. Soc. L.

61. Sermon on Death. Oxford, 1839.

BUCKLAND (W.) and CONYBEARE (W. D.).

1. On the South-western Coal district of England.—Trans. Geol. Soc. Lond. ser. 2, I. p. 210.—Isis, 1823, IV. Litt. Anz. pp. 178, 191.—*Leonh.* Zeitschr. 1826, II. p. 94.

2. Sectional Views of the North-east Coast of Ireland.—Trans. Geol. Soc. Lond. ser. 1, V. p. 111.

3. Illustrations of the Landslip on the Coast of Devonshire. Lond. 1840, fol.

BUCKLAND (W.) and DE LA BECHE (H. T.).

1. On the Geology of the Neighbourhood of Weymouth and the adjacent Parts of the Coast of Dorset.—Trans. Geol. Soc. Lond. ser. 2, IV. p. 1.—Proc. Geol. Soc. I. p. 217.—Phil. Mag. ser. 2, VII. p. 454.

BUCKLAND (W.) and GREENOUGH (G. B.).

1. On Vitreous Tubes in Sand-hills near Drig in Cumberland.— Trans. Geol. Soc. Lond. VII. p. 528.

BUCKLAND (W.) and SYKES (W. H.).

1. On the Interior of the Dens of living Hyænas.—Edinb New Phil. Journ. II. p. 377.—*Féruss.* Bull. 1828, XV. p. 59.

BUCKLER (Charles).

1. On the Occurrence of the Red-breasted Tanager near Chelten-ham.—Zoologist, p. 444.

BUCKLEY (S. B.).

1. On the Discovery of a nearly complete Skeleton of the Zyglodon of *Owen* (Basilosaurus of *Harlan*) in Alabama.—Americ. Journ. Science, Apr. 1843, p. 409.

BUCKMAN (J.).

1. On the occurrence of the remains of Insects in the Upper Lias of the County of Gloucester.—Proc. Geol. Soc. IV. p. 211.—Ann. and Mag. N. Hist. XIV. p. 73.

2. Geological Chart of the Oolitic Strata of the Cotteswold Hills, and the Lias of the Vale of Gloucester. fol. Cheltenham, 1843.

BUCQUET ( ).

1. Introduction à l'Étude des Corps naturels tirés du Règne miné-ral. Paris, 1771, 2 vols. 12°.—*Böhm.* Bibl. IV. 1, p. 72.

BÜCQUOY (Jac. de).

1. Reisebeschreibung durch einen Theil des Schlesischen Gebirges. Bünzl. 1783, 8°.—*Böhm.* Bibl. I. 1, p. 578.

2. Aanmerkelyke Ontmædingen in de sestien jaarige Reyse naar de

Indien. Harl. 1745, 4°.—(Germ.) Leipz. 1771, 8°.—*Böhm.* Bibl.
I. 1, p. 672.

BUDA (J. L.).

1. Vulcania Lithosylloge Ætnæ, in classes digesta. Catan. 1789, 8°.

BUDDLE (John).

1. An Account of the Explosion which took place in Jarrow Colliery, Aug. 3, 1830. Newcastle, 1831, 4°.—Tr. Nat. Soc. Northumb.—Cat. R. Soc. L.

2. Synopsis of the several Seams of Coal in the Newcastle district. 4°. Newc. 1831.—Tr. Nat. Soc. Northumb.—Cat. R. Soc. L.

3. On a Whin Dyke lately discovered in the Fenham Division of Benwell Colliery.—Trans. Nat. Hist. Soc. Newcastle, I. p. 9.

4. On Subsidences produced by working Beds of Coal.—Trans. Geol. Soc. ser. 2, VI. p. 165.—Proc. Geol. Soc. III. p. 147.—Phil. Mag. ser. 3, XVI. p. 146.

5. On the Great Fault, called the Horse, in the Forest of Dean Coalfield.—Trans. Geol. Soc. ser. 2, VI. p. 215.—Proc. Geol. Soc. III. p. 287.—Phil. Mag. ser. 3, XVIII. p. 229.

6. On the Newcastle Coal-field.—Rep. Brit. Assoc. 1838, Sect. p. 74.

BUDGE (Jul.).

1. Beitrag zur Lehre von den Sympathien.—*Müll.* Arch. 1839, p. 389.

2. Untersuchungen über das Nervensystem. I.-II. Heft. Frankf. 1840–42, 8°.

3. Die Lehre vom Erbrechen; nach Erfahrungen u. Versuchen. (Vorrede v. *Nasse.*) Bonn, 1840, 8°.— *Val.* Repert. VI. p. 42.

4. Mémoire sur les Sympathies qui existent entre diverses parties du Corps.—C. R. X. p. 820.

5. Ueber den Verlauf der Nervenfasern im Rückenmarke des Frosches.—*Müll.* Arch. 1844, p. 160, t. 6–8.

BUFALINI (Maur.).

1. Saggio sulla Dottrina della Vita. Forli, 1813, 8°.

BUFALINI ( ), etc.

1. Giornale di Patologia; *vide sup.* Pars I. p. 73.

BUFF ( ).

1. Observations sur la probabilité d'un Dépôt salifère en Westphalie.—*Karst.* Arch. XVII. 1, p. 97.—*Féruss.* Bull. XXV. p. 153.

2. Observations géognostiques sur le Terrain de Craie du comté de la Marck et du duché de Westphalie, et sur les Sources salées qu'il paraît renfermer.—*Nögger.* Das Gebirg in Rheinl.-Westphal. III. p. 42.—*Leonh.* Tasch. 1824, III. p. 718.— *Féruss.* Bull. 1824, II. p. 323.

3. Ueber Gang-Bildungen, welche eine Lager-artige Entstehung zu haben scheinen.—*Karst.* Arch. VI. p. 439.—*L.* u. *Br.* N. Jahrb. 1834, p. 688.

4. Observations géognostiques sur le Gîte de l'Antimoine dans la mine Caspori, à Wintrop, etc. en Westphalie.—*Karst.* Arch. XVI. 1, p. 54.—*Féruss.* Bull. XXV. p. 155.

BUFFON (G. L. Leclerc de).

1. Traité des Minéraux. Paris, 1774, 9 vols. 12º.

2. Discours sur la nature des Animaux. Genève, 1754, 12º.— *Böhm.* Bibl. II. 1, p. 38.

3. Sämmtliche Werke, übers. v. *Rave.* 9 vols. 8º. Düsseld. 1836–40, fig.—Id. übers. v. *Schaltenbrandt,* sammt den Ergänzungen nach *Cuvier.* Köln, 1836–1840, 300 Liefer.

4. Naturgeschichte des Menschen. Aus d. Franz. v. *Allmenstein.* Berl. 1805, 8º.

5. Les Époques de la Nature. Par. 1778, 2 vols. 8º. fig.—Ibid. 1790, 2 vols. 12º. fig.—(Germ. *J. F. Hackmann*): *Epochen der Natur.* Pet. u. Leipz. 1781, 2 vols. 8º. fig.—(Pol.) Warsch. 1787, 8º.—(Angl.) 1793, 8º.—(Angl. *Smellie*) 1812, 20 vols. 8º. fig.

6. Lettres à l'Abbé *Bexon.*—Conserv. I. an. VIII.—Biogr. Un. VI. p. 241.

7. Naturgeschichte der vierfüssigen Thiere aus dem Franz. v. *Martini* u. *Otto.* Berl. 1772–1802, 23 part. fig.

8. Œuvres choisies. Paris, 1837, 6 vols. 8º.—N. Bibl. class. XXV.-XXX.

9. Œuvres complètes. 36 vols. 4º. Paris, 1774–1804; 40 vols. 8º. Paris, 1824–32 (par *Lamoureux* et *Desmarest*); 17 vols. 8º. Paris, 1817–19 (par *Lacépède*); 26 vols. 8º. Paris, 1820–22; 34 vols. 8º. Paris, 1801; 32 vols. 8º. Paris, 1825–28 (par *M. A. Richard*); 55 vols. 8º. Paris, 1828–35 (par *G. Cuvier*); 42 vols. 8º. Paris, 1828–31 (par *F. Cuvier*); 80 vols. 18º. Paris, 1829–34; 6 vols. 4º. Paris, 1834 (par *A. Compte*); 9 vols. 8º. Paris, 1837; 20 vols. 8º. Paris, 1833–34 (par *A. Richard*); 6 vols. 8º. Paris, 1837–39; 5 vols. 8º. Paris, 1837–38 (par *Geoffroy St. Hilaire*); 9 vols. 8º. Paris, 1844.—(Ital.) 40 vols. 8º. Venezia, 1820–24.

Buffon (G. L. L. de), Daubenton (L. J. M.), Montbeillard (P. G. de) et Lacépède (B. G. E. de).

1. Histoire naturelle, générale et particulière, avec la description du
cabinet du Roi. 44 vols. 4°. Paris, 1749–1804 ; 90 vols. 12°.
Paris, 1752–1805; 13 vols. 12°. Paris, 1769 ; 129 vols. 8°. Paris,
1798–1807 (rédigé par *C. N. S. Sonnini*) ; 11 vols. 8°. Paris,
1804 (par *B. Bernard*); 36 vols. 18°. Paris, 1825–26 (par *G.
Cuvier*); 76 vols. 18°. Paris, 1799–1802 (par *Lacépède*); 54
vols. 8°. Deux Ponts, 1785–90; 13 vols. 8°. Berne, 1784–86.—
(Germ. *Kästner*) 4°. Hamb. 1750.—(Id. *Haller*) 16 vols. 4°.
Hamb. 1750–74.—(Id. *Martini*) 7 vols. 8°. Berlin, 1771–74.—
(Id. *Funke*) 4 vols. 8°. Hamb. 1803–1808.—(Angl. *Kenrick*) 6 vols.
London, 1755.—(Id. *W. Smellie*) 9 vols. 8°. Edinb. 1781–85 ;
1791 ; 20 vols. 8°. London, 1812.—(Ital.) 80 vols. 18°. Livorno,
1829.—(Hispan. *Castel*) 22 vols. 12°. Madrid, 1802.

Buffon (G. L. L. de) et Daubenton (L. J. M.).

1. Planches enluminées. 5 vols. 4°. Paris.

Buffon (G. L. L. de) et Montbeillard (P. G. de).

1. Histoire naturelle des Oiseaux. 10 vols. 4°. Paris, 1770–86 ; 26
vols. 8°. Paris, 1806 (par *Sonnini*).—(Angl. *W. Smellie*) 9 vols.
8°. London, 1793.—(Ital.) 6 vols. 8°. Milan, 1774–77.

Bugnion (Ch.).

1. Description de quatre nouvelles espèces de Lépidoptères de la
Syrie et de l'Égypte.—Ann. Soc. Ent. Fr. VI. p. 439, fig.
2. Note sur le *Satyrus Styx*, Escher.—Ann. Soc. Entom. III.
p. 337.—*Wiegm.* Arch. 1835, II. p. 57.

Bugs (Egb.).

1. Nieuw en vollkommen Woordenbock van Konsten en Wetten-
schappen. Amst. 1769–72, 8°. I.–IV.—*Böhm.* Bibl. I. 1, p. 313.

Buhle (C. A. A.).

1. Das Fischbuch, oder Beschr. u. Abb. mehrerer Fische. 8°. Halle,
1812.
2. Naturgeschichte der Schädlichen Feldmaus. 8°. Leipz. 1819,
fig.
3. Naturgeschichte des Hamsters. 8°. Leipz. 1821 ; 1841, fig.
4. Naturgeschichte des Maulwurfes. 8°. Leipz. 1835, fig.
5. Die Maulwurfsgrille. 8°. Leipz. 1835.
6. Handbuch der Naturgeschichte des Thierreichs. Halle, 1804, fig.

7. Die Wasserratte. 8°. Leipz. 1835.

8. Naturgeschichte der domesticirten Thiere, in ökonomischer und technischer Hinsicht. Heft I. II. Stuttg. 1842, 8°. fig.

9. Raupen- u. Schmetterlingskalender der deutschen Falter. 4°. Leipz. 1837.

10. Die Tag- u. Abendschmetterlinge Europas. 4°. Leipz. 1837.

BUHLE (J. G.).

1. De fontibus, unde Albertus Magnus libris suis XXV. de animalibus materiem hauserit.—Comm. Soc. Gotting. XII. pt. 3, p. 94.

BÜHLMANN (Fr.).

1. Ueber eine eigenthümliche, auf den Zähnen des Menschen vorkommende Substanz.—*Müll.* Arch. 1840, p. 442, fig.

BUISCH (J. G.).

1. Bemerkungen auf einer Reise durch einen Theil Schwedens. Hamb. 1783, 8°.—*Böhm.* Bibl. I. 1, p. 618.

BUISSIÈRE (P.).

1. Lettre sur un Œuf trouvé dans la trompe de Fallope d'une femme, avec des remarques sur la Génération.—J. des Sav. Sept. 1695.—Biogr. Un. VI. p. 246.

2. Description anatomique du Cœur des Tortues de terre. 1700; 12°. Paris, 1713.—Biogr. Un. VI. p. 246.—Phil. Trans. XXVII. p. 170 (*Anatomical Descript. of the Heart of the Land-Tortoises from America*).

3. Examen des faits observés par *M. Duverney*, du Cœur de la Tortue de terre.—Mém. Ac. Sc. 1703.—Biogr. Un. VI. p. 246.

BUISSON (M. Fr. Régis).

1. Anatomie et Physiologie des annexes du Fœtus. Paris, 8°.—*Müll.* Arch. 1835, p. 238.

2. De la division la plus naturelle des Phénomènes physiologiques considérés dans l'Homme, avec un Précis historique sur *Xav. Bichat.* Paris, 1802, 8°.—Biogr. Un. VI. p. 247.

BUIST (G.).

1. On the Geology of Perthshire.—Prize Essays Highland Soc. XIII. p. 1.

Buist (H.).

1. On the Pupa of *Necrodes littoralis.*—Mag. Nat. Hist. ser. 2, III. p. 600.

Bujack (J. G.).

1. Ueber die Zeit des Verschwindens der Biber in Preussen.—Pr. Provinzialbl. XVI. p. 160.—*Wiegm.* Arch. 1837, II. p. 174.

2. *Bos Urus*, L.—Pr. Provinzialbl. XV. p. 425.—*Wiegm.* Arch. 1837, II. p. 186.

3. Naturgeschichte der höhern Thiere, mit besonderer Berücksichtigung der *Fauna Prussica.* Königsb. 1837, 8°. fig.

4. Naturgeschichte des Elchwildes oder Elens, mit Rücksicht auf die neueren Beobachtungen in den Forsten Ost-Preussens. 8°. Königsb. 1837.—Preuss. Provinzialbl. 1837, XVIII. p. 33.

Bülffinger (G. B.).

1. An Aër sanguini pulmones transeunti misceatur, per experimenta quæsivit.—Comm. Ac. Petr. 1728, III. p. 230.

Bulfinch ( ).

1. Report on the Casts of Fossil Footsteps from Cheshire, England. —Proc. Bost. Soc. N. H. 1841, p. 45.

Bulifon (Ant.).

1. Lettera nella quale si da distincto Ragguaglio dell' Incendio del Vesuvio succedato d'Aprile 1694, con una breve Notizia degli Incendi antecedenti. Nap. 1694, 12°. fig.—*Böhm.* Bibl. IV. 2, p. 379.

2. Compendio istorico del Monte Vesuvio, in cui si ha piena Notizia di tutti gl' Incendi, etc. Nap. 1698, 1701, 12°. fig.—*Böhm.* Bibl. IV. 2, p. 379.

Bullivant (B.).

1. On some Natural Observations made in New England.—Phil. Trans. 1693, p. 167.

Bullmann (J. C.).

1. Meinungen über die Natur und Entstehung des fliegenden Sommers.—N. Schr. d. Naturf. zu Halle, I. p. 1.—Bibl. Ent. I. p. 56.

Bullock (W.).

1. *Virginia examinata*, or a Natural and Political Description of this Country. Lond. 1649, 4°.—*Böhm.* Bibl. I. 1, p. 745.

BULLOCK (W. II.).

1. Companion to the London Museum. 2 vols. Lond. 1816, 12°.—
Tr. Linn. Soc. Lond. XII. 2, p. 590.

2. An Account of four rare species of British Birds.—Tr. Linn.
Soc. Lond. XI. 1, p. 175.

3. Ueber die Colibris.—Resid. in Mexico.—*Brehm*, Ornis, III. p. 98.

BÜLOW-RIETH ( ).

1. Neue Beobachtungen über den Kiefernspinner (*Phalæna Bombyx pini*). Stettin, 1828, 8°.

2. Neue Beobachtungen über die Nonne (*Phalæna monacha*). 8°.
Stettin, 1831.

BULWER (James).

1. On the *Isocardia Cor* of the Irish Sea.—Zool. Journ. II. p. 357.
—*Féruss.* Bull. 1828, XIII. p. 254.

BULWER (John).

1. *Pathomyotomia* (Dissection des Muscles qui indiquent les affec-
tions de l'âme). Lond. 1649, 12°.—Biogr. Un. VI. p. 263.

BUNBURY (C. J. F.).

1. On some remarkable Fossil Ferns from Frostburg, Maryland.—
Journ. Geol. Soc. II. 82.

2. On Fossil Plants from the Coal-field near Richmond, Virginia.—
Journ. Geol. Soc. III. Part I. p. 281.—*Sill.* Am. Journ. ser. 2, IV.
p. 114.

3. On the Fossil Plants communicated by Mr. *Dawson* from Nova
Scotia.—Journ. Geol. Soc. II. 136.

4. On Fossil Plants from the Coal Formation of Cape Breton.—
Journ. Geol. Soc. III. p. 423.

BUNBURY (Sir H.).

1. On the Strata observed in boring at Mildenhall in Suffolk.—
Trans. Geol. Soc. ser. 2, I. p. 379.—*Féruss.* Bull. 1826, VIII.
p. 323.

BUNEL (H.).

1. Apercus géologiques et paléontologiques. Caen, 1836.

BUNIVA (Mich.).

1. Mémoire sur la plupart des Insectes les plus remarquables qui

attaquent les Végétaux dont les Hommes tirent de la nourriture en Piémont.—Mém. Acad. Tur. XVI. p. LXXVIII.

2. Intorno agli Insetti nocivi. Torino, 1804, 12°.

3. Mémoire sur la Physiologie et la Pathologie des Poissons, suivi d'un Tableau indiquant l'Ichthyographie subalpine.—Mém. Ac. Tur. XII. p. 78.

4. Observationes et Experimenta ad recognoscenda bubulæ speciei potissimùm in Subalpinâ Regione infesta Animalia, horumque nocendi modum detegendum.—Mem. Accad. Tor. XI. p. 215.

5. Animalia Bobus infesta. Dissert. 1797, 4°.

BUNKER (J. M.).

1. Vegetable Origin of Anthracite.—*Sill.* Am. Journ. XXIV. 1, p. 172.

BUNSEN ( ).

1. Auszug einer Schreibens aus *J. J. Berzelius.* 1846.

2. Ueber den Zusammenhang der pseudovulcanischen Erscheinungen Islands.—*Liebig's* Annalen, 1847.

3. Beitrag zur Kenntniss des Islands Tuffgebirgen.—*Liebig's* Annalen, 1847.

BUNSEN (R. W.) u. BERTHOLD (A. A.).

1. Das Eisenoxydhydrat, ein Gegengift der arsenigten Säure. Götting. 1833, 4°; 1834, 8°.—N. Act. Nat. Cur. XVII. p. xx.—Isis, 1834, VIII. p. 871.

BÜNTINGEN (J. Ph.).

1. *Sylva subterranea,* oder vortreffliche Nutzbarkeit des unterirrdischen Waldes der Steinkohlen. Halle, 1693, 12°.—*Böhm.* Bibl. IV. 1, p. 484.

BUONAMICI (Fr.).

1. Sulle Glossopetre, gli Occhi di Serpe, i Bastoncelli di M. *Paolo,* ed altre Pietre figurate delle Isole di Malta e Gozzo.—Opusc. Sc. XII.

BUONANNI. *Vide* BONANNI.

BUONUOMO (Cosmo).

1. Osservazioni intorno a Pellicelli del Corpo umano. Fir. 1687, 4°.—*Mod.* B. H.

BUQUET (Lucien).

1. Description de quelques Coléoptères longicornes appartenant

aux genres *Calocomus, Stenaspis* et *Galissus.*—Rev. Zool. 1840, p. 142.

2. Note sur quelques Coléoptères nouveaux d'Algérie et particulièrement de Constantine.—Rev. Zool. 1840, p. 240.

3. Description d une nouvelle espèce de *Buprestide,* du genre *Anthaxia?*, Eschsch.—Rev. Zool. 1841, p. 194.

4. Description d'une nouvelle espèce de *Trachydéride,* appartenant au genre *Ozodera.*—Rev. Zool. 1840, p. 110.

5. Description de quelques Coléoptères nouveaux, de la famille des Carabiques, appartenant aux genres *Colliuris,* Latr., *Diaphorus,* Dej., *Agra,* Fabr., *Cymindis,* Latr., *Calleida,* Dej., *Lebia,* Latr., *Coptodera,* Dej., *Helluo* et *Anchomenus,* Bonelli.—Ann. Soc. Entom. IV. p. 603.—*Wiegm.* Arch. 1836, II. p. 307.

6. Description de onze espèces nouvelles du genre *Lebia,* rapportées de Cayenne par *M. Leprieur.*—Ann. Soc. Entom. III. p. 673. —*Wiegm.* Arch. 1835, II. p. 24.

7. Description de quelques espèces nouvelles de Lamellicornes appartenant aux genres *Goliathus, Macronota, Gnathocera* et *Macroma.*—Ann. Soc. Entom. V. p. 201, fig.—*Erichs.* in *Wiegm.* Arch. 1837, II. p. 296.

8. Description d'un Coléoptère nouveau, du genre *Goliathus.*—Ann. Soc. Entom. IV. p. 135, fig.—*Burm.* in *Wiegm.* Arch. 1836, II. p. 307 (*Goliathus Daphnis*).

9. Description de deux Insectes nouveaux du genre *Oodes.*—Ann. Soc. Ent. Fr. III. p. 47.

10. *Oxycheila acutipennis, Cicindela guttula, Graphipterus femoratus, Lebia quadrinotata.*—Mag. de Zool. IX. fig.—*Burm.* in *Wiegm.* Arch. 1836, II. p. 306.

11. Notice sur un nouveau genre de Longicornes de la tribu des Cérambycins.—Rev. Zool. 1838, p. 253

12. Notice sur deux Coléoptères longicornes de la tribu des Lamiaires et appartenant au genre *Phacellus* de *M. Dejean.*—Rev. Zool. 1838, p. 254.

13. Notice sur un genre nouveau de Longicornes, de la tribu des Cérambycins.—Ann. Soc. Ent. Fr.—Rev. Zool. 1840, pp. 286, 292.

14. Notice supplémentaire sur un genre de Longicornes (le *G. Pteroplatus*).—Ann. Soc. Ent. Fr.—Rev. Zool. 1841, p. 206.

15. Espèce nouvelle du genre *Hexodon.*—Rev. Zool. 1840, p. 212.

16. Coléoptères nouveaux.—Rev. Zool. 1840, p. 172.

17. Note sur sept espèces algériennes du genre *Rhizotragus.*—Rev. Zool. 1840, p. 171.

18. Scarabée Jupiter, insecte nouveau.—Rev. Zool. 1840, p. 42.

19. Note sur le genre *Trochoideus,* et description d'une nouvelle espèce.—Rev. Zool. 1840, p. 173.

BUQUOY (G. von).

1. Zusammenstellung einiger Hauptmomente aus der Geotomie, Phytotomie u. Zootomie. 4°. Leipz. 1826.

2. Ideelle Verherrlichung des empirisch erfassten Naturlebens. 2 vols. 8°. Leipz. 1826.

3. Anregungen für philosophisch-wissenschaftliche Forschung und dichterische Begeisterung, u. s. w. Leipz. 1825.—N. Act. Nat. Cur. XII. p. XI.

4. Skizzen zu einem Gesetzbuche der Natur, zu einer sinnigen Auslegung desselben, und zu einer hieraus hervorgehenden Characteristik der Natur. Leipz. 1817–26, 4°. fig.—N. Act. Nat. Cur. XI. p. LXVII.—Isis, 1819, VII. p. 1168.

5. Die Fundamentalgesetze an den Erscheinungen der Wärme empirisch begründet, und deren Bedeutung nach dynamisch-mathematischen Ansichten im Geiste hervorgerufen, ohne Annahme eines Wärmestoffes. Erster Nachtrag zu dem Werke : Skizzen zu einem Gesetzbuche der Natur. Leipz. 1819.—N. Act. Nat. Cur. XI. p. LXVIII; XII. p. XI.

6. Aphorismen für Meditation und Naturdichtung.—Isis, 1829, V. p. 470.

7. Eigenthümliche Darstellung der Hauptzüge der Physiologie.— Isis, 1835, VI. p. 481 ; VII. p. 577 ; VIII. p. 673 ; IX. p. 762 ; X. p. 841 ; XI. p. 921 ; XII. p. 1001.

8. Ueber das Wesen und die Bedeutung der Muskelbewegung.— Isis, 1826, IV. p. 419.

9. Prodromus zu einer einstmaligen physiologischen Darstellung der successiven Erdbildung, diese mit der Thier- und Pflanzenentwickelung als Eines betrachtet.—Isis, 1826, IV. p. 397.— *Féruss.* Bull. 1826, VIII. p. 315 (*Prodr. d'un Exposé physiologique de la formation successive de la croûte terrestre,* etc.).

10. Betrachtungen über die Formation der Erdoberfläche, dargestellt im Lichte der Organogenie.—Isis, 1834, VIII. p. 761.

11. Beitrag zu den bisherigen Bemühungen für sachgemässe Classification in der Zoologie, blos den Haupt- nicht den Unterabtheilungen nach.—Isis, 1833, VIII. p. 756.

12. Beschreibung einer neuen Art der Gattung *Pterodactylus,* Cuv., *Ornithocephalus,* Sömmer.—N. Act. Nat. Cur. XV. p. 49, fig.

BURAT (Améd.).

1. Description des Terrains volcaniques de la France centrale. 8°.

Paris, 1833, fig.—Cat. R. Soc. L.—*L.* u. *Br.* N. Jahrb. 1833, p. 559.

2. Traité de Géognosie, ou Exposé des Connaissances actuelles sur la Constitution physique et minérale du Globe terrestre, etc. Par. 1834, 1835, 8°. 3 vols.

3. Géologie appliquée, ou Traité de la recherche et de l'exploitation des Minéraux utiles. Paris, 1843, 8°. fig.—(Germ. *Krause* et *Hochmuth*) : *Angewandte Geognosie, oder das Auffinden und der Bau nutzbarer Mineralien.* Berl. 1844, 8°. fig.—Journ. Geol. Soc. I. p. 133.

BURAT (Jul.).

1. Des Puits artésiens. Notions générales de Géologie appliquée à la recherche des Eaux souterraines. 8°.

BURCHARD (E. Fr.).

1. Programma de Terrâ Cœmiterii *B. Gerthrudæ.* Rost. 1747, 4°.—*Böhm.* Bibl. IV. 1, p. 195.

2. Epistola ad *C. Linnæum* de *Cocco polonico.*—Act. Ups. 1742, p. 53.—N.´ Hamb. Mag. XXIII. p. 481 ; XXIV. p. 499.—Bibl. Ent. I. p. 56.

BURCHARD (J. A.).

1. De Tumore cranii recens Natorum sanguineo Symbolæ. Vratisl. 4°.—*Val.* Repert. III. p. 20.

BURCHELL (W. J.).

1. Travels in the Interior of South Africa, ann. 1811–1815. Lond. 2 vols. 4°. fig. 1822-24.—Isis, 1818, IV. p. 618; 1821, XII. Beil. No. 21 ; 1823, III. Litt. Anz. p. 129.

2. Sur une nouvelle espèce de Rhinocéros, et Observations de *M. de Blainville* sur les différentes espèces de ce genre.—J. de Phys. LXXXV. p. 163.—Isis, 1817, IX. p. 1318.

3. Description of *Malaconotus atro-coccineus.*—Zool. Journ. I. p. 461.

BURCKHARDT (Aug.).

1. Obs. anat. de Uteri vaccini fabrica. 4°. Basel. 1834.

BURCKHARDT (Chr.).

1. Ueber den *Palinurus Sueurii* des Muschelkalks.—Verh. Basl. Naturf. Ges. IV. p. 78.—*L.* u. *Br.* N. Jahrb. 1841, p. 740.

2. Ostindianische Reisebeschreibung, oder Abriss von Ostindien,

zusammt derselben Lebensart, einiger Thiere, Gewächse, Edelgesteine, etc. Halle u. Leipz. 1693, 12º.—*Böhm.* Bibl. I. 1, p. 662.

BURCKHARDT (J. L.).

1. Travels in Nubia and in the interior of N. E. Africa, performed in 1813. Lond. 1819, 4º.—Biogr. un. Suppl. LIX. p. 443.

BURDACH (C. Fr.).

1. Vom Baue und Leben des Gehirns und Rückenmarks. Leipz. 1819, 3 vols. 4º. fig.—Bibl. med.-ch. p. 70.—*Féruss.* Bull. 1823, III. p. 435 (*De la structure et de la vie de l'Encéphale*).

2. Tabellarische Uebersicht der Hyologie des menschlichen Körpers, zum Gebrauch bei Vorlesungen. Königsb. 1835, fol.—Bibl. med.-ch. p. 70.

3. Handbuch der neuesten Entdeckungen in der Heilmittellehre. Leipz. 1806, 8º.—Bibl. Ent. I. p. 56.

4. Die Physiologie als Erfahrungswissenschaft. Leipz. 1828–1835, 8º. 5 Bde. (Mit Beitr. v. *Mayer, Rathke, Valentin,* u. and.)— Bibl. med.-ch. p. 70.—*Val.* Repert. I. pp. 7, 10.—*Müll.* Arch. 1838, p. cxciv.—(2te Aufl.) Leipz. 1838, mit Beitr. v. *Hayn* u. *Moser.*—6ter Bd. mit Beitr. v. *E. Burdach* u. *J. F. Dieffenbach,* 1840.—(Gall. *Jourdan*): *Traité de Physiologie considérée comme Science d'Observation, avec des additions,* etc. Par. 1837–41, 9 vols. 8º.

5. Ueber den Schlag und Schall des Herzens. Wien, 1832, 4º.— Bibl. med.-ch. p. 70.

6. Der Mensch nach den verschiedenen Seiten seiner Natur, oder Anthropologie für das gebildete Publikum. Stuttg. 1836, 8º. fig. (1te–4te Abth.)—*Val.* Repert. II. p. 20.—Bibl. med.-ch. p. 70.

7. Berichte von der kön. anatomischen Anstalt zu Königsberg. Leipz. 1818–1824, 8º.—Bibl. med.-ch. p. 69.

8. Beiträge zur nähern Kenntniss des Gehirns, in Hinsicht auf Physiologie, Medizin und Chirurgie. Leipz. 1806, 8º. 2 Th.— Bibl. med.-ch. p. 70.

9. De Fœtu humano adnotationes anatomicæ, etc. Lips. 1828, fol. fig.—Bibl. med.-ch. p. 70.

10. Blicke ins Leben. 2 vols. 8º. Leipz. 1842.

BURDACH (E.).

1. Bericht von der königl. anatomischen Anstalt zu Königsberg. 1835, 8º.—*Val.* Repert I. p. 8.

2. Beitrag zur microscopischen Anatomie der Nerven. Königsb.

1837, 4°.—*Val.* Repert. III. p. 14.—Ann. Sc. n. IX. pp. 96, 247 (*Mém. sur l'Anatomie microscopique des Nerfs*).

3. Untersuchung über das Verhalten der Vasa vasorum im Gegensatz zu den ernährenden Gefässen des Herzens.—Ber. kön. anatom. Anst. Königsb.—*Müll.* Arch. 1836, p. xxvii.

4. Beitrag zur vergleichenden Anatomie des Affen.—*Rathke*, N eunt. Ber. v. d. k. anat. Anst. zu Königsb. 1838, 8°.

5. Quelques Considérations sur le *Nidamentum*, ou Enveloppe des Œufs.—Ann. Sc. n. XXVII.

BÜRDE (Fr.).

1. Abbildungen merkwürdiger Säugethiere (I. u. IIte Lief. monogr. bearbeitet von *Brandt* u. *Wiegmann*). fol. u. 4°. Berl. 1831–32. —Isis, 1833.—*Wiegm.* Arch. 1837, p. 186.

BURDIN (C. J.).

1. Cours d'études médicales, ou Exposition de la Structure de l'Homme comparée à celle des Animaux. 5 vols. 8°. Paris, 1803. —(Angl.) *Course of Medical Studies*, etc. 3 vols. 8°. Lond. 1803. —(Germ. v. *Reuss*.) Stuttg. 1803, 8°. 2 Th.—Bibl. med.-ch. p. 71.

BUREAU ( ).

1. De novâ Vermium intestinalium Specie.—*VanderM*. Rec. IV. p. 341.—*Mod.* B. Helm.

BURG (P. van der).

1. Erste grondbeginselen der Natuurkunde. 8°. Gouda, 1844.

BURGATZKY ( ).

1. Diss. de Vespertilionibus quibusdam gravidis. 4°. Tubing. 1817.

BÜRGER (Mich.).

1. Dissertatio de Lumbricis. Regiom. 1706, 4°.

BURGESS (Th. H.).

1. The Physiology and Mechanism of Blushing, illustrative of the influence of mental emotion on the capillary circulation. 8°. Lond. 1838.

BURGGRAEVE (Ad.).

1. Histologie ou Anatomie de Texture. Gand, 1843, 1 vol. 8°. fig.

2. Cours théorique et pratique d'Anatomie, etc. I. Gand, 1840, 8°.

3. Précis de l'Histoire de l'Anatomie, comprenant l'Examen comparatif des Ouvrages des principaux Anatomistes, anciens et modernes. Gand, 1840, 1 vol. 8°.

BURGGRAV (Fr. Ph.).

1. Bedenken von dem Werke der Erzeugung. Frankf. 1737, 4°.—
Böhm. Bibl. II. 1, p. 131.

2. De Aëre, Aquis et Locis. Francof. 1751, 8°.—*Böhm.* Bibl. I.
1, p. 591.

BURGHARD (Ern. Fr.).

1. Arenaria Reichenbacensis.—Med. Sil. Satyr. Spec. I. p. 36, fig.
—*Mod.* B. Helm.

BURGHARD (God. H.).

1. *Iter Sabothicum,* oder Beschreibung einiger 1733 u. d. folg.
Jahre auf den Zobtenberg gethanen Reisen. Bresl. u. Leipz.
1736, 8°. fig.

2. Libellæ, seu Perlæ sudelicæ descriptio.—Med. Sil. Satyr. Spec.
V. p. 28, fig.—Bibl. Ent. I. p. 56.

BURGOS (Vinc. de).

1. Historia natural, do se tratan las propriedades de todas las coses.
Toledo, 1529, fol.—*Böhm.* Bibl. I. 1, p. 227.

BURGSDORF (Fr. A. L. v.).

1. Beiträge zur Naturgeschichte des Rothhirsches (*Cervus Elephas,*
Linn.).—Schr. Berl. Naturf. VI. p. 411.—Biogr. Un. VI. p. 316.

2. Von den verschiedenen Knoppern, als ein Beitrag zur Naturgeschichte der Eichen und Insecten (*Cynips chalcis*).—Berl. Naturf.
IV. p. 1.—Bibl. Ent. I. p. 56.

3. Versuch einer vollständigen Geschichte von den Holzarten in
systemat. Abhandl. Berl. 1783–1800, 4°.—Bibl. Ent. I. p. 56.

4. Von den Wölfen in Neu Ost-Preussen.—N. Schr. Gesellsch.
Naturf. Fr. Berl. III. p. 570.

5. Von dem Luchs in Neu Ost-Preussen.—Neue Schr. Gesellsch.
Naturf. Fr. Berl. III. p. 569.

6. Naturhistorische Beschreibung Neu Ost-Preussens.—N. Schr.
Gesellsch. Naturf. Fr. Berl. III. p. 567.

7. Bemerkungen auf einer Reise nach dem Unterharz, dessgleichen
nach Destedt, Helmstädt und Harbke im August 1783.—Schr.
Berl. Gesellsch. Naturf. Fr. V. p. 148.

BURGUNDUS vel BOURGOINGNE (Vinc.).

1. Speculum quadruplex naturale, doctrinale, historiale, etc. Duaci,
1624, 4 vols. fol.—*Mod.* B. Helm.

2. Lapidum Historia. Duaci, 1622, fol.—*Böhm.* Bibl. IV. 1, p. 215.

3. Speculum majus, quod alia Specula complectitur, naturale, doc-trinale, morale, historiale. Argentinæ, 1473 et 1476, fol. 6 vols. —(Titulo: Bibliotheca Mundi). Basil.—Venet. 1591, fol. 4 vols. —Duaci, 1624, fol. 4 vols. (Studio Theologorum in Acad. Duac. aucta)—*Böhm.* Bibl. I. 1, p. 223.

BURKART (J.).

1. Geognostische Bemerkungen über die Berge von Santiago östlich v. Zacatecas.—*Karst.* Arch. VI. p. 413.—*L.* u. *Br.* N. Jahrb. 1835, p. 482.

2. Zusammen-Vorkommen von Granit u. Basalt in der Schneegrube im Riesengebirge (Brief an *Nöggerath*) (*Extrait d'un lettre à* Nöggerath: *Présence du Granit accompagné de Basalte sur le Schneegrube dans le Riesengebirge*).—Min. Taschenb. 1823, IV. p. 831.—*Féruss.* Bull. I. 1824, p. 108.

3. Geognostische Beobachtungen auf Reisen in Mexico gesammelt (*Observations géognostiques faites pendant un voyage au Mexique*). —Zeitschr. f. Mineral. 1826, II. p. 1, fig.—*Féruss.* Bull. 1828, XIII. p. 40.

4. Geognostische Skizze der Gebirgs-Bildungen des Kreisen Kreuz-nach, etc. (*Esquisse géognostique des formations du Cercle de Kreutznach et de quelques contrées environnantes du Palatinat*). —*Nögger.* Das Gebirge im Rheinl. Westphal. IV. p. 142.—*Féruss.* Bull. 1828, XIII. p. 22.

5. Rapports géologiques des Mines d'Angangeo au Mexique.— Zeitschr. f. Mineral. 1827, XI. et XII. p. 401.—*Féruss.* Bull. 1829, XVII. p. 41.

6. Aufenthalt und Reisen in Mexiko in den Jahren 1825–1834; Bemerkungen über Land, Produkte, Leben u. Sitten der Ein-wohner ü. Beobachtungen aus dem Gebiete der Mineralogie, Geognosie, etc. Stuttg. 1836, 2 vols. 8°. fig.

7. Geognostische Verhältnisse der Silber-Bergwerke von Veta grande in der Prov. Zacatecas in Mexiko.—*Karst.* Arch. VI. p. 319.— *L.* u. *Br.* N. Jahrb. 1836, p. 609.

8. Geognostische Bemerkungen auf einer Reise, etc. im Staate von Michoacan.—*Karst.* Arch. V. p. 158.—*L.* u. *Br.* N. Jahrb. 1833, p. 211.

9. Geognostische Bemerkungen auf einer Reise zwischen Ramos u. Catoree.—*Karst.* Arch. VI. p. 422.—*L.* u. *Br.* N. Jahrb. 1834, p. 589.

10. Ueber die Ausbruche des Jorullo u. des Tustla.—*L.* u. *Br.* N. Jahrb. 1835, p. 36.

BURKE (W.).

1. The Mineral Springs of Western Virginia, their Use, etc.   New York, 1842, 12°.

BURKHARDT (Aug.).

1. Observationes anatomicæ de Uteri vaccini fabricâ.   Basil, 1834, 4°. fig.—*Müll.* Arch. 1836, p. LIX.

BURKHART (J. Rod.).

1. Ueber das Blut und das Athmen, etc.   Basel, 1828, 8°.

BURMANN (Nic. Laur.).

1. Series Zoophytorum Indicorum (edit. cum ejus Flora Indica). Lugd. Bat. 1768, 4°.—*Böhm.* Bibl. II. 2, p. 498.—*Mod.* B. H.

BURMEISTER (H.).

1. Die Respirationsorgane von *Julus* und *Lepisma.*—Isis, 1834, II. p. 134, fig.

2. *Distomum globiporum* Rud. ausführlich beschrieben.—*Wiegm.* Arch. 1835, II. p. 187.—*Müll.* Arch. 1836, p. 237.

3. Ueber den Bau der Augen bei *Branchiopus paludosus.*—*Müll.* Arch. 1835, pp. 529, 613, fig.; 1836, V. p. CI.

4. Handbuch der Entomologie.   I. u. II. Bd. 8°. Berl. 1832, 1835, fig. 4°. III. Bd. ibid. 1842.—Isis, 1834, III. p. 315.—N. Act. Nat. Cur. XVII. p. XXI.—(Trad. angl.) *A Manual of Entomology, translated from the German by* W. E. Shuckard. Lond. 1836, 8°. —*Val.* Repert. II. p. 19.—*Wiegm.* Arch. 1836, II. p. 294.

5. De Insectorum Systemate naturali.   Hall. 1830, 8°.—*Eis.* Insect. p. 143.

6. Bericht über die Fortschritte der Entomologie im Jahre 1834. —*Wiegm.* Arch. 1835, II. p. 7.

7. Bericht über die Fortschritte der Entomologie im Jahre 1835. —*Wiegm.* Arch. 1836, II. p. 293.

8. Die Verwandlungsgeschichte von *Chlamys monstrosa.*—*Wiegm.* Arch. 1835, II. p. 245, fig.

9. Beschreibung der Raupe und Puppe v. *Plusia consona* und *amethystina.*—*Thon* Arch. II. 1, p. 36.—*Eis.* Insect. p. 212.

10. Ueber die Gattung *Nomatocera* Meig. *Hexatoma* Latr.—*Thon* Arch. II. 1, p. 35, fig.—*Eis.* Insect. p. 235.

11. Mémoire sur la division naturelle des Punaises terrestres (*Geocores*), etc.—*Wiegm.* Arch. 1835, II. p. 187.—*Silb.* Rev. Entom. II. 1834, p. 1, fig.—Bibl. Ent. I. p. 57.

12. Ueber die Gattung *Aclysia.*—Isis, 1834, II. p. 138, fig.—*Wiegm.* Arch. 1835, I. p. 353.

13. Beschreibung einiger neuen oder weniger bekannten Schmarotzerkrebse, nebst allgemeinen Betrachtungen über die Gruppe, welcher sie angehören.—N. Act. Nat. Cur. XVII. p. 269, fig.— *Wiegm.* Arch. 1836, II. p. 221.

14. Beiträge zur Naturgeschichte der Rankenfüsser (*Cirripedia*). Berlin, 1834, 4°. fig.—*Wiegm.* Arch. 1835, I. p. 344.—*Müll.* Arch. 1835, pp. 74, 80.

15. Genera Insectorum, Iconibus illustravit et descripsit. I. Rhynchota. Berl. 1838–40, 8°. fig. col.—*Erichs.* in *Wiegm.* Arch. 1838, II. p. 262.—Ann. of Nat. H. III. p. 52.—Rev. Zool. 1841, p. 284.

16. Lehrbuch der Naturgeschichte. Halle, 1830, 8°.—Isis, 1831, IV. p. 343.

17. Handbuch der Naturgeschichte. Berl. 1837.

18. Grundriss der Naturgeschichte, für Gymnasien und höhere Bürgerschulen. 2te verbess. Aufl. Berl. 1835, 8°.

19. Ueber den innern Bau der Carabenlarven. Zergliederung der Larve des *Calosoma sycophanta.*—Trans. Entom. Soc. Lond. I. p. 235, fig.—*Erichs.* in *Wiegm.* Arch. 1837, II. p. 293.

20. Zur Naturgeschichte der Gattung *Calandra*, u. Beschreib. einer neuen Art (*C. Sommeri*). Berl. 1837, 4°. fig.

21. Zoologischer Handatlas. Berl. 1835–39, 4°. fig. (Lief. I.–VI.)

22. Geschichte der Schöpfung. Eine Darstellung des Entwickelungsganges der Erde und ihrer Bewohner. Leipz. 1843, 8°.

23. Canbophorarum species enumeratæ.—*Silbermann*, Revue Entom. I. 1833, p. 227, fig.

24. On the genus *Myocoris*, of the family *Reduvini.*—Trans. Entom. Soc. II. p. 102.

25. Die Organization der Trilobiten, aus ihren lebenden verwandten entwickelt. 4°. Berlin, 1843.—Journ. Geol. Soc. I. p. 129.— (Angl.) RAY SOCIETY. fol. London, 1847.

26. Zoologische Zeitung. Halle, 1847–48.

BURMESTER (J. H.).

1. Dissert. II. de Margâ ejusque Historiâ naturali. Lugd. Bat. 1754, 4°.

BURN (Al.).

1. On the Habits of Blister-flies.—Entom. Soc. Aug. 1840.—Ann. and Mag. N. H. VII. p. 147.

BURNABY (Andr.).

1. Travels through the middle Settlements in N. America in the years 1759–60, with Observations upon the State of the Colonies. Lond. 1775, 8°.—(Germ.) *Reisen durch die mittlern Colonien der Engländer in Nordamerika,* etc. Hamb. 1776, 8°.—(Gall.) Laus. 1778, 8°.—*Böhm.* Bibl. I. 1, p. 736.

BURNELL (E. H.).

1. List of Lepidopterous Insects found in the neighbourhood of Witham, Essex.—Mag. Nat. Hist. ser. 2, p. 601.

BURNES (A.).

1. On the Salt Mines of the Punjaub.—J. A. S. B. I. p. 145.

2. On the Geology of the Chari Range in Cutch.—J. A. S. B. VI. p. 159.

3. On Peshawar Coal.—J. A. S. B. II. p. 267.

4. On the Geology of the Banks of the Indus, the Indian Caucasus, and the Plains of Tartary, to the Shores of the Caspian.—Trans. Geol. Soc. ser. 2, III. 491.—Proc. Geol. Soc. II. 8.—Phil. Mag. ser. 3, IV. p. 225.—*L. u. Br.* N. Jahrb. 1836, p. 224.

BURNET (Gilb.).

1. Some Lettres containing an Account of what seemed most re-marquable in Switzerland, Italy, Germany, France. Roterd. 1686 et 1687, 8°.—(Gall.) Roterd. 1687 ; 1688 ; 1718, 12°.—(Germ.) Leipz. 1687 ; 1688 ; 1693, 12°.—*Böhm.* Bibl. I. 1, p. 450.

BURNET (Th.).

1. An Answer to the late Objections made by *E. Warren* against the Theory of the Earth. Lond. 1690, 4°.—Act. Erud. 1691, p. 329.

2. The Theory of the Earth ; the two last books concerning the Burning of the World, and the new Heaven and the new Earth. Lond. 1697, fol.—Cat. Bibl. Turic.

3. The Sacred Theory of the Earth, containing an Account of its Creation and of the Changes which it hath undergone, or is to undergo. Lond. 1759, 2 vols. 8°.—Cat. R. S. L.

4. Telluris theoria sacra, Orbis nostri originem et mutationes ge-nerales, quas aut jam subiit, aut olim subiturus est, complectens, etc. Lond. 1681, 4° ; 1684 (Libr. I. et II. de Diluvio et Para-diso).—Lond. 1689, 4°. (Libr. III. et IV. de Flagratione Mundi

et futuro rerum statu).—*Böhm.* Bibl. IV. 2, p. 231.—(Germ. *J. J. Zimmermann.*) Hamb. 1698, 4º.

BURNETT (G. T.).

1. Illustrations of the Manupeda, or Apes and their Allies.—Journ. Roy. Inst. XXVI. p. 300.—Isis, 1833, X. p. 938.

2. Illustrations of the Alipeda, or Bats and their Allies.—Journ. Roy. Inst. XXVII. p. 261.—Isis, 1833, X. p. 939.

3. On the Teeth of the *Erinaceus europæus*, Urchin, or Common Hedgehog.—Journ. Roy. Inst. XXVIII. p. 332.

4. Illustrations of the Quadrupeda.—Journ. Roy. Inst. XXVIII. p. 336.—Isis, 1833, X. p. 941.

5. Illustrations of the Cetotheræ, including the Loripeda, Semipeda and Pinnipeda.—Journ. Roy. Inst. XXIX. p. 355.

6. Illustrations of the Herpornitheræ.—Journ. Roy. Inst. XXIX. p. 362.

BURNEY (Col.).

1. On Fossil Bones from Ava.—J. A. S. B. III. p. 365.

2. On a fossil *Hippopotamus* from Ava.—J. A. S. B. VI. p. 1099.

BUROW (C. F.).

1. De Vasis Sanguiferis Ranarum. Regiom. 183...—*Müll.* Arch. 1835, p. 62.

2. Mémoire sur le Système vasculaire du Phoque ; trad. de l'Allem. par *Höfer.*—Ann. d'Anat. II. p. 292.

3. Echinorhynchi strumosi Anatome. Regiom. 1836, 8º.—*Wiegm.* Arch. 1837, II. p. 258.—*Müll.* Arch. 1838, p. CXLVII.

4. Ueber das Gefässsystem der Robben.—*Müll.* Arch. 1838, p. 230, fig.

5. Ueber das Menstrualblut.—*Müll.* Arch. 1840, p. 36.

6. Ueber den Bau der Macula lutea des menschlichen Auges.— *Müll.* Arch. 1840, p. 38, fig.

7. Beitrag zur Gefässlehre des Fötus.—*Müll.* Arch. 1838, p. 44, fig.

8. Beiträge zur Physiologie und Physik des menschlichen Auges. Berl. 1841, 8º. fig.—*Val.* Repert. VII. p. 48.

BURR (F.).

1. Elements of Practical Geology as applicable to Mining, Engi-

neering, Architecture, etc., with a comprehensive view of the geological structure of Great Britain.   Lond. 1839, 8°. (new edit.)

2. On the Geology of the line of the proposed Birmingham and Gloucester Railway.—Proc. Geol. Soc. II. p. 593.—Phil. Mag. ser. 3, XII. p. 573.

3. On the Geology of Aden, on the Coast of Arabia.—Trans. Geol. Soc. ser. 2, VI. p. 499.—Proc. Geol. Soc. III. p. 355.—Phil. Mag. ser. 3, XIX. p. 174.

BURRELL (J.).

1. A Catalogue of Insects found in Norfolk.—Lond. Ent. Soc. 1812, p. 101.—Bibl. Ent. I. p. 57.

2. On the *Lygeus micropterus.*—Lond. Ent. Soc. 1812, p. 73.—Bibl. Ent. I. p. 57.

3. Remarks on *Staphylinus tricornis.*—Lond. Ent. Soc. 1812, p. 310. —Bibl. Ent. I. p. 57.

BURRIEL (A. Marc.).

1. Noticia de la California, y de su conquista temporal y spiritual, etc.   Madr. 1758, 4°. 3 vols.—(Angl.) *A natural and civil History of California, transl. of Mig. Venegas.*   Lond. 1759, 2 vols. 8°.—(Belg.) *Natuurlijke Historie van Californien.*   Harl. 1761, 4°.—(Gall.) *Histoire nat. et civ. de la Californie,* etc., trad. de *l'Angl. par E. F.*   Paris, 1767, 8°. 3 vols.—(Germ.) *Natürliche u. bürgerliche Geschichte von Californien, aus dem Engl. v. J. Chr. Adelung.*   Lemgo, 1769, 4°. 3 vols.—*Böhm.* Bibl. I. 1, p. 741.

BURROW (E. J.).

1. Description of *Mus castorides,* a new species.—Tr. Linn. Soc. Lond. XI. p. 167.

2. Elements of Conchology, according to the Linnæan System, illustrated by 28 plates drawn from Nature.   Lond. 1829, 8°.— (Ital. *Baldassini.*) 8°. Milan, 1836.—Tr. Linn. Soc. Lond. XII. 2, p. 590.—Mag. Zool. and Bot. 1838, II. p. 238.—*Féruss.* Bull. 1829, p. 118.

BURROW (G. M.).

1. Account of two cases of Death from eating Mussels ; with some general Observations on Fish Poison.   Lond. 1815, 8°.—Tr. Linn. Soc. Lond. XII. 2, p. 590.

BURSER (Joach.).

1. Introductio ad Scientiam naturalem, in Acad. Soranâ discentibus exhibita.   Amst. 1652, 8°.—*Böhm.* Bibl. I. 1, p. 243.

2. De Fontium origine Tractatus. Hafn. 1639, 8º.—*Böhm*. Bibl. V. p. 63.

BURT (A.).

1. On the Dissection of the Pangolin.—Asiat. Res. II. p. 353.

BURT (T. Seymour).

1. Observations on the Curiosities of Nature, by the late *W. Burt.* 8º. Lond. 1836.—Cat. R. Soc. L.

BURTIN (Fr. X. de).

1. Mémoire sur les Révolutions et l'âge du Globe terrestre.—Verh. Teyler's Genoot. VIII. 1790, 4º.—Biogr. Un. Suppl. LIX. p. 455.

2. Voyage minéralogique de Bruxelles, par Wavre, à Court-St.-Etienne. Harl. 1781, 8º.—Mém. Ac. Brux. V. p. 123.—Biogr. Un. Suppl. LIX. p. 456.

3. Oryctographie de Bruxelles, ou description des Fossiles, tant naturels qu'accidentels, découverts jusqu'à ce jour dans les environs de cette ville. Brux. 1784, fol. fig. col.—Biogr. Un. Suppl. LIX. p. 455.

4. Des Bois fossiles découverts dans les différentes parties des Pays-Bas. Harl. 1781, 8º.—Biogr. Un. Suppl. LIX. p. 456.

BURTON (E.)

1. Observations on the Natural History and Anatomy of the *Pelecanus aquilus* of Linnæus (*Tachypetes aquila*, Vieillot).—Trans. Linn. Soc. XIII. p. 1.

2. On a *Ratelus* from India.—Proc. Zool. Soc. III. p. 113.—*Wiegm.* Arch. 1836, II. p. 281.

3. On a new species of *Pipra.*—Proc. Zool. Soc. IV. p. 113.—*Wiegm.* Arch. 1837, II. p. 207.

4. On a species of *Agriopus*, Cuv.—Proc. Zool. Soc. III. p. 116.—*Wiegm.* Arch. 1836, II. p. 242.

5. Characters of several Birds from the Himalaya Mountains.—Proc. Zool. Soc. III. p. 152.—*Wiegm.* Arch. 1836, II. pp. 262, 266, 270.

6. Eine neue Art von *Euprepes*, unter dem Namen *Tiliqua Fernandi.*—Proc. Zool. Soc. p. 62.—Lond. and Edinb. Phil. Mag. Suppl. p. 514.—*Wiegm.* Arch. 1837, II. p. 229.

7. On a new Kingfisher (*Ceyx microsoma*), and a female specimen of *Caprimulgus monticolus.*—Proc. Zool. Soc. V. p. 89.—Ann. of Nat. H. I. p. 227.

BURTON (H.).

1. Characters of a new species of the genus *Monacanthus*, Cuv.—
Proc. Zool. Soc. II. p. 121.—*Wiegm.* Arch. 1835, II. p. 270.

BURY (C. A.).

1. On the Birds of the Isle of Wight.—Zoologist, pp. 516, 634, 915,
970; Remarks on Mr. Waterton's Essay on the Oil-gland, p. 751;
On the Mammalia of the Isle of Wight, p. 776; Reptiles of the
Isle of Wight, p. 1027.

BUSCH (D. W. H.).

1. Das Geschlechtsleben des Weibes, in physiologischer, pathologischer u. therapeutischer Hinsicht dargestellt. Leipz. 1839–40,
2 vols. 8°.

BUSCH (J. D.).

1. Grundriss einer zootomischen Beschreibung der landwirthschaftlichen Thiere. Heidelb. 1796.

2. Dell' uso dell' Intestino cieco e dell' appendice vermiforme.—
Ann. Med. Stran. VII.

BUSCH ( ), GRÄFE ( ), HUFELAND ( ), etc.

1. Encyklop. Wörterb.; *vide sup.* Pars I. p. 13.

BÜSCHING (Ant. Fr.).

1. Vollständige Topographie der Mark Brandenburg. Berl. 1775,
4°.—*Böhm.* Bibl. I. 1, p. 604.

2. Eigene Gedanken u. gesammlete Nachrichten von der Tarantel,
zur gänzlichen Vertilgung der Vorurtheiles von der Schädlichkeit
ihrer. Berl. 1772, 8°.—*Böhm.* Bibl. II. 2, p. 355.

3. Versuch, die Kenntniss der Natur den Kindern leicht u. fasslich
zu machen. Berl. 1772, 8°.—(Hernach unter dem Titel): *Unterricht in der Naturgeschichte, für diejenigen, welche noch wenig,
oder gar nichts davon wissen.* Berl. 1775, 8°; 1776, fol.; 1780,
8°. fig.—Bibl. Ent. I. p. 55.—(Dan.) *Underviisning i Natur-
historien.* Othin. 1778, 8°.—Copenh. 1779, 8°.—*Böhm.* Bibl. I.
1, p. 295.

BUSH (J.).

1. Hibernia curiosa. Lond. 1767, 8°.

BUSHELL ( ).

1. Theory of Minerals. Lond. 1690, 4°.—*Böhm.* Bibl. IV. 1, p. 38.

BUSHNAN (J. S.).

1. Nature, structure and economical use of Fishes. 12°. Edinb. 1840.—*Jard.* Nat. Lib. Ichth. II.

2. Philosophy of Instinct and Reason. 8°. Edinb. 1837.

3. Introduction to the Study of Nature, illustrative of the Attributes of the Almighty. 8°. London, 1834.

BUSK (Geo.).

1. On the Young of a species of *Ixodes* from Brazil.—Trans. Microsc. Soc. I. p. 88.

2. Observations on the Structure and Nature of the *Filaria Medinensis* or Guinea Worm.—Trans. Microsc. Soc. II. p. 65.

3. On the Anatomy of *Trichocephalus dispar.*—Microsc. Soc. March 1841.—Ann. and Mag. N. H. VII. p. 212.

4. Some observations on the Natural History of the Echinococcus. —Trans. Microsc. Soc. II. p. 10.

BUSSIÈRE (D.).

1. Epistola de Substantiâ, vas aliquod pulmonale referente, tussi rejectâ.—Excerpt. Philos. Angl. Apr. 1700, p. 534, fig.—Act. Erudit. XX. p. 231.

2. Epistola de triplici Vesicâ.—Philos. Trans. 1701, Jan. p. 752, fig. —Act. Erudit. XXI. p. 27.

BUSSIÈRE (Paul).

1. Anatomical Description of the Heart of Land Tortoises from America.—Philos. Trans. XXVII. p. 170.—Dict. Sc. n. XV. p. 248.

BUSSUEIL (   ).

1. Sur l'Hétéradelphe de Chine (Extrait d'une Note recueillie à Macao).—Mém. Mus. XV. p. 407.

BUSTAMANTIUS (Jac.).

1. De Reptilibus animantibus seu Animalibus Scripturæ S. Libri VI. duobus tomis comprehensi. Compluti, 1591; 1595, 4°.—Lugd. 1620, 8°; 1658, 8°.—Alcala de Henarez, 1595, 4°.—*Dum.* et *Bib.* I. p. 310.—*Böhm.* Bibl. II. 1, p. 72.

BUTEUX ( ).

1. Mémoire sur la Géologie d'une partie du Dép^t. de la Somme. Par. 1835, 8°. fig.—Bull. Soc. Géol. Fr. VI. p. 103.

BUTINI ( ).

1. Mémoire sur le *Tænia* à anneaux courtes, ou ver solitaire.—Mém. Soc. Roy. Méd. 1782 et 1783, p. 285.

BUTLER (C.).

1. *Apum Historia*, seu Feminine Monarchy, or the History of the Bees. Oxf. 1609, 8°. Lond. 1623, 4°.—(Latinè) Lond. 1634, 4°; 1673, 8°. fig.—(Angl.) Oxf. 1692, 8°. Lond. 1704, 8°.— (Franç.) La Haye, 1740, 8°.—*Eis.* Insect. p. 224.—*Böhm.* Bibl. II. 2, p. 281.

BUTTE (W.).

1. Die Biotomie des Menschen, oder die Wissenschaft der Natureintheilungen des Lebens als Mensch, als Mann und als Weib, nach seinen auf- und absteigenden Linien, etc.   Bonn, 1829, 8°. fig.—Bibl. med.-ch. p. 72.

2. Uebersicht der anthropologischen Biotomie, und Andeutung der klimatologischen Geotomie.   Köln, 1829, 8°.—Bibl. med.-ch. p. 72.—Isis, 1830, VI. p. 641.

3. Zusammenhang der anthropologischen Biotomie mit dem Systeme klimatologischen Geotomie.—Isis, 1830, VI. p. 663.

BUTTER (J.).

1. On the Change of Plumage exhibited by many species of Female Birds at an advanced period of life.—Mem. Wern. Soc. III. p. 183. —Isis, 1818, IV. p. 586.

BÜTTNER (Chr. G.).

1. Liste des Noms d'Animaux usités dans l'Asie méridionale. 1780 (publ. p. *Ekkard*).—Biogr. Un. VI. p. 402.

2. In vielen Jahren gesammelte anatomische Wahrnehmungen. Königsb. 1768, 4°. fig.—Bibl. med.-ch. p. 73.

3. Sechs besondere anatomisch-chirurgische Wahrnehmungen. Königsb. 1774, 4°. fig.—Bibl. med.-ch. p. 73.

4. Observations sur quelques espèces de *Tænia*. 1774.—Biogr. Un. VI. p. 402.

Büttner (Dan. Sig.).

1. *Rudera Diluvii testes*; d. i. Zeichen u. Zeugen der Sündfluth, in Ansehung des jetzigen Zustandes unserer Erd- u. Wasserkugel, insonderheit der darinnen im Querfurthischen Revier Angetroffenen, ehemals verschwemmten Thiere u. Gewächse, etc. Leipz. 1710, 4°. fig.—*Böhm.* Bibl. IV. 2, p. 234.—*Mod.* B. Helm.— Biogr. Un. VI. p. 398.

2. Coralliographia subterranea, etc. Lips. 1714, 4°. fig.—*Böhm.* Bibl. IV. 2, p. 311.—*Mod.* B. H.

3. Sur quelques Nautilites (Allem.). Querfurth, 4°.

4. Diss. de Scientiæ naturalis constitutione in Gymnasio Hamburg. Hamb. 1690, 4°.—*Böhm.* Bibl. I. 1, p. 255.

5. Diss. de Nive. 4°. Hamb. 1690.—*Böhm.* Bibl. V. p. 54.

6. De Lapidibus tàm pretiosis quàm vulgaribus. Hamb. 1694, 4°. —*Böhm.* Bibl. IV. 1, p. 217.

7. Diss. de Pluviâ. Hamb. 1690, 4°.—*Böhm.* Bibl. V. p. 52.

Büttner (J. G.).

1. Sur les restes du Monde primitif trouvés en Courlande.—Verh. Kurland. Ges. f. Litt. u. K. I. p. 195.—*Féruss.* Bull. 1826, VII. p. 91.

2. Vermischte Bemerkungen über einige Käferarten.—*Germ.* Mag. Ent. III. 1818, p. 245.—Bibl. Ent. I. p. 55.

Butty (J.).

1. Essay towards a Natural History of the County of Dublin. 1772. 12 vols. 8°.

Buxbaum (J. C.).

1. De Plantis Submarinis Observationes.—Comm. Acad. Petrop. IV. p. 279.

Buxtorf (Joh.).

1. Lumbrici teretes ex ulcere inguinis dextri prodeuntes.—Act. Helv. VII. p. 106.

Buzzi ( ).

1. Nuove Esperienze fatte sull' Occhio umano.—Opusc. Scelt. V. p. 95.

2. Storia anatomica di una varieta di Uomo bianco eliofobo.— Opusc. Scelt. VIII. p. 85.

BYERLEY (J.).

1. On the Natural History of Sierra Leone.—Trans. Geol. Soc. ser. 2, I. p. 418.

2. Attempt to explain the Phænomena of Geology by the Precession of the Equinoxes.—Mag. Nat. Hist. ser. 1, IV. p. 308 ; V. pp. 172, 201.

BYLANDT (E. de.).

1. Disquisitio circa Telam cellulosam anatomica, physiologica et pathologica. Berol. 1838, 8°.— Val. Repert. IV. p. 17.

BYLANDT-PALSTERCAMP (A. de).

1. Théorie des Volcans. Paris, 1836, 3 vols. 8°. Atl. fol.

2. Résumé préliminaire d'un Ouvrage sur la Théorie des Volcans. 8°. Naples, 1833.—Paris, 1834.—Cat. R. Soc. L.

BYRES (R. W.).

1. On the Traces of the Action of Glaciers at Porth-Treiddyn in Caernarvonshire.—Journ. Geol. Soc. I. 153.

BYRON (Lord).

1. Voyage to the Sandwich Islands. 4°. London, 1826.

BYTEMEISTER (H. J.).

1. Bibliothecæ Appendix, s. Catalogus Apparatûs curiosorum, artificialium et naturalium, subjunctis experimentis. Helmst. 1731, 8° ; Juliæ, 1745, 4°. fig.—Tr. Linn. Soc. Lond. VII. p. 243.— Klein, Disp. Echinod. p. ix.—Böhm. Bibl. I. 1, p. 391.—Mod. B. H.

BYWATER (J.).

1. On Animalcules, particularly on the Polypes.—Phil. Mag. XLIX. p. 283.

END OF VOL. I.

PRINTED BY RICHARD AND JOHN E. TAYLOR,
RED LION COURT, FLEET STREET.